Computed Tomography

Principles, Design, Artifacts, and Recent Advances

SECOND EDITION

Computed Tomography

Principles, Design, Artifacts, and Recent Advances

SECOND EDITION

Jiang Hsieh

Contents

Preface

Since the release of the first edition of this book in 2003, x-ray computed tomography (CT) has experienced tremendous growth thanks to technological advances and new clinical discoveries. Few could have predicted the speed and magnitude of the progress, and even fewer could have predicted the diverse nature of the technological advancement. The second edition of this book attempts to capture these advances and reflect on their clinical impact.

The second edition provides significant changes and additions in several areas. The first major addition is a new chapter on radiation dose. In the last few years, significant attention has been paid to this subject by academic researchers, radiologists, the general public, and the news media. An increased awareness of the impact of radiation dose on human health has led to the gradual adoption of the "as low as reasonably achievable" (ALARA) principle, the implementation of American College of Radiology (ACR) accreditation and other dose reference levels, and the development of many advanced dose-saving features for CT scanners. The new Chapter 11 briefly describes some of the known biological effects of radiation dose, then presents different dose definitions and measurements, and concludes with an illustration of various dose-reduction techniques.

At the time the first edition was published, the term "multislice" CT was an accurate description of state-of-the-art scanners. Sixteen-slice scanners had just been introduced commercially, and their clinical utilities and advantages had just begun to be discovered. Since then, the "slice war" has continued, and now 64-, 128-, 256-, and 320-slice scanners are the new state of the art. These scanners can be easily labeled as "cone-beam" CT. They require not only a detector with wider coverage, but also other technologies such as new calibration techniques and reconstruction algorithms. Chapter 10 has been significantly expanded to discuss the technologies associated with these scanners and the new image artifacts created by them.

Since the first edition, CT advancement has not been limited to the technology. Advances also have been made in many areas of clinical applications, including the rapid development of cardiac CT imaging and new applications inspired by the reintroduction of dual-energy CT. Chapter 12 presents these advances and the fundamental physics and technologies behind them.

Image artifacts have accompanied x-ray CT ever since its birth over 30 years ago. Some artifacts are caused by the characteristics of the physics involved, some are caused by technological limitations, some are created by new technologies, some are related to the patient, some result from suboptimal design, and some are introduced by the operator. Chapter 7 has been expanded to reflect the ever-evolving nature of these artifacts and various efforts to overcome them.

Historically, CT advances were driven by the development of new hardware. However, it has become increasingly clear that hardware alone cannot solve all of the technical and clinical problems that CT operators face. The second edition includes significant updates to the section on statistical iterative reconstruction technology and presents some of the exciting new developments in this area.

To enhance readers' understanding of the material and to inspire creative thinking about these subjects, a set of "problems" concludes each chapter. Many problems are open-ended and may not have uniquely correct solutions. Hopefully readers will find these problems useful and will develop new problems of their own.

At the time of this writing, the world is experiencing an unprecedented financial crisis—that some call a financial "tsunami." It is impossible to estimate or predict the impact of this crisis on the market for x-ray CT. However, CT technology is unlikely to remain stagnant. Many new exciting advances will take place in both the technology and its clinical applications.

Acknowledgments

Many of the ideas, principles, results, and examples that appear in this book stem from thoughts provoked by other books and research papers, and the author would like to take this opportunity to acknowledge those sources. The author would like to express his appreciation to Prof. Jeffrey A. Fessler of the University of Michigan for his review of this text. His expert critical opinions have significantly strengthened and enhanced the manuscript. The author owes a debt to two people for supplying materials for both editions of this book: Dr. Ting-Yim Lee of the Robarts Research Institute for providing reference materials on CT perfusion, and Mr. Nick Keat of the ImPACT group in London for supplying historical pictures on early CT development. The author would also like to thank Dr. T. S. Pan of the M.D. Anderson Cancer Research Center for providing some of the positron emission computed tomography (PET-CT) images, and Dr. P. Kinahan of the University of Washington for providing research results on patient motion artifacts. The author would like to thank many current and former colleagues at GE Healthcare Technologies and the GE Global Research Center for useful discussions, joint research projects, inspiration, and many beautiful images. Finally, the most significant acknowledgment of all goes to the author's spouse, Lily J. Gong, for her unconditional support of the project, and to his children, Christopher and Matthew, for their forgiveness of the missed vacation.

Jiang Hsieh
August 2009

Nomenclature and Abbreviations

2D:	two-dimensional
3D:	three-dimensional
AAPM:	American Association of Physicists in Medicine
ACR:	American College of Radiology
ALARA:	as low as reasonably achievable
ART:	algebraic reconstruction technique
ASIC:	application-specific integrated circuit
BMD:	bone mineral density
bpm:	beats per minute (heart rate)
CAC:	coronary artery calcification
CAI:	coronary artery imaging
CAT:	computer-aided tomography
CBF:	cerebral blood flow
CBV:	cerebral blood volume
CDRH	Center for Devices and Radiological Health (FDA)
CG:	conjugate gradient
COPD:	chronic obstructive pulmonary disease
CT:	computed tomography
CTDI:	CT dose index
DAS:	data acquisition system
DECT:	dual-energy CT
DFT:	discrete Fourier transform
DLP:	dose-length product
DSP:	digital signal processing
EBCT:	electron-beam computed tomography
EBT:	electron-beam tomography
EC:	European Commission
ECG / EKG:	electrocardiogram
FBP:	filtered backprojection
FDA:	US Food and Drug Administration
FDK:	Feldkamp-Davis-Kress (cone beam reconstruction algorithm)
FFT:	fast Fourier transform
FOV:	field of view
FWHM:	full width at half maximum
FWTM:	full width at tenth maximum
GDE:	geometric detection efficiency

GPU: graphic processor unit
HCT: helical computed tomography
HU: Hounsfield unit
IAC: inner auditory canal
IAEA: International Atomic Energy Agency
ICD: iterative coordinate decent
ICRP: International Commission on Radiological Protection
IFFT: inverse fast Fourier transform
IR: iterative reconstruction
LCD: low-contrast detectibility
LSF: line spread function
MIP: maximum intensity projection
minMIP: minimum intensity projection
ML: maximum likelihood
MPR: multiplanar reformation
MSAD: multiple-scan average dose
MTF: modulation transfer function
MTT: mean transit time
NPS: noise power spectrum
OS: ordered subset
PET: positron emission computed tomography
PSF: point spread function
QA: quality assurance
QDE: quantum detection efficiency
rad: radiation absorbed dose
RCA: right coronary artery
rem: Roentgen equivalent man
ROI: region of interest
SPECT: single-photon-emission computed tomography
SSP: slice sensitivity profile
Sv: sieverts
TAT: transverse axial tomography
VR: volume rendering
WL: (display) window level
WW: (display) window width

Chapter 1
Introduction

According to Webster's dictionary, the word "tomography" is derived from the Greek word "tomos" to describe "a technique of x-ray photography by which a single plane is photographed, with the outline of structures in other planes eliminated."[1] This concise definition illustrates the fundamental limitations of the conventional radiograph: superposition and conspicuity due to overlapping structures. In conventional radiography, the three-dimensional (3D) volume of a human body is compressed along the direction of the x ray to a two-dimensional (2D) image, as shown in Fig. 1.1(a). All underlying bony structures and tissues are superimposed, which results in significantly reduced visibility of the object of interest. Figure 1.1(b) shows an example of a chest x-ray study. The superposition of the ribs, lungs, and heart is quite evident. Consequently, despite the image's superb spatial resolution (the ability to resolve closely placed high-contrast objects), it suffers from poor low-contrast resolution (the ability to differentiate a low-contrast object from its background). A recognition of this limitation led to the development of conventional tomography.

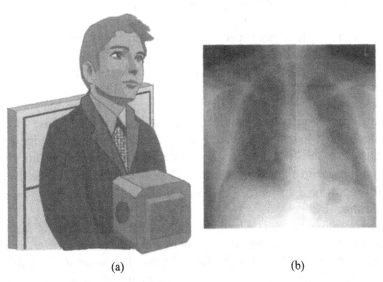

(a)   (b)

Figure 1.1 Illustration of conventional x ray. (a) Acquisition setup and (b) example of a chest study.

1

1.1 Conventional X-ray Tomography

Conventional tomography is also known as planigraphy, stratigraphy, laminagraphy, body section radiography, zonography, or noncomputed tomography.[2] One of the pioneers of conventional tomography was A. E. M. Bocage.[3] As early as 1921, Bocage described an apparatus to blur out structures above and below a plane of interest. The major components of Bocage's invention consisted of an x-ray tube, an x-ray film, and a mechanical connection to ensure synchronous movement of the tube and the film. The principle of conventional tomography is illustrated in Fig. 1.2. For convenience, consider two isolated points, A and B, located inside a patient. Point A is positioned on the focal plane and point B is off the focal plane. The shadows cast on the x-ray film by points A and B are labeled A1 and B1, respectively, as shown in Fig. 1.2(a). The image produced on the film at this instant is not at all different from a conventional radiograph. Next, we move both the x-ray source and the x-ray film synchronously in opposite directions (for example, the x-ray source moves to the left and the film moves to the right, as shown in the figure) to reach a second location. We want to make sure that the shadow A2 produced by the stationary point A overlaps with the shadow A1 produced by point A in the first position. This can be easily accomplished by setting the distance traveled by the x-ray source and the x-ray film to be proportional to their respective distances to point A, as shown in Fig. 1.2(b). However, the shadow B2 produced by the stationary point B at the second position does not overlap B1. This is due to the fact that point B is off the focal plane, and the distance ratio from point B to the x-ray source and to the film deviates significantly from that of point A. When the x-ray tube and the film move continuously along a straight line (in opposite directions, of course), the shadow produced by point B forms a line segment. This property holds for any point located above or below the focal plane. Note that the intensities of the shadows produced by the off-focus points are reduced, since the shadows are distributed over an extended area. On the other hand, any point located at the focal plane retains its image position on the film. Its shadow remains a point and the corresponding intensity is not degraded.

Conventional tomography has several problems. Although the focal plane in conventional tomography is theoretically a true plane, planes close to the focal plane undergo little blurring. If we use the amount of blurring to judge whether a point belongs to the focal plane, the slice thickness based on this definition depends on the sweep angle α, as shown in Fig. 1.3. In fact, the slice thickness is inversely proportional to $\tan(\alpha/2)$. Clearly, α must be fairly large to obtain a reasonable slice thickness.

Another problem associated with conventional tomography is the fact that little blurring takes place in the direction perpendicular to the movement of the x-ray source and the film. The net effect is that for structures parallel to the direction of the source motion, the sharpness of the shadow boundaries is not significantly reduced as desired. These structures appear to be elongated only

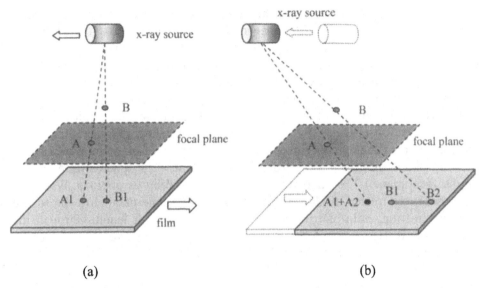

(a) (b)

Figure 1.2 Illustration of conventional tomography principle. (a) X-ray source and film produce shadows A1 and B1 of points A and B at a first position. (b) X-ray source and film are moved reciprocally such that shadow A2 of point A overlaps shadow A1, but shadow B2 of point B does not overlap B1.

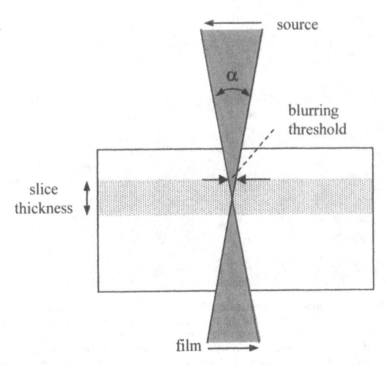

Figure 1.3 Illustration of slice thickness as a function of the scan angle.

along the direction of motion. This effect is illustrated with a computer simulation shown in Fig. 1.4. The imaged object was made of two long ellipsoids and two spheres with a 2:1 density ratio in favor of the ellipsoids. The goal was to enhance the visibility of the spheres. In the first simulation, the ellipsoids were placed such that their long axes were perpendicular to the direction of the source motion. For comparison, a conventional radiography image (stationary source and detector) is shown in Fig. 1.4(a) and a conventional tomography image is shown in Fig. 1.4(b). Compared to the conventional radiograph, the ellipsoids in the conventional tomography image were blurred by the source motion because the ellipsoids were placed away from the focal plane. (The spheres were located on the focal plane.) The improvement in sphere visibility is obvious. When the ellipsoids were rotated 90 deg so their long axes became parallel to the source motion direction, little blurring took place, since the path lengths through the ellipsoids at different source locations were virtually unchanged. Consequently, no improvement in sphere visibility was obtained.

To partially compensate for the lack of tomographic effect in certain directions, pluridirectional tomography has been proposed.[2] For these devices, the x-ray source and the film synchronously undergo more complicated motion

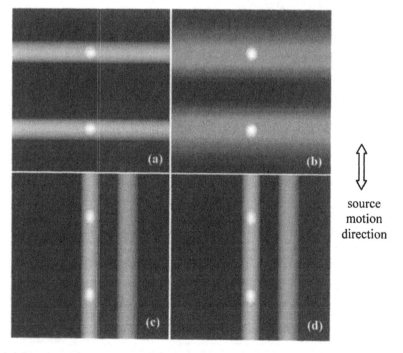

source
motion
direction

Figure 1.4 Simulated images of conventional tomography. The phantoms are made of two long ellipsoids and two spheres. The top row depicts the scenario in which the long axes of the ellipsoids are perpendicular to the direction of source-detector motion, and the bottom row depicts the scenario in which the ellipsoids' long axes are parallel to the motion. (a) and (c) show conventional radiography images of the phantoms; (b) and (d) show conventional tomography images of the phantoms.

patterns, such as circular, ellipsoidal, sinusoidal, hypocycloidal, or spiral. Figure 1.5 depicts an example of an elliptical motion pattern that produces more uniform blurring of the structures outside the focal plane. Disadvantages of pluridirectional tomography include higher cost, increased procedure time, and a larger x-ray dose to the patient.

Instead of forming a focal plane parallel to the patient long axis, axial transverse tomography (also known as transverse axial tomography or TAT) defines a cross-sectional plane that is perpendicular to the patient long axis, as shown in Fig. 1.6. In this apparatus, the x-ray source is stationary and is oriented at a shallow angle θ with respect to the x-ray film. Both the patient and the film rotate about their vertical axes synchronously at an identical direction and speed. Because the geometric relationship between the x-ray source, the patient, and the film remains unchanged, the magnification factor for each point located inside the tomographic plane is constant (the magnification factor is defined as the distance between the source and the shadow on the film over the distance between the source and the point in the tomographic plane).

During the imaging process, structures inside the tomographic plane remain in sharp focus, since structures inside the plane remain in the field of view (FOV) at all times, and the shadow locations produced by these structures do not change relative to the film. On the other hand, structures outside the tomographic plane do not always stay inside the FOV, and their shadows move around relative to the film during the scan. Thus, these shadows do not appear as sharp. Strictly speaking, the tomographic plane is actually a volume. The thickness of the volume decreases with the angle θ between the center ray of the source and the film. Since θ is limited by many practical factors, the minimum thickness of the tomographic volume is also limited. For example, for an extremely small θ, the

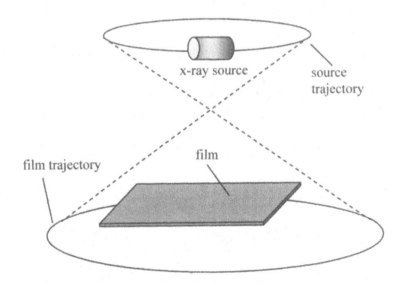

Figure 1.5 Illustration of pluridirectional tomography.

amount of x-ray flux detected by the film is severely limited, and the image quality is degraded by the quantum noise.

Although all of these tomographic techniques are somewhat successful in producing images at the plane of interest, they all suffer from two fundamental limitations: these techniques do not increase the object contrast, and they cannot completely eliminate other structures outside the focal plane. Note that conventional tomography blurs the overlying structure. The visibility of the structures inside the focal plane may be enhanced, but the contrast between different structures inside the focal plane is not enhanced. In addition, the blurred overlying structures superimposed on the tomographic image significantly degrade the quality of the image. Combined with the larger x-ray dose to the patient, conventional tomography has found limited use in clinical applications.

With the development of digital flat-panel technology, the combination of digital processing techniques and conventional tomographic acquisitions has found renewed interest.[4-8] The combined technique is often called *tomosynthesis*. Compared to conventional tomography where the tomographic effect is achieved through analog means, tomosynthesis has the advantage of producing tomographic images at different depths with a single data acquisition and additional improvements in image quality using image processing technologies. It offers the potential of providing low-dose alternatives to x-ray CT for certain clinical applications, such as breast cancer or lung cancer screening. Figure 1.7 shows an example of the improved visualization of a stone in a ureter (b) compared to the conventional radiograph (a). Note that the intensities of the spine and ribs in the reconstructed tomosynthesis image are significantly reduced but not completely eliminated.

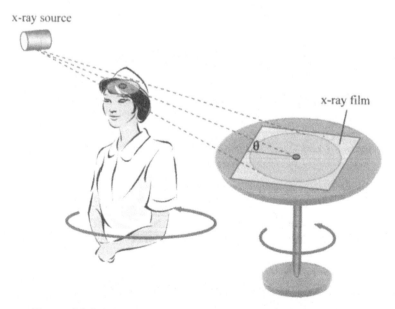

Figure 1.6 Schematic diagram of axial transverse tomography.

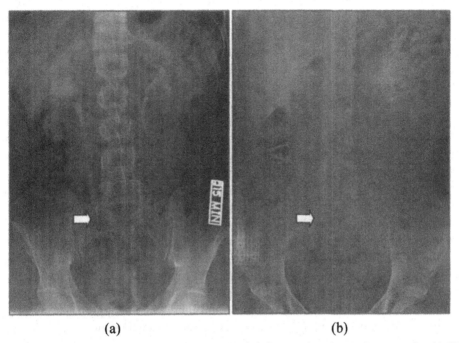

(a)   (b)

Figure 1.7 Clinical example of tomosynthesis. (a) Conventional anterior-posterior (A-P) radiography. (b) Tomosynthesis image showing a stone in the right ureter.

1.2 History of Computed Tomography

It is quite remarkable that image reconstruction from projections was attempted as early as 1940.[9] Needless to say, these attempts were made without the benefits of modern computer technology. In a patent granted in 1940, Gabriel Frank described the basic idea of today's tomography.[10] The patent includes drawings of equipment to form sinograms (representations of measurement data as linear samples versus view samples) and optical backprojection techniques to reconstruct images. Backprojection can be roughly described as a process in which the intensity of a sample is distributed uniformly along the path that formed the sample. A more detailed discussion on the subject can be found in Chapter 3. Although images generated from the proposed approach suffer from blurring, the patent clearly envisioned the fundamental requirements for tomographic devices.

Twenty-one years later, William H. Oldendorf, an American neurologist from Los Angeles, performed a series of experiments based on principles similar to those later used in CT.[11] The objective of his work was to determine whether internal structures within dense structures could be identified by transmission measurements. His experimental apparatus is schematically diagrammed in Fig. 1.8. The phantom consists of two concentric rings of iron nails embedded in a plastic block of $10 \times 10 \times 4$ cm to represent the cranial vault. Another iron nail and an aluminum nail were placed 1.5 cm apart near the center of the ring. The phantom was placed on a model train. The train was pulled by a clock motor along the track at a slow speed (88 mm per hour). A relatively faster rotational motion was provided by placing the entire apparatus on a turntable rotating at 16 revolutions per minute (rpm). A collimated radioiodine (^{131}I) source provided a pencil beam gamma ray to irradiate the phantom, and the gamma ray beam always passed through the center of rotation. The signal was detected by a sodium iodine scintillation crystal and a photomultiplier.

To understand the nature of the measurement, consider the intensity modulation of the measured beam. Every nail in the phantom passed the gamma ray exactly twice per rotation. At a rotation rate of 16 rpm, the relatively rapid variation in the transmitted beam intensity caused by the peripheral nails was a high-frequency signal. The translation of the central nails near the center of rotation caused a relatively slow variation in the transmitted beam, resulting in low-frequency signals. A low-frequency signal can be separated from a high-frequency signal through filtering (Oldendorf used an electronic low-pass filter with a 30-sec time constant). For this experiment, only one line passing through the center of rotation was reconstructed. The reconstruction of additional lines would require shifting the phantom relative to the center of rotation. Because each scan required one hour to complete and there was no appropriate means of storing the data, no attempt was made to reconstruct the 2D structure.

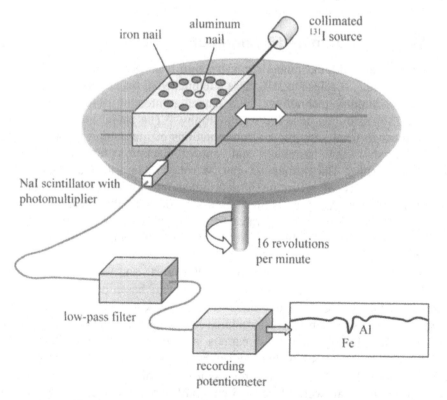

Figure 1.8 Simplified diagram of Oldendorf's experiment.

In 1963, David E. Kuhl and Roy Q. Edwards introduced transverse tomography using radioisotopes, which was further developed and evolved into today's emission computed tomography.[12] A sequence of scans was acquired at uniform steps and regular angular intervals with two opposing radiation detectors. At each angle, the film was exposed to a narrow line of light moving across the face of a cathode ray tube with a location and orientation corresponding to the detectors' linear position. This is essentially the analog version of the backprojection operation. The process was repeated at 15-deg angular increments (the film was rotated accordingly to sum up the backprojected views). In later experiments, the film was replaced by a computer-based backprojection process. What was lacking in these attempts was an exact reconstruction technique.

The mathematical formulation for reconstructing an object from multiple projections dates back to the Austrian mathematician J. Radon. In 1917, Radon demonstrated mathematically that an object could be replicated from an infinite set of its projections. The concept was first applied by R. N. Bracewell in 1956 to reconstruct a map of solar microwave emissions from a series of radiation measurements across the solar surface.[13] Between 1956 and 1958, several Russian papers accurately formulated the tomographic reconstruction problem as an inverse Radon transform.[14–16] These papers discussed issues associated with the implementation and proposed methodologies of performing reconstructions with television-based systems. Although these algorithms were somewhat inefficient, they offered a satisfactory performance.[17]

In 1963, Allan M. Cormack reported the findings from investigations of perhaps the first CT scanner actually built.[18] His work could be traced back to 1955 when he was asked to spend one and a half days a week at Groote Schuur Hospital attending to the use of isotopes after the resignation of the hospital physicist (Cormack was the only nuclear physicist in Capetown). While observing the planning of radiotherapy treatments, Cormack came to realize the importance of knowing the x-ray attenuation coefficient distribution inside the body. He wanted to reconstruct attenuation coefficients of tissues to improve the accuracy of radiation treatment. During a sabbatical at Harvard University in late 1956, he derived a mathematical theory for image reconstruction. When he returned to South Africa in 1957, he tested his theory in a laboratory simulation using a 5-cm-thick disk that was 20 cm in diameter. The disk consisted of a central cylinder of pure aluminum (1.13 cm in diameter) surrounded by an aluminum alloy annulus. This was surrounded in turn by an oak annulus. A collimated ^{60}Co isotope was used as the radiation source and a Geiger counter as a detector. The disc was translated through the gamma ray in 5-mm steps to form a linear scan. Because of circular symmetry, the disc was scanned at only one angular position (the linear scans obtained from other angles would be identical). The attenuation coefficients were calculated for aluminum and wood with the use of his reconstruction technique.

In 1963, Cormack repeated his experiment at Tufts University (where he became a member of the physics department in 1957) with a circularly unsymmetrical phantom of aluminum and plastic. The phantom consisted of an outer ring of aluminum to represent the skull, a Lucite filling to represent soft tissue, and two aluminum disks within the Lucite to represent tumors. Again, a collimated gamma ray beam was used as the source. Twenty-five linear scans were performed at 7.5-deg angular increments over a 180-deg angle. The results of two experiments were published in 1963 and 1964, respectively. Unfortunately, little attention was paid at the time to his work due to the time and difficulty of performing the necessary calculations. Cormack remarked in his 1979 Nobel lecture that "There was virtually no response. The most interesting request for a reprint came from the Swiss Center for Avalanche Research. The method would work for deposits of snow on mountains if one could get either the detector or the source into the mountain under the snow!" [19]

The development of the first clinical CT scanner began in 1967 with Godfrey N. Hounsfield at the Central Research Laboratories of EMI, Ltd. in England. While investigating pattern recognition techniques, he deduced, independent of Cormack, that x-ray measurements of a body taken from different directions would allow the reconstruction of its internal structure.[20] Preliminary calculations by Hounsfield indicated that this approach could attain a 0.5% accuracy of the x-ray attenuation coefficients in a slice. This is an improvement of nearly a factor of 100 over the conventional radiograph. For their pioneering work in CT, Cormack and Hounsfield shared the 1979 Nobel Prize in Physiology and Medicine.

The first laboratory scanner was built in 1967, as shown in Fig. 1.9. Linear scans were performed on a rotating specimen in 1-deg steps (the specimen remained stationary during each scan). Because of the low-intensity americium gamma source, it took nine days to complete the data acquisition and produce a picture. Unlike the reconstruction method used by Cormack, a total of 28,000 simultaneous equations had to be solved by a computer in 2.5 hours (a more detailed discussion on this reconstruction technique can be found in Section 3.4). The use of a modified interpolation method, a higher-intensity x-ray tube, and a crystal detector with a photomultiplier reduced the scan time to nine hours and improved the accuracy from 4% to 0.5%.

The first clinically available CT device was installed at London's Atkinson-Morley Hospital in September 1971, after further refinement on the data acquisition and reconstruction techniques. Images could be produced in 4.5 minutes. On October 4, 1971, the first patient, who had a large cyst, was scanned and the pathology was clearly visible in the image.[21] Figure 1.10 depicts a patient's head scan performed with the first clinical CT device, and Fig. 1.12(a) depicts one of the first clinical head images obtained from a CT scanner.

Humans are not the sole beneficiaries of this wonderful invention. Over the years, CT scanners have been used to scan trees, animals, industrial parts, mummies, and just about everything that can fit inside a CT gantry. As an example, Fig. 1.11 shows two images of scanning procedures on big cats from zoos. Given the fact that CT is often called a CAT (computer-aided tomography) scanner, scanning cats by CAT seems to be quite natural!

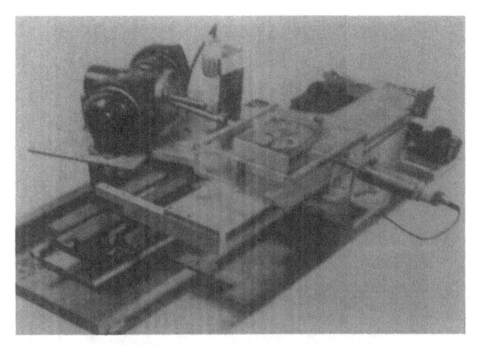

Figure 1.9 Original lathe and scanner used in an early CT experiment by Hounsfield at EMI Central Research Laboratory in 1967.

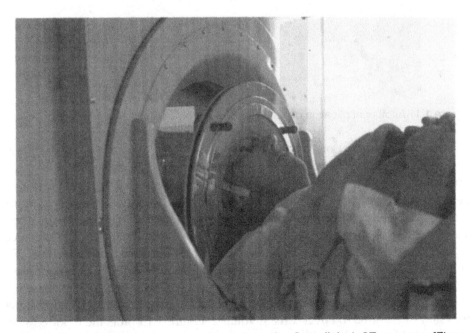

Figure 1.10 A patient head scan performed on the first clinical CT scanner. [Figure supplied by ImPACT (www.impactscan.org), St George's Hospital NHS Trust, London, and reprinted with permission.]

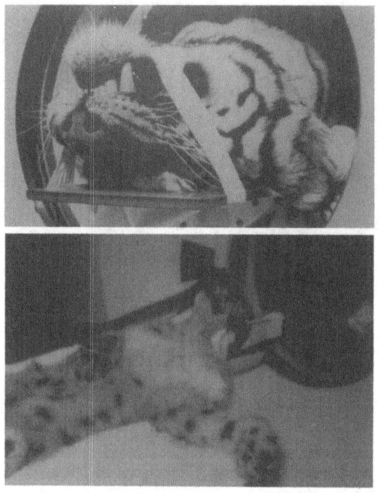

Figure 1.11 Another application of a medical CT: use of CAT to scan big cats.

Since the introduction of the first clinical scanner, tremendous advancement has been made in CT technology. For illustration, Fig. 1.12(b) depicts a CT head image obtained with a state-of-the-art scanner. Improvements in spatial and low-contrast resolution are evident. As another simple illustration, we can examine the progress made on one performance benchmark: time to acquire a single slice. To cover a volume, consecutive slices must be acquired with an interslice spacing that is roughly equal to the slice thickness. Therefore, the scan time of a single slice includes the actual data acquisition time as well as interscan delays between slices. Although other benchmarks such as gantry speed, reconstruction time, or tube power can be considered, the scan time per slice is a more direct indicator of the time to cover a fixed volume. Figure 1.13 plots the reported scan time over 30 years of CT history. For clarity, a logarithmic scale is used. A straight line (in logarithmic scale) was fitted through the samples depicted by the diamonds. The reduction in scan time (and therefore, the increase in volume coverage speed) follows an exponential relationship over time. Based on the slope of the fitted straight line, we can calculate the time it takes to reduce the scan time to half; the result is 2.3 years. This means that CT scanners have doubled their acquisition power every 2.3 years over the last 30 years! This performance matches Moore's law for electronics.

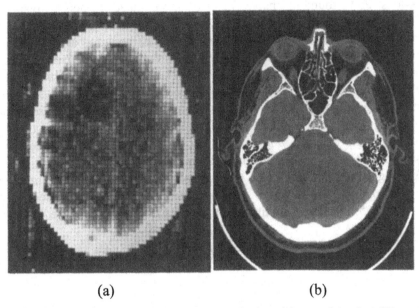

(a) (b)

Figure 1.12 Comparison of CT head images acquired on (a) one of the first CT scanners and (b) the GE LightSpeed VCT 2005.

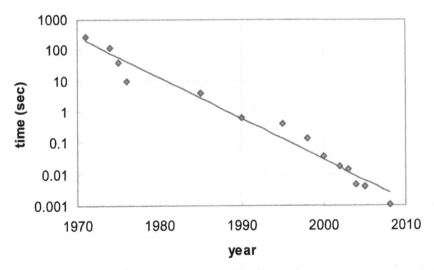

Figure 1.13 Scan time per slice as a function of time (excluding electron-beam computed tomography, discussed in Section 1.3). The scan time/slice decreased exponentially over time and was reduced by a factor of 1.34 per year.

1.3 Different Generations of CT Scanners

The previous section presented a brief review of CT history. This section describes the evolution of CT technology over the last 30 years. Because of the nature of the overview, detailed comparisons of the pros and cons of different types of scanners are omitted. We will provide more in-depth coverage on this subject in later sections of the book, wherever appropriate. For example, many discussions can be found in Chapter 7 as part of the image artifacts analysis.

The type of scanner built by EMI in 1971 is called the first-generation CT. In a first-generation scanner, only one pencil beam is measured at a time. In the original EMI head scanner, the x-ray source was collimated to a narrow beam of 3 mm wide (along the scanning plane) and 13 mm long (across the scanning plane). The x-ray source and detector were linearly translated to acquire individual measurements. The original scanner collected 160 measurements across the scan field. After the linear measurements were completed, both the x-ray tube and the detector were rotated 1 deg to the next angular position to acquire the next set of measurements, as shown in Fig. 1.14.

Although clinical results from the first-generation scanners were promising, there remained a serious image quality issue associated with patient motion during the 4.5-min data acquisition.[22] The data acquisition time had to be

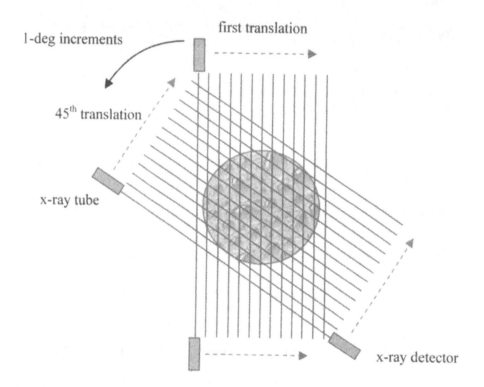

Figure 1.14 First-generation CT scanner geometry. At any time instant, a single measurement is collected. The x-ray tube and detector translate linearly to cover the entire object. The entire apparatus is then rotated 1 deg to repeat the scan.

reduced. This need led to the development of the second-generation scanner illustrated in Fig. 1.15. Although this was still a translation-rotation scanner, the number of rotation steps was reduced by the use of multiple pencil beams. The figure depicts a design in which six detector modules were used. The angle between the pencil beams was 1 deg. Therefore, for each translation scan, projections were acquired from six different angles. This allowed the x-ray tube and detector to rotate 6 deg at a time for data acquisition, representing a reduction factor of 6 in acquisition time. In late 1975, EMI introduced a 30-detector scanner that was capable of acquiring a complete scan under 20 sec.[9] This was an important milestone for body scanning, since the scan interval fell within the breath-holding range for most patients.

One of the most popular scanner types is the third-generation CT scanner illustrated in Fig. 1.16. In this configuration, many detector cells are located on an arc concentric to the x-ray source. The size of each detector is sufficiently large so that the entire object is within each detector's FOV at all times. The x-ray source and the detector remain stationary to each other while the entire apparatus rotates about the patient. Linear motion is eliminated to significantly reduce the data acquisition time.

In the early models of the third-generation scanners, both the x-ray tube power and the detector signals were transmitted by cables. Limitations on the length of the cables forced the gantry to rotate both clockwise and

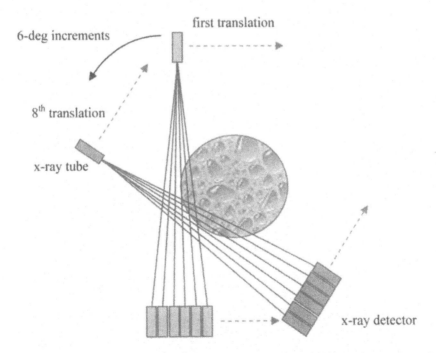

Figure 1.15 Second-generation CT scanner geometry. At each time instant, measurements from six different angles are collected. Although the x-ray source and detector still need to be linearly translated, the x-ray tube and detector can rotate every 6 deg.

counterclockwise to acquire adjacent slices. The acceleration and deceleration of the gantry, which typically weighed several hundred kilograms, restricted the scan speed to roughly 2 sec per rotation. Later models used slip rings for power and data transmission. Since the gantry could rotate at a constant speed during successive scans, the scan time was reduced to 0.5 sec or less. The introduction of slip ring technology was also a key to the success of helical or spiral CT (Chapter 9 is devoted to this topic). Because of the inherent advantages of the third-generation technology, nearly all of the state-of-the-art scanners on the market today are third generation.

Several technology challenges in the design of the third-generation CT, including detector stability and aliasing, led to investigations of the fourth-generation concept depicted in Fig. 1.17. In this design, the detector forms an enclosed ring and remains stationary during the entire scan, while the x-ray tube rotates about the patient. Unlike the third-generation scanner, a projection is formed with signals measured on a single detector as the x-ray beam sweeps across the object. The projection, therefore, forms a fan with its apex at the detector, as shown by the shaded area in Fig. 1.17 (a projection in a third-generation scanner forms a fan with the x-ray source as its apex). One advantage of the fourth-generation scanner is the fact that the spacing between adjacent samples in a projection is determined solely by the rate at which the measurements are taken. This is in contrast to third-generation scanning in which

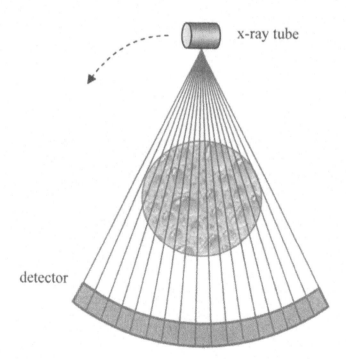

Figure 1.16 Third-generation CT scanner geometry. At any time instant, the entire object is irradiated by the x-ray source. The x-ray tube and detector are stationary with respect to each other while the entire apparatus rotates about the patient.

the sample spacing is determined by the detector cell size. A higher sampling density can eliminate potential aliasing artifacts. In addition, since at some point during every rotation each detector cell is exposed directly to the x-ray source without any attenuation, the detector can be recalibrated dynamically during the scan. This significantly reduces the stability requirements of the detector.

A potential drawback of the fourth-generation design is scattered radiation. Because each detector cell must receive x-ray photons over a wide angle, no effective and practical scatter rejection can be performed by a post-patient collimator. Although other scatter correction schemes, such as the use of a set of reference detectors or software algorithms, are useful, the complexity of these corrections is likely to increase significantly with the introduction of multislice or volumetric CT.

A more difficult drawback to overcome is the number of detectors required to form a complete ring. Since the detector must surround the patient at a fairly large circumference (to maintain an acceptable source-to-patient distance), the number of detector elements and the associated data acquisition electronics become quite large. For example, a recent single-slice fourth-generation scanner required 4800 detectors. The number would be much higher for multislice scanners. Thus, for economical and practical reasons, fourth-generation scanners are likely to be phased out.[9]

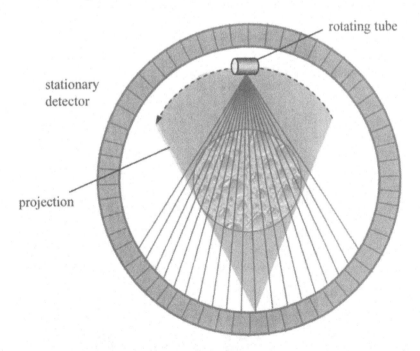

Figure 1.17 Geometry of a fourth-generation CT scanner. At any time instant, the x-ray source irradiates the detectors in a fan-shaped x-ray beam, as shown by the solid lines. A projection is formed using measurement samples from a single detector over time, as depicted by the shaded fan-shaped region.

The electron-beam scanner, sometimes called the fifth-generation scanner, used in electron-beam computed tomography (EBCT), or electron-beam tomography (EBT), was built between 1980 and 1984 for cardiac applications.[23] To "freeze" cardiac motion, a complete set of projections must be collected within 20 to 50 ms. This is clearly very challenging for conventional third- or fourth-generation types of scanners due to the enormous centripetal force placed on the x-ray tube and the detector. In the electron-beam scanner, the rotation of the source is provided by the sweeping motion of the electron beam (instead of the mechanical motion of the x-ray tube). Figure 1.18 shows a simplified schematic diagram of an electron-beam scanner. The bottom arc (210 deg) represents an anode with multiple target tracks. A high-speed electron beam is focused and deflected by carefully designed coils to sweep along the target ring, similar to a cathode ray tube. The entire assembly is sealed in a vacuum. Fan-shaped x-ray beams are produced and collimated to a set of detectors, represented by the top arc of 216 deg. The detector ring and the target ring are offset (noncoplanar) to make room for the overlapped portion. When multiple target tracks and detector rings are used, a coverage of 8 cm along the patient long axis can be obtained for the heart. Since the system has no mechanical moving parts, scan times as fast as 50 ms can be achieved. However, for noise considerations, multiple scans are often averaged to produce the final image. A more complete description of the electron-beam scanner can be found in Ref 24.

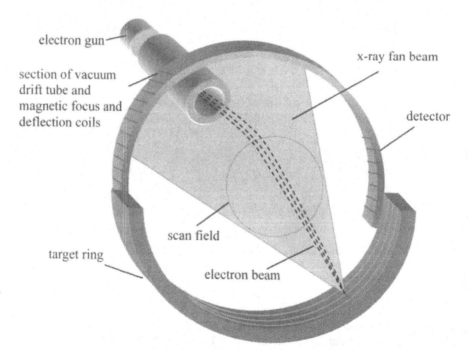

Figure 1.18 Geometry of an electron-beam scanner.

1.4 Problems

1-1   Section 1.1 discussed conventional tomography and the many drawbacks and shortcomings of this technique. However, recently many research and development activities have combined conventional tomography with advanced reconstruction techniques. This technology is called tomosynthesis, as illustrated by a clinical example in Fig. 1.7. To understand the rationale behind its comeback, list at least three advantages of tomosynthesis over CT.

1-2   Figure 1.4 illustrates one limitation of conventional tomography. Pluridirectional tomography was invented to overcome this shortcoming. What are the limitations of this approach? Design a phantom to demonstrate its limitations.

1-3   In a breast tomosythesis device similar to the configuration shown in Fig. 1.2, the film is replaced by a digital detector that remains stationary during data acquisition. Only the x-ray tube translates during data acquisition. What are the design considerations that ensure a good resolving capability of such a device to suppress overlapping structures across the entire volume?

1-4   In a tomosynthesis device similar to the one shown in Fig. 1.2, a digital detector replaces the film and is capable of translational motion, as shown in the figure. Is the focal plane uniquely defined by the speed of the x-ray tube and the detector as in the case of conventional tomography? If not, how is the focal plane defined, assuming the digital detector is capable of high-speed sampling?

1-5   One application of x-ray CT technology is luggage scanning for airport security. Luggage is transported over a conveyer belt, and cross-sectional images of the luggage are generated and analyzed for security threats. List at least three similarities and three differences between a medical CT device and a luggage CT device in terms of technical design considerations.

1-6   List three advantages and three disadvantages of the fourth-generation CT scanner over the third-generation CT scanner.

1-7   In many micro-CT scanners designed for small-animal imaging, the CT gantry remains stationary while the animal spins about its long axis, while the gantry in a human clinical CT scanner rotates about the patient. Describe the pros and cons of each design.

References

1. D. B. Guralnik, Ed., *Webster's New World Dictionary of the American Language*, 2nd college edition, William Collins + World Publishing Co., Cleveland (1974).

2. C. L. Morgan, *Basic Principles of Computed Tomography*, University Park Press, Baltimore (1983).

3. A. E. M. Bocage, "Procede et dispositifs de radiographie sur plaque en mouvement," French Patent No. 536,464 (1921).

4. J. T. Dobbins III and D. J. Godfrey, "Digital x-ray tomosynthesis: current state of the art and clinical potential," *Phys. Med. Biol.* **48**, 65–106 (2003).

5. R. J. Warp, D. J. Godfrey, and J. T. Dobbins III, "Applications of matrix inversion tomosynthesis," *Proc. SPIE* **3977**, 376–383 (2000).

6. G. M. Stevens, R. Fahrig, and N. J. Pelc, "Filtered backprojection for modifying the impulse response of circular tomosynthesis," *Med. Phys.* **28**(3), 372–380 (2001).

7. B. Nett, S. Leng, and G.-H. Chen, "Plannar tomosynthesis reconstruction in a parallel-beam framework via. Virtual object reconstruction," *Proc. SPIE* **6510**, 651028 (2007).

8. T. Deller, K. Jabri, J. Sabol, X. Ni, G. Avinash, R. Saunders, and R. Uppaluri, "Effect of acquisition parameters on image quality in digital tomosynthesis, *Proc. SPIE* **6510**, 65101L (2007).

9. L. W. Goldman, "Principles of CT and the evolution of CT technology," in *Categorical Course in Diagnostic Radiology Physics: CT and US Cross-sectional Imaging*, L. W. Goldman and J. B. Fowlkes, Eds., Radiological Society of North America, Inc., Oak Brook, IL (2000).

10. S. Webb, "Historical experiments predating commercially available computed tomography," *British J. of Radiology* **65**, 835–837 (1992).

11. W. H. Oldendorf, "Isolated flying spot detection of radiodensity discontinuities: displaying the internal structural pattern of a complex object," *IEEE Trans. Biomed. Electr.* **8**, 68–72 (1961).

12. D. E. Kuhl and R. Q. Edwards, "Reorganizing data from transverse section scans of the brain using digital processing," *Radiology* **91**, 975–983 (1968).

13. R. N. Bracewell, "Strip integration in radiation astronomy," *Australian J. of Physics* **9**, 198–217 (1956).

14. S. I. Tetel'baum, "About the problem of improvement of images obtained with the help of optical and analog instruments," *Bull. Kiev Polytech. Inst.* **21**, 222 (1956) (in Russian).

15. S. I. Tetel'baum, "About a method of obtaining volume images with the help of x-rays," *Bull. Kiev Polytech. Inst.* **22**, 154–160 (1957) (in Russian).

16. B. I. Korenblyum, S. I. Tetel'baum, and A. A. Tyutin, "About one scheme of tomography," *Bull. Inst. Higher Educ. Radiophys.* **1**, 151–157 (1958).

17. H. H. Barrett, W. G. Hawkins, and M. L. G. Joy, "Historical note on computed tomography," *Radiology* **147**, 172 (1983).

18. A. M. Cormack, "Representation of a function by its line integrals, with some radiological applications," *J. Appl. Phys.* **34**, 2722–2727 (1963).

19. A. M Cormack, "Early two-dimensional reconstruction and recent topics stemming from it," Nobel Lecture, December 9, 1979.

20. G. N. Hounsfield, "Historical notes on computerized axial tomography," *J. of the Canadian Assoc. of Radiologists* **27**, 135–142 (1976).

21. J. Ambrose, "A brief review of the EMI scanner," *Proc. Br. Inst. Radiol.* **48**, 605–606 (1975).

22. D. Schellinger, G. Di Chiro, S. Axelbaum, H. L. Twigg, and R. S. Ledley, "Early clinical experience with the ACTA scanner," *Radiology* **114**, 257–261 (1975).

23. D. P. Boyd, R. G. Gould, J. R. Quinn, R. Sparks, J. H. Stanley, and W. B. Herrmannsfeldt, "A proposed dynamic cardiac 3D densitometer for early detection and evaluation of heart disease," *IEEE Trans. Nucl. Sci.* **26**, 2724–2727 (1979).

24. C. H. McCollough, "Principles and performance of electron beam CT," in *Medical CT and Ultrasound: Current Technology and Applications*, L. W. Goldman and J. B. Fowlkes, Eds., Advanced Medical Publishing, Madison, WI, 487–518 (1995).

Chapter 2
Preliminaries

This chapter covers two major topics: the fundamentals of the mathematics and the fundamentals of the physics involved in x-ray physics. The objective is to provide a general review of the important mathematics tools and background knowledge of x-ray physics that are used throughout the book. The chapter serves mainly as a refresher. Readers who have not been previously exposed to these topics will find a list of recommended reference materials at the end of the chapter so that detailed studies can be performed.

Although these topics can be integrated into other chapters when they are encountered, many of the topics appear multiple times and at different locations. The goal is to provide a convenient and quick reference point to the readers by consolidating these subjects in a separate chapter.

2.1 Mathematics Fundamentals

2.1.1 Fourier transform and convolution

The one-dimensional (1D) Fourier transform of a function $f(x)$ is defined as

$$F(u) = \int_{-\infty}^{\infty} f(x)e^{-j2\pi ux}dx,\qquad(2.1)$$

where $j = \sqrt{-1}$. In this equation, $e^{-j2\pi ux} = \cos 2\pi ux - j\sin 2\pi ux$. Substituting this expression into Eq. (2.1), the 1D Fourier transform can also be expressed as

$$F(u) = \int_{-\infty}^{\infty} f(x)\cos 2\pi ux dx - j\int_{-\infty}^{\infty} f(x)\sin 2\pi ux dx.\qquad(2.2)$$

When $f(x)$ is real, the real part of Eq. (2.2) is an even function of frequency u, and the imaginary part is an odd function of frequency u. Based on the definition of a complex conjugate, the Fourier transform of any real function possesses the following property:

$$F(-u) = F^*(u),\qquad(2.3)$$

where * denotes a complex conjugate. This property is often called *Hermitian symmetry*. The Fourier transform of the measured CT projections always possesses Hermitian symmetry, because the projection measurement $p(x)$ is always a real function. An interesting special case occurs when $f(x)$ is a real and even function of x. It can be shown that its Fourier transform $F(u)$ is also a real and even function of u.

The inverse Fourier transform is defined as

$$f(x) = \int_{-\infty}^{\infty} F(u)e^{j2\pi ux} du .$$ (2.4)

One can show that Eqs. (2.1) and (2.4) form a transformation pair. Several properties of the 1D Fourier transform pair can be readily obtained from these definitions. Interested readers can refer to several textbooks for details.[1-3] These properties will be presented below without proof:

(1) *Linearity*: If $f(x) = af_1(x) + bf_2(x)$, then $F(u) = aF_1(u) + bF_2(u)$. Here, $F_1(u)$ and $F_2(u)$ are the Fourier transforms of $f_1(x)$ and $f_2(x)$, respectively. This property indicates that the linear combination of two functions leads to a linear combination of their Fourier transforms.

(2) *Scaling*: The Fourier transform of $f(ax)$ is $(1/a) F(u/a)$.

(3) This property states that scaling the size of a function leads to a compression in frequency and an amplification in magnitude. A simple example is shown in Fig. 2.1. In the discrete implementation of the Fourier transform (called the discrete Fourier transform or DFT), one technique that is often used to increase the sampling density of a spatial (or time) domain signal is to perform zero padding in the Fourier domain. The process is as follows: First, the Fourier transform of the original signal is performed. When the original signal contains N samples, its corresponding DFT also contains N samples. If the original sampling density must be increased by a factor of K, we pad $(K-1)N$ zeroes to the Fourier transform before the inverse Fourier transform is performed. Thus, the original signal is expanded to KN samples, as shown in Fig. 2.2.

(4) *Shift property*: The Fourier transform of $f(x-x_0)$ is $e^{-j2\pi ux_0} F(u)$. This indicates that a shift of a function in the spatial domain is equivalent to a phase shift in the Fourier domain. This property is quite useful in data resampling or interpolation. It is often preferred over spatial domain interpolations. Figure 2.3 shows an example of shifting the original data by 3.2 sampling intervals.

(5) *Derivative*: The Fourier transform of the derivative of $f(x)$, $df(x)/dx$, is $j2\pi uF(u)$.

(6) *Conservation of energy*: $\int_{-\infty}^{\infty} |f(x)|^2 dx = \int_{-\infty}^{\infty} |F(u)|^2 du$.

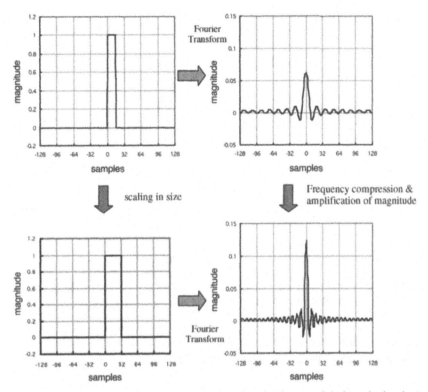

Figure 2.1 Scaling the size of the original function in the spatial domain leads to the compression of frequency and amplification of magnitude in the frequency domain.

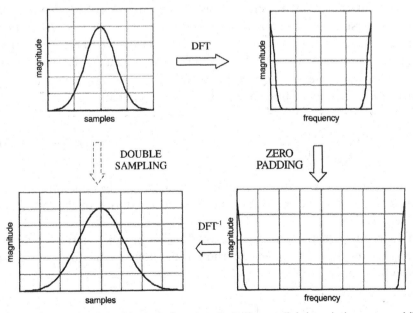

Figure 2.2 Illustration of double sampling density in the spatial domain by zero padding in the frequency domain.

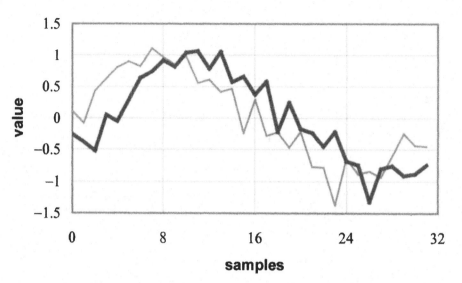

Figure 2.3 Data resampling with phase shift in the Fourier domain. Gray line: original samples; black line: shifted by a 3.2 sampling interval.

Another useful property of the Fourier transform is its application to the convolution process. The convolution of two functions, $f_1(x)$ and $f_2(x)$, is defined by

$$g(x) = f_1(x) * f_2(x) = \int_{-\infty}^{\infty} f_1(x')f_2(x-x')dx' . \qquad (2.5)$$

The convolution operation can be conceptually performed by a sequence of operations. First we flip the function $f_2(x')$ at $x' = 0$ to arrive at a function $f_2(-x')$, as shown in Fig. 2.4. The function is then continuously shifted along the x axis by an amount x to obtain $f_2(x-x')$. The area under the product of the two functions, $f_1(x')f_2(x-x')$, is assigned as the convolution of the two functions at location x.

If we take the Fourier transform of Eq. (2.5), we obtain

$$G(u) = \int_{-\infty}^{\infty}\left[\int_{-\infty}^{\infty} f_1(x')f_2(x-x')dx' \right] e^{-j2\pi ux}dx = F_1(u)F_2(u) . \qquad (2.6)$$

Equation (2.6) states that the Fourier transform of the convolution of two functions is equal to the product of the Fourier transforms of the two functions taken separately.

If the original spatial domain function $f(x, y)$ is 2D, the Fourier transform can be easily extended to 2D by the inclusion of the Fourier transform in the second dimension:

$$F(u,v) = \int_{-\infty}^{\infty}\int_{-\infty}^{\infty} f(x,y)e^{-j2\pi(ux+vy)}dxdy . \qquad (2.7)$$

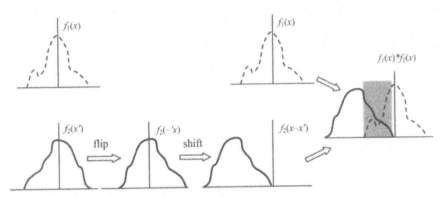

Figure 2.4 Illustration of the convolution process between functions $f_1(x)$ and $f_2(x)$. $f_2(x)$ is first flipped and shifted with respect to its x axis. The resulting convolution represents the area under the product of the two functions.

The inverse Fourier transform can be defined in a similar manner:

$$f(x,y) = \int_{-\infty}^{\infty} \int_{-\infty}^{\infty} F(u,v) e^{j2\pi(ux+vy)}\, du\, dv . \tag{2.8}$$

All properties presented for the 1D Fourier transform can be extended to the two dimensional case.

2.1.2 Random variables

For a random variable x', its cumulative distribution function $P(x)$ is defined as a function whose value at each point x is the probability that a random observation of x' will be less than or equal to x. Mathematically, it is represented by

$$\Pr\{x' \le x\} = P(x) . \tag{2.9}$$

The probability density function $p(x)$ is defined as

$$p(x) = \frac{dP(x)}{dx} . \tag{2.10}$$

Clearly, the cumulative distribution function $P(x)$ is simply the integration of the probability density function $p(x)$ from $-\infty$ to x:

$$P(x) = \int_{-\infty}^{x} p(x')dx' . \tag{2.11}$$

The probability that the random variable x' will fall in any interval between a and b is the area under the probability density function $p(x)$ from a to b:

$$\Pr(a < x' \le b) = \int_{a}^{b} p(x)dx = P(b) - P(a) . \tag{2.12}$$

The pictorial representation is shown in Fig. 2.5. The shaded area under the probability density function [Fig. 2.5(a)] equals the difference between two points on the cumulative distribution function [Fig. 2.5(b)]. Because a random observation x must have some value, we have

$$P(\infty) = \int_{-\infty}^{\infty} p(x)dx = 1.$$

The expected value or the mean of a random variable $\overline{x}$ is defined as

$$E(x) = \overline{x} = \int_{-\infty}^{\infty} xp(x)dx. \qquad (2.13)$$

The variance of a random variable σ^2 is defined as the second moment of the probability density function with respect to the mean:

$$\sigma^2 = E(x - \overline{x})^2 = \int_{-\infty}^{\infty} (x - \overline{x})^2 p(x)dx. \qquad (2.14)$$

σ is often called the standard deviation of x. One widely used distribution function is the normal distribution (also called the Gaussian distribution), which has the probability density function

$$p(x) = \frac{1}{\sigma\sqrt{2\pi}} e^{-(x-\mu)^2/2\sigma^2}, \qquad (2.15)$$

where μ is the mean and σ is the standard deviation. This distribution is of particular interest in both applied and theoretical statistics. One application is the estimation of confidence interval defined by the following probability:

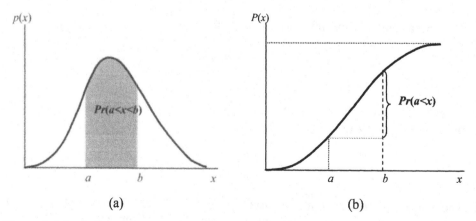

(a)   (b)

Figure 2.5 Illustrations of (a) the probability that random variable x falls in an interval from a to b using probability density function $p(x)$, and (b) cumulative distribution function $P(x)$.

$$\Pr(x_1 < x < x_2) = \int_{x_1}^{x_2} \frac{1}{\sigma\sqrt{2\pi}} e^{-(x-\mu)^2/2\sigma^2} dx = 1-\alpha, \qquad (2.16)$$

where $(1-\alpha)100\%$ is called the confidence interval, which states that a selected random sample will fall inside the region (x_1, x_2) with a $1-\alpha$ probability. If $\bar{x}$ is the value of the mean of N random samples from a normal distributed population with variance σ^2, $(1-\alpha)$ 100% confidence interval for the population mean μ is

$$\bar{x} - z_{\alpha/2}\left(\frac{\sigma}{\sqrt{N}}\right) < \mu < \bar{x} + z_{\alpha/2}\left(\frac{\sigma}{\sqrt{N}}\right), \qquad (2.17)$$

where $z_{\alpha/2} = 1.282, 1.645, 1.960, 2.326,$ and 2.576 for 80%, 90%, 95%, 98%, and 99% confidence intervals, respectively, and $\sigma/\sqrt{N}$ is the standard deviation of $\bar{x}$. It should be pointed out that the confidence interval is not unique. For example, the one-sided $(1-\alpha)$ 100% confidence interval for population mean μ is defined by

$$\mu < \bar{x} + z_{\alpha}\left(\frac{\sigma}{\sqrt{N}}\right), \qquad (2.18)$$

where $z_{\alpha} = 1.282, 1.645, 1.960, 2.326,$ and 2.576 for 90%, 95%, 97.5%, 99%, and 99.5% confidence intervals, respectively. Note that when $N \geq 30$, the distribution of $\bar{x}$ approximately follows a normal distribution with the mean μ and the standard deviation of $\sigma/\sqrt{N}$, regardless of the shape of the population sampled. This is the so-called central limit theorem. In Chapter 5, this theorem will be used to estimate the low-contrast resolution of a CT system.

For statistics related to x-ray photons, one often uses the Poisson distribution. Unlike in the previous discussion, Poisson is a discrete distribution in that the variable takes on only non-negative integers. For example, in x-ray photon statistics, the Poisson distribution represents the probability of observing x number of photons in a fixed time interval. The distribution is represented by

$$p(x) = \frac{m^x e^{-m}}{x!}, \qquad (2.19)$$

where $x = 0, 1, 2, \dots$. It can be shown that both the mean and the variance of the Poisson distribution equal m. Figure 2.6 shows examples of the Poisson distribution for $m = 10, 20,$ and 30. This section discusses only a few selected topics with a limited scope on statistics, so interested readers can refer to Refs. 4–7 for more detailed treatments of the subject.

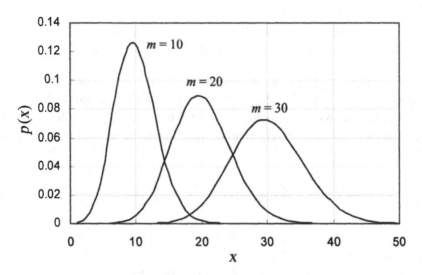

Figure 2.6 Poisson distributions for *m* = 10, 20, and 30.

2.1.3 Linear algebra

An *m* × *n* matrix **A** is a rectangular array of numbers arranged in *m* rows and *n* columns:

$$\mathbf{A} = \begin{bmatrix} a_{11} & a_{12} & \cdots & a_{1n} \\ a_{21} & a_{22} & \cdots & a_{2n} \\ \vdots & & & \vdots \\ a_{m1} & a_{m2} & \cdots & a_{mn} \end{bmatrix}. \tag{2.20}$$

The transpose of a matrix **A**, denoted by $\mathbf{A}^T$, is an *n* × *m* matrix obtained by writing the rows of **A**, in order, as columns:

$$\mathbf{A}^T = \begin{bmatrix} a_{11} & a_{12} & \cdots & a_{1n} \\ a_{21} & a_{22} & \cdots & a_{2n} \\ \vdots & & & \vdots \\ a_{m1} & a_{m2} & \cdots & a_{mn} \end{bmatrix}^T = \begin{bmatrix} a_{11} & a_{21} & \cdots & a_{m1} \\ a_{12} & a_{22} & \cdots & a_{m2} \\ \vdots & & & \vdots \\ a_{1n} & a_{m2} & \cdots & a_{mn} \end{bmatrix}. \tag{2.21}$$

One special case of a matrix is a square matrix (*m* = *n*). The square matrix has many applications in medical imaging. For example, most CT images are reconstructed into 512 × 512 pixels and can be represented by a square matrix.

A diagonal matrix is a square matrix whose nondiagonal entries are all zero:

$$\mathbf{A} = \begin{bmatrix} a_{11} & 0 & \cdots & 0 \\ 0 & a_{22} & \cdots & 0 \\ \vdots & & & \vdots \\ 0 & 0 & \cdots & a_{nn} \end{bmatrix}. \tag{2.22}$$

If $a_{ii} = 1$ for $i = 1, 2, \ldots, n$ of a diagonal matrix, this matrix is called a unit or identity matrix and is often denoted by $\mathbf{I}_n$.

Another special case of an $m \times n$ matrix $\mathbf{A}$ is a matrix with $m = 1$:

$$\mathbf{A} = \begin{bmatrix} a_1, & a_2, & \cdots, & a_n \end{bmatrix}. \tag{2.23}$$

This matrix is often called a row vector. The transpose of a row vector forms a column vector. A set of projection measurements that forms a view in a CT scan can be considered a vector.

The sum of two $m \times n$ matrices $\mathbf{A} = [a_{i,j}]$ and $\mathbf{B} = [b_{i,j}]$ is another $m \times n$ matrix, $\mathbf{C} = [c_{i,j}]$:

$$c_{i,j} = a_{i,j} + b_{i,j}. \tag{2.24}$$

A zero matrix, or additive identity matrix, is a matrix that consists of m rows and n columns of zeroes. If $\mathbf{C}$ is a zero matrix in Eq. (2.24), matrix $\mathbf{B}$ is an additive inverse of $\mathbf{A}$.

If $\mathbf{A}$ is an $m \times n$ matrix and $\mathbf{B}$ is an $n \times p$ matrix, the product, $\mathbf{C} = \mathbf{AB}$, is an $m \times p$ matrix:

$$c_{i,j} = \sum_{k=1}^{n} a_{ik} b_{kj}. \tag{2.25}$$

When $\mathbf{A}$ is a row vector and $\mathbf{B}$ is a column vector ($m = 1$ and $p = 1$), the product is a single element matrix or a scalar:

$$c = \sum_{k=1}^{n} a_k b_k. \tag{2.26}$$

The product of a scalar r by a matrix $\mathbf{A}$ is called the scalar multiplication:

$$r\mathbf{A} = \begin{bmatrix} ra_{11} & ra_{12} & \cdots & ra_{1n} \\ ra_{21} & ra_{22} & \cdots & ra_{2n} \\ \vdots & & & \vdots \\ ra_{m1} & ra_{m2} & \cdots & ra_{mn} \end{bmatrix}. \tag{2.27}$$

A set of properties that is often useful in matrix manipulations is shown below without proof. Interested readers can easily prove these properties by examining the entries of the resulting matrices.

1. associative law: $(\mathbf{AB})\mathbf{C} = \mathbf{A}(\mathbf{BC})$
2. left distributive law: $\mathbf{A}(\mathbf{B} + \mathbf{C}) = \mathbf{AB} + \mathbf{AC}$
3. right distributive law: $(\mathbf{B} + \mathbf{C})\mathbf{A} = \mathbf{BA} + \mathbf{CA}$
4. if r is a scalar: $r(\mathbf{AB}) = (r\mathbf{A})\mathbf{B} = \mathbf{A}(r\mathbf{B})$
5. for an n-square matrix $\mathbf{A}$: $\mathbf{AI}_n = \mathbf{I}_n\mathbf{A} = \mathbf{A}$
6. $(\mathbf{AB})^{\mathrm{T}} = \mathbf{B}^{\mathrm{T}}\mathbf{A}^{\mathrm{T}}$
7. $(\mathbf{A+B})^{\mathrm{T}} = \mathbf{A}^{\mathrm{T}} + \mathbf{B}^{\mathrm{T}}$

For the convenience of a computer implementation of the algorithms, we can define a set of matrix element-by-element algebraic operations that are typically not defined in a linear algebra textbook. We call these the elemental addition, multiplication, and division operations, denoted by $\oplus, \otimes$, and $-$, respectively. If r is a scalar and $\mathbf{A}$ is an $m \times n$ matrix, the elemental addition is defined as

$$r \oplus \mathbf{A} = \mathbf{A} \oplus r = \begin{bmatrix} r+a_{11} & r+a_{12} & \dots & r+a_{1n} \\ r+a_{21} & r+a_{22} & \dots & r+a_{2n} \\ \vdots & & & \vdots \\ r+a_{m1} & r+a_{m2} & \dots & r+a_{mn} \end{bmatrix}. \tag{2.28}$$

If $\mathbf{A}$ and $\mathbf{B}$ are $m \times n$ matrices, their elemental product $\mathbf{C}$ is also an $m \times n$ matrix:

$$\mathbf{C} = \mathbf{A} \otimes \mathbf{B} = \begin{bmatrix} a_{11}b_{11} & a_{12}b_{12} & \dots & a_{1n}b_{1n} \\ a_{21}b_{21} & a_{22}b_{22} & \dots & a_{2n}b_{2n} \\ \vdots & & & \vdots \\ a_{m1}b_{m1} & a_{m2}b_{m2} & \dots & a_{mn}b_{mn} \end{bmatrix}. \tag{2.29}$$

Similarly, we define the elemental division of a scalar r by a matrix $\mathbf{A}$ as

$$\frac{r}{\mathbf{A}} = \begin{bmatrix} r/a_{11} & r/a_{12} & \dots & r/a_{1n} \\ r/a_{21} & r/a_{22} & \dots & r/a_{2n} \\ \vdots & & & \vdots \\ r/a_{m1} & r/a_{m2} & \dots & r/a_{mn} \end{bmatrix}. \tag{2.30}$$

The elemental division of two matrices $\mathbf{A}$ and $\mathbf{B}$ is defined as

$$\frac{\mathbf{A}}{\mathbf{B}} = \begin{bmatrix} a_{11}/b_{11} & a_{12}/b_{12} & \cdots & a_{1n}/b_{1n} \\ a_{21}/b_{21} & a_{22}/b_{22} & \cdots & a_{2n}/b_{2n} \\ \vdots & & & \vdots \\ a_{m1}/b_{m1} & a_{m2}/b_{m2} & \cdots & a_{mn}/b_{mn} \end{bmatrix}. \tag{2.31}$$

In the same spirit, we can define a set of matrix trigonometry functions such as $\sin(\mathbf{A})$, $\cos(\mathbf{A})$, and $\tan(\mathbf{A})$. For example,

$$\tan(\mathbf{A}) = \begin{bmatrix} \tan(a_{11}) & \tan(a_{12}) & \cdots & \tan(a_{1n}) \\ \tan(a_{21}) & \tan(a_{22}) & \cdots & \tan(a_{2n}) \\ \vdots & & & \vdots \\ \tan(a_{m1}) & \tan(a_{m2}) & \cdots & \tan(a_{mn}) \end{bmatrix}. \tag{2.32}$$

It can be argued that the definitions of these operators are unnecessary, since the same results can be achieved through the conventional linear algebraic operators. For example, the elemental addition can be achieved by the following series of operations:

$$r \oplus \mathbf{A} = (r\mathbf{U}) + \mathbf{A},$$

where $u_{i,j} = 1$. The computer implementation of the second notation implies that we must first generate an intermediate matrix $\mathbf{U}$, multiply it by the scalar r, and perform matrix addition. In reality, many image processing software packages, such as MATLAB® and IDL™, have the elemental addition operator defined. Therefore, explicitly defining these notations allows the user to program these operations more efficiently. This will become clear as we discuss the matrix implementation of parallel-beam and fan-beam backprojection operations in Chapter 3.

Because the focus of this chapter is to provide a refresher course on the subject, many theories and properties of linear algebra cannot be covered. Fortunately, many linear algebra textbooks are available, and interested readers should refer to these books for further study.[8,9]

2.2 Fundamentals of X-ray Physics

2.2.1 Production of x rays

An x ray is an electromagnetic waveform, as are microwaves, infrared, visible light, ultraviolet, and radio waves. The wavelength of the x ray ranges from a few picometers to a few nanometers. The energy of each x-ray photon E is proportional to its frequency ν, and is described by the following expression:

$$E = h\nu = \frac{hc}{\lambda},$$ (2.33)

where h is Planck's constant and equals 6.63×10^{-34} J•s, c is the speed of light and equals 3×10^8 m/s, and λ is the wavelength of the x ray. Therefore, x-ray photons with longer wavelengths have lower energies than the photons of shorter wavelengths. For convenience, x-ray energy is usually expressed in the unit of eV (1 eV = 1.602×10^{-19} J). This is the amount of kinetic energy with which an electron is accelerated across an electrical potential of 1 V (the actual energy equals the product of the electron charge times the voltage). Since x-ray photons are produced by striking a target material with high-speed electrons (the kinetic energy is transformed into electromagnetic radiation), the maximum possible x-ray photon energy equals the entire kinetic energy of the electron. Equation (2.30) can then be converted to the unit of eV:

$$E = \frac{1.24 \times 10^3 \, eV \times nm}{\lambda}.$$ (2.34)

An x ray with a wavelength in the range of 10 nm (124 eV) to 0.1 nm (12.4 keV) is often called a soft x ray due to its inability to penetrate the thicker layers of materials. These x rays have little value in radiology. The wavelength of diagnostic x-rays varies roughly from 0.1 nm to 0.01 nm, corresponding to an energy range of 12.4 keV to 124 keV. Although x rays with a much shorter wavelength are highly penetrating, they provide little low-contrast information, and therefore are of little interest to diagnostic imaging. It is worth pointing out that x rays are at the high-energy (short wavelength) end of the electromagnetic spectrum. For illustration, Fig. 2.7 depicts the electromagnetic spectrum with different types of electromagnetic waves labeled.

X-ray photons are produced when a substance is bombarded by high-speed electrons. When a high-speed electron interacts with the target material, different types of collisions take place. The majority of these encounters involve small energy transfers from a high-speed electron to electrons that are knocked out of the atoms, leading to ionization of the target atoms. This type of interaction does not produce x rays but gives rise to delta rays and eventually heat. For a typical x-ray tube, over 99% of the input energy is converted to heat.

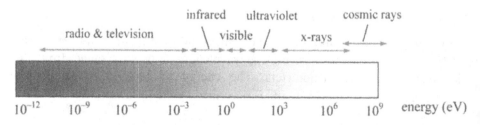

Figure 2.7 Illustration of the electromagnetic spectrum.

The more interesting interactions are the following three types of interactions. The first occurs when an electron approaches close to the nucleus of an atom and suffers a radiation loss, as shown in the upper-left part of Fig. 2.8. The bottom portion of Fig. 2.8 shows a typical x-ray spectrum produced by an x-ray tube operating at a 120-kV potential with additional filtration to remove the low-energy photons. The high-speed electron travels partially around the nucleus due to the attraction between the positive nucleus and the negative electron. The sudden deceleration of the electron produces bremsstrahlung radiation. The energy of the resulting radiation depends on the amount of incident kinetic energy that is given off during the interaction. If the energetic electron barely grazes the atomic coulomb field, the resulting x-ray has relatively low energy. As the amount of interaction increases, the resulting x-ray energy increases. This type of radiation is responsible for white radiation, which covers the entire range of the energy spectrum.

X rays can also be generated by other charged particles besides high-speed electrons, such as protons or alpha particles. The total intensity of the bremsstrahlung radiation I that results from a charged particle of mass m and charge ze incident on the target nuclei with charge Ze is proportional to[10]

$$I \propto \frac{Z^2 z^4 e^6}{m^2}. \qquad (2.35)$$

This equation indicates that an electron is over 3 million times more efficient at generating bremsstrahlung than a massive particle such as a proton or alpha particle. This is one of the major reasons that high-speed electrons are the practical choice for the production of x rays. Equation (2.35) also indicates that the bremsstrahlung production efficiency increases rapidly as the atomic number of the target Z increases.

The second type of interaction occurs when a high-speed electron collides with one of the inner shell electrons of the target atom (the upper-middle part of Fig. 2.8 depicts a collision with a K-shell electron) and liberates the inner shell electron. When the hole is filled by an electron from an outer shell, characteristic radiation is emitted. In the classic Bohr model of an atom, electrons occupy orbits with specific quantized energy levels. The energy of the characteristic x ray is the difference between the binding energies of two shells. For example, the binding energies of the K, L, M, and N shells of tungsten are 70, 11, 3, and 0.5 keV, respectively. When a K-shell electron is liberated and the hole is filled with an L-shell electron, a 59-keV x-ray photon is generated. Similarly, when an M-shell electron moves to the K-shell, the produced x-ray photon has 67 keV energy. This is illustrated by the characteristic peaks in the x-ray spectrum shown in Fig. 2.8. Note that each element in the periodic table has its own unique shell binding energies, so the energies of characteristic x rays are unique to each atom.

The third type of interaction occurs when an electron collides directly with a nucleus and its entire energy appears as bremsstrahlung. The x-ray energy produced by this interaction represents the upper energy limit in the x-ray spectrum. For example, for an x-ray tube operating at 120 kVp, this interaction produces x-ray photons with 120-keV energy, as shown in Fig. 2.8. The probability of such collisions is low, as shown by its near-zero magnitude in the spectrum.

Before concluding this section, we want to emphasize the distinction between kV (electric potential) and keV (energy). The unit kVp is often used to prescribe a CT scan. For instance, a 120-kVp CT scan indicates that the applied electric potential across the x-ray tube is 120 kV. Under this condition, the electrons that strike the target have 120 keV of kinetic energy. The highest-energy photons that can be produced by such process are 120 keV.

2.2.2 Interaction of x rays with matter

The typical energy range of the x-ray photons generated for medical CT is roughly between 20 keV and 140 keV. In this energy range, there are three fundamental ways in which x rays interact with matter: the photoelectric effect, the Compton effect, and coherent scattering.

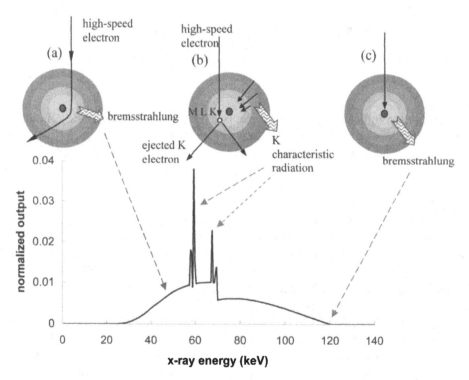

Figure 2.8 Illustration of electron interaction with a target and its relationship to the x-ray tube energy spectrum. (a) Bremsstrahlung radiation is generated when high-speed electrons are decelerated by the electric field of the target nuclei. (b) Characteristic radiation is produced when a high-speed electron interacts with a target electron and ejects it from its electronic shell. When outer-shell electrons fill in the vacant shell, characteristic x rays are emitted. (c) A high-speed electron hits the nucleus directly, and the entire kinetic energy is converted to x-ray energy. For the x-ray spectrum shown in the figure, the target material is tungsten, and additional filtration is used to remove low-energy x rays.

The photoelectric effect describes a situation in which the x-ray photon energy is greater than the binding energy of an electron, and the incident x-ray photon gives up its entire energy to liberate an electron from a deep shell of an atom, as shown in Fig. 2.9(a). The free electron is often called a photoelectron. The photon then ceases to exist. The hole created in the deep shell is filled by an outer-shell electron. Since the outer-shell electron is at a higher energy state than the inner shell, a characteristic radiation results. Thus, the photoelectric effect produces a positive ion (the affected atom lacks an electron to be electrically neutral), a photoelectron, and a photon of characteristic radiation. This effect was first explained by Albert Einstein in 1905, for which he received the Nobel Prize in Physics in 1921.

For tissue-like materials, the binding energy of the K-shell electrons is very small (roughly 500 eV). Hence, the photoelectron acquires essentially the entire energy of the x-ray photon. In addition, at such low energy, the characteristic x rays produced in the interaction do not travel very far (less than the dimensions of a typical human cell) before being attenuated. Even for materials such as calcium (a constituent of bone), the K-shell binding energy is only 4 keV. Since the mean free path of a 1-keV x-ray photon in muscle tissue is about 2.7 µm, we can safely assume that all of the characteristic x rays produced in patients by the photoelectric effect are reabsorbed.

The more tightly bound electrons are more important in bringing about photoelectric absorption. For example, the two K electrons in lead are more than five times as effective as the eight L electrons in a photoelectric interaction. In addition, the maximum absorption occurs when the incident x-ray photon has just enough energy to eject the bound electron. In fact, the probability of the photoelectric interaction $P_{\text{photoelectric}}$ is roughly inversely proportional to the cubic of the excess photon energy[11]:

$$P_{\text{photoelectric}} \propto E^{-3}. \qquad (2.36)$$

In addition, the probability of interaction $P_{\text{photoelectric}}$ is proportional to the cube of the atomic number Z (Ref. 11):

$$P_{\text{photoelectric}} \propto Z^{3}. \qquad (2.37)$$

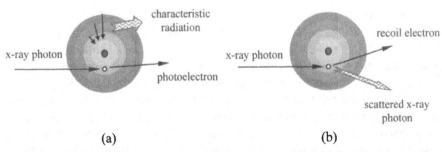

(a)   (b)

Figure 2.9 Illustrations of (a) photoelectric interaction and (b) Compton interaction.

Consequently, tissues with small differences in atomic numbers produce greater differences in the probabilities of photoelectric effects. This, in turn, results in different absorption rates of x-ray photons and leads to greater contrast between different tissues. Equations (2.36) and (2.37) indicate that lower-energy x-ray photons are important for the low-contrast differentiation of tissues.

The second way in which x rays interact with matter, the Compton effect, is named after Arthur Holly Compton, who received the Nobel Prize in 1927 for its discovery. This is an important interaction mechanism in tissue-like materials.[12] In this interaction, the energy of the incident x-ray photon is considerably higher than the binding energy of the electron. An incident x-ray photon strikes an electron and frees the electron from the atom. The incident x-ray photon is deflected or scattered with partial loss of its initial energy, as shown in Fig. 2.9(b). Thus, a Compton interaction produces a positive ion, a "recoil" electron, and a scattered photon. The scattered photon may be deflected at any angle from 0 to 180 deg. Low-energy x-ray photons are preferentially backscattered (with a deflection angle larger than 90 deg), whereas high-energy photons have a higher probability of forward scattering (a deflection angle smaller than 90 deg). Because of the wide deflection angle, the scattered photon provides little information about the location of interaction and the photon path.

After a Compton interaction, most of the energy is retained by the photon. The deflected photon may undergo additional collisions before exiting the patient. Since only a small portion of the photon energy is absorbed, the energy (or radiation dosage) absorbed by the patient is considerably less than the photoelectric effect. The probability of a Compton interaction depends on the electron density of the material, not the atomic number Z. The lack of dependence on the atomic number provides little contrast information between different tissues (the electron density difference between different tissues is relatively small). Consequently, nearly all medical CT devices try to minimize the impact of the Compton effect by either postpatient collimation or algorithmic correction.

The third (and least important to researchers in clinical CT) way in which x rays interact with matter is coherent scattering (also known as Rayleigh scattering). In this interaction, no energy is converted into kinetic energy, and ionization does not occur. The process is identical to what occurs in the transmitter of a radio station. An electromagnetic wave with an oscillating electric field sets the electrons in an atom into momentary vibration. These oscillating electrons emit radiation of the same wavelength. Because the scattering is a cooperative phenomenon, it is called coherent scattering.

Coherent scattering occurs mainly in the forward direction to produce a slightly broadened x-ray beam. Since no energy is transferred to kinetic energy, this process has historically limited interest to CT. Laboratory research, however, has shown that coherent scattering can be used for bone characterization.[13, 14]

To understand the relative importance of different types of interactions, Fig. 2.10 shows the percent of interactions in water as a function of the photon energy (in the diagnostic energy range). The graph is based on Table 5-5 of Ref. 12. Although the graph is for interactions in water, the same graph can be used with little error for interactions in soft tissues. Note that the percent of photoelectric interaction decreases quickly with the increase in x-ray energy, while the percent of the Compton interaction increases quickly with the increase in x-ray energy.

A more important measurement is the percent of the energy transferred as a function of x-ray photon energy for different processes, as shown in Fig. 2.11. Note that at low x-ray energies, the majority of the energy is transferred by the photoelectric process. For example, although at 30 keV only 36.3% of the interactions are photoelectric, the energy transferred by the process is more than 93%. This phenomenon can be understood from the fact that more energy is transferred in a photoelectric interaction than a Compton interaction. Based on Fig. 2.11, we can divide the diagnostic x-ray energy range into three different zones. The first zone covers the energy range up to 50 keV, as depicted by the light-shaded region in Fig. 2.11. In this zone, photoelectric absorption dominates. In the next zone, represented by the dark-shaded region (between 50 and 90 keV), both the photoelectric and Compton interactions are important. The third zone is the remaining region between 90 and 150 keV. The Compton interaction clearly dominates. Many investigations and developments on dual-energy CT imaging are based on the dominating interactions in different x-ray energy zones. This topic will be discussed in more detail in Chapter 12.

The net effect of these interactions (photoelectric, Compton, and coherent scattering) is that some of the photons are absorbed or scattered. In other words, x-ray photons are attenuated when they pass through a material. The attenuation can be expressed by an exponential relationship for a monochromatic (monoenergetic) incident x-ray beam and a material of uniform density and atomic number:

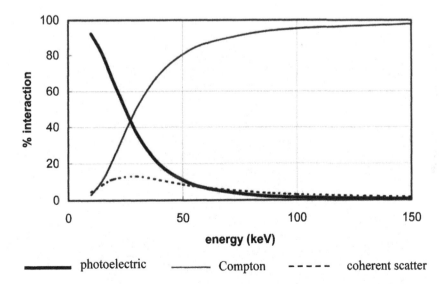

Figure 2.10 Percentages of different types of interactions as a function of energy in water.

$$I = I_0 e^{-(\tau+\sigma+\sigma_r)L} , \qquad\qquad (2.38)$$

where I and I_0 are the incident and transmitted x-ray intensities; L is the thickness of the material; and τ, σ, and σ_r are the attenuation coefficients of the photoelectric, Compton, and coherent scattering interactions of the material, respectively. Equation (2.38) is often expressed as

$$I = I_0 e^{-\mu L} , \qquad\qquad (2.39)$$

where μ is the linear attenuation coefficient of the material. This is often called the Beer–Lambert law. Clearly, μ is a function of the incident x-ray photon energy. Figure 2.12 plots μ as a function of x-ray energy for iodine, bone, soft tissue, and water (μ is calculated based on a table provided by the National Institute of Standards and Technology (NIST), which can be found at: http://physics.nist.gov/PhysRefData/XrayMass-Coef/cover.html.

Several observations can be made. First, in the diagnostic x-ray energy range, water and soft tissue have nearly identical attenuation coefficients. This is not surprising because a large portion of soft tissue is made up of water. Therefore, water phantoms are often used in the CT calibration process to ensure accurate CT numbers in soft tissues.

Second, iodine has a much higher attenuation coefficient (and therefore is more attenuating) than soft tissue or water. Thus, iodine is often injected into patients intravenously to be used as a contrast agent so that blood vessels appear more opaque in x rays. For example, in CT angiographic studies, iodine contrast

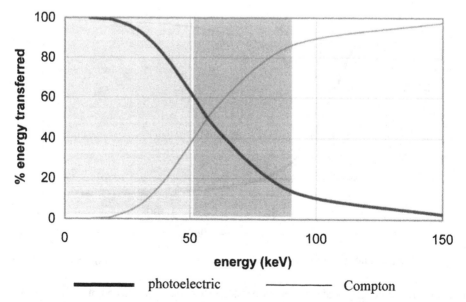

Figure 2.11 Percentages of energy transfer of different interactions in water.

injection is followed by CT scans to investigate the integrity of blood vessels and identify the presence of stenosis. A closer inspection of the iodine attenuation curve shows that a sudden change in the attenuation coefficient occurs at 33.2 keV, which is the binding energy of the K-shell electron. At the x-ray energy slightly lower than 33.2 keV, the attenuation coefficient of iodine is 32.31 cm^{-1}, while the attenuation jumps to 176.59 cm^{-1} for x-ray energies slightly higher than 33.2 keV. This phenomenon is called the K edge of iodine. The K edges are different for different materials. For example, the K edge is 69.5 keV for tungsten and 50.2 keV for gadolinium.

Third, the attenuation coefficient of bone is higher and the shapes of the attenuation curves are significantly different from those of soft tissue. This often leads to CT imaging artifacts due to the broad x-ray spectrum of the x-ray flux produced by the x-ray tubes. Closer inspection of the bone attenuation curves also shows that a discontinuity occurs at 4.0 keV, which is the K edge of calcium. This is not surprising, since a major portion of the bone is made of calcium. This will be discussed in more detail in Chapter 7.

For an x-ray tube operating at 120 kVp, the average photon energy is roughly 70 keV. At the average energy, the μ values for water and muscle are 0.1928 cm^{-1} and 0.1916 cm^{-1}, respectively. Clearly, the difference is quite small. To enhance small differences between different tissue types, the intensity scale (called the CT number) used in the reconstructed CT image is defined by

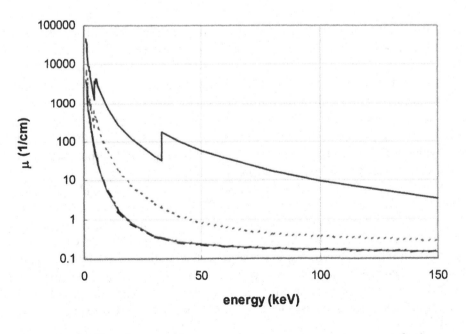

Figure 2.12 Linear attenuation coefficients for different materials.

$$\text{CT number} = \frac{\mu - \mu_{\text{water}}}{\mu_{\text{water}}} \times 1000, \tag{2.40}$$

where μ_{water} is the linear attenuation coefficient of water. This unit is often called the Hounsfield unit (HU), honoring the inventor of CT. By definition, water has a CT number of zero. The CT number for air is -1000 HU, since $\mu_{\text{air}} = 0$. Soft tissues (including fat, muscle, and other body tissues) have CT numbers ranging from -100 HU to 60 HU. Cortical bones are more attenuating and have CT numbers from 250 HU to over 1000 HU.

2.3 Measurement of Line Integrals and Data Conditioning

Chapter 1 explained that CT relies on x-ray flux measurements from different angles to form an image. At each angle, the measurement is essentially identical to that of a conventional x ray. We record the x-ray flux impinging on the x-ray detector after attenuation by a patient. If we assume that the input x-ray photons are monoenergetic, the x-ray intensities measured on the entrance and exit sides of a uniform material, shown in Fig. 2.13(a), follow the Beer–Lambert law, as discussed in Section 2.2.2:

$$I = I_0 e^{-\mu \Delta x}. \tag{2.41}$$

In this equation, I_0 is the entrance x-ray intensity, I is the exit x-ray intensity, Δx is the thickness, and μ is the linear attenuation coefficient of the material. In general, μ changes with the x-ray energy and varies with the selection of the material.

From Eq. (2.41), it is clear that objects with higher μ values are more attenuating to the x-ray photons than objects with lower μ values. For example, the μ value of bone is higher than that of soft tissues, indicating that it is more difficult for x-ray photons to penetrate bones than soft tissues. On the other hand, the μ value for air is nearly zero, indicating that the input and output x-ray flux is virtually unchanged when passing through the air ($e^0 = 1$).

Now consider the case of a nonuniform object (an object made of multiple materials with different attenuation coefficients). The overall attenuation characteristics can be calculated by dividing the object into smaller elements, as shown in Fig. 2.13(b). When the size of each element is sufficiently small, each element can be considered as a uniform object. Equation (2.41) is now valid to describe the entrance and exit x-ray intensities for each element. Since the exit x-ray flux from an element is the entrance x-ray flux to its neighbor, Eq. (2.41) can be repeatedly applied in a cascade fashion. Mathematically, it is expressed as the following:

$$I = I_0 e^{-\mu_1 \Delta x} e^{-\mu_2 \Delta x} e^{-\mu_3 \Delta x} \cdots e^{-\mu_n \Delta x} = I_0 e^{-\sum_{n=1}^{N} \mu_n \Delta x}. \tag{2.42}$$

If we divide both sides of Eq. (2.42) by I_0 and take the negative logarithm of the quantity, we obtain

$$p = -\ln\left(\frac{I}{I_0}\right) = \sum_{n=1}^{N}\mu_n\Delta x . \tag{2.43}$$

When Δx approaches zero, the above summation becomes an integration over the length of the object:

$$p = -\ln\left(\frac{I}{I_0}\right) = \int_L \mu(x)dx . \tag{2.44}$$

In CT, p is the projection measurement. Equation (2.44) states that the ratio of the input x-ray intensity over the output x-ray intensity after a logarithm operation represents the line integral of the attenuation coefficients along the x-ray path. The reconstruction problem for CT can now be stated as follows: Given the measured line integrals of an object, how do we estimate or calculate its attenuation distribution?

Before trying to find a solution to the image reconstruction problem, detailed discussions of Eq. (2.44) are necessary. The relationship presented in this equation is valid only under "ideal" conditions, but the measurements obtained on any medical CT scanner rarely satisfy these conditions. As a result, the data obtained on the scanner must be conditioned or preprocessed before any reconstruction formula can be applied. In fact, the mathematical formulas known to many people for CT reconstruction represent only a small fraction of the total amount of computation required to obtain an image. To separate the data-

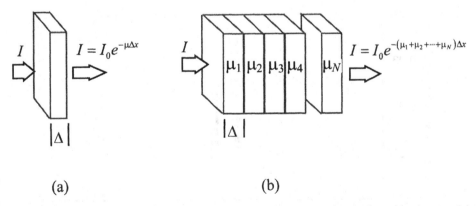

(a)   (b)

Figure 2.13 Illustration of material attenuation for a monochromatic x-ray beam. (a) Attenuation of monoenergetic x rays by a uniform object follows the Beer–Lambert law. (b) Any nonuniform object can be subdivided into multiple elements. Within each element, a uniform attenuation coefficient can be assumed. The Beer–Lambert law can then be applied in a cascade fashion.

conditioning portion of the reconstruction from the "ideal data" reconstruction, we call the algorithms presented in the next chapter the "textbook" reconstructions to refer to the fact that they have been presented in many textbooks and publications.[15–18] To appreciate the complexity of the preprocessing steps, we will briefly outline some of the major sources of error that are an integral part of the data collection process.

One basic assumption of Eq. (2.44) is the monoenergetic nature of the input x-ray beam. This assumption requires all of the x-ray photons emitting from an x-ray source to have the same energy. In "real world" clinical CT, this condition is rarely satisfied. The output energy spectrum of an x-ray tube is quite broad. For example, when we select 120 kVp on a CT scanner, the output x-ray photon energies vary between 10 kVp and 120 kVp, as shown in Fig. 2.14. For the majority of the materials, the μ value varies significantly with the x-ray energy, as previously illustrated in Fig. 2.12. Because of the polyenergetic nature of the x-ray beam and the energy dependency of μ, Eq. (2.44) is no longer valid. The linear relationship between the measured projection and the thickness of the object does not exist. This is the well-known beam-hardening problem. Beam hardening can cause cupping, shading, or streaking artifacts in reconstructed images.[19–21] A detailed discussion of this subject can be found in Chapter 7.

Another challenging problem that affects the accuracy of the projection measurement is scattered radiation. Not all of the x-ray photons that reach the detector are primary photons. A significant portion of the signal is generated from scattered radiation. Scattered radiation adds a low-frequency bias to the true attenuation measurements. The minus logarithm operation that is necessary to convert x-ray photon flux to the line integrals [Eq. (2.44)] transforms the linear summation to a nonlinear operation. This can be easily understood from the fact that $\log(x+y) \neq \log(x) + \log(y)$. Unlike conventional radiography, scattered

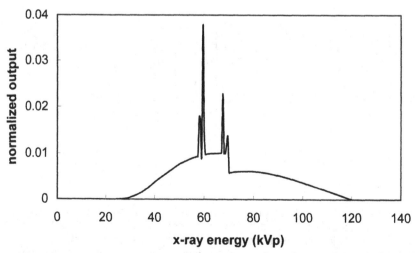

Figure 2.14 Example of the x-ray energy spectrum of an x-ray tube operating at 120 kVp.

radiation not only causes inaccuracy in the reconstructed CT numbers, but also produces shading and streaking image artifacts.[19,22,23] The problem becomes more pronounced with the recent introduction of multislice scanners, since in general scattered radiation increases with irradiated volume. Examples of the scattered radiation impact are included in Chapter 7.

A third source of measurement error comes from the nonlinearity of the detector or data acquisition systems. For example, it is well known that dark currents are present in all electronics equipment. Dark currents produce a channel-dependent and temperature-dependent bias to the measured x-ray flux. Another example of nonlinearity is the gain variation of the detector scintillation material. Some scintillators, such as $CdWO_4$, are known to exhibit hysteresis or radiation damage phenomenon. The detector gain becomes dependent on its radiation exposure history. Although much of the radiation damage phenomenon can be self-recovered over an extended period of time, it can present significant problems for continuous patient scans or repeated scans.

A fourth source of measurement inaccuracy comes from the scanned object itself. In most clinical applications, the object of interest is a living patient. Even the most advanced CT scanners take seconds to complete their coverage of an organ, so voluntary and involuntary patient motions are inevitable. As a result, projections taken at different moments in time do not represent the attenuation line integrals of the same object, a circumstance that deviates from the basic tomographic assumption. It is not difficult to understand that a violation of inconsistency conditions can lead to image artifacts.[19,24]

Other error sources can also lead to image artifacts and measurement inaccuracies. A partial list of these error sources includes off-focal radiation of the x-ray source, the presence of metal objects in the scanning plane, x-ray photon starvation, CT gantry misalignment, x-ray tube arcing, a deficiency in the projection sampling, a partial volume effect, focal spot drift, mechanical vibration, projection truncation, x-ray tube rotor wobble, and special data acquisition modes. A detailed treatment of these topics is left for Chapter 7.

It is clear from the above discussion that CT reconstruction is not limited to the problem of image formation using ideal projections. In fact, the preprocessing and postprocessing steps used to overcome nonideal data collection often outweigh, in terms of the number of operations, "textbook" tomographic reconstruction. Figure 2.15 depicts a representative flow diagram of the CT image formation process. For simplicity, we placed the "preprocessing" block ahead of the "minus logarithm" operation. In reality, however, some of the preprocessing steps are performed after the logarithm step. Many of the preprocessing and postprocessing steps are proprietary in nature and therefore are not available in the public domain literature. On the other hand, most textbook reconstruction methods are nonproprietary. The focus of the next chapter is on textbook tomographic reconstruction. Other pre- and postprocessing steps will be discussed in detail in Chapters 7 and 10.

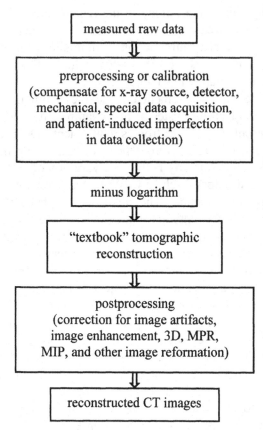

Figure 2.15 Flow diagram of the CT image formation process. For simplicity, all preprocessing steps are placed in a block ahead of the minus logarithm step. In reality, many of the preprocessing steps are performed after the logarithm operation.

2.4 Sampling Geometry and Sinogram

For the ease of future reference, we will define the terminologies used throughout this book. First, we will define the projection sampling geometry. Chapter 1 described the different generations of CT scanners. Data collected in either first- or second-generation scanners have samples in a single projection that form a set of parallel rays, as shown in Fig. 2.16(a). This type of sampling is generally called *parallel projection*.

Data collected in third- or fourth-generation scanners share the same property in that samples of a single projection focus to a point. This type of data collection is called a *fan-beam projection*, as shown in Fig. 2.16(b). For a third-generation scanner, measurements from all the detector channels collected roughly at the same time instant constitute a projection, as illustrated by Fig. 1.16. The focal point, therefore, is naturally the x-ray source. However, for a fourth-generation scanner, formation of a projection is slightly more complicated. The roles of the x-ray tube and detector are reversed, and the focal point is a detector channel. A single projection is formed by taking measurements on a particular detector at different x-ray tube positions, as shown by the shaded fan-shaped region in Fig. 1.17.

The third data collection mode is called *cone-beam projection*. Although samples from a cone-beam projection still focus to a single point, multiple fan-beam planes are collected simultaneously to cover a volume, as shown in Fig. 2.16(c). Note that of all the fan-beam planes, only one is perpendicular to the axis of rotation. All of the other fan-beam planes are tilted with respect to this axis. We will discuss in Chapter 10 how the tilting of the fan-beam plane is a major source of image artifacts when the scanning trajectory (the loci of the x-ray source) is a single circle. A simple comparison of the three sampling geometries shows that the sampling pattern becomes increasingly complicated as we move from a parallel-beam geometry to a cone-beam geometry. To enable readers to easily comprehend the basic principles of CT reconstruction, we will start our discussion on parallel-beam geometry.

Consider a projection dataset taken over a 2π angular range with a parallel geometry. (Theoretically, only projection data over a π angular range is needed for parallel-beam reconstruction, but 2π projection data are used here to cover the case of the fan beam as well). Clearly, the projection dataset can be presented in many ways. The most popular presentation is the so-called sinogram. In the sinogram space, the horizontal axis represents the detector channels, and the vertical axis represents the projection angle. Therefore, a single projection is represented in the sinogram as a set of samples located along a horizontal line, as shown in Fig. 2.17. The data collected over all of the projection angles form a 2D image with its intensities representing the magnitude of the projection samples. Now examine the projection of a point in the object [specified by its polar coordinate (r, ϕ)] in the sinogram space. To calculate the loci of the point on the detector plane, we define a rotating coordinate system (x', y') with its y' axis parallel to the x-ray beam path, as shown in Fig. 2.17. The x' coordinate of the point (and therefore its location on the projection) follows the following relationship:

$$x' = r \cdot \cos(\phi - \beta). \qquad (2.45)$$

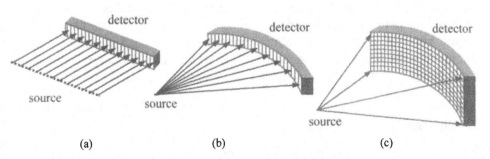

Figure 2.16 Illustration of different data sampling geometries: (a) parallel beam, (b) fan beam, and (c) cone beam.

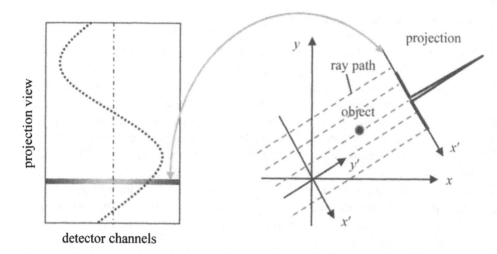

detector channels

Figure 2.17 Illustration of mapping between the object space and the sinogram space (left). A sinogram is formed by stacking all of the projections of different views, so that a single projection is represented by a horizontal line in the sinogram. The projection of a single point forms a sinusoidal curve in the sinogram space.

In this equation, β is the projection angle formed with the x axis. This equation indicates that if we plot the projection of a single point as a function of the projection angle, we obtain a sinusoidal curve. Since any object can be approximated by a collection of points located in space, we expect its projection to be a set of overlapped sine or cosine curves in the sinogram space. For illustration, Fig. 2.18 shows the cross section of a body phantom (b) and its sinogram (a). The high-intensity curves near the middle (left to right) of the sinogram correspond to the projections formed by the phantom itself, and the low-intensity curves across the left-right span of the sinogram are formed by the projections of the table underneath the phantom. The two air pockets located near the center of the phantom (b) are clearly visible in the sinogram (a) as two dark sinusoidal curves near the center, and the five high-density ribs near the periphery of the phantom (b) are depicted as bright sinusoidal curves (a).

A sinogram is a useful tool for analyzing projection data. It is often used to detect abnormalities in a CT system. For example, a defective detector channel manifests itself as a vertical line in a sinogram, since data collected by a single detector channel maintain a fixed distance from the iso-center over all projection angles. Similarly, a temporary malfunction of the x-ray tube produces a disrupted sinogram with horizontal lines, since each horizontal line in the sinogram corresponds to a particular projection view, and therefore relates to a particular time instance.

2.5 Problems

2-1   Derive the linearity, scaling, and shift properties of the Fourier transform.

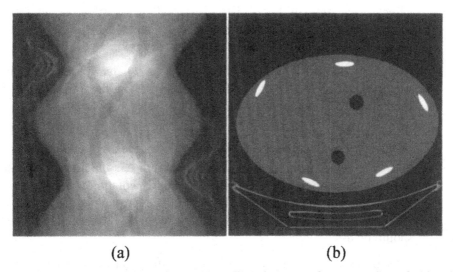

(a)   (b)

Figure 2.18 (a) Sinogram and (b) the object. The sinogram of any complicated object is formed with overlapping sinusoidal curves, since any object can be considered a collection of many small points.

2-2   Derive the conservation of energy property of the Fourier transform.

2-3   The Fourier zero-padding technique described in Section 2.1 can be used to increase the sampling densities of a spatial domain function. Spatial domain interpolation techniques can also be used to increase the sampling density. Discuss the pros and cons of the Fourier domain technique versus spatial domain techniques.

2-4   To take the derivative of a function $f(x)$, one can use the Fourier transform property and take the inverse Fourier transform of $j2\pi uF(u)$. For a discretely sampled function $f(n)$, one can also take the difference between two adjacent samples, $f(n) - f(n-1)$, to produce the derivative signal. Are these two approaches equivalent? Discuss the advantages and disadvantages of each approach.

2-5   If a random sample of size $N = 40$ from a normal population with $\sigma = 20$ has the mean $\bar{x} = 50$, find the value of x_1 such that the population mean μ falls inside $(-\infty, x_1)$ with a 95% confidence interval.

2-6   Based on the central limit theorem, the distribution of the mean $\bar{x}$ approximates a normal distribution when $N \geq 30$. For population number 1, the mean μ_1 and the standard deviation σ_1 of $\bar{x}$ are 20 and 2, respectively. For population number 2, the corresponding values are $\mu_2 = 25$ and $\sigma_2 = 2$. Are the two populations different based on a 95% confidence interval?

2-7   Show that the mean and standard deviation of Eq. (2.19) is m.

2-8   Direct calculation of Eq. (2.19) may result in overflow or underflow errors. Derive a recursive formula for the calculation of the Poisson distribution. That is, express $p(x+1)$ in terms of $p(x)$.

2-9   If the Poisson distribution calculation must be efficiently carried out in a vector operation (e.g., MATLAB® or IDL™), how do you structure the calculation to avoid overflow or underflow?

2-10   A 2D image is denoted by $f(i, j)$, where $i = 1, 2, ..., N$ and $j = 1, 2, ..., N$. In an x-y coordinate system, however, the indexes i and j represent discrete x values from $-x_1$ to x_2 at an interval of Δx, and discrete y values from $-y_1$ to y_2 at an interval of Δy. Use matrix notation to write a set of operations that produces an image $g(i, j)$ representing $f(i, j)$ rotated clockwise by β in the x-y coordinate system. Bilinear interpolation should be used.

2-11   Repeat the above exercise to generate an image $g(i, j)$ that represents a magnified $f(i, j)$ by a factor of α and then shifted by (a, b).

2-12   From Eq. (2.37), it is clear that the photoelectric effect enhances the low-contrast differentiation of materials. In Fig. 2.10, the photoelectric effect is more dominant in the low x-ray energy range and is desirable to enhance low-contrast objects. In CT designs, what are the factors that potentially limit our use of low-energy (soft) x rays?

2-13   Figure 2.12 shows linear attenuation coefficients of water, iodine, and bone. For x-ray photons with energy higher than 40 keV, what kind of iodine-water mixture will produce attenuation similar to that of bone?

2-14   Assume the input x-ray spectrum is uniformly distributed between 4 keV and 10 keV, and the attenuation coefficient for water changes linearly within this range with $\mu(4 \text{ keV}) = 81.1 \text{cm}^{-1}$ and $\mu(10 \text{ keV}) = 5.33 \text{cm}^{-1}$. What fraction of the x-ray photons will reach the detector after passing through 2 cm of water?

2-15   Equation (2.39) shows the relationship between incident and transmitted x-ray intensities for a monochromatic x-ray source. If the input x-ray spectrum is $q(E)$ where E denotes the x-ray energy, derive the relationship between the incident and transmitted x-ray intensities, assuming an attenuation coefficient of $\mu(E)$.

2-16   Assume polychromatic x-ray photons $q(E)$ travel along the y axis. Derive an equation that describes the relationship between incident and transmitted x-ray intensities if the object is nonuniform and with the attenuation coefficient = $\mu(E, y)$.

2-17   In Fig. 2.12, μ_{water} is a function of the x-ray photon energy E. However, the CT number defined by Eq. (2.40) uses μ_{water} as a reference. In a CT system, do you need to store the curve of $\mu_{water}(E)$ in order to generate an accurate CT number?

2-18   Section 2.3 mentioned that one source of error is the nonideal detector response. Assume the input signal $p(n)$ is a simple step function

$$p(n) = \begin{cases} 1.0, & n = 0,1,...5 \\ 0.01, & n = 6,7,...100 \end{cases}$$

where n is the sampling index. For simplicity, assume the detector's impulse response is 0.2^n, where $n = 0, 1, \ldots$. Calculate the detector output for $n = 10, 20$, and 100.

2-19   Repeat the above calculation for the case where the detector's impulse response is input dependent. For simplicity, assume that the impulse response $h(n)$ is

$$h(n) = \left[0.2 - 0.01 \sum_{k=0}^{n-1} p(k) \right]^n .$$

2-20   Assume that the average input x-ray flux is $I_0 = 10^6$ and the average measured transmitted x-ray fluxes are $I = 10^1$, 10^3, and 10^5 with a Poisson distribution. What are the signal-to-noise ratios of p using Eq. (2.44)?

2-21   Equation (2.45) shows the loci of a point (r, ϕ) projected onto a detector for parallel geometry. Derive an equation for the loci of a point in fan-beam geometry. Assume that the detector is an arc concentric to the x-ray focal spot, the source-to-detector distance is D, the source-to-iso distance is S, and d represents the curved distance on the detector measured from the detector iso-channel, as shown in Fig. 2.19.

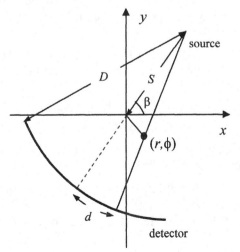

Figure 2.19 Illustration of fan-beam geometry.

References

1. J. W. Goodman, *Introduction to Fourier Optics*, McGraw-Hill, San Francisco (1968).

2. A. Papoulis, *The Fourier Integral and its Applications,* McGraw-Hill, New York (1962).

3. W. D. Stanley, G. R. Dougherty, and R. Dougherty, *Digital Signal Processing*, Prentice-Hall Company, Reston, VA (1984).

4. E. L. Crow, F. A. Davis, and M. W. Maxfield, *Statistics Manual*, Dover Publications, Inc., New York (1960).

5. P. R. Bavington, *Data Reduction and Error Analysis for the Physical Sciences*, McGraw-Hill, New York (1969).

6. A. Popoulis, *Probability, Random Variables, and Stochastic Processes*, McGraw-Hill, New York (1984).

7. J. E. Freund and R. E. Walpole, *Mathematical Statistics*, Prentice Hall, Inc., Upper Saddle River, NJ (1980).

8. S. Lipschutz, *Linear Algebra*, McGraw-Hill, New York (1968).

9. P. Lowman and J. Stokes, *Introduction to Linear Algebra*, Harcourt Brace Jovanovich Publishers, San Diego (1991).

10. J. M. Boone, "X-ray production, interaction, and detection in diagnostic imaging," in *Handbook of Medical Imaging*, Vol. 3, J. Beutel, H. L. Kundel, and R. L. VanMetter, Eds., SPIE Press, Bellingham, WA, p. 78 (2000).

11. C. L. Morgan, *Basic Principles of Computed Tomography*, University Park Press, Baltimore (1983).

12. H. E. Johns and J. R. Cunningham, *The Physics of Radiology*, Charles C. Thomas Publisher, Ltd., Springfield, IL (1983).

13. D. L. Batchelar, W. Dabrowski, and I. A. Cunningham, "Tomographic imaging of bone composition using coherently scattered x rays," *Proc. SPIE* **3977**, 353–361 (2000).

14. M. S. Westmore, A. Fenster, and I. A. Cunningham, "Tomographic imaging of the angular-dependent coherent-scatter cross section," *Med. Phys.* **24** (1), 3–10 (1997).

15. G. T. Herman, *Image Reconstruction from Projections: The Fundamentals of Computerized Tomography*, Academic Press, New York (1980).

16. A. C. Kak and M. Slaney, *Principles of Computerized Tomographic Imaging*, IEEE Press, Piscataway, NJ (1988).

17. S. Napel, "Computed tomography image reconstruction," in *Medical CT and Ultrasound: Current Technology and Applications*, L. W. Goldman and J. B. Fowlkes, Eds., Advanced Medical Publishing, Madison, WI, 603–626 (1995).

18. J. Hsieh, "CT image reconstruction," in *RSNA Categorical Course in Diagnostic Radiology Physics: CT and US Cross-Sectional Imaging 2000*, L. W. Goldman and J. B. Fowlkes, Eds., Radiological Society of North America, Inc., Oak Brook, IL, 53–64 (2000).

19. J. Hsieh, "Image artifacts, causes, and correction," in *Medical CT and Ultrasound: Current Technology and Applications*, L. W. Goldman and J. B. Fowlkes, Eds., Advanced Medical Publishing, Madison, WI, 488–518 (1995).

20. P. M. Joseph and R. D. Spital, "A method for correcting bone induced artifacts in CT scanning," *J. Comp. Assist. Tomogr.* **12**, 100–108 (1978).

21. J. Hsieh, R. C. Molthen, C. A. Dawson, and R. H. Johnson, "An iterative approach to the beam hardening correction in cone beam CT," *Med. Phys.* **27**(1), 23–29 (2000).

22. P. M. Joseph and R. D. Spital, "The effects of scatter in x-ray computed tomography," *Med. Phys.* **9**(4), 464–472 (1982).

23. A. C. Kak, "Tomographic imaging with diffracting and non-diffracting sources," in *Array Signal Processing*, S. Haykin, Ed., Prentice-Hall, Englewood Cliffs, NJ (1985).

24. J. Hsieh, "Three-dimensional artifact induced by projection weighting and misalignment," *IEEE Trans. Med. Imaging* **18**(4), 1375–1384 (1999).

Chapter 3
Image Reconstruction

3.1 Introduction

As discussed in Chapter 2, the object that we try to reconstruct can be considered a 2D distribution of some kind of function. For CT, this function represents the linear attenuation coefficients of the object. The problem imposed on the tomographic reconstruction can be stated as the following: Suppose we have collected a set of measurements, and each measurement represents the summation or line integral of the attenuation coefficients of the object along a particular ray path. These measurements are collected along different angles and different distances from the iso-center. To avoid redundancy in the data sampling, assume that the measurements are taken in the following sequence. First, we take a set of measurements along parallel paths that are uniformly spaced, as shown by the solid lines in Fig. 3.1. These measurements form a "view" or a "projection." Then we repeat the same measurement process at a slightly different angle, as shown by the dotted lines in Fig. 3.1. This process continues until the entire 360 deg (theoretically, only 180 deg of parallel projections are necessary) is covered. During the entire process, the angular increment between adjacent views remains constant, and the scanned object remains stationary. For CT reconstruction, the question is how to estimate the attenuation distribution of the scanned object based on these measurements.

Before we begin rigorous discussions on the mathematical principles of CT, we will present a simple "thinking" experiment to demonstrate how CT could work. Assume that we try to guess the internal structure of a semitransparent object. The object is formed with five spheres embedded inside a cylinder, as shown in Fig. 3.2(a). We are not allowed to view the object in a top-down fashion (looking directly into the page); we can only examine it from the side. If we examine the object only at the angle shown in the figure, two of the spheres are partially blocked (since they are semitransparent) by the sphere in front, and we see only one overlapped sphere. Although the opacity of the overlapped sphere is expected to be higher than the nonoverlapped spheres, we cannot infer from the opacity the number of spheres that overlap, because we do not know the opacity of each sphere. Based on a single view, we may erroneously conclude that the cylinder contains only three spheres. If we rotate the object and observe it from a different angle, as shown in Fig. 3.2(b), the spheres that were blocked become visible. If we examine the object from multiple angles, we correctly conclude that the cylinder contains five smaller objects, but we also conclude that these objects are indeed spheres, since the size and the intensity do not change with the viewing angle. This experiment demonstrates that by examining a semi-opaque object from multiple angles, we can estimate the internal structure of the object. If we replace the light source with an x-ray source, our eyes with an x-ray detector, and our brain with a computer, we will create a CT system. The two key points are the observer's ability to examine the object from multiple view angles, and the object's property of semi-opaqueness. This second condition is as important as the first, because if the object is totally opaque, an estimation of the internal structure is no longer possible, regardless of the number of angles at which the observer can examine the object. Fortunately, the majority of the materials under medical CT examination (soft tissue, bone, and a contrast agent) are semitransparent to x rays.

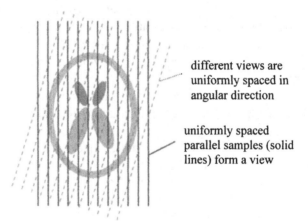

different views are uniformly spaced in angular direction

uniformly spaced parallel samples (solid lines) form a view

Figure 3.1 Data sampling pattern. Samples within a view are parallel to each other and are uniformly spaced. Views are also uniformly spaced in the angular direction.

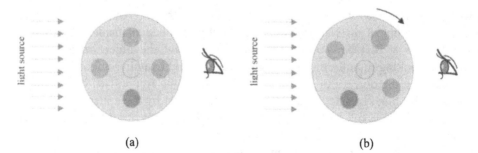

(a)   (b)

Figure 3.2 Illustration of the general concept of tomography. (a) When viewing a semitransparent object from one angle, some internal spheres can overlap. Since we do not know *a priori* the opaqueness of each sphere, it is impossible to estimate the shape, intensity, and number of spheres. (b) After the object is rotated, the shadows produced by the spheres are no longer overlapped, and each sphere can be viewed individually. When more viewing angles are used, the shape, intensity, and number of spheres can be accurately estimated.

The subject of CT image reconstruction is interesting and complicated. Its mathematical formulation can be traced back to 1917 when J. Radon first developed a solution for the reconstruction of a function from its line integrals. With the development of clinically viable CT scanners in the late 1970s and early 1980s, research activities in this area experienced tremendous growth. Numerous research papers, conference proceedings, book chapters, and even textbooks were dedicated to this topic.[1-14] Many techniques have been proposed that make various tradeoffs among computational complexity, spatial resolution, temporal resolution, noise, clinical protocols, flexibility, and artifacts. Because of the complexity of the subject, it is impossible to cover all aspects of image reconstruction in a single chapter. The objective is to provide an overview of the topic, and to give readers a basic understanding of the underlying principles of reconstruction. Although mathematical formulations are provided in detail, our emphasis is not on the rigorous derivation of mathematical equations. Whenever appropriate, we provide intuitive explanations or derivations to these mathematical solutions. We hope that this approach will help readers to gain easier access to the domain of image reconstruction. We also hope that this book will serve as a source of reference to interested readers for further exploration on the subject.

3.2 Several Approaches to Image Reconstruction

To understand some of the methodologies employed in CT reconstruction, we will begin with an extremely simplified case in which the object is formed with four small blocks. The attenuation coefficients are homogeneous within each block and are labeled μ_1, μ_2, μ_3, and μ_4, as shown in Fig. 3.3. We will further consider the scenario where line integrals are measured in the horizontal, vertical, and diagonal directions. Five measurements in total are selected in this example. It can be shown that the diagonal and three other measurements form a set of independent equations. For example,

$$
\begin{aligned}
p_1 &= \mu_1 + \mu_2, \\
p_2 &= \mu_3 + \mu_4, \\
p_3 &= \mu_1 + \mu_3, \\
p_4 &= \mu_1 + \mu_4.
\end{aligned}
\tag{3.1}
$$

Here, four independent equations are established for the four unknowns. From elementary algebraic knowledge, we know that there is a unique solution to the problem, since the number of equations equals the number of unknowns. If we generalize the problem to the case where the object is divided into N by N small elements, we could easily reach the conclusion that as long as enough independent measurements (N^2) are taken, we can always uniquely solve the attenuation coefficient distribution of the object.

Many techniques are readily available to solve linear sets of equations. Direct matrix inversion was the method used on the very first CT apparatus in 1967. Over 28,000 equations were simultaneously solved. When the object is divided into finer and finer elements (corresponding to higher spatial resolutions), the task of solving simultaneous sets of equations becomes quite a challenge, even with today's computer technology. In addition, to ensure that enough independent equations are formed, we often need to take more than N^2 measurements, since some of the measurements may not be independent. A good example is the one given in Fig. 3.3. Assume that four measurements, p_1, p_2, p_3, and p_5, are taken in the horizontal and vertical directions. It can be shown that these measurements are not linearly independent ($p_5 = p_1 + p_2 - p_3$). A diagonal value must be added to ensure their orthogonality. When the number of equations exceeds the number of unknowns, a straightforward solution may not always be available. This is even more problematic when we consider the inevitable possibility that errors exist in some of the measurements. Therefore, different reconstruction techniques need to be explored. Despite its limited usefulness, the linear algebraic approach proves the existence of a mathematical solution to the CT problem.

One possible remedy to solve this problem is the so-called iterative reconstruction approach. For the ease of illustration, we again start with an over-simplified example (a more rigorous discussion on this topic can be found in Section 3.6). Consider the four-block object problem discussed previously. This time, we assign specific attenuation values for each block, as shown in Fig. 3.4(a). The corresponding projection measurements are depicted in the same figure. We will start with an initial guess of the object's attenuation distribution. Since we have no *a priori* knowledge of the object itself, we assume that it is homogeneous. We can start with an initial estimate using the average of the projection samples. The sum of the projection samples ($3 + 7 = 10$ or $4 + 6 = 10$) evenly distributed over the four blocks results in an average value of 2.5 ($10/4 = 2.5$). Next, we calculate the line integrals of our estimated distribution along the same paths as the original projection measurement. For example, we can calculate the projection samples along the horizontal direction and obtain the calculated projection values of 5 ($2.5 + 2.5$) and 5, as shown in Fig. 3.4(b). By

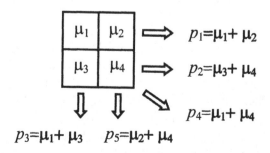

Figure 3.3 A simple example of an object and its projections.

comparing the calculated projections against the measured values of 3 and 7 [Fig. 3.4(a)], we observe that the top row is overestimated by 2 (5 − 3) and the bottom row is underestimated by 2 (5 − 7). Since we have no *a priori* knowledge of the object, we again assume that the difference between the measured and the calculated projections must be split evenly among all pixels along each ray path. Therefore, we decrease the value of each block in the top row by 1 and increase the bottom row by 1, as shown in Fig. 3.4(c). The calculated projections in the horizontal direction are now consistent with the measured projections. We repeat the same process for projections in the vertical direction and reach the conclusion that each element in the first column must be decreased by 0.5 and each element in the second column increased by 0.5, as shown in Fig. 3.4(d). The calculated projections in all directions are now consistent with the measured projections (including the diagonal measurement), and the reconstruction process stops. The object is correctly reconstructed. This reconstruction process is called the algebraic reconstruction technique (ART). For a more rigorous treatment of the subject, readers should refer to Section 3.6 or consult the references listed at the end of this chapter.[1-5]

Based on the above discussion, it is clear that iterative reconstruction methods are computationally intensive because forward projections (based on the estimated reconstruction) must be performed repeatedly. This is in addition to the updates required of the reconstructed pixels based on the difference between the measured projection and the calculated projection. All of the iterative reconstruction algorithms require several iterations before they converge to the desired results. Given the fact that the state-of-the-art CT scanner can acquire a complete projection data set in a fraction of a second, and each CT examination typically contains several hundred images, the routine clinical usage of ART is still a long way from reality. Despite its limited utility, ART does provide some insight into the reconstruction process. Recall the image-update process that was used in Fig. 3.4. When there is no *a priori* knowledge of the object, we always assume that the intensity of the object is uniform along the ray path. In other words, we distribute the projection intensity evenly among all pixels along the ray path. This process leads to the concept of backprojection.

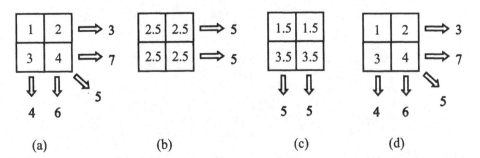

Figure 3.4 Illustration of iterative reconstruction. (a) Original object and its projections, (b) initial estimate of the object and its projection, (c) updated estimation of object and its projection, and (d) final estimation and projections.

Consider a simple case in which the object of interest is an isolated point. The corresponding projection is an impulse function with its peak centered at the location where a parallel ray intersects the point, as shown in Fig. 3.5. Similar to the reasoning used for ART, we do not know, *a priori*, the location of the point other than the fact that it is located on that line. Therefore, we have to assume a uniform probability distribution for its location. We paint the entire ray path with the same intensity as the measured projection, as shown in Fig. 3.5(a). In this example, the first projection is oriented vertically. The next projection is again an impulse function, and again we paint the entire ray path that intersects the impulse with the same intensity as the measurement. This time, however, the ray path is slightly rotated relative to the first one because of the difference in projection angles. This process is repeated for all projection samples. Figures 3.5(b)–(i) depict the results obtained over different angular ranges in 22.5-deg increments. Note that the painting procedure essentially reverses the projection process and formulates a 2D object from a set of 1D line integrals. As a result, this process is called backprojection, and is one of the key image reconstruction steps used in many commercial CT scanners. From Fig. 3.5(i), it is clear that by backprojecting over the range of 0 to 180 deg, a rough estimate of the original object (a point) can be obtained. By examining the intensity profile of the reconstructed point (Fig. 3.6), we conclude that the reconstructed point

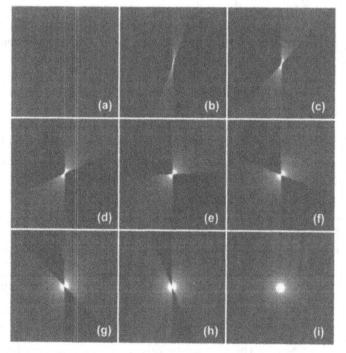

Figure 3.5 Backprojection process of a single point. (a) Backprojected image of a single projection. (b)–(i) Backprojection of views covering: (b) 0 to 22.5 deg; (c) 0 to 45 deg; (d) 0 to 67.5 deg; (e) 0 to 90 deg; (f) 0 to 112.5 deg; (g) 0 to 135 deg; (h) 0 to 157.5 deg; and (i) 0 to 180 deg.

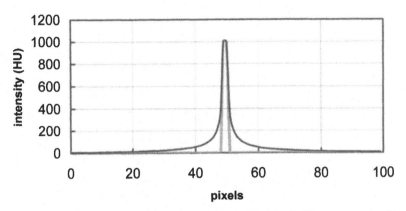

Figure 3.6 Profile of a reconstructed point. Solid black line: reconstruction with backprojection; thick gray line: ideal reconstruction.

(black line) is a blurred version of the true object (gray line). Degradation of the spatial resolution is obvious. From the linear system theory, we know that Fig. 3.5(i) is essentially the impulse response of the backprojection process. Therefore, we should be able to recover the original object by simply deconvolving the backprojected images with the inverse of the impulse response. This approach is often called the backprojection-filtering approach.

3.3 The Fourier Slice Theorem

The intuitive illustration of the image reconstruction approaches above can help explain the theory behind the tomographic reconstruction process used in many of today's commercial CT scanners. For the ease of future reference, we will denote by $f(x, y)$ the object being reconstructed, and by $p(t, \theta)$ a parallel projection of $f(x, y)$ taken at angle θ, as shown in Fig. 3.7. In this notation, t represents the distance of the projection ray to the iso-center (center of rotation). The theory that governs tomographic reconstruction is generally known as the Fourier slice theorem.[2] To many researchers, it is also known as the central slice theorem, and can be stated as follows:

> The Fourier transform of a parallel projection of an object $f(x, y)$ obtained at angle θ equals a line in a 2D Fourier transform of $f(x, y)$ taken at the same angle.

The proof of the theorem is straightforward. We will consider the case where a projection of $f(x, y)$ is taken parallel to the y axis, as shown in Fig. 3.8. The projection $p(x, 0)$ is related to the original function $f(x, y)$ by the following equation:

$$p(x,0) = \int_{-\infty}^{\infty} f(x,y)dy. \tag{3.2}$$

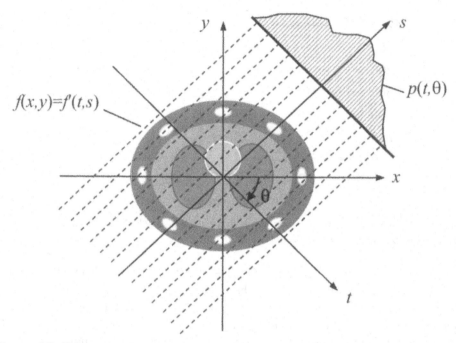

Figure 3.7 Schematic diagram of the original coordinate system and the rotated coordinate system.

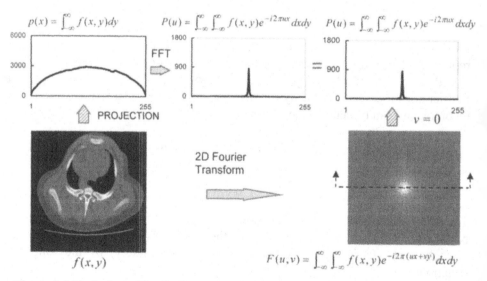

Figure 3.8 Illustration of the Fourier slice theorem for a projection at an angle of 0 deg.

If we take the Fourier transform with respect to x on both sides of Eq. (3.2), we obtain

$$P(u) = \int_{-\infty}^{\infty} p(x,0) e^{-j2\pi ux} dx = \int_{-\infty}^{\infty} \int_{-\infty}^{\infty} f(x,y) e^{-j2\pi ux} dx dy . \qquad (3.3)$$

Next, we will consider the 2D Fourier transform of the original function $f(x, y)$ evaluated at $v = 0$:

$$F(u,v)\Big|_{v=0} = \int_{-\infty}^{\infty}\int_{-\infty}^{\infty} f(x,y) e^{-j2\pi(ux+vy)} dx dy \Big|_{v=0} = \int_{-\infty}^{\infty}\int_{-\infty}^{\infty} f(x,y) e^{-j2\pi ux} dx dy . \qquad (3.4)$$

Comparing the right-hand side of Eq. (3.3) to Eq. (3.4), we conclude that the two equations are identical. This states that the Fourier transform of an object's 0-deg projection is the same as the $v = 0$ line in the 2D Fourier transform of the same object. The conclusion is depicted pictorially in Fig. 3.8. Because the coordinate system is selected arbitrarily, the above conclusion is valid in any rotated coordinate system. In other words, the Fourier transform of an object's projection at any angle equals a line taken in the same orientation of the 2D Fourier transform of the same object.

Alternatively, the Fourier slice theorem can be derived directly as follows. We will select a rotated coordinate system such that one axis s is parallel to the x-ray path for the projection with angle θ, as shown in Fig. 3.9. The object $f(x, y)$ can be represented by $f'(t, s)$ in the rotated coordinate system, where the two coordinate systems are related by the following set of equations:

$$\begin{cases} t = x\cos\theta + y\sin\theta, \\ s = -x\sin\theta + y\cos\theta. \end{cases} \qquad (3.5)$$

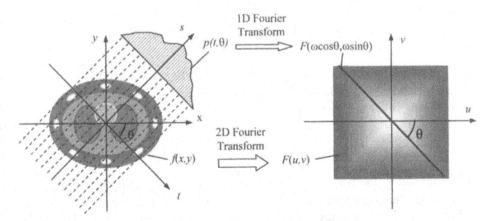

Figure 3.9 Illustration of the Fourier slice theorem.

The projection $p(t, \theta)$ is then simply the integration of the function $f'(t, s)$ along the s axis:

$$p(t,\theta) = \int_{-\infty}^{\infty} f'(t,s)ds . \tag{3.6}$$

If we denote by $P(\omega, \theta)$ the Fourier transform of $p(t, \theta)$ over variable t, we obtain

$$P(\omega,\theta) = \int_{-\infty}^{\infty} \int_{-\infty}^{\infty} f'(t,s)ds e^{-i2\pi\omega t} dt . \tag{3.7}$$

Next we will perform coordinate transformation on the right-hand side of Eq. (3.7). In a general theory of calculus, we know that the differential area of the two coordinate systems is related by the following equation:

$$dsdt = Jdxdy = \begin{vmatrix} \dfrac{\partial t}{\partial x} & \dfrac{\partial s}{\partial x} \\ \dfrac{\partial t}{\partial y} & \dfrac{\partial s}{\partial y} \end{vmatrix} dxdy , \tag{3.8}$$

where J is the Jacobian determinant. Combining Eqs. (3.6), (3.7), and (3.8), we obtain

$$P(\omega,\theta) = \int_{-\infty}^{\infty} \int_{-\infty}^{\infty} f(x,y)e^{-i2\pi\omega(x\cos\theta+y\sin\theta)} dxdy . \tag{3.9}$$

To relate the Fourier transform of a projection $P(\omega, \theta)$ to the Fourier transform of the original function $f(x, y)$, examine the 2D Fourier transform of $f(x, y)$, $F(u, v)$. Based on the definition of the Fourier transform, we obtain

$$F(u,v) = \int_{-\infty}^{\infty} \int_{-\infty}^{\infty} f(x,y)e^{-i2\pi(xu+yv)} dxdy . \tag{3.10}$$

Note the similarity between the right-hand sides of Eqs. (3.9) and (3.10). If we let $u = \omega\cos\theta$ and $v = \omega\sin\theta$, the two equations become identical. Mathematically, the following relationship exists:

$$F(\omega\cos\theta, \omega\sin\theta) = P(\omega,\theta). \tag{3.11}$$

In Fourier space, the two variables ($u = \omega\cos\theta$ and $v = \omega\sin\theta$) define a straight line through the origin that forms an angle θ with respect to the u axis, as shown in Fig. 3.10. Thus, we have shown that the Fourier transform of the parallel projection of an object $f(x, y)$ is a slice in the 2D Fourier transform of the object.

The slice is taken at the same angle as the projection. We have completed the proof of the Fourier slice theorem.

Consider the implication of the Fourier slice theorem. From each projection, we obtain a line in the 2D Fourier transform of the object by performing the Fourier transform on the projection. If we collect a sufficient number of projections over the range from 0 to π, we can fill the entire Fourier space of the object being reconstructed. Once the object's Fourier transform is obtained, we can recover the object itself by the inverse Fourier transform. The tomographic reconstruction process becomes a series of 1D Fourier transforms followed by a 2D inverse Fourier transform.

3.4 The Filtered Backprojection Algorithm

Although the Fourier slice theorem provides a straightforward solution for tomographic reconstruction, it presents some challenges in actual implementation. First, the sampling pattern produced in the Fourier space is non-Cartesian. The Fourier slice theorem states that the Fourier transform of a projection is a line through the origin in 2D Fourier space. As a result, samples from different projections fall on a polar coordinate grid, as shown in Fig. 3.10. To perform a 2D inverse Fourier transform, these samples must be interpolated or regridded to a Cartesian coordinate. Interpolation in the frequency domain is not as straightforward as interpolation in real space. In real space, an interpolation error is localized to the small region where the pixel is located. This property does not hold, however, for interpolation in the Fourier domain, since each sample in a 2D Fourier space represents certain spatial frequencies (in the horizontal and vertical directions). Therefore, an error produced on a single sample in Fourier space affects the appearance of the entire image (after the inverse Fourier transform).

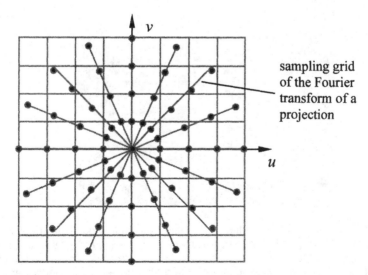

sampling grid
of the Fourier
transform of a
projection

Figure 3.10 Sampling pattern in Fourier space based on the Fourier slice theorem.

To illustrate the sensitivity of Fourier domain interpolation, we performed the following simple experiment. A shoulder phantom was scanned and reconstructed in a 512×512-pixel matrix denoted by $f(x, y)$, where $x = 0, 1, \ldots, 511$, and $y = 0, 1, \ldots, 511$. The reconstructed image is shown in Fig. 3.11(a). (We clipped the top and bottom portions of the image in the display, since they contain only air in the background. At the selected display window level, it is completely black.) Next, we performed a 2D discrete Fourier transform of the image and obtained the function $F(u, v)$, where $u = 0, 1, \ldots, 511$, and $v = 0, 1, \ldots, 511$. Note that $F(u, v)$ is a 512×512 complex array. In this matrix, $F(0, 0)$ represents the dc component of the image. If we simply take the discrete inverse Fourier transform of $F(u, v)$, we should arrive at the original image, $f(x, y)$. Note that the function $F(u, v)$ is the quantity that we are trying to estimate using the parallel projections (the Fourier slice theorem).

We can simulate a case in which we make an error in the estimation of $F(u, v)$. Instead of the original value of $F(0, 1)$, we replace this with the average of its two neighbors, $F(0, 0)$ and $F(0, 2)$. This is a logical step, since interpolation simply uses the neighboring values to estimate the missing quantity. Then we take the inverse Fourier transform of the modified function and arrive at an estimated image, $f'(x, y)$, shown in Fig. 3.11(b). Because of the overall shift in image intensity, we have to lower the display window level by 300 HU in order to visually inspect the image. This indicates that a modification of a single value in $F(u, v)$ has created a CT number shift. In addition, a significant shading artifact can be observed in the image. This phenomenon can be easily understood by examining the representation of $F(0, 1)$. $F(0, 1)$ represents the dc component

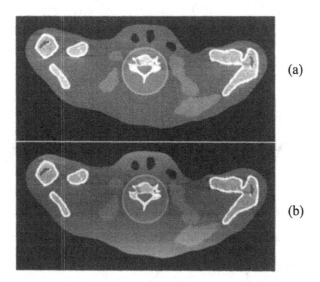

(a)

(b)

Figure 3.11 Illustration of the sensitivity of Fourier domain interpolation where WW = 650 HU. (a) Original image. (b) Modified original image by replacing $F(0, 1)$ with the average of $F(0, 0)$ and $F(0, 2)$ in the Fourier transform of (a). The display WL for (b) is 300 HU lower than that of (a), indicating a significant grayscale shift. In addition, shading artifacts are clearly visible in (b).

of the image, $f(x, y)$, in the horizontal direction and the first harmonic in the vertical direction. Therefore, an error in the estimation of $F(0, 1)$ produces an intensity shift and a single cycle sinusoidal shading in the vertical direction. Although this example illustrates the difficulties in performing interpolation in Fourier space, it does not imply that accurate Fourier space interpolation is impossible. Many advanced interpolation algorithms have been developed to perform such tasks.

Another disadvantage of direct Fourier space reconstruction is the difficulty of performing targeted reconstruction. Targeted reconstruction is a technique used quite often in CT to examine fine details of a small region in the object. To illustrate the targeted reconstruction, Fig. 3.12(a) shows a reconstructed image of a human skull at 25-cm FOV. The FOV is adequate when we try to examine the overall structures and large-sized objects in the image. But if we examine detailed structures of the sinus (shown by the dotted circle), for example, the image size becomes inadequately small. The fine structures that are closely spaced become difficult to differentiate. If we can somehow "focus" the reconstruction on only the area of interest, the fine details of the object can be better visualized, as shown in Fig. 3.12(b). Using the direct Fourier reconstruction approach, we pad $F(u, v)$ with a large number of zeroes to essentially perform frequency domain interpolation. The size of the inverse Fourier transform is inversely proportional to the size of the targeted region of interest (ROI). For a very small ROI, the matrix size becomes unmanageably large (detailed discussions on how the targeted reconstruction is performed will be presented later in the chapter). Although other techniques can be used to overcome some of the difficulties, the implementation of these techniques is still not straightforward. Consequently, it is desirable to explore alternative implementations of the Fourier slice theorem. The most popular implementation is the so-called filtered backprojection (FBP) algorithm.[1,2]

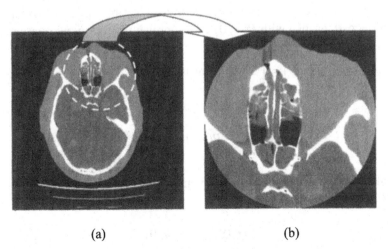

(a)   (b)

Figure 3.12 Concept of targeted reconstruction. (a) Human skull phantom reconstructed with a 25-cm FOV. (b) The same scan reconstructed with a 10-cm FOV centered at the sinus.

3.4.1 Derivation of the filtered backprojection formula

We start with the well-known fact that the Fourier transform and the inverse Fourier transform are conjugate operators. The image function $f(x,y)$ can be recovered from its Fourier transform $F(u, v)$ by the inverse Fourier transform:

$$f(x,y) = \int_{-\infty}^{\infty} \int_{-\infty}^{\infty} F(u,v) e^{j2\pi(ux+vy)} du\,dv . \qquad (3.12)$$

Similar to the coordinate transformation that we performed in the derivation of the Fourier slice theorem, we switch from a Cartesian coordinate (u, v) to a polar coordinate (ω, θ). The purpose of the coordinate switch is to express the quantity $F(u, v)$ in the more natural form in which the data are collected (the Fourier transform of each parallel projection falls on a polar grid, as shown in Fig. 3.10). The coordinate transformation is as follows:

$$\begin{cases} u = \omega\cos\theta \\ v = \omega\sin\theta \end{cases} \qquad (3.13)$$

and

$$dudv = \begin{vmatrix} \partial u / \partial\omega & \partial u / \partial\theta \\ \partial v / \partial\omega & \partial v / \partial\theta \end{vmatrix} d\omega d\theta = \omega d\omega d\theta . \qquad (3.14)$$

Substituting Eqs. (3.12) and (3.13) into (3.14), we obtain

$$f(x,y) = \int_{0}^{2\pi} d\theta \int_{0}^{\infty} F(\omega\cos\theta, \omega\sin\theta) e^{j2\pi\omega(x\cos\theta+y\sin\theta)} \omega d\omega . \qquad (3.15)$$

If we make use of the Fourier slice theorem described in Eq. (3.11), we can replace $F(\omega\cos\theta, \omega\sin\theta)$ with $P(\omega, \theta)$ and establish the following relationship:

$$\begin{aligned} f(x,y) &= \int_{0}^{2\pi} d\theta \int_{0}^{\infty} P(\omega,\theta) e^{j2\pi\omega(x\cos\theta+y\sin\theta)} \omega d\omega \\ &= \int_{0}^{\pi} d\theta \int_{0}^{\infty} P(\omega,\theta) e^{j2\pi\omega(x\cos\theta+y\sin\theta)} \omega d\omega \\ &\quad + \int_{0}^{\pi} d\theta \int_{0}^{\infty} P(\omega,\theta+\pi) e^{-j2\pi\omega(x\cos\theta+y\sin\theta)} \omega d\omega . \end{aligned} \qquad (3.16)$$

For parallel sampling geometry, a convenient symmetry property exists among the projection samples:

$$p(t,\, \theta+\pi) = p(-t,\, \theta). \qquad (3.17)$$

This property can be easily understood by examining the sampling geometry of a set of parallel beams that are 180 deg apart, as shown in Fig. 3.13. The two sets of projections represent exactly the same set of ray paths. Based on the properties of the Fourier transform, a similar relationship exists for the corresponding Fourier transformed pair:

$$P(\omega, \theta+\pi) = P(-\omega, \theta). \tag{3.18}$$

Substituting Eq. (3.23) into Eq. (3.21), we arrive at the following relationship:

$$f(x,y) = \int_0^\pi d\theta \int_{-\infty}^\infty P(\omega,\theta)|\omega|e^{j2\pi\omega(x\cos\theta+y\sin\theta)}d\omega. \tag{3.19}$$

If we express the above equation in the rotated coordinate system (s,t) and use the relationship denoted in Eq. (3.5), we arrive at the following equation:

$$f(x,y) = \int_0^\pi d\theta \int_{-\infty}^\infty P(\omega,\theta)|\omega|e^{j2\pi\omega t}d\omega. \tag{3.20}$$

Here, $P(\omega,\theta)$ is the Fourier transform of the projection at angle θ. The inside integral is the inverse Fourier transform of the quantity $P(\omega,\theta)|\omega|$. In the spatial domain, it represents a projection filtered by a function whose frequency domain response is $|\omega|$, and is therefore called a "filtered projection."

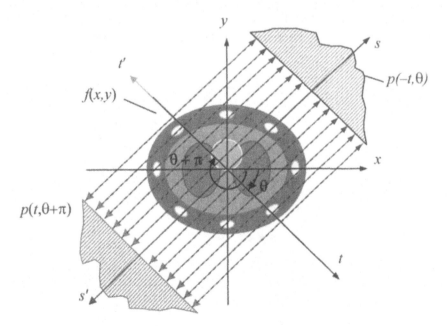

Figure 3.13 Illustration of the symmetry property of parallel projection. Projections that are 180 deg apart sample the same ray paths.

If we denote the filtered projection at angle θ by $g(t,\theta)$ represented by the inside integral of Eq. (3.20), then

$$g(t,\theta) = g(x\cos\theta + y\sin\theta) = \int_{-\infty}^{\infty} P(\omega,\theta)|\omega| e^{j2\pi\omega(x\cos\theta+y\sin\theta)} d\omega. \quad (3.21)$$

Equation (3.20) can be rewritten in the form

$$f(x,y) = \int_0^\pi g(x\cos\theta + y\sin\theta) d\theta. \quad (3.22)$$

The variable $x\cos\theta + y\sin\theta$ is simply the distance of the point (x, y) to a line that goes through the origin of the coordinate system and forms an angle θ with respect to the x axis, as shown in Fig. 3.14. Equation (3.22) states that the reconstructed image $f(x, y)$ at location (x, y) is the summation of all filtered projection samples that pass through that point. Alternatively, we can focus on a particular filtered projection sample and examine its contribution to the reconstructed image. Since $x\cos\theta + y\sin\theta$ represents a straight line that overlaps the ray path that produces the projection sample, the intensity of $g(x\cos\theta + y\sin\theta)$ is added uniformly to the reconstructed image along the straight line, as shown by the gray line in Fig. 3.14. Consequently, the value of the filtered projection sample is "painted" or "superimposed" along the entire straight-line path. This is indeed the "backprojection" process (of the filtered projection) described in Section 3.2.

For readers who are less interested in the detailed mathematical derivation, we will present an intuitive explanation of the filtered backprojection approach. Based on the Fourier slice theorem, the 2D Fourier transform of an object is obtained by patching together multiple 1D Fourier transforms. Ideally, if we

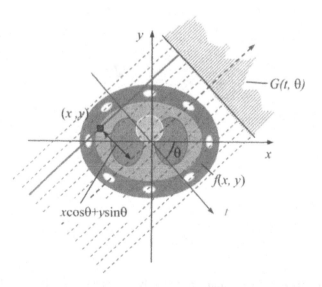

Figure 3.14 Illustration of the backprojection process.

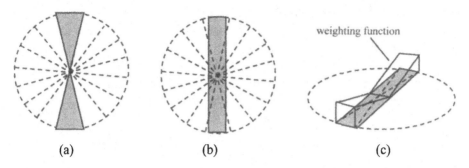

Figure 3.15 Illustration of the filtered backprojection concept. (a) Ideal frequency data from one projection, (b) actual frequency data from one projection, and (c) weighting function in the frequency domain to approximate ideal conditions.

assume that the Fourier transform of a projection is shaped as a sliced pie [shown in Fig. 3.15(a)], we can simply insert each wedge into its proper place to obtain a 2D Fourier transform of the object. Unfortunately, in frequency space, the Fourier transform of each projection is shaped as a strip, as shown in Fig. 3.15(b). If we simply sum up the Fourier transforms of all projections that are uniformly spaced over 2π, the center region is artificially enhanced and the outer regions are underrepresented. To approximate the pie-shaped region with the strip-shaped region, we can multiply the strip-shaped Fourier transform by a function that has a lower intensity near the center and a higher intensity near the edge. For example, we can multiply the Fourier transform of the projection by the width of the pie-shaped wedge at that frequency, as shown in Fig. 3.15(c). If we assume N projections evenly spaced over 180 deg, the width of each wedge is $\pi|\omega|/N$ at frequency ω. The net effect of the weighting function is that the summation of the weighted strips maintains the same "mass" as the summation of the pie-shaped wedges.

3.4.2 Computer implementation

Equation (3.19) cannot be implemented directly in its present form. To understand, consider the interpretation of Eq. (3.21). Based on the Fourier transform property, we know that the multiplication of two functions in the Fourier domain is equivalent to the convolution of two corresponding spatial-domain functions. The corresponding function of $P(\omega, \theta)$ in the spatial domain is the measured parallel projection $p(t,\theta)$. The spatial domain function (or the impulse response) $\xi(t)$ that corresponds to the filter $|\omega|$ is simply the inverse Fourier transform of the function

$$\xi(t) = \int_{-\infty}^{\infty} |\omega| e^{j2\pi\omega t} d\omega . \tag{3.23}$$

$\xi(t)$ does not exist. For illustration, consider the value of $\xi(0)$ by setting $t = 0$ in the above equation. $\xi(0)$ represents the area under the curve $|\omega|$. When $\omega \to \infty$, then $\xi(0) \to \infty$. Therefore, Eq. (3.19) cannot be implemented in its present form and an alternative approach must be explored. One such approach is to introduce a band-limiting function to the equation.

Assume that the Fourier transform of the projection is band limited. In other words, zero energy is contained outside the frequency interval $(-\Gamma, \Gamma)$. Under this assumption, Eq. (3.21) can be expressed in the following form:

$$g(t,\theta) = \int_{-\Gamma}^{\Gamma} P(\omega,\theta)|\omega|e^{j2\pi\omega t}d\omega . \qquad (3.24)$$

Equation (3.24) indicates that in order to calculate the filtered projection $g(t, \theta)$, we need to take the Fourier transform of the projection $p(t,\theta)$ to arrive at $P(\omega,\theta)$, multiply it by $|\omega|$ in the range of $(-\Gamma, \Gamma)$, and then perform the inverse Fourier transform. Unfortunately, this seemingly simple problem is complicated by two factors: the discretization of the truncated filter kernel, and the nature of circular convolution. To fully understand the filter kernel issue, we must first derive the ideal kernel in the spatial domain. To ensure aliasing-free sampling, the projection bandwidth Γ must satisfy the Nyquist sampling criterion:

$$\Gamma = \frac{1}{2\delta} \text{ cycles/mm,} \qquad (3.25)$$

where δ is the projection sampling interval in millimeters. Under this condition, the original ramp function $|\omega|$ is effectively multiplied by a window function $q(\omega)$:

$$H(\omega) = |\omega|q(\omega) , \qquad (3.26)$$

where

$$q(\omega) = \begin{cases} 1, & |\omega| < \Gamma \\ 0, & \text{otherwise.} \end{cases}$$

The filter function $H(\omega)$ is depicted in Fig. 3.16. The impulse response of the filter can be described by

$$h(t) = \int_{-\Gamma}^{\Gamma} |\omega|e^{j2\pi\omega t}d\omega = \frac{1}{2\delta^2}\left(\frac{\sin 2\pi\Gamma t}{2\pi\Gamma t}\right) - \frac{1}{4\delta^2}\left(\frac{\sin \pi\Gamma t}{\pi\Gamma t}\right)^2 . \qquad (3.27)$$

It is interesting to note that since $H(\omega)$ is a real and even function of ω, the corresponding impulse response $h(t)$ is a real and even function of t. This property of the Fourier transform is discussed in Section 2.1.1.

The projection is sampled at an interval of $\delta = (2\Gamma)^{-1}$. By the convolution theorem, Eq. (3.20) can be written as

$$f(x,y) = \int_0^\pi d\theta \int_{-t_m}^{t_m} p(t',\theta)h(t-t')dt', \tag{3.28}$$

where t_m is the value of t' for which $p(t',\theta) = 0$ and $\forall |t'| > t_m$. Here, we make use of the fact that the scanned object has finite spatial support. In the discrete implementation of the filtered projection, we are interested in only the filter values at the integer multiples of δ. By substituting $t = n\delta$ into Eq. (3.27), we obtain

$$h(n\delta) = \begin{cases} \dfrac{1}{4\delta^2}, & n = 0, \\ 0, & n = \text{even}, \\ -\dfrac{1}{(n\pi\delta)^2}, & n = \text{odd}. \end{cases} \tag{3.29}$$

The impulse response of the filter function is depicted in Fig. 3.17. In this figure, we set $\delta = 1$. If we denote the discrete samples of the projection at angle θ by $p(k\delta,\theta)$ ($k = 0, \ldots, N-1$), the filtered projection described in Eq. (3.21) can be expressed as a spatial domain convolution:

$$g(n\delta,\theta) = \delta \sum_{k=0}^{N-1} h(n\delta - k\delta)p(k\delta,\theta), \ n = 0, 1, \ldots, N-1. \tag{3.30}$$

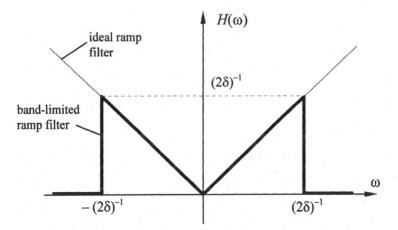

Figure 3.16 Frequency representation of the band-limited ramp filter.

Here, we make use of the fact that each projection has finite support in space; that is, $p(k\delta,\theta)$ is zero outside the index range. This implies that to determine $g(n\delta,\theta)$, we use only $h(m\delta)$ in the range $-(N-1)\le m \le (N-1)$.

Although the discrete convolution implementation of Eq. (3.30) is straightforward to arrive at the filtered projection, it is often more efficient to carry out the operation in the frequency domain when N is large [using fast Fourier transform (FFT) operations]. For a typical CT scanner today, the number of samples N in a single projection is close to a thousand. Therefore, we want to arrive at a frequency domain version of the $h(m\delta)$ sequence. The discrete Fourier transform $H'(\omega)$ of $h(m\delta)$ in the finite range is different from $H(\omega)$ described by Eq. (3.26), as illustrated by Fig. 3.18 (which was generated by again setting $\delta = 1$). The primary difference between the two is the dc component. Although the difference is quite small, its impact on the CT number accuracy of the reconstructed image is not negligible. To illustrate this effect, we reconstructed an oval phantom using the discrete Fourier transform of the kernel described in Eq. (3.29), and the intensity profile along a vertical line is plotted as a black line in Fig. 3.19. Then we set the dc component of the kernel (in Fourier space) to zero and reconstructed the same scan. The intensity profile is depicted by the gray line, indicating a bias in the image intensity. Although the magnitude of the kernel's dc component is small (Fig. 3.18), it resulted in a nearly 20-HU drop in the reconstructed image intensity.

Next, consider the issue of circular convolution. The original filtering operation described in Eq. (3.21) requires an aperiodic convolution. When this operation is performed in the frequency domain, only periodic or circular convolution is possible. If we implement directly the sequence of operations described previously, interference artifacts may result. This is the so-called "wrap-around" effect or interperiod interference. To avoid the artifacts, we must pad each projection with a sufficient number of zeroes prior to the Fourier

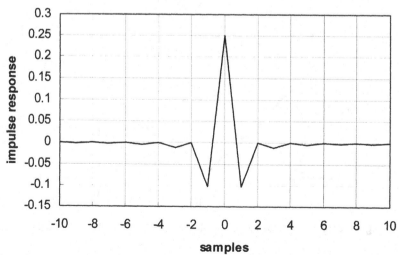

Figure 3.17 Impulse response of the ramp filter.

transform and filtering operation.[2] The minimum number of zeroes must equal the number of samples in the original projection minus one ($N - 1$). To illustrate the interperiod interference effect, we scanned and reconstructed an oval phantom with and without proper zero padding. The results are shown in Figs. 3.20(a) and (b). A close examination of the regions near the edge of the FOV indicates a shading artifact for the image produced without zero padding. The difference image shown in Fig. 3.21 clearly shows the artifact's appearance. For a quantitative measurement, we can plot a vertical profile through the center of the image (Fig. 3.22). A nearly 20-HU drop in the CT number near the peripheries of the image is observed. For further theoretical analysis on this topic, interested readers should refer to Ref. 2.

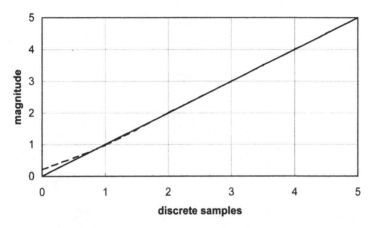

Figure 3.18 Comparison of $|\omega|$ function (solid line) and the Fourier transform of the band-limited ramp filter (dashed line).

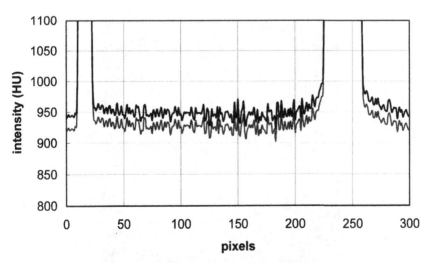

Figure 3.19 Impact of the dc term in the convolution filter design. Black line: reconstruction profile by filter derived from band-limited truncated kernel; gray line: reconstruction profile obtained by setting the dc term to zero.

After this lengthy discussion of the filter implementation process, it is important to understand the properties of the filter kernel and their impact on the reconstructed images. The characteristics of the ramp filter $H'(\omega)$ shown in Fig. 3.18 indicate that $H'(\omega)$ prefers the high-frequency contents of the projection over the low-frequency contents. In fact, the ramp filter behaves roughly as a derivative operator. This comes from the Fourier transform property that was discussed in Chapter 2. Therefore, we could consider the filtering operation as a

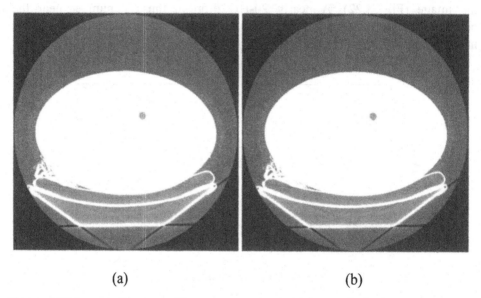

(a)   (b)

Figure 3.20 Impact of zero padding on reconstructed images. Projections are (a) zero padded and (b) not zero padded prior to the filtered backprojection.

Figure 3.21 Difference image (non-zero-padded image minus zero-padded image).

deconvolution process that removes the blurring caused by backprojection. For illustration, Figs. 3.23(a) and (b) plot the profiles of a projection of a uniform cylinder and its filtered version. The edge-enhancement characteristics of the filter are clearly shown by the sharp negative side lobes of the filtered projection near the edges of the object.

Equation (3.26) used a simple rectangular window function to band limit the filter kernel. An additional modification can be applied to the window function to shape the filter's frequency response. In practice, the window function is often used as a tool to modify the noise characteristics of the reconstructed images, thus achieving a desirable balance between the spatial resolution and image noise. For example, we can multiply the original filter by a Hanning or sinc

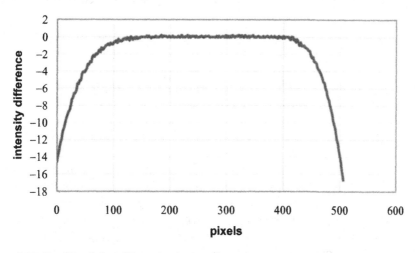

Figure 3.22 Profile of the difference image (image reconstructed without zero padding minus image reconstructed with proper zero padding of an oval phantom).

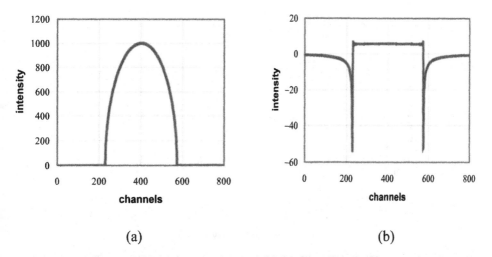

(a)   (b)

Figure 3.23 (a) A projection and (b) the filtered projection.

window function. Mathematically, the general Hanning window $\Pi_H(\omega)$ is expressed by

$$\Pi_H(\omega) = \begin{cases} 1, & |\omega| \le \omega_L \\ 0.5 + 0.5\cos\left(\dfrac{\pi(|\omega| - \omega_L)}{\omega_H - \omega_L}\right), & \omega_L < |\omega| \le \omega_H, \\ 0, & |\omega| > \omega_H \end{cases} \qquad (3.31)$$

where ω_H and ω_L are the upper and lower limits of the frequency to be modified, respectively. The general sinc window $\Pi_S(\omega)$ can be described by the following equation:

$$\Pi_S(\omega) = \begin{cases} 1, & |\omega| \le \omega_L \\ \dfrac{\sin\left(\pi(|\omega| - \omega_L)/(\omega_H - \omega_L)\right)}{\pi(|\omega| - \omega_L)/(\omega_H - \omega_L)}, & \omega_L < |\omega| \le \omega_H. \\ 0, & |\omega| > \omega_H \end{cases} \qquad (3.32)$$

Similarly, ω_H and ω_L are the upper and lower limits of the frequency to be modified, respectively. To illustrate the impact of different window functions on the reconstructed image, we reconstructed an oval phantom with a rectangular and a sinc function (the window functions are shown in Fig. 3.24). The reconstructed images are shown in Fig. 3.25. For convenience, the symbols "WW" and "WL" denote the display window width and window level in all figures. A comparison between Figs. 3.25(a) and (b) shows a significant noise reduction plus a slight reduction in spatial resolution [note the outline of the

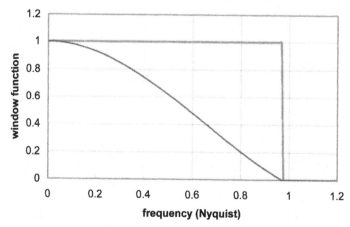

Figure 3.24 Different window functions for the reconstruction filter. Thick gray line: rectangular window; thin black line: sinc window.

objects in the difference image of Fig. 3.25(c)]. This is not surprising if the characteristics of the two window functions are examined. Compared to the rectangular window, the sinc function suppresses the high-frequency contents preferentially. Since most of the quantum noise in the measured projection is high frequency in nature, a noise reduction is achieved with the sinc window.

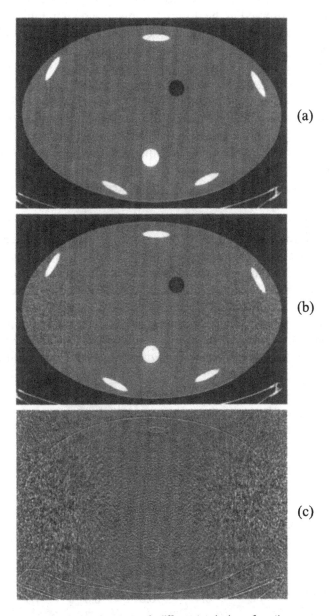

(a)

(b)

(c)

Figure 3.25 Illustration of the impact of different window functions on reconstructed image. (a) Reconstructed with the sinc function shown in Fig. 3.24 (WW = 400). (b) Reconstructed with the rectangular window function shown in Fig. 3.24 (WW = 400). (c) Difference image (rectangular-sinc) (WW = 100).

In addition to the shape-of-the-window function, we can also modify the cutoff frequency ω_H of the window function to achieve the desired noise and spatial resolution. For illustration, Fig. 3.26 shows two sinc functions with different cutoff frequencies. The corresponding images reconstructed with these filters are shown in Figs. 3.27(a) and (b). Although the impacts on image noise and spatial resolution are not as dramatic as in the previous example, the effect can still be clearly observed in the difference image shown in Fig. 3.27(c).

The simple examples given above illustrate the flexibility of the reconstruction filter design. When reconstruction filters are generated for commercial CT scanners, other enhancement techniques are also used to emphasize or de-emphasize certain frequency contents for different clinical applications. For example, the LightSpeed™ CT scanners (GE Healthcare, Milwaukee, Wisconsin) have six different types of filters: soft, standard, detail, lung, bone, and edge. The standard kernel is often used for routine whole-body scans, since it produces reasonably quiet images. The bone kernel, on the other hand, is often used in inner auditory canal (IAC), lung, and spine studies to examine fine structures. Naturally, the bone kernel provides a higher spatial resolution at the expense of image noise. For the newer vintage scanner Discovery™ CT750 HD (GE Healthcare, Milwaukee, Wisconsin), additional reconstruction kernels are added to take advantage of the increased inherent spatial resolution of the scanner. For example, for cardiac imaging, five kernels are added to optimize the appearance of stent, calcified plaques, soft plaques, or stenosis. The design of the reconstruction kernels is usually driven by specific clinical applications rather than uniform coverage over a range of frequency space. The filter design process is quite complicated and requires extensive clinical testing and validation. Unfortunately, the design details of these filters are generally considered proprietary and are not available in the public domain.

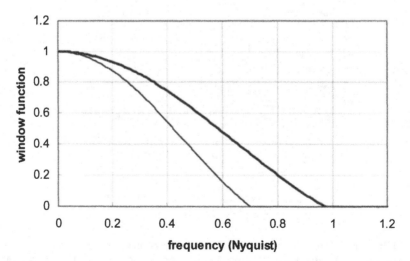

Figure 3.26 Window function with different cutoff frequencies. Thick line: sinc function with cutoff frequency at 97% Nyquist; thin line: sinc function with cutoff frequency at 68% Nyquist.

Having confronted several filter implementation issues, it is time to discuss implementation issues related to the backprojection process. As stated previously, the backprojection process maps the value of a filtered projection sample to a straight line in the reconstructed image, as shown in Fig. 3.28(a). This figure ignores the discrete nature of the backprojection processing—that is, it assumes the projection is a 1D continuous function and the reconstructed image is a 2D continuous function. In any computer implementation, however, neither the projection nor the reconstructed image is continuous. The projection waveform is sampled by the detector into a discrete set of samples (the samples remain in discrete form after the filtering process). Similarly, the reconstructed image is formed on a discrete matrix, as shown in Fig. 3.28(b). For backprojection, interpolation is required.

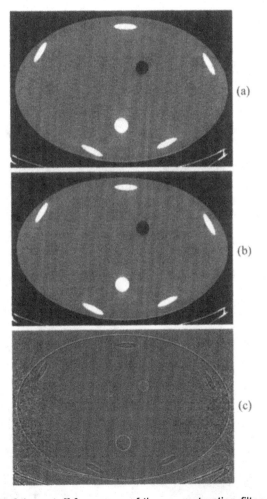

Figure 3.27 Impact of the cutoff frequency of the reconstruction filter window function on the reconstructed image. (a) Sinc function with cutoff frequency at 97% of Nyquist (WW = 200). (b) Sinc function with cutoff frequency at 68% of Nyquist (WW = 200). (c) Difference image (97% – 68% window) (WW = 50).

The interpolation can be performed either on the filtered projection samples or on the reconstructed images. Different interpolation processes are shown in Fig. 3.29. Figure 3.29(a) shows a backprojection process that starts at a discrete projection sample. The filtered sample is traced along the x-ray path at a fixed increment. At each stop along the ray path, the intensity of the sample is distributed to the neighboring pixels. The distribution process can be either linear or nonlinear. For example, one could use nearest-neighbor interpolation in which the entire sample is added to the pixel that is closest to the stop. One could also use bilinear interpolation in which the intensity is distributed to the nearest four neighboring pixels based on their distance to the stop, as shown by Fig. 3.29(a). Of course, other higher-order interpolation schemes can be used. The interpolation is, in general, 2D. This type of backprojection is commonly called "ray-driven" backprojection.

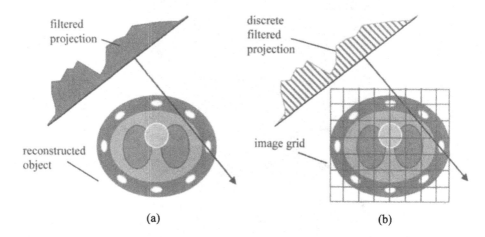

(a)   (b)

Figure 3.28 Difference between (a) a continuous backprojection and (b) a discrete backprojection.

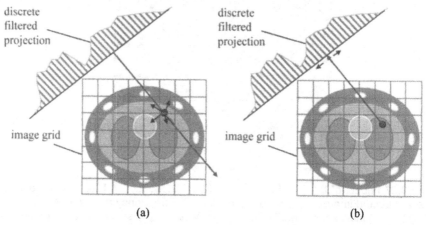

(a)   (b)

Figure 3.29 Illustrations of (a) ray-driven backprojection, and (b) pixel-driven backprojection.

Alternatively, we can start at an image pixel location and calculate the intensity contribution from the projection, as shown in Fig. 3.29(b). In this approach, we follow the ray path that passes through the center of the pixel and locate its intersection with the filtered projection. In general, the location of the intersection does not line up with the discrete samples of the projection. As a result, some form of interpolation is needed to estimate the filtered projection sample at the location of intersection. Either nearest neighbor or linear interpolation can be used. Because the interpolation takes place in the projection space, only 1D interpolation is needed. This backprojection type is commonly called "pixel-driven" backprojection. Because of its computational savings (1D versus 2D), pixel-driven backprojection is often preferred over a ray-driven backprojection.

To improve the accuracy of the pixel-driven interpolation, one often uses higher-order interpolations (e.g., Lagrange or cubic-spline interpolation). For a large image matrix size (e.g., 512×512), the interpolation can be computationally expensive. To speed up the backprojection process and preserve high-frequency information contents, the filtered projection is often pre-interpolated to a finer sampling grid with high-order interpolation schemes prior to the backprojection. During backprojection, simpler interpolation schemes such as nearest-neighbor or linear interpolation can then be used. To illustrate the impact of different interpolation schemes on spatial resolution and image noise, we scanned a 0.08-mm tungsten wire phantom. The filtered projection data were then preinterpolated to double their size (the distance between the interpolated samples was half of the original sampling distance). Two interpolation algorithms were used: frequency domain interpolation and linear interpolation. The interpolated projection was then backprojected using a pixel-driven algorithm with linear interpolation.

To quantitatively analyze the impact of different interpolation algorithms on the spatial resolution of the reconstructed images, the modulation transfer functions (MTF) of the two images are plotted in Fig. 3.30. The MTF measures a

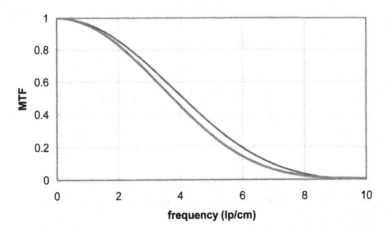

Figure 3.30 Impact of interpolation algorithm on spatial resolution in a backprojection. Thin black line: frequency domain interpolation; thick gray line: linear interpolation.

system's response to different frequencies. An "ideal" system has a flat MTF curve such that the system response is independent of the input frequency. A detailed description of MTF can be found in Section 5.1. In Fig. 3.30, the thin black line is the MTF of the frequency domain interpolation and the thick gray line is the MTF of the linear interpolation. Clearly, frequency domain interpolation produces better spatial resolution, as expected. A quantitative measurement of the MTF curves shows a 0.34 line pair per centimeter (lp/cm) increase at the 50% point and a 0.48 lp/cm increase at the 10% point (a line pair is a pair of equal-sized black-white bars; see Section 5.1). On the other hand, the image produced by the linear interpolation results in a less-noisy image compared to the frequency-domain interpolation (a standard deviation of 4.97 HU versus 5.76 HU).

In many high-level language software systems for numerical computations and graphics, such as MATLAB® (The MathWorks, Natick, Massachusetts) or IDL® (Research Systems, Inc., Boulder, Colorado), vectors or matrices can be represented simply as variables. Various operators are also defined for the vectors (as opposed to the element-by-element manipulation in C programming or Fortran). In such an environment, the implementation of parallel backprojection can be quite easy. For illustration, consider an example of pixel-driven backprojection. We denote by $g(k,\theta)$ the filtered projection at angle θ, where $k = 0, 1, ..., K-1$ is the index of the filtered parallel samples in a view; the iso-ray is located at c_g (c_g is not necessarily an integer); v is the reconstruction FOV in millimeters; n is the number of rows and columns of the reconstructed image (typically, $n = 512$); c is the center of the image matrix and equals $(n-1)/2$ (typically, $c = 255.5$); and (x_c, y_c) is the center of reconstruction in millimeters [the iso-center is at $(0, 0)$]. The x coordinates $\mathbf{X_0}$ and y coordinates $\mathbf{Y_0}$ of the reconstructed image matrix at $\theta = 0$ (in millimeters) are

$$\mathbf{X}_0 = (v/n)\mathbf{R} \oplus x_c \qquad (3.33)$$

and

$$\mathbf{Y}_0 = (v/n)\mathbf{R}^T \oplus y_c, \qquad (3.34)$$

where $\mathbf{R}$ is an $n \times n$ matrix with $r_{i,j} = j - c$, $i = 0,..., n-1$ and $j = 0,...,n-1$, and $\oplus$ represents elemental addition as defined in Chapter 2. The intersection of all of the pixel rays through the image matrix at angle θ with the filtered projection is simply the x coordinates of the image matrix rotated by $-\theta$ (rays are parallel to the y axis when $\theta = 0$):

$$\mathbf{X}_\theta = \cos\theta\, \mathbf{X}_0 - \sin\theta\, \mathbf{Y}_0. \qquad (3.35)$$

Since $\mathbf{X}_\theta$ is specified in millimeters, we need to convert it to the filtered projection sample index:

$$\mathbf{T}_\theta = (1/\delta)\mathbf{X}_\theta \oplus c_g. \qquad (3.36)$$

If $\mathbf{T'}_\theta = \text{int}(\mathbf{T}_\theta)$ is the integer portion of $\mathbf{T}_\theta$, $\mathbf{L}$ is the $n \times n$ reconstructed image matrix (with its elements initialized to 0), M is the number of views in the projection, δ is the filtered projection sampling spacing in millimeters, θ_0 is the projection angle of the first view, and $\Delta\theta$ is the angular increment, then the backprojection processing can be described by the following loop:

For $m = 0$, M–1, we perform the following set of operations:
$\theta = \theta_0 + m\,\Delta\theta$
$\mathbf{T}_\theta = (1/\delta)(\cos\theta\;\mathbf{X}_0 - \sin\theta\;\mathbf{Y}_0) \oplus c_g$
$\mathbf{T'}_\theta = \text{int}(\mathbf{T}_\theta)$
$\alpha_\theta = \mathbf{T}_\theta - \mathbf{T'}_\theta$
$\mathbf{L} = \mathbf{L} + [1 \oplus (-\alpha_\theta)] \otimes g(\mathbf{T'}_\theta, \theta) + \alpha_\theta \otimes g[(\mathbf{T'}_\theta \oplus 1), \theta]$
End of loop

Of course, the final result must be scaled by π/M. The above implementation assumes that the filtered projections are preinterpolated and only linear interpolation is necessary for the final interpolation.

The entire reconstruction process for parallel-beam geometry can be summarized as follows. For each measured projection (after preprocessing or preconditioning of the data), the projection is padded with sufficient zeroes to avoid interperiod interference. A Fourier transform is applied to the zero-padded projection, and the transformed projection is multiplied by a filtering function. The result is then inverse Fourier transformed to arrive at a filtered projection. This projection is backprojected (either pixel driven or ray driven) to the image matrix. To improve the spatial resolution, the filtered projection is often pre-interpolated prior to the backprojection process. The entire process is repeated for every view in the projection data set. A flowchart in Fig. 3.31 describes the reconstruction process for parallel-beam projections.

3.4.3 Targeted reconstruction

The typical image matrix size for medical CT is 512 × 512. For a 50-cm reconstruction FOV, each pixel is roughly 1 mm in size. According to the Nyquist sampling theory, the highest-frequency contents supported by such a sampling density is 5 lp/cm. If we wish to examine anatomical structures with higher spatial frequency contents, the sampling distance between image pixels must be reduced by either increasing the reconstructed image matrix size or by reducing the reconstruction FOV. An increased image size will impact not only the reconstruction speed (since the number of pixels to be reconstructed scales quadraticly with the matrix size), but also increases the storage. For example, if we wish to examine an object with the highest frequency contents of 20 lp/cm, the image matrix size needs to be at least 2048 × 2048. This is a 16-fold increase in the reconstruction complexity and storage. Consequently, this approach is rarely used in practice.

An alternative approach is to reduce the reconstruction FOV. For example, if we reduce the FOV from 50 to 12.5 cm, the same 512 × 512 image matrix size can now support frequencies up to 20 lp/cm (with an image pixel size that is roughly 0.25 mm). Since most high-resolution applications require the examination of only a small area (e.g., the inner auditory canal or vertebral bones), the reduced FOV is not a limitation. This approach is often called a "targeted" or "zoomed" reconstruction because the reconstruction is targeted to a smaller area.

The reconstruction processing of a targeted reconstruction is similar to a full FOV reconstruction. Once a filtered projection is obtained (the filtering process is identical to the full FOV process), the backprojection maps the reduced FOV to the projection. For example, consider a pixel-driven backprojection that is performed for a reconstruction FOV of v mm centered at (x_c, y_c) mm from the system iso-center. Assume that the image matrix size is $n \times n$, and the center of the image matrix is denoted by i_c, j_c. An image pixel at location (i, j) can be mapped to a point (x, y) in the original coordinate system centered at the iso-center by the following set of equations:

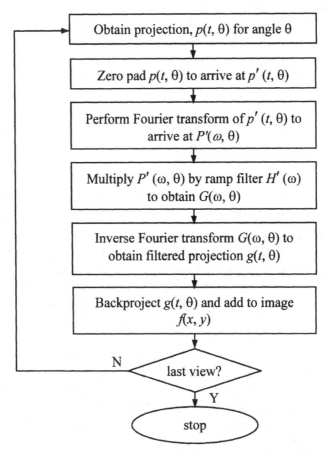

Figure 3.31 Flowchart of the reconstruction process for parallel projections.

$$\begin{cases} x = \dfrac{(i - i_c)v}{n} + x_c, & i = 1, \dots, n \\[2mm] y = \dfrac{(j - j_c)v}{n} + y_c, & j = 1, \dots, n. \end{cases} \tag{3.37}$$

Since we know how to perform backprojection for a point located at (x, y) for the full FOV reconstruction, the filtered projection samples can be located, interpolated, and added to the reconstructed image in a fashion similar to that of the full FOV reconstruction. In fact, the reconstruction loop outlined in the previous section can be applied directly without modification.

To illustrate the impact of the targeted reconstruction, we performed the following experiment. A GE Performance^{TM} phantom was scanned and reconstructed with a 50-cm FOV. The reconstructed image is shown in Fig. 3.32(a). The same scan was then reconstructed with a reconstruction algorithm targeted on the center 10-cm FOV; the reconstructed image is shown in Fig. 3.32(b). A comparison between Figs. 3.32(a) and (b) shows that the targeted reconstruction reveals more of the fine details of the phantom than the full FOV reconstruction shows. For example, the thin wire that is barely visible in Fig. 3.32(a) can be clearly seen in Fig. 3.32(b). To demonstrate the fact that the same effect cannot be achieved by image space interpolation, Fig. 3.32(a) was interpolated using a 2D interpolation to produce an image representing the center 10-cm region of the original image, as shown in Fig. 3.32(c). By comparing the bar patterns in Figs. 3.32(b) and (c), we can conclude that the targeted reconstruction produced a much improved spatial resolution than did the interpolated image. This simple experiment clearly demonstrates the value of targeted reconstruction.

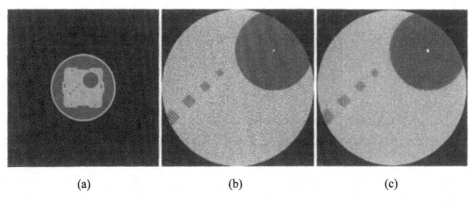

(a)   (b)   (c)

Figure 3.32 Impact of reconstruction FOV and interpolation on spatial resolution (WW = 300). (a) Image reconstructed with a 50-cm FOV and a high-resolution kernel. (b) Image reconstructed with a 10-cm FOV and a high-resolution kernel. (c) Magnified view of (a) with bilinear interpolation.

3.5 Fan-Beam Reconstruction

So far, our discussion has been limited to the case of parallel-beam geometry. In nearly all of the commercially available third-generation CT scanners, the x-ray source is approximately a single point and a projection is collected by sampling nearly 1000 detector channels in a short time window. Even in the case of a fourth-generation scanner, a projection is formed by collecting a series of samples on a fixed detector channel while a point x-ray source travels along a circle. In both cases, the rays that form a single projection are not parallel lines; instead, they focus to a single point. Consequently, the formula derived previously cannot be applied directly.

To solve the reconstruction issue, we can take a closer look at fan-beam geometry. In practice, two of the most popular third-generation design choices use equiangular fan beam or equal-spaced fan beam geometries. The difference is driven mainly by different detector designs. In the case of an equiangular fan beam, a detector is made of a large number of detector modules. These modules are placed on an arc that is concentric to the x-ray source, as shown in Fig. 3.33(a). Because the width of each module is quite small compared to the module-to-source distance, the angle formed by each detector cell with the x-ray source can be assumed constant. For example, for HighSpeed^{TM} scanners, the source-to-module distance is roughly 1100 mm, while the detector cell width is roughly 1 mm. A module contains 16 detector cells. The maximum difference between the angles formed by the detector cells within the module (the boundary

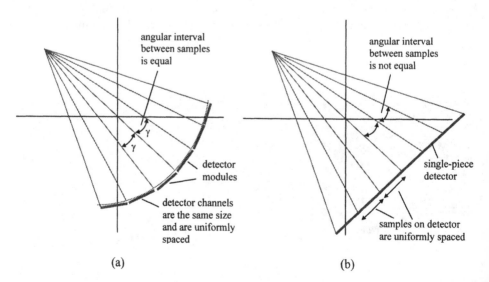

(a)   (b)

Figure 3.33 Illustrations of (a) equiangular and (b) equal-spaced sampling for fan-beam geometry. In equiangular sampling, a detector is made of many small modules that are located on an arc concentric to the x-ray source. The detector cells within each module are uniformly spaced. The angular interval between adjacent samples is constant throughout the detector. In equal-spaced geometry, the detector surface is flat, and sampling elements are uniformly spaced along the surface. The angular interval between adjacent samples changes from the center to the periphery.

channel versus the center channel) is only 0.0000024 deg, so for all practical purposes, the difference can be safely ignored. Because each module is aligned to the x-ray focal spot, the angular error is the same for all modules. The advantage of this design is its improved manufacturability. Since all modules are interchangeable, there is no need to design different geometries for modules at different detector locations. In addition, a defective channel in a detector can be corrected by replacing a single module.

The second fan-beam geometry is the equal-spaced design. This design is popular for detectors made of a single piece. For example, the detector can be an image intensifier (II) or a digital flat panel. In both cases, the detector is formed on a flat surface instead of a curved surface. The sampling is uniformly spaced, as shown in Fig. 3.33(b). For this type of design, the angular interval between adjacent sampling elements with the x-ray source changes gradually from center to peripheries.

3.5.1 Reconstruction formula for equiangular sampling

Ever since the concept of the third-generation CT scanner geometry was developed, researchers were motivated to derive a reconstruction formula similar to that of parallel-beam geometry. These efforts led to the development of a fan-beam reconstruction formula.[1,2,6] For the ease of derivation, we will define some terminologies. Any fan-beam ray can be uniquely specified by two parameters γ and β, where γ is the angle formed by the ray with the iso-ray (an imaginary ray that connects the x-ray source with the iso-center), and β is the angle of the iso-ray formed with the y axis, as shown in Fig. 3.34. β is called the projection angle and indicates which projection view is used. γ is called the detector angle and specifies the location of a ray within the fan. Recalling the discussion on parallel projection reconstruction, we can specify a ray uniquely by a parameter pair t and θ, where t is the distance of the ray to the iso-center and θ is the projection angle. A projection sample $q(\gamma, \beta)$ in a fan-beam projection would belong to a projection sample $p(t, \theta)$ in a parallel projection, if the following conditions are satisfied:

$$\begin{cases} \theta = \beta + \gamma \\ t = D\sin\gamma, \end{cases} \tag{3.38}$$

where D is the distance between the x-ray source and the iso-center. For parallel projections, the reconstruction formula was derived in Eq. (3.28):

$$f(x,y) = \int_0^\pi d\theta \int_{-t_m}^{t_m} p(t',\theta)h(t-t')dt'$$

$$= \int_0^\pi d\theta \int_{-t_m}^{t_m} p(t',\theta)h(x\cos\theta + y\sin\theta - t')dt'. \tag{3.39}$$

We can modify Eq. (3.39) to include all projections over 2π:

$$f(x,y) = \frac{1}{2} \int_0^{2\pi} d\theta \int_{-t_m}^{t_m} p(t',\theta)h(x\cos\theta + y\sin\theta - t')dt'. \qquad (3.40)$$

Expressing this equation in polar coordinates (r, ϕ), we obtain

$$f(r,\varphi) = \frac{1}{2} \int_0^{2\pi} d\theta \int_{-t_m}^{t_m} p(t',\theta)h(r\cos(\theta-\varphi) - t')dt'. \qquad (3.41)$$

Here, we make use of the relationship $r\cos(\theta-\varphi) = r\cos\varphi\cos\theta + r\sin\varphi\sin\theta$. We want to perform variable substitutions so that the entire expression is written in terms of (γ, β) instead of (t, θ). Substituting Eq. (3.38) into Eq. (3.41) and observing the fact that $dt d\theta = D\cos\gamma\,d\gamma\,d\beta$, we obtain

$$f(r,\varphi) = \frac{1}{2} \int_{-\gamma}^{2\pi-\gamma} d\beta \int_{-\gamma_m}^{\gamma_m} q(\gamma,\beta)h(r\cos(\beta+\gamma-\varphi) - D\sin\gamma)D\cos\gamma\,d\gamma, \qquad (3.42)$$

where γ_m is the maximum γ beyond which $q(\gamma, \beta) = 0$. Naturally, $\gamma_m = \sin^{-1}(t_m/D)$. Since all functions of β are periodic with a period of 2π, the integration limit $-\gamma$ and $2\pi-\gamma$ in Eq. (3.42) may be replaced by 0 and 2π, respectively:

$$f(r,\varphi) = \frac{1}{2} \int_0^{2\pi} d\beta \int_{-\gamma_m}^{\gamma_m} q(\gamma,\beta)h(r\cos(\beta+\gamma-\varphi) - D\sin\gamma)D\cos\gamma\,d\gamma. \qquad (3.43)$$

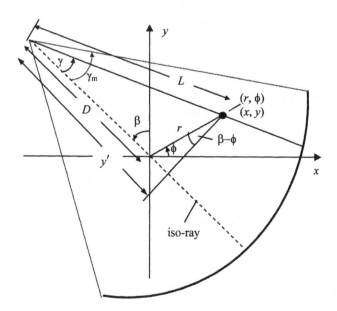

Figure 3.34 Fan-beam geometry. The reconstructed pixel is expressed by (x, y) or (r, ϕ). The detector angle γ is the angle between the source-to-pixel line and the iso-ray, and L is the distance from the x-ray source to the pixel (x, y).

Clearly, the independent variable for the filtering function h is quite complicated and cannot be easily implemented on a computer. Note that the expression is not in the form of a convolution. We can rewrite the argument in the following form:

$$r\cos(\beta + \gamma - \varphi) - D\sin\gamma = r\cos(\beta - \varphi)\cos\gamma - [r\sin(\beta - \varphi) + D]\sin\gamma. \quad (3.44)$$

If we denote the distance from the x-ray source to the point of reconstruction (r, φ) by L and the detector angle of the ray that passes through (r, φ) by γ', we have

$$\begin{cases} L\cos\gamma' = D + r\sin(\beta - \varphi) \\ L\sin\gamma' = r\cos(\beta - \varphi) \end{cases}. \quad (3.45)$$

Combining Eqs. (3.44) and (3.45), we obtain

$$r\cos(\beta + \gamma - \varphi) - D\sin\gamma = L\sin(\gamma' - \gamma). \quad (3.46)$$

Here, we make use of the property that $L\sin(\gamma' - \gamma) = L\sin\gamma'\cos\gamma - L\cos\gamma'\sin\gamma$. By substituting this equation into Eq. (3.43), we produce a simplified reconstruction formula:

$$f(r, \varphi) = \frac{1}{2}\int_0^{2\pi} d\beta \int_{-\gamma_m}^{\gamma_m} q(\gamma, \beta) h(L\sin(\gamma' - \gamma)) D\cos\gamma \, d\gamma. \quad (3.47)$$

In this equation, the only quantity that is not well defined is the argument for the filter function h. Based on the original definition of the filter function, we have

$$h(L\sin\gamma) = \int_{-\infty}^{\infty} |\omega| e^{j2\pi\omega L\sin\gamma} d\omega$$

$$= \left(\frac{\gamma}{L\sin\gamma}\right)^2 \int_{-\infty}^{\infty} |\omega'| e^{j2\pi\omega'\gamma} d\omega'$$

$$= \left(\frac{\gamma}{L\sin\gamma}\right)^2 h(\gamma). \quad (3.48)$$

In the derivation, we used the transformation $\omega' = \dfrac{\omega L\sin\gamma}{\gamma}$. Since variable L does not depend on γ, and (x, y) and (r, φ) are interchangeable, the reconstruction formula (3.47) may be written as

$$f(x, y) = \int_0^{2\pi} L^{-2} d\beta \int_{-\gamma_m}^{\gamma_m} q(\gamma, \beta) h''(\gamma' - \gamma) D\cos\gamma \, d\gamma, \quad (3.49)$$

where

$$h''(\gamma) = \frac{1}{2}\left(\frac{\gamma}{\sin\gamma}\right)^2 h(\gamma).$$ (3.50)

If we assume that the projection $q(\gamma, \beta)$ is sampled in γ with an interval ν, the discrete form of $h''(\gamma)$ is simply

$$h''(n\delta) = \begin{cases} \dfrac{1}{8\nu^2}, & n = 0, \\ 0, & n = \text{even}, \\ -\dfrac{1}{2}\left(\dfrac{1}{\pi\sin n\nu}\right)^2, & n = \text{odd}. \end{cases}$$ (3.51)

Next, compare the reconstruction formula for parallel-beam projection [Eq. (3.28)] with the formula for fan-beam reconstruction [Eq. (3.49)]. For convenience, we will duplicate Eq. (3.28) below:

$$f(x, y) = \int_0^\pi d\theta \int_{-t_m}^{t_m} p(t', \theta) h(t - t') dt'.$$ (3.52)

Structurally, Eqs. (3.49) and (3.52) are quite similar. Therefore, the discussions on computer implementation issues for parallel projection can be readily applied to the fan-beam case. For example, to avoid interperiod interference, the projection must be padded with sufficient zeroes prior to the filtering process, assuming the Fourier transform is used. Similarly, the window function for the filter can be modified by additional functions to shape its frequency response. If pixel-driven backprojection is used, the filtered projection can be preinterpolated to a finer sampling grid to improve spatial resolution and reduce computational complexity.

Now examine the difference between parallel-beam and fan-beam reconstructions. Unlike parallel reconstruction, the fan-beam projection needs to be multiplied by $D\cos\gamma$ prior to performing the filtering operation. Because this function is independent of projection angle β, it can be precalculated and stored prior to the start of the reconstruction. The second difference between the two reconstruction algorithms is the fact that weighted backprojection is used for fan-beam reconstruction. The backprojection is carried out along the fan. The scaling factor L^{-2} changes from pixel to pixel. Note that L represents the distance from the reconstructed pixel to the x-ray source. Because L depends on both γ and β, the amount of computation is substantial. More efficient and less time-consuming implementations of the weighted backprojection have been investigated. To ensure adequate reconstruction speed, some CT manufacturers have designed application-specific integrated circuit (ASIC) chips to perform the fan-beam backprojection operation.

To illustrate the difference between a fan-beam backprojection and a parallel-beam backprojection, we performed a backprojection operation on a single filtered projection of a GE Performance™ phantom. The backprojected images are shown in Figs. 3.35(a) and (b). A comparison of the two images shows that the fan-beam backprojection (b) not only changes its shape as a function of its distance to the x-ray source (located in the 6 o'clock position in this example), but also changes its intensity as a distance to the source (the image is brighter toward the bottom of the figure). For a better understanding of the fan-beam reconstruction process, a reconstruction flow chart is shown in Fig. 3.36.

Although the backprojection equation described by Eq. (3.49) is acceptable for computer implementations, it requires the calculation of the term L^{-2}, where L is the distance from the x-ray source to the point of reconstruction. This needs to be performed for each reconstructed pixel and each projection view. It is often desirable to simplify the expression so that the new quantity depends only on the rotated coordinate y' of each pixel. In this case, the distance calculation can be avoided by simply noting that

$$y' = L\cos\gamma, \tag{3.53}$$

where y' is the projected distance (along the y' axis) from the reconstructed point to the x-ray source, as shown in Fig. 3.34. If the filtered projection in Eq. (3.49) is first multiplied by $\cos^2\gamma$, the L^{-2} term becomes $(y')^{-2}$. Since the multiplication by $\cos^2\gamma$ is performed only once for each view, the total amount of computation of the backprojection is reduced.

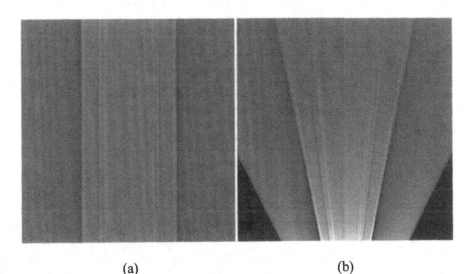

(a)   (b)

Figure 3.35 (a) Parallel backprojection and (b) fan-beam backprojection of a single filtered projection of a GE Performance™ phantom.

The implementation for fan-beam backprojection is slightly more complicated than the parallel-beam case. The filtered projection is denoted by $q'(k, \beta)$, where $k = 0, 1, \ldots, K-1$ is the index along the detector angle, and β is the projection angle. The image matrix coordinates calculation should be the same as the one described in Eqs. (3.33) and (3.34), since the image matrix does not change with the projection sampling pattern. Note that the projection taken at angle β is equivalent to the projection taken at angle 0 while rotating the image matrix by an angle $-\beta$:

$$\mathbf{X}_\beta' = \cos\beta \, \mathbf{X}_0 - \sin\beta \, \mathbf{Y}_0 \qquad (3.54)$$

$$\mathbf{Y}_\beta' = \sin\beta \, \mathbf{X}_0 + \cos\beta \, \mathbf{Y}_0. \qquad (3.55)$$

Next, we will denote the source-to-iso distance in millimeters by D. The scaling factor $\mathbf{S}_\beta$ (assuming the filtered projection has been prescaled by $\cos^2 \gamma$) is

$$\mathbf{S}_\beta = \frac{1}{D \oplus \left(-\mathbf{Y}_\beta'\right)}. \qquad (3.56)$$

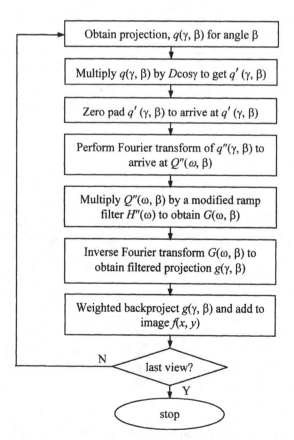

Figure 3.36 Flowchart of equiangular fan-beam reconstruction processing.

The symbols $\oplus$ and $-$ denote elemental addition and division as defined in Chapter 2. The detector angle Γ_β corresponding to the reconstructed image pixels can be described by the following equation:

$$\Gamma_\beta = \tan^{-1}\left(\mathbf{X}'_\beta \otimes \mathbf{S}_\beta\right). \tag{3.57}$$

Note that the denominators for Eqs. (3.56) and (3.57) are the same and can be calculated only once to save computation time. Since Γ_β is specified in radians, we must convert it to the filtered projection sample index:

$$\mathbf{T}_\beta = (1/\Delta\gamma)\Gamma_\beta \oplus c_g , \tag{3.58}$$

where c_g is the location of the iso-ray of the filtered projection, and $\Delta\gamma$ is the angular spacing between adjacent samples. If we denote the integer portion of $\mathbf{T}_\beta$ by $\mathbf{T}'_\beta = \text{int}(\mathbf{T}_\beta)$, the $n \times n$ reconstructed image matrix (with its elements initialized to 0) by $\mathbf{L}$, the number of views in the projection by M, the projection angle of the first view by β_0, and the view angular increment by $\Delta\beta$, then the backprojection processing can be described by the following loop:

For $m = 0$, M–1, we perform the following set of operations:
$$\beta = \beta_0 + m\,\Delta\beta$$
$$\mathbf{X}_\beta' = \cos\beta\,\mathbf{X}_0 - \sin\beta\,\mathbf{Y}_0$$
$$\mathbf{Y}_\beta' = \sin\beta\,\mathbf{X}_0 + \cos\beta\,\mathbf{Y}_0$$
$$\mathbf{S}_\beta = \frac{1}{D \oplus \left(-\mathbf{Y}_\beta'\right)}$$
$$\mathbf{T}_\beta = \left(\frac{1}{\Delta\gamma}\right)\tan^{-1}\left(\mathbf{X}'_\beta \otimes \mathbf{S}_\beta\right) \oplus c_g$$
$$\mathbf{T}'_\beta = \text{int}(\mathbf{T}_\beta)$$
$$\alpha_\beta = \mathbf{T}_\beta - \mathbf{T}'_\beta$$
$$\mathbf{L} = \mathbf{L} + \mathbf{S}_\beta \otimes \mathbf{S}_\beta \otimes \left\{\left[1 \oplus \left(-\alpha_\beta\right)\right] \otimes g'(\mathbf{T}'_\beta,\beta) + \alpha_\beta \otimes g'\left[(\mathbf{T}'_\beta \oplus 1),\beta\right]\right\}$$
End of loop

The final result must be scaled by $2\pi/M$. For computational efficiency, we assume that the filtered projections are preinterpolated prior to the backprojection operation and that only linear interpolation is necessary during the backprojection process.

3.5.2 Reconstruction formula for equal-spaced sampling

Similar to the equiangular fan-beam projection, a projection sample in an equal-spaced fan-beam projection can be specified uniquely with the parameter set (s, β), where s is the distance between the iso-center and the intersection of a ray with an imaginary detector, as shown in Fig. 3.37. The imaginary detector is parallel to the actual detector and intersects the iso-center. The use of an imaginary detector is mainly for the convenience of reconstruction formulation. β is the projection angle defined in the same fashion as the equiangular fan beam.

The derivation of the equal-spaced fan-beam reconstruction algorithm is similar to that of the equiangular fan beam. Again, we start with the reconstruction formula of a parallel-beam projection and establish a geometric relationship between a fan-beam sampling and a parallel-beam sampling. Because of the similarity of the two types of reconstruction, here we will present only the reconstruction formula of the equal-spaced fan-beam algorithm without derivation. Interested readers can refer to Ref. 2 for a detailed derivation of the formula. The equal-spaced fan-beam projection is denoted by $q(s, \beta)$. The reconstructed image $f(r, \phi)$ can be shown to be:

$$f(r,\varphi) = \frac{1}{2} \int_0^{2\pi} U^{-2}(r,\varphi,\beta)d\beta \int_{-\infty}^{\infty} \left(\frac{D}{\sqrt{D^2+s^2}} \right) q(s,\beta)h(s'-s)ds , \quad (3.59)$$

where

$$U(r,\varphi,\beta) = \frac{D+r\sin(\beta-\varphi)}{D} .$$

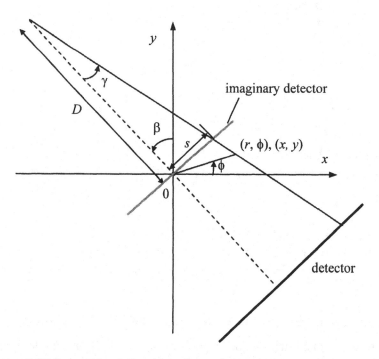

Figure 3.37 Geometric relationships of an equal-spaced fan-beam projection.

Here, D is the distance between the x-ray source and iso-center, as shown in Fig. 3.37. A comparison between Eqs. (3.49) and (3.59) reveals a striking similarity. Both processes require multiplication of the projection by a function prior to the filtering process, and both algorithms call for weighted backprojection. A closer examination of the scaling function $D / \sqrt{D^2 + s^2}$ reveals that the function represents the cosine of the projection fan angle γ, as shown in Fig. 3.37. Equation (3.59) states that the original fan-beam projection must be scaled by the detector fan angle prior to performing the filtering operation. Examine the physical meaning of the weighting function $U(r, \phi, \beta)$ used in the backprojection. The quantity in the numerator represents the projected distance between the x-ray source and the reconstructed point (r, ϕ) onto the iso-ray. $U(r, \phi, \beta)$ is the ratio of the projected distance over the source-to-iso distance. Similar to the case of the equiangular fan-beam reconstruction, the weighting function changes from pixel to pixel in the reconstructed image. As was discussed previously, the computational complexity for weighted backprojection is considerably higher than that of parallel-beam backprojection.

3.5.3 Fan-beam to parallel-beam rebinning

In Section 3.4, reconstruction algorithms for parallel-beam geometry were derived and their implementations discussed in detail. If we reformat the fan-beam projections into a set of parallel-beam projections, the previously discussed reconstruction algorithms can be applied directly. The existence of such a reformation should not be a surprise, since the fan-beam reconstruction algorithms presented in the previous sections were derived from the parallel-beam case based on the relationship of Eq. (3.38). In this section, we try to establish a mechanism to convert fan-beam projections to parallel-beam projections. This process is often called *rebinning*.

To understand the principle behind rebinning, first examine the sampling pattern of a set of parallel projections in sinogram space. Figure 3.38 depicts a 2D graph with its horizontal axis representing the distance of a ray to the iso-center, and the vertical axis representing the angle of the ray formed with the x axis. A projection ray is mapped to a single point in the sinogram space. For parallel projections, all samples fall onto a uniformly spaced rectangular grid, as shown in Fig. 3.39(a). Each parallel projection is represented by a single row of dots (indicated by the dotted rectangle in the figure). If we map a set of fan-beam samples (either equiangular or equal-spaced) to the same sinogram, they appear as dots that generally do not fall onto the rectangular grid, as shown by Fig. 3.39(b). Each fan-beam projection is mapped to a slanted row (not necessarily a straight line) of dots indicated by the dotted rectangle, because the angles formed by the fan-beam rays with the x axis increase (or decrease) from one end of the detector to the other in a single projection. Therefore, these samples have different θ values. Similarly, the distance of the fan-beam rays to the iso-center does not increase (or decrease) in a uniform step. For equiangular fan-beam sampling, the relationship between the angle θ and the sampling index i is linear:

$$\theta = (i - i_{\text{iso}})\Delta\gamma, \tag{3.60}$$

where i_{iso} is the index for the iso-channel, and $\Delta\gamma$ is the angle of two adjacent detector channels formed with the x-ray source. Here, "iso-channel" is an imaginary detector channel that lines up perfectly with the iso-ray, and i_{iso} is not necessarily an integer. As will be discussed in chapter 7, in many CT systems the detector is purposely "misaligned" so that none of the real detector channels lines up with the iso-ray. The distance d_i of the fan beam ray i to the iso-center can be described by

$$d_i = r \cdot \sin\left(i - i_{\text{iso}}\right)\Delta\gamma, \tag{3.61}$$

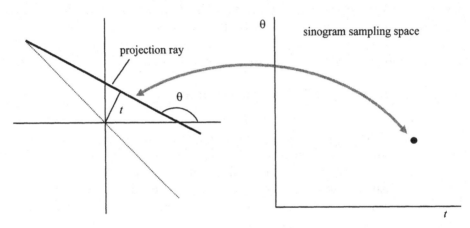

Figure 3.38 Projection sample in real space mapped to a point in the sinogram space.

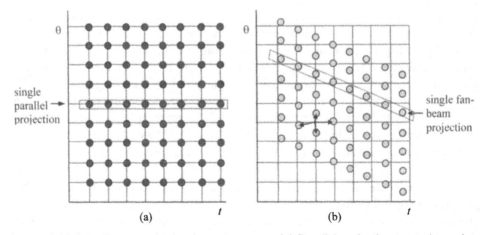

Figure 3.39 Sampling patterns in sinogram space. (a) Parallel projections are shown by the solid dots and (b) fan-beam projections are shown by the circles. These samples can be rebinned or regridded to the parallel sampling pattern.

where r is the source-to-iso-center distance. Figure 3.39(b) and Eqs. (3.60) and (3.61) indicate that the fan-beam samples (shown by the gray dots) can be mapped to the parallel-beam sampling grid by interpolation. For example, if we wish to estimate a parallel projection ray located at the intersection of the grid, we can simply interpolate the intensities of the neighboring fan-beam samples. Figure 3.39(b) shows an example in which four fan-beam samples are used to estimate a parallel sample. Following the discussions rendered in Section 3.4.2, the interpolation process can also be either linear or nonlinear.

A more convenient way to perform fan-beam to parallel-beam rebinning is a two-step process. In the first step, interpolation is carried out only across projection views to arrive at a set of nonuniformly spaced parallel samples. That is, interpolation is carried out among projection samples taken from the same detector channel. Based on Eq. (3.60), the location of a parallel-beam projection forms a straight line in the fan-beam sinogram. If we use the iso-channel angle as the basis for the parallel-beam projection angle, the rebinned nonuniformly spaced parallel projection $p(i, j)$ is related to the fan-beam projection $q(i, j')$ by

$$p(i,j) = q\left(i, j - \frac{(i - i_{iso})\Delta\gamma}{\Delta\beta} \right), \qquad (3.62)$$

where $\Delta\beta$ is the angular increment between views. The uniformly spaced parallel-beam projections are obtained by a second interpolation, based on Eq. (3.61), along the detector channel direction i. Once a set of parallel projections is obtained, parallel-beam reconstruction algorithms can be applied directly to produce an image. Since the two interpolations are independent, the order of interpolation can be reversed. In other words, one could first interpolate the equiangular fan-beam projections into a set of fan-beam projections with a uniform distance to the iso-center (interpolation along i) and then interpolate the uniform-distance fan beam to a set of uniformly spaced parallel projections (interpolation along j).

The two-step interpolation approach also provides flexible data handling. For example, although interpolation to a nonuniformly spaced parallel projection can be carried out with computer software, we can use channel-dependent sampling delay to directly obtain a set of parallel projections. Based on Eq. (3.62), a projection sample i becomes parallel to the iso-channel sample if it is taken with a time delay of τ_i relative to the iso-channel (in milliseconds):

$$\tau_i = \frac{(i - i_{iso})\Delta\gamma}{\Delta\beta}\Delta t, \qquad (3.63)$$

where Δt is the view sampling interval (in milliseconds) of the original fan-beam data acquisition. Since this approach does not require any interpolation, it is likely to yield a better spatial resolution. The drawback of this approach, however, is the increased complexity and cost of the data acquisition electronics, since an elaborate delay-control mechanism must be built with each system.

The parallel projections obtained by view resampling (either by interpolation or sampling delay) are not uniformly spaced. Although the resulting projections can be interpolated again in i to produce uniformly spaced samples, modified parallel reconstruction algorithms for nonuniformly spaced samples can be applied directly to the projections for reconstruction.[7]

The motivation to pursue fan-to-parallel rebinning for reconstruction goes beyond computational efficiency, since the speed of computer hardware has improved dramatically over the years. Some vendors have designed ASIC chips specifically for fan-beam reconstruction. From a computing resources point of view, the backprojection operation is not the bottleneck for image reconstruction speed. Careful design of the reconstruction engine ensures that the preprocessing, filtering, and backprojection operation speeds are balanced.

One of the key advantages of the fan-to-parallel rebinning approach is the improved noise uniformity in the reconstructed image. In the fan-beam backprojection process, each filtered projection sample is multiplied by the L^{-2} factor as it is backprojected across the image, where L is the distance from the x-ray source to the backprojected point. For simplicity, consider the backprojection process for a particular image pixel where the standard deviation of the corresponding filtered projection sample at projection angle β is denoted by $\sigma_p(\beta)$. The variance of the reconstructed pixel σ_I^2 is

$$\sigma_I^2 = \sum_{k=1}^{N} w^2(k)\sigma_p^2(k), \qquad (3.64)$$

where $w(\beta)$ is the weighting factor applied during the backprojection process, and N is the number of projection views. For the parallel backprojection process, the weighting factor is a constant, $1/N$. If we assume a constant projection noise $\sigma_p(\beta) = \sigma_c$, the variance of the image pixel from the parallel backprojection is

$$\sigma_{para}^2 = \frac{\sigma_c^2}{N}. \qquad (3.65)$$

For fan-beam backprojection, however, the weighting factor contains the L^{-2} term, which constantly changes for image pixels away from the iso-center. Consider a pixel $(0, y)$ located on the y axis; the reconstructed variance for this pixel is

$$\sigma_{fan}^2 = \left(\frac{\sigma_c}{N}\right)^2 \sum_{k=1}^{N} \frac{D^4}{\left(D^2 + y^2 - 2Dy\sin\theta_i\right)^2}. \qquad (3.66)$$

Because of the rotational symmetry of CT sampling geometry, this formula is not limited to a single point on the y axis and is applicable to all image pixels that are at a distance y to the iso-center. Thus, the variance ratio of the fan-beam backprojection over the parallel-beam backprojection is

$$\frac{\sigma_{fan}^2}{\sigma_{para}^2} = \sum_{k=1}^{N} \frac{D^4}{\left(D^2 + y^2 - 2Dy\sin\theta_i\right)^2} . \tag{3.67}$$

Figure 3.40 shows the variance ratio as a function of the distance to the iso-center. It is clear from the figure that for pixels located away from the iso-center, the noise is higher for fan-beam reconstruction. Although this analysis is extremely simplified and assumes a constant projection noise, it provides insight to the impact of different reconstruction algorithms on noise.

Before concluding this section, we want to point out that the development of the fan-beam reconstruction algorithm is far from complete. Recent research activities show that new formulations and derivations of fan-beam reconstruction are continuously being discovered.[8-12] These developments shed new light onto a three-decade-old problem and may potentially offer better solutions in fan-beam CT reconstruction.

3.6 Iterative Reconstruction

All of the reconstruction algorithms discussed so far are analytical in nature. Each projection data point is weighted, filtered, and backprojected. When the last projection view is processed, the reconstruction is complete, and reconstructed

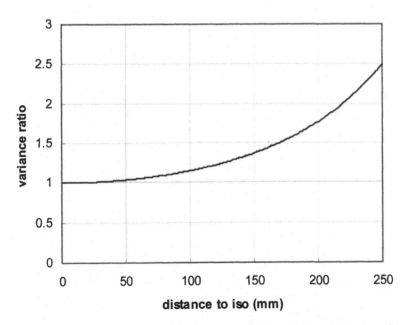

Figure 3.40 Variance ratio of fan-beam over parallel-beam reconstruction (assuming constant noise).

images are generated. Although these reconstruction algorithms are efficient and have served us well for the past 30 years, the question always remains whether the reconstructed image quality can be further improved with other classes of reconstruction algorithms.

3.6.1 Mathematics versus reality

To answer this question, we will first examine the CT system and the assumptions being made to derive the filtered backprojection (FBP) algorithm. Figure 3.41 is a schematic diagram of a CT scanner with a significantly enlarged x-ray source, detector, and reconstructed image pixel for illustration purposes. In the FBP algorithm derivation, we assume that the x-ray source (focal spot) is infinitely small and can be approximated by a point. Although we consider the detector spacing in the derivation of the reconstruction filters, the shape and dimension of each detector cell are ignored, and we assume that all of the x-ray photon interactions take place at a point located at the geometric center of the detector cell. These assumptions lead to the formation of a pencil beam that represents the line integral of the attenuation coefficient along the path, as illustrated by the figure. We ignore the shape and size of the reconstruction image pixels and assume that they are infinitely small points located on a square grid. In addition, we assume that each of our measurements is accurate and not influenced by statistical fluctuations. These simplified assumptions were made for a good reason—to make the mathematics more manageable—and allow us to derive a set of closed-form analytical equations that enable the efficient reconstruction of CT images.

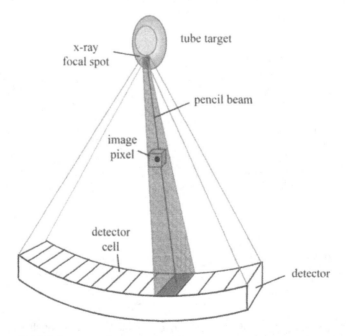

Figure 3.41 Schematic diagram of a CT system and sampling geometry.

Despite the rationale behind these simplifications, the CT system model on which the FBP algorithms were developed does not represent reality. As will be discussed in Chapter 6, the size of the x-ray focal spot is not negligible; the typical focal spot size is about 1×1 mm. For a detector cell spacing of slightly larger than 1 mm, the width of the active detector area is roughly 80% of the spacing and is roughly 1 mm. The reconstructed image pixel size depends on the selection of the reconstruction FOV and slice thickness. For example, for an image with a 512×512-matrix size representing a 50-cm FOV and a 5-mm slice thickness, the reconstructed pixel is $0.98 \times 0.98 \times 5$ mm. In addition, because each projection sample is acquired at a finite sampling interval (less than a millisecond on modern scanners) and the x-ray tube output has a limited output flux, the measurement received at the detector has limited photon statistics (a limited number of x-ray photons) after the x rays are attenuated by the patient. Given the finite size of the x-ray focal spot, x-ray detector, reconstructed image pixel, and photon statistics, there is little doubt that the simplifications made in the FBP derivation should have an impact on the reconstructed image quality. Therefore, we must ask ourselves the following key questions: To what degree is the image quality impacted by these assumptions? If the impact is significant, what approaches can we take to incorporate a realistic set of physical models into the reconstruction process? If the solution exists, will the computational intensity prevent its application in clinical settings?

3.6.2 The general approach to iterative reconstruction

One way of overcoming the intractability of an analytical solution is to use a technique called iterative reconstruction (IR). IR algorithms have been studied extensively for many years.[15-32] For simplicity, we will limit our discussion to a 2D object and its projection measurements (the same discussion can be easily extended to the 3D case). The scanned object is denoted by a 2D vector μ and its measured projections (after all calibration steps) by $\mathbf{p}$. We want to establish a relationship that maps from the object μ to its projections $\mathbf{p}$. It can be shown that the two variables are linked by a system matrix $\mathbf{A}$ and an error vector $\mathbf{e}$:

$$\mathbf{p} = \mathbf{A}\mu + \mathbf{e}. \qquad (3.68)$$

For an ideal object and its ideal set of projections, an element in the system matrix $\mathbf{A}$ contains simply the amount of contribution of a particular object pixel to a particular projection sample, and $\mathbf{e}$ is zero. For real systems, the system matrix $\mathbf{A}$ can be determined based on the system geometry, focal spot shape, detector response, and many other significant geometrical and physical parameters of a CT system. The error vector $\mathbf{e}$ accounts for any measurement bias and additive noise. The reconstruction process is used to estimate μ given a measurement vector $\mathbf{p}$, and can be formulated in the Bayesian framework of maximizing the posterior probability $\Pr(\mu|\mathbf{p})$:

$$\hat{\mu} = \arg\max_{\mu}\left[\Pr(\mu\,|\,\mathbf{p})\right], \qquad (3.69)$$

where $\hat{\mu}$ is the optimal estimation of μ based on $\mathbf{p}$. Based on the Bayes rule, we can express $\Pr(\mu|\mathbf{p})$ in the following form:

$$\Pr(\mu\,|\,\mathbf{p}) = \frac{\Pr(\mathbf{p}\,|\,\mu)\Pr(\mu)}{\Pr(\mathbf{p})}. \qquad (3.70)$$

When no *a priori* information about the measured projection is available, the optimization problem of Eq. (3.70) is equivalent to

$$\hat{\mu} = \arg\max_{\mu}\left[\log\Pr(\mathbf{p}\,|\,\mu) + \log\Pr(\mu)\right]. \qquad (3.71)$$

The first term in Eq. (3.71) is the log likelihood that links the estimated object μ to the measured projection $\mathbf{p}$. The second term is related to the rules that we place on the estimated object, such as the amount of variation allowed in a small region.

First we must maximize the first term. Based on Eq. (3.68), the object estimation is performed by requiring μ and $\mathbf{e}$ to satisfy specified optimization criteria. The estimation process is an iterative procedure in which we produce a sequence of vectors, $\mu^{(0)}$, $\mu^{(1)}$, ..., $\mu^{(n)}$ such that the sequence converges to $\hat{\mu}$. For each iteration j, we calculate the quantity $\mathbf{p}^{(j)}$ as

$$\mathbf{p}^{(j)} = \mathbf{A}\mu^{(j)} + \mathbf{e}. \qquad (3.72)$$

Based on the difference between the calculated projection $\mathbf{p}^{(j)}$ and the measured projection $\mathbf{p}$, we modify our estimation $\mu^{(j)}$ such that the difference between the two quantities is reduced. Typically, certain constraints are placed on $\mu^{(j)}$ in the estimation process. One of the most commonly used constraints is positivity. This comes from the basic physics property of linear attenuation coefficients. Negative linear attenuation coefficients imply that the x-ray intensity increases as it passes through the object, which is a clear violation of fundamental physics.

Figure 3.42 provides a simple pictorial illustration of the iterative process. In this example, parallel beams are used for simplicity (an extension to fan-beam geometry is straightforward and will not be discussed separately). For a particular view angle, the calculated projection is obtained by performing forward projection on the image of the estimated object. Forward projection essentially sums up the intensities of all pixels along a particular ray path and assigns this value as the estimated projection. The calculated projection $\mathbf{p}^{(j)}$ is then compared to the actual measurement $\mathbf{p}$, and the estimation $\mu^{(j)}$ is modified based on the difference. For example, we can distribute the correction $\mathbf{p} - \mathbf{p}^{(j)}$ evenly among all elements in $\mu^{(j)}$ that contribute to the calculation of $\mathbf{p}^{(j)}$. This distribution is carried out on a ray-by-ray basis, and the step continues for all projection views. The entire process is repeated if certain criteria are not met. In the above calculations, we concentrate mainly on the geometric factors to produce projections $\mathbf{p}^{(j)}$, and update our estimated reconstruction $\mu^{(j)}$. This type of iterative algorithm is called the algebraic reconstruction technique (ART).[1-5]

Some advantages of IR over FBP are its ability to suppress metal artifacts, better handling of truncated projections, and improved performance for limited-angle tomography. When the structure of the scanned object is sparse, iterative algorithms are capable of producing good image quality even with a small number of projections. For illustration, Fig. 3.43(a) shows a subtracted projection of an x-ray angiograph pre- and post-contrast injection. Because of the good image registration, overlay structures such as ribs and spines are completely removed. Since the vascular structure is sparse, only a limited number of projections are needed to perform the image reconstruction. In this particular example, 44 subtracted projections were collected and an ART reconstruction algorithm was used over three iterations. The volume-rendered version of the reconstructed images is shown in Fig. 3.43(b).

3.6.3 Modeling of the scanner's optics and physics

There are several ways to use the system matrix $\mathbf{A}$ to model the optics of a CT system. The most straightforward method is to cast many pencil-beam rays through the x-ray focal spot, the image pixels, and the detector. That is, instead of using one pencil beam to represent the line integral between an x-ray focal spot and the pixel in the forward projection process, we use many rays that mimic

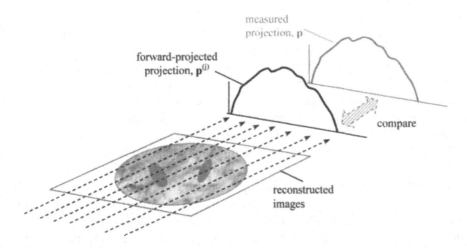

Figure 3.42 Illustration of the iterative reconstruction approach. The estimated image is forward projected to obtain an estimated projection. The estimated projection is compared to the measured projection. The estimated image is modified based on certain criteria to reduce the difference between two projections.

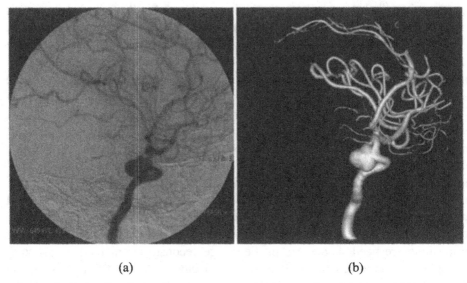

(a)   (b)

Figure 3.43 Example of image reconstruction using the ART algorithm. (a) Subtracted projection angiograph before and after contrast injection (512 × 512). (b) Volume-rendered image produced by the ART algorithm using 44 projections.

different x-ray photons interacting with the object, as depicted in Fig. 3.44. The final forward-projected value is the weighted summation of all rays. One strategy that ensures unbiased coverage of different areas on the focal spot and different subregions inside an image pixel is to subdivide the focal spot and pixel into equal-sized small elements called "lets." For example, we can divide the 1 × 1-mm focal spot area into 3 × 3 source-"lets," divide each image pixel into 3 × 3 × 3 small pixel-lets, and compute forward projections for each "lets" pair (Fig. 3.45). Needless to say, this process is very time consuming.

An alternative approach models the "shadow" cast by each image pixel onto the detector (Fig. 3.46).[15] In this approach, only one ray is cast through the centers of the focal spot and image pixel to generate the intersection point with the detector. A response function is then used to distribute the image pixel's contribution to different detector cells during the forward-projection process. The response function is nonstationary and changes with the image location to account for different magnifications and orientations. Compared to the ray casting approach, this method is computationally more efficient.

So far, we have not addressed the statistical nature of the problem. We assume that the measured projections **p** represent the true line integrals of the attenuation coefficients. Because of the statistical nature of the measurements, the measured projections (after calibration) deviate from their expected values. In other words, the algebraic approach is appropriate only if we are given a set of noiseless projections **p**. In the real world, this is impossible to accomplish. One way to overcome this difficulty is to incorporate statistical models into our calculation. This type of reconstruction is typically called *statistically based*

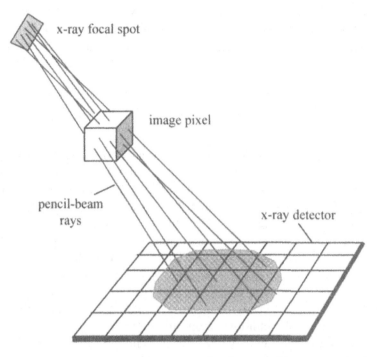

Figure 3.44 Illustration of modeling of a realistic focal spot and image voxel using pencil-beam rays.

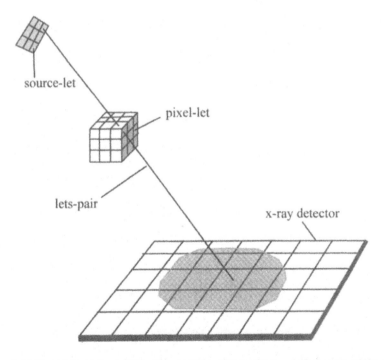

Figure 3.45 Illustration of modeling the system optics with source-voxel let-pairs.

iterative reconstruction.[15,16,21-32] For example, if we use the second-order Taylor series expansion and the log of Poisson probability function, the log likelihood term of Eq. (3.71), log Pr($\mathbf{p}|\boldsymbol{\mu}$), can be approximated by the following expression:[15]

$$\log \Pr(\mathbf{p}\,|\,\boldsymbol{\mu}) \approx -\frac{1}{2}(\mathbf{p}-\mathbf{A}\boldsymbol{\mu})^T D(\mathbf{p}-\mathbf{A}\boldsymbol{\mu})+f(\mathbf{p}),\qquad(3.73)$$

where D is a diagonal matrix with d_i proportional to the detector x-ray photons, and f is a function of the data. However, the incorporation of complete statistical models into CT reconstruction is not straightforward. There are major difficulties in building such models. For example, the noise in the measured projections without calibration (with electronic dark current removed) may not be accurately modeled as a simple Poisson distribution due to the data acquisition electronic noise and energy-dependent signals. One study has shown that noise at this stage can be modeled as compound Poisson plus Gaussian noise.[33] After the preprocessing and calibration steps, the statistical properties of $\mathbf{p}$ are not well understood and cannot be easily modeled. The operations in the calibration steps are generally not simple additions or multiplications. The complexity is further compounded by the logarithm operator, which cannot be avoided in CT.

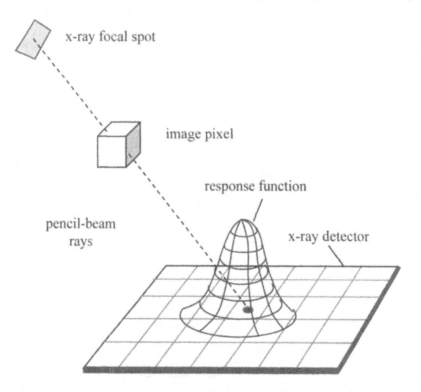

Figure 3.46 Illustration of modeling the system optics with the response function.

One way to use the statistical model in IR is to assign a confidence level associated with each projection measurement. During each iteration, the quantity $\mathbf{p} - \mathbf{p}^{(j)}$ is used to update the reconstructed image. For those elements of $\mathbf{p}$ that are noisy, we can place a lower confidence weighting to the correction so that their contributions to the image update are less than those samples with lower noise.

We are now ready to discuss the maximization of the second term in Eq. (3.71). The nature of human anatomy is such that the attenuation characteristics of a small element inside a body are not completely isolated from and independent of the surrounding elements. That is, the CT number of an image pixel should fall within some reasonable range of the CT numbers of its neighbors. If the CT number of a pixel is far from its neighbors' CT numbers, it should be penalized and modified. This process helps to reduce random noise in the image, since the value of the random noise is unrelated to its neighbors. This process is often called the *regularization* step in IR. The model that determines the amount of penalty placed on a "run away" pixel value is called *prior*. If we develop a penalty or cost function associated with local intensity fluctuations, we can maximize the second term in Eq. (3.71) by minimizing the cost function. Figure 3.47 shows examples of several cost functions. As the normalized difference between a pixel and its neighbors increases, the penalty increases. The cost function is typically nonlinear in order to optimize the tradeoffs between noise suppression and edge preservation.

3.6.4 Updating strategy

The image update strategy from $\mu^{(j)}$ to $\mu^{(j+1)}$ has two approaches: concurrent and sequential. In the concurrent strategy, the calculation of the modification $\Delta\mu_i$ for the ith projection is performed for all pixels inside the entire FOV before moving

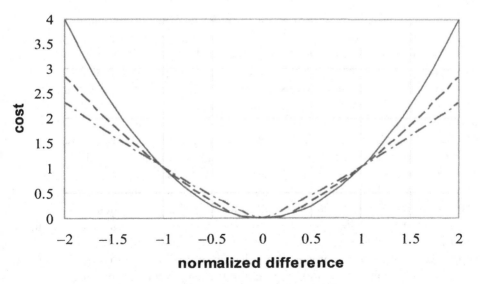

Figure 3.47 Examples of three cost (penalty) functions.

on to the next projection. The pixel update takes place concurrently for all pixels. Examples of algorithms that use this update strategy include the maximum-likelihood (ML) algorithm and the conjugate-gradient (CG) algorithm.

Early concurrent algorithms performed the image update step after examining all of the projections. Clearly, this process is computationally expensive. To speed up the reconstruction process, algorithms were proposed that performed the image update based on subsets of the projection dataset.[24,26,27] This approach is the so-called ordered-subset (OS) algorithm. The image update process starts with one subset and continues until all projection subsets are used. These iterative algorithms are used extensively in positron emission tomography (PET) and single-photon-emission computed tomography (SPECT).

On the other hand, the sequential update strategy focuses on the generation of the update value for a single pixel or a subset of pixels. In this process, a pixel or a subset of pixels is selected at random. The update value or values are calculated using all projection views, and the update is carried out on the selected pixels while the rest of the pixels remain unchanged. In the next round of updating, another set of pixels is selected and updated. An example of an algorithm that uses this update strategy is the iterative coordinate descent (ICD) algorithm.

A comprehensive survey of statistical methods for transmission tomography can be found in Ref. 31. The vast majority of IR methods for transmission tomography are based on a simple monoenergetic model rather than a comprehensive physics model of the CT system. In recent years, algorithms have been proposed that model polychromatic energy characteristics.[29,32] This is a good first step toward better modeling of x-ray physics. Since the system matrix $\mathbf{A}$ is capable of modeling the physics properties of the detector, x-ray source, and scanned object in addition to the scanner optics, the advantages of using the raw measurements prior to calibration (instead of the calibrated projection $\mathbf{p}$) should be fully explored.

The potential advantages of IR over the FBP algorithm are improved metal artifact reduction, better handling of truncated projections, limited-angle tomography, and improved noise performance. For illustration, Fig. 3.48 shows two coronal images reconstructed from the same projection dataset. The patient was scanned with a super-low-dose protocol to test the dose-reduction capability of the IR algorithm. A comparison of the two images clearly shows that the noise performance of the IR is significantly better than that of the FBP-based reconstruction. Anatomical details that are completely masked by noise in the FBP image [Fig. 3.48(a)] are clearly visible in the IR image [Fig. 3.48(b)].

Because the system matrix $\mathbf{A}$ accurately models the system optics, spatial resolution degradation due to the finite focal spot size, detector aperture, and pixel size can be reduced when the IR algorithm is applied. To test this hypothesis, a patient head scan was reconstructed with both an FBP and an IR algorithm. Inspection of Figs. 3.49(a) and (b) shows that the IR not only suppressed noise in the image, but also improved the spatial resolution, as demonstrated by the fine details depicted in the temporal bone regions.

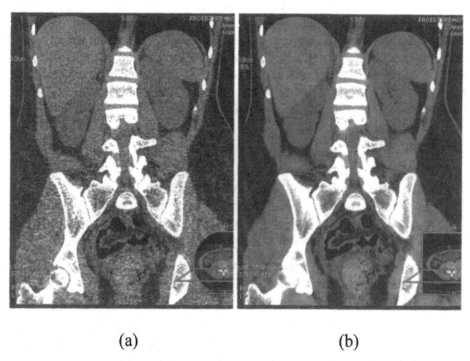

(a)   (b)

Figure 3.48 Coronal images of a super-low-dose patient CT scan (a) reconstructed with FBP and (b) reconstructed with iterative reconstruction.

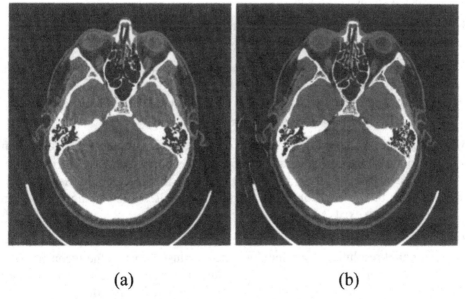

(a)   (b)

Figure 3.49 Axial images of a patient head scan (a) reconstructed with FBP, and (b) reconstructed with iterative reconstruction.

The investigation of statistically based iterative algorithms for CT is still in its infancy, and other clinical applications of IR are likely to be discovered. Historically, one of the most important factors that has prevented the application of iterative algorithms to CT is computational complexity. Although the same issue is still present today, the magnitude of the obstacle has been significantly reduced and is likely to be removed in the near future with advanced computer hardware and improved algorithm efficiency.

3.7 Problems

3-1   For the simplified iterative reconstruction approach shown in Fig. 3.4, set the initial conditions of the reconstruction [Fig. 3.4(b)] to 1, 4 (top row), and 4, 2 (bottom row). Repeat the iterative process as illustrated in Section 3.2. How many iterations are needed before the final solution is reached? What observation can you make based on this example?

3-2   Implement a parallel-beam reconstruction algorithm that takes advantage of the vector form, using MATLAB®, IDL®, or another vector-based software package.

3-3   A set of parallel projections was acquired with a CT gantry rotated clockwise over 2π. A student developed a parallel-beam reconstruction algorithm and made one mistake by assuming the gantry rotated counterclockwise. Would the student obtain a reconstructed image that somewhat resembled the true object? What is the relationship, if any, between the two images?

3-4   Repeat the exercise of problem 3-3 for the case of a projection set acquired with fan-beam geometry.

3-5   Equation (3.28) states that for parallel-beam geometry, projection views only in the angular range from 0 to π are needed to accurately reconstruct an image. What are the possible reasons that we would sometimes want to collect projection data over the range of 0 to 2π?

3-6   Because of the large number of projection samples ($N \approx 1000$), an operation used to generate a filtered projection is typically carried out in Fourier space during the reconstruction, as discussed in Section 3.5. Does this conclusion hold when N is small? What is the break-even point, if it exists?

3-7   Section 3.4 illustrates the impact of a targeted reconstruction on image spatial resolution. Can one draw the conclusion that as the reconstruction FOV is reduced, the spatial resolution always increases? What other limiting factors may lead to a plateau of diminished return?

3-8   For an evenly spaced view sampling pattern, the projection view angle can be specified by $\theta = \Delta\theta \times i$, $i = 0, 1,..., N-1$, where θ is the projection angle, $\Delta\theta = 2\pi/N$, and i is the view index. A student designed a first-generation CT scanner to collect parallel-beam samples. The projection sampling is nonuniform and follows the equation

$$\theta = 2\pi \left(\frac{i}{N} \right)^{1.2}, i = 0, 1, \ldots, N-1.$$

What modifications to the parallel-beam reconstruction formula are needed, if any, to ensure good image quality?

3-9   Equation (3.67) provides a variance ratio of fan-beam backprojection over parallel-beam backprojection for a 2π projection reconstruction when the filtered projection standard deviation is constant. Calculate the variance ratio if 0-to-π projections are used, assuming the source-to-detector distance is 950 mm, and the source-to-iso distance is 550 mm (for this problem, ignore other complexities in the fan-beam reconstruction process when less than a full rotation of projections is used). Compared to the result of a 2π projection, what observation can you make?

3-10   For a backprojection process carried out over a 2π projection angle, will the variance of the parallel backprojected image always be less than or equal to the variance of the fan-beam backprojected image, assuming that the variances of the filtered parallel projection samples equal the variances of the corresponding filtered fan-beam projection samples (but not constant)?

3-11   One of the main drivers for targeted reconstruction is to ensure that high-frequency signals are properly supported. If the detector cell is 1 mm in width, the source-to-detector distance is 950 mm, and the source-to-iso distance is 550 mm, what is the size of the reconstruction FOV to support the highest frequency of the system near the iso-center, assuming that the focal spot size is 0 mm and the image matrix size is 512 × 512?

3-12   Repeat Exercise 3-11 if the loss of spatial resolution in the interpolation process during the backprojection is 1 lp/cm.

3-13   The limiting resolution of a CT system is 20 lp/cm, based on the system geometry, detector size, focal spot size, and sampling frequency. Derive an equation to express the filter cutoff frequency as a function of the reconstruction FOV from 10 to 50 cm in diameter, assuming the image matrix size is 512 × 512.

3-14   For fan-parallel rebinning, how do you generate the rebinned parallel samples for θ near 0 or 2π, assuming the original fan-beam data are collected in a step-and-shoot mode over 0 to 2π?

3-15   Describe the pros and cons of performing 2D interpolation versus two 1D interpolations in the fan-to-parallel rebinning process.

3-16   Derive the reconstruction formula for the equal-spaced fan-beam geometry shown in Eq. (3.59).

3-17   Derive a fan-to-parallel beam rebinning formula for equal-spaced fan-beam geometry.

3-18   Compared to a parallel reconstruction, the noise uniformity of a fan-beam reconstruction is degraded, as shown in Fig. 3.40, and the computational efficiency is worse for a fan-beam-based reconstruction. What are the potential advantages of the fan-beam algorithm over the parallel reconstruction?

3-19   The x-ray output intensity is not uniform across the focal spot. If we assume that the x-ray flux intensity has a Gaussian shape with its full width at half maximum (FWHM) equal to the width of the spot, how can one incorporate this information into the system optics model shown in Fig. 3.46?

3-20   One major challenge of iterative reconstruction is computational complexity. One approach to improving reconstruction speed is to use multiple computers or central processing units (CPUs) to perform parallel processing. Discuss parallel processing approaches for both concurrent and sequential update strategies.

3-21   In Fig. 3.47, all three cost functions are symmetrical, i.e., the penalty value is the same for two normalized differences with opposite signs but identical magnitude. Does the cost function need to be symmetrical?

3-22   In the IR discussion, the image pixel is shaped as a rectangular cube. Discuss the pros and cons if the image pixel has a spherical shape.

3-23   In Section 3.4, targeted reconstruction with FBP is described. Outline the steps to perform targeted reconstruction for an iterative algorithm, assuming the reconstruction FOV is significantly smaller than the scanned object.

3-24   During the implementation of ordered-subset IR, one student chooses to divide the original projection dataset from 0 to 2π into 8 subsets, with the projections of one subset being 0 to 0.25π, the projections of a second subset being 0.25π to 0.5π, etc. Another student sets the views to 1, 9, 17, ... as one subset, and the views of another subset to 2, 10, 18, ..., etc. Which approach is better?

References

1. G. T. Herman, *Image Reconstruction from Projections: The Fundamentals of Computerized Tomography*, Academic Press, New York (1980).
2. A. C. Kak and M. Slaney, *Principles of Computerized Tomographic Imaging*, IEEE Press, Piscataway, NJ (1988).

3. R. Gordon, R. Bender, and G. T. Herman, "Algebraic reconstruction techniques (ART) for three-dimensional electron microscopy and x-ray photography," *J. Theor. Biol.* **29**, 471–481 (1970).

4. G. T. Herman, A. Lent, and S. Rowland, "ART: Mathematics and applications, a report on the mathematical functions and on the applicability to real data of algebraic reconstruction technique," *J. Theor. Biol.* **42**(1), 1–32 (1973).

5. G. N. Hounsfield, "A method of and apparatus for examination of a body by radiation such as x or gamma radiation," Patent Specification No. 1283915, The Patent Office, London (1972).

6. B. K. P. Horn, "Fan-beam reconstruction methods," *Proc. IEEE* **67**, 1616–1623 (1979).

7. B. K. P. Horn, "Density reconstructions using arbitrary ray sampling schemes," *Proc. IEEE* **66**, 551–562 (1978).

8. G. Chen, "A new framework of image reconstruction from fan-beam projections," *Med. Phys.* **30**, 1151–1161 (2003). •

9. Y. Wei, J. Hsieh, and G. Wang, "General formula for fan-beam computed tomography," *Phys. Rev. Lett.* **95**, 258102 (2005).

10. Y. Wei, G. Wang, and J. Hsieh, "Relation between filtered backprojection algorithm and backprojection algorithm in CT," *IEEE Signal Process. Lett.* **12**(9), 633–636 (2005).

11. Y. Wei, G. Wang, and J. Hsieh, "An intuitive discussion on the ideal ramp filter in computed tomography," *Comput. Math. Applic.* **49**, 731–740 (2005).

12. G. Chen, R. Tokalkanahalli, T. Zhuang, B. E. Nett, and J. Hsieh, "Development and evaluation of an exact fan-beam reconstruction algorithm using an equal weighted scheme via locally compensated filtered backprojection (LCFBP), *Med. Phys.* **33**(2), 475–481 (2006).

13. S. Napel, "Computed tomography image reconstruction," in *Medical CT and Ultrasound: Current Technology and Applications*, L. W. Goldman and J. B. Fowlkes, Eds., Advanced Medical Publishing, Madison, WI, 311–327 (1995).

14. J. Hsieh, "CT image reconstruction," in *RSNA Categorical Course in Diagnostic Radiology Physics: CT and US Cross-Sectional Imaging 2000*, L. W. Goldman and J. B. Fowlkes, Eds., Radiological Society of North America, Inc., Oak Brook, IL, 53–64 (2000).

15. J.-B. Thibault, K. D. Sauer, C. A. Bouman, and J. Hsieh, "A three-dimensional statistical approach to improved image quality for multislice helical CT," *Med. Phys.* **34**(11), 4526–4544 (2007).

16. Z. Yu, J.-B. Thibault, K. Sauer, C. Bouman, and J. Hsieh, "Accelerated line search for coordinate descent optimization," in *Proc. IEEE Nuclear Science Symposium and Medical Imaging Conf.* **6498**, 2841–2844 (2006).

17. K. Sauer and C. A. Bouman, "A local update strategy for iterative reconstruction from projections," *IEEE Trans. Signal Process* **41**, 553–548 (1993).

18. P. Sukovic and N. H. Clinthorne, "Penalized weighted least-squares image reconstruction in single and dual energy x-ray computed tomography," *IEEE Trans. Med. Imag.* **19**, 1075–1091 (2000).

19. J. Nuyts et al., "Iterative reconstruction for helical CT: a simulation study," *Phys. Med. Biol.* **43**, 729–737 (1998).

20. Y. Trousset, C. Picard, C. Ponchut, R. Romeas, R. Campagnolo, S. Croci, J. M. Scarabin, and M. Amiel, "3D x-ray angiography: from numerical simulations to clinical routine," in *Proc. 1995 Int. Meeting on Fully 3D Imag. Recon. in Radio. and Nucl. Med.*, 3–10 (1995).

21. L. A. Shepp and Y. Vardi, "Maximum likelihood reconstruction for emission tomography," *IEEE Trans. Med. Imag.* **MI-1**(2), 113–122 (1982).

22. J. A. Browne and T. J. Holmes, "Developments with maximum likelihood x-ray computed tomography," *IEEE Trans. Med. Imag.* **11**(1), 40–52 (1992).

23. K. Lange and R. Carson, "EM reconstruction algorithm for emission and transmission tomography," *J. Comput. Assist. Tomogr.* **8**(2), 306–316 (1984).

24. H. Erdogan and J. A. Fessler, "Ordered subsets algorithms for transmission tomography," *Phys. Med. Biol.* **44**, 2835–2851 (1999).

25. B. DeMan, "Iterative reconstruction for reduction of metal artifacts in computed tomography," Ph.D. thesis, Katholieke Universiteit Leuven, Belgium (2001).

26. S. H. Manglos, G. M. Gange, A. Krol, F. D. Thomas, and R. Narayanaswamy, "Transmission maximum-likelihood reconstruction with ordered subsets for cone beam CT," *Phys. Med. Biol.* **40**, 1225–1241 (1995).

27. C. Kamphius and F. J. Beekman, "Accelerated iterative transmission CT reconstruction using an ordered subset convex algorithm," *IEEE Trans. Med. Imag.* **17**(6), 1101–1105 (1988).

28. C. A. Bouman and K. Sauer, "A unified approach to statistical tomography using coordinate descent optimization," *IEEE Trans. Imag. Process.* **5**, 480–492 (1996).

29. I. A. Elbakri and J. A. Fessler, "Statistical image reconstruction for polyenergetic x-ray computed tomography," *IEEE Trans. Med. Imag.* **21**, 89–99 (2002).

30. G. Wang, M. W. Vannier, and P. C. Cheng, "Iterative x-ray cone beam tomography for metal artifact reduction and local region reconstruction," *Microsc. Microanal.* **5**, 58–65 (1999).

31. J. A. Fessler, "Statistical image reconstruction methods for transmission tomography," in *Handbook of Medical Imaging, Vol. 2, Medical Image Processing and Analysis*, M. Sonka and J. M. Fitzpatrick, Eds., SPIE Press, Bellingham, WA, 1–70 (2000).

32. B. De Man et al., "Reduction of metal artifacts in x-ray computed tomography using a transmission maximum a posteriori algorithm," *IEEE Trans. Nucl. Sci.* **47**(3), 977–981 (2000).

33. B. R. Whiting, "Signal statistics in x-ray computed tomography," *Proc. SPIE* **4682**, 53–60 (2002).

Chapter 4
Image Presentation

4.1 CT Image Display

For radiologists, the most important output from a CT scanner is the image itself. As was discussed in Chapter 2, although the reconstructed images represent the linear attenuation coefficient map of the scanned object, the actual intensity scale used in CT is the Hounsfield unit (HU). For convenience, we replicate the mapping function below:

$$CT\ number = \frac{\mu - \mu_{water}}{\mu_{water}} \times 1000 . \tag{4.1}$$

The linear attenuation coefficient is magnified by a factor over 1000 (note the division by μ_{water}). When specified in HU, air has the value of -1000 HU; water has the value of 0 HU; and bones, contrast, and metal objects have values from several hundred to several thousand HU. Because of the large dynamic range of the CT number, it is impossible to adequately visualize it without modification on a standard grayscale monitor or film. Typical display devices use eight-bit grayscales, representing 256 (2^8) different shades of gray. If a CT image is displayed without transformation, the original dynamic range of well over 2000 HU must be compressed by a factor of at least 8. Figure 4.1(a) shows a reconstructed image of a head phantom in which the minimum and maximum pixel intensities are -1000 HU and 1700 HU, respectively. When this dynamic range is linearly mapped to the dynamic range of the display device (0 to 255), the original grayscale is so severely compressed that little intensity variation can be visualized inside the human skull. This is clearly unacceptable. To overcome this difficulty, a CT image is typically displayed with a modified grayscale:

$$
p_w(x, y) = \begin{cases} 0, & p(x, y) \leq L - \dfrac{W}{2} \\[2em] \dfrac{p(x, y) - \left(L - \dfrac{W}{2}\right)}{W} I_{max}, & L - \dfrac{W}{2} < p(x, y) \leq L + \dfrac{W}{2} \\[2em] I_{max}, & p(x, y) > L + \dfrac{W}{2} \end{cases} \quad (4.2)
$$

Here, L and W are called the display window level and display window width, respectively, and I_{max} is the maximum intensity scale of the display device (for an 8-bit display device, $I_{max} = 255$). In essence, the process maps the original intensity scale between $(L–W/2, L+W/2)$ to the full scale of the display device, as shown in Fig. 4.2. Figure 4.1(b) shows the same head image displayed at a window width of 100 HU and a window level of 20 HU. The gray and white matter inside the brain can be clearly visualized.

Within the display window of $(L–W/2, L+W/2)$, the display process performs a linear transformation between the reconstructed image intensity and the displayed image intensity. In many cases, however, it is desirable to perform a nonlinear mapping between the two to change the appearance of certain features

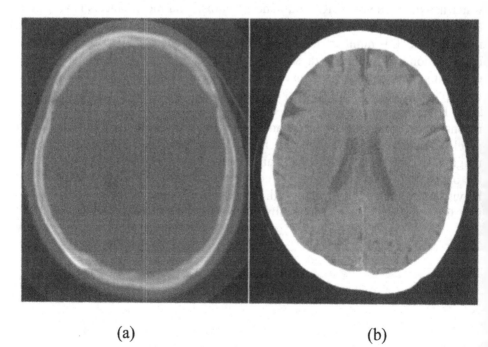

(a)   (b)

Figure 4.1 Reconstructed image of a patient head scan. (a) Image displayed with the full dynamic range of the image (2700 HU). (b) Image displayed with a window width (WW) of 100 HU and window level (WL) of 20 HU.

in the image. By adjusting the mapping function, these features can be enhanced or de-emphasized. In general, the mapping function can be expressed by the following equation:

$$q(x, y) = \begin{cases} G[p_w(x, y)], & t_L \le p_w(x, y) < t_H \\ p_w(x, y), & \text{otherwise,} \end{cases} \quad (4.3)$$

where $q(x, y)$ is the grayscale after the mapping, G is the mapping function, $p_w(x, y)$ is the reconstructed CT value after the original window width and window level transformation, and t_L and t_H are parameters of the mapping function. Figure 4.3 depicts three different mapping functions: the identity function and two grayscale enhancement functions. The enhancement functions are designed so that the contrast near the CT number of 30 is increased due to the larger-than-unity slope of the two curves. Figure 4.4 shows the appearance of a CT brain image displayed with three different mapping functions. Figure 4.4(a) shows the identity mapping function in which the original reconstructed CT numbers are faithfully displayed. Although the gray and white matters in the image can be identified, their differentiation is not clearly visible. To enhance the difference between the gray and white matters in the brain, we can display the same image using the enhanced mapping functions depicted in Fig. 4.3; the results are shown in Figs. 4.4(b) and (c). However, the enhanced gray-white matter differentiation is achieved at the price of increased noise, as shown by the larger fluctuations in the uniform regions.

The concept of using different mapping functions to enhance the appearance of images is not limited to CT; it was first used in x-ray radiographs to adjust the

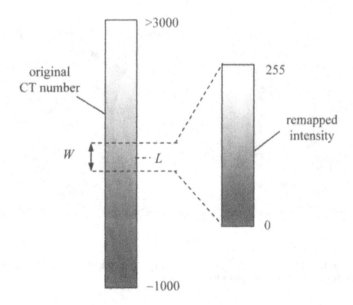

Figure 4.2 Illustration of the display WW and WL.

film contrast characteristic known as the gamma curve. The gamma curve describes the relationship between the x-ray exposure and the optical density of the film. By adjusting the average gradient of the gamma curve, the contrast of the imaged object can be modified. The advantage of the digital mapping function is its flexibility in defining an arbitrary function shape that is sometimes difficult to accomplish with the analog method.

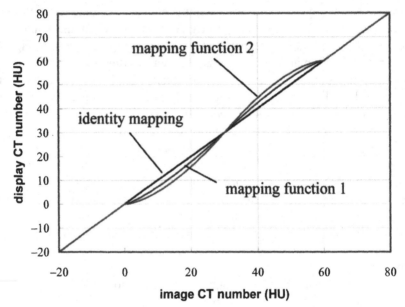

Figure 4.3 Illustration of grayscale mapping functions.

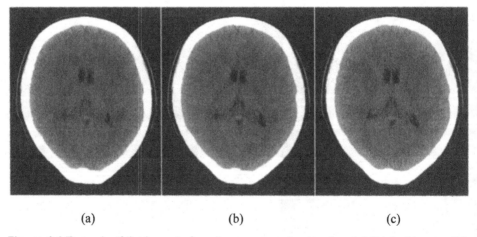

Figure 4.4 Example of the impact of nonlinear grayscale mapping. (a) Original image, (b) mapping function 1, and (c) mapping function 2.

4.2 Volume Visualization

Traditionally, CT images are viewed in a slice-by-slice mode. A series of reconstructed CT images are placed on films, and radiologists are trained to form, in their heads, the volume information from multiple 2D images. Although the ability to generate 3D images by computer has been available nearly since the beginning of CT,[1] the use of this capability has only recently become popular. This change is mainly due to three factors. The first is related to the quality and efficiency of 3D image generation. Because of the slow acquisition speed of the early CT scanners, thicker slices were typically acquired in order to cover the entire volume of an organ. The large mismatch between the in-plane and cross-plane resolution produced undesirable 3D image quality and artifacts. In addition, the amount of time needed to generate 3D images from a 2D dataset was quite long, due to computer hardware limitations and a lack of advanced and efficient algorithms.

The second factor is the greater productivity of radiologists. Historically, CT scans were "organ centric"; each CT examination covered only a specific organ, such as the liver, lung, or head, with either thick slices or sparsely placed thin slices. The number of images produced by a single examination was well below 100. With the recent advances in CT technology (helical and multislice), a large portion of the human body can be easily covered with thin slices. For example, some of the CT angiographic applications typically cover a 120-cm range from the celiac artery to a foot with thin slices.[2] The number of images produced by a single examination can be several hundred to over 1000. To view these images sequentially would be an insurmountable task.

The third factor that influences the image presentation format is related to clinical applications. CT images have become increasingly useful tools for surgical planning, therapy treatment, and other applications outside the radiology department. It is more convenient and desirable to present these images in a format that can be understood by people who do not have radiological training. The solution to all of these clinical demands is the advanced computer graphical techniques.

4.2.1 Multiplanar reformation

One popular graphical technique that can be performed along a flat plane or a curved surface is multiplanar reformation (MPR).[3] To perform reformation, we first establish a patient-based coordinate system using three orthogonal axes: left-right, anterior-posterior (A-P), and superior-inferior, as shown in Fig. 4.5. Reformation planes are defined relative to these axes. The sagittal plane is parallel to both the A-P and superior-inferior axes. The coronal plane is parallel to both the left-right and superior-inferior axes. Images formed parallel to the sagittal or coronal planes are called sagittal or coronal images, respectively, as depicted by a pair of examples shown in Figs. 4.6(a) and (b). For CT images

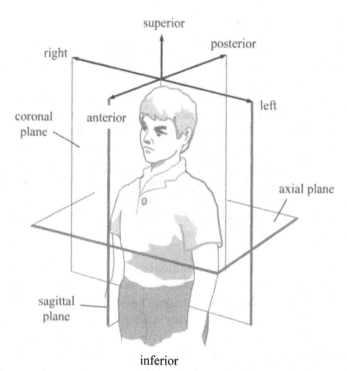

Figure 4.5 Illustration of a patient-based coordinate system and different plane orientations. The coronal plane is parallel to both the left-right and superior-inferior axes. The sagittal plane is parallel to both the anterior-posterior and superior-inferior axes. Without gantry tilt, both planes are perpendicular to the axial plane of the reconstructed images.

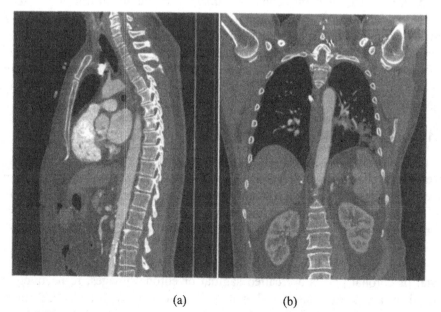

(a)   (b)

Figure 4.6 Planar reformatted images of CT scans (WW = 800). (a) Sagittal image and (b) coronal image.

generated without gantry tilt, the reconstructed image (axial plane) is perpendicular to both the sagittal and coronal planes. Any plane that is not parallel to these planes is called an oblique plane.

To form MPR images, the reconstructed images are arranged sequentially in a "stack" as shown in Fig. 4.7. Each CT image has a certain slice thickness, and the distance between consecutive images represents the slice spacing. Reformatted images are produced by interpolating the image volume in different orientations. If there is no gap between images (where slice spacing is less than or equal to the slice thickness), the reformation process is fairly simple. Even under this condition, an image pixel in the reformatted image often requires the interpolation of neighboring pixels, since it is unlikely that the reformatted pixel is located exactly at the original grid of the axial images. The interpolation can be carried out in any of one, two, or three dimensions. For example, for a sagittal or coronal reformation, interpolation is often needed only across the image slices.

The selection of different interpolation algorithms has a significant impact on spatial resolution, noise, artifact, and speed. Tradeoffs often must be made among these image performance parameters. Figure 4.8 depicts two sagittal images generated on the same reconstructed volume with two different interpolation algorithms: a linear interpolation and a four-point Lagrange interpolation. Closer inspection of the images shows that the sagittal image produced by the Lagrange interpolator is slightly sharper than its counterpart produced by linear interpolation.

In the production of reformatted images, special attention must be paid to the slice thickness of the original CT images. It is desirable to start with a stack of axial images that have isotropic spatial resolution—that is, the spatial resolutions in the x, y, and z directions should be approximately equal. When this rule of thumb is significantly violated, the result is degraded image quality or artifacts. Figure 4.9 shows a set of sagittal images generated with reconstructed images of

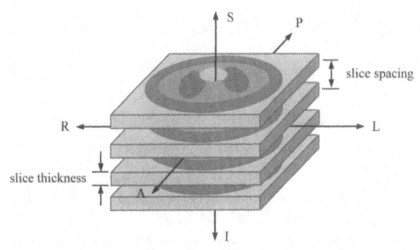

Figure 4.7 Geometric relationships of the reformation process.

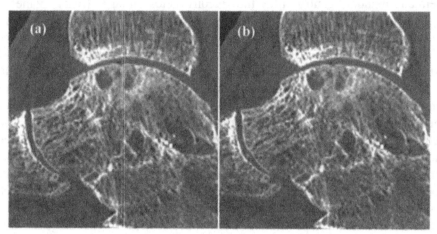

Figure 4.8 Illustration of the interpolation algorithm on MPR (0.625 mm at 1.25-mm spacing). (a) Linear interpolation and (b) Lagrange interpolation (WW = 1500).

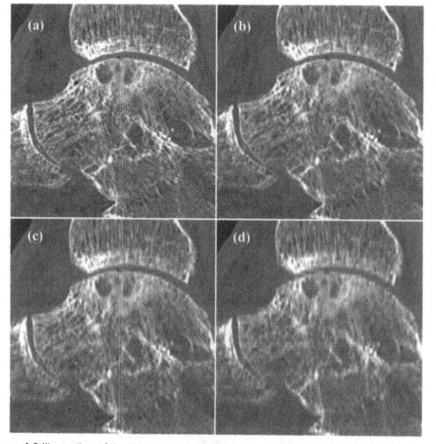

Figure 4.9 Illustration of the impact of slice thickness on MPR (WW = 1500). (a) 0.625 mm at 0.625-mm spacing, (b) 1.25 mm at 1.25-mm spacing, (c) 1.875 mm at 1.875-mm spacing, and (d) 2.5 mm at 2.5-mm spacing.

different slice thicknesses: 0.625 mm, 1.25 mm, 1.875 mm, and 2.5 mm. As the slice thickness increases, the fine structures in the sagittal image become less visible, and sawtooth-type artifacts start to appear along the bony edges. This was a key factor that prevented the widespread use of reformatted images before the introduction of multislice CT, since most clinical practices routinely used thicker slices to reduce the total acquisition time, increase the volume coverage, and avoid tube cooling delays.

For oblique image reformation, interpolation must be carried out within each image slice as well as across slices. For the best image quality, an overlapped reconstruction is needed in which the image spacing is significantly smaller than the slice thickness, due to the Nyquist sampling requirement: The original signal can be faithfully recovered if it is sampled at twice the highest-frequency contents in the signal. It is also important to ensure that the slice thickness of the reconstructed image is roughly equal to its in-plane resolution. A significant difference between the two can produce stair-step artifacts or pixilated structures.

Sometimes the reformation must be performed along a curved plane or surface, as illustrated by Fig. 4.10. For example, for dental scans it is more convenient to examine the reformatted image along the curvature of the jaw so that the teeth and the supporting bony structure can be examined at the same

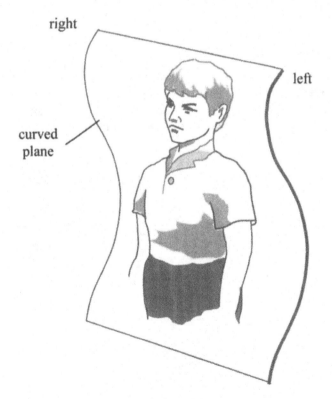

Figure 4.10 Illustration of curved-plane reformation and oblique-plane reformation.

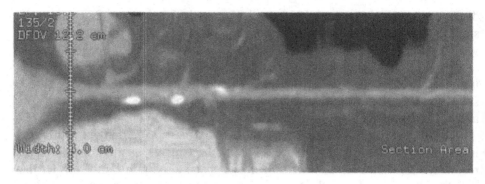

Figure 4.11 Curve plane reformatted image of the left anterior descending artery.

time. A similar situation exists when examining the carotid arteries, the aorta, and the iliac arteries. The reformation must follow the curvature of the vessel to ensure easy identification of stenoses or vessel narrowing. However, the effectiveness of a curved reformation depends heavily on the precise definition of the vessel center. When the curved plane is off the vessel center for some locations, the reformatted images may produce a false vessel narrowing. Figure 4.11 shows an example of the reformatted left anterior descending artery. Although the artery itself is curved, it appears as a straight vessel in the reformatted image.

4.2.2 MIP, minMIP, and volume rendering

Another popular display technique that compresses a 3D volume into a 2D image is volume rendering (VR).[4–6] The first step in VR is to select an observer's viewing orientation and a hypothetical screen, as shown in Fig. 4.12. A set of imaginary rays is then cast through the volume along the viewing orientation.

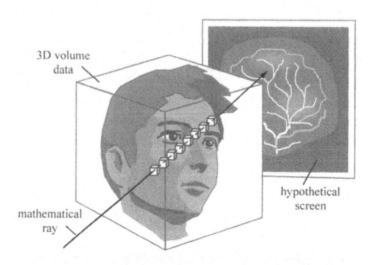

Figure 4.12 Geometric relationships of volume rendering.

The rays can be either divergent or parallel. For computational efficiency, parallel rays are typically used for medical applications. The intensity or color of a pixel on the hypothetical screen is determined by all of the pixels in the 3D volume that intersect the ray through a mapping function. For computational efficiency, the original 3D volume is often reinterpolated so that the rows or columns of the reoriented volume are parallel to the ray.

A special VR case is maximum intensity projection (MIP). The shade of gray of an MIP pixel is chosen to be the brightest pixel intensity in a 3D volume encountered by the ray. Figure 4.13 shows an example of an MIP image of a renal study. Compared to the regular projection image (similar to a conventional x-ray radiograph), MIP offers superior contrast of the vascular structures.

Because MIP intensity is determined by the maximum value along a projection path, one should be aware of some MIP characteristics. Unlike other operations in image manipulation, the intensity of the MIP varies with the projection path length, even for a uniform object,[3] as illustrated by Fig 4.14. Consider two rays in an MIP image generation process. One ray passes through the object with a longer path length (path 1) and the other with a shorter path length (path 2). Assume that the sequences of samples along both paths have an identical mean and standard deviation (they come from the same population). Since the samples taken along path 1 contain a longer sequence (more samples), the probability of encountering a sample with high intensity is higher than the

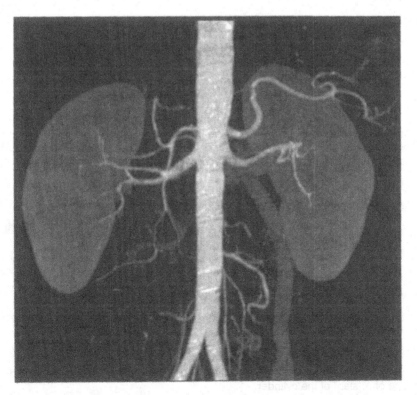

Figure 4.13 Maximum intensity projection image of a renal study.

path 2 samples (although the probability of path 1 encountering a lower-intensity pixel is also higher than that of path 2, the MIP process automatically removes the low samples). Consequently, the corresponding MIP intensity is higher (on average) for a longer path than a shorter one. Figure 4.15(a) shows an MIP image of a uniform cylindrical phantom. The phantom axis is placed vertically, so each horizontal line in the MIP image represents maximum intensity projections across a uniform disc. It is clear from the figure that the intensity of the produced MIP image is not uniform across the image; the intensity is higher in the middle than in the peripheries. This image provides the false impression that the object has a nonuniform CT number across the FOV. To quantify our analysis, we plot the average MIP intensity (averaged along the vertical direction) as a thick gray line in Fig. 4.16. The intensity is clearly nonuniform. Also, the average intensity does not match the average intensity of the cylinder phantom that produced the MIP image. The average CT number of the phantom is only 50 HU.

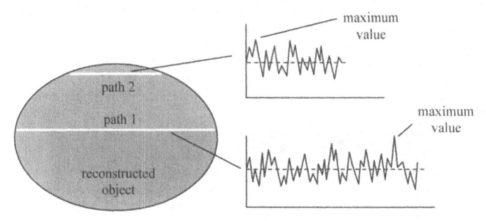

Figure 4.14 Illustration of MIP intensity variation with path length. For a uniform object with identical noise distribution, the MIP value for the longer path length (path 1) is likely to be higher than the MIP value for the shorter path (path 2), because the probability of encountering a higher sampling value increases with the increase in sampling size.

(a)   (b)

Figure 4.15 MIP images of a uniform cylinder. (a) MIP image of the entire cylinder. (b) MIP image of a "slab" of the cylinder.

One way to overcome the intensity variation over the path length is to generate MIP images over a thin slab of the object, as shown in Fig. 4.17. Because the thickness of the slab is selected to be much smaller than the object's size, the path length across the object is nearly constant. This helps to eliminate the intensity variation problem that we observed previously. Indeed, Fig. 4.15(b) shows a slab MIP image of the same cylindrical phantom. Its intensity is quite uniform. As expected, the average intensity of the slab MIP image is lower than the regular MIP. A quantitative analysis can be performed using the average slab MIP intensity profile shown by the black line in Fig. 4.16. The slab MIP intensity is still much higher than the true CT number of the cylinder (50 HU). Therefore, one should not use MIP as a quantitative analysis tool.

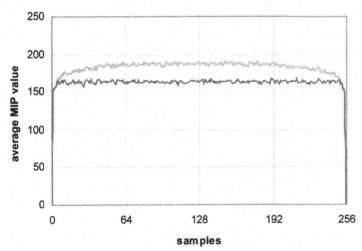

Figure 4.16 Average MIP intensity profile (averaged over different slices). Thick gray line: MIP profile of the entire cylinder; thin black line: MIP profile of a "slab" of cylinder.

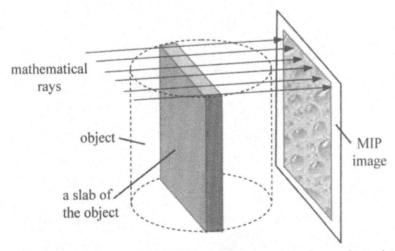

Figure 4.17 Illustration of the "slab" MIP. Instead of searching for maximum intensities over the entire object volume, only a section of the volume is used. Typically, the section or slab is selected to be perpendicular to the mathematical rays for MIP calculation.

Similar to the MIP process, minimum intensity projection (minMIP) also finds its use in many clinical settings. As its name implies, minMIP calculates the minimum value encountered by a ray and assigns this value as the minMIP value at the intersection of the ray with the projection plane. One application of minMIP is the investigation of the air pathway in the lung, since air has the lowest attenuation coefficient of zero. MinMIP images will reveal the integrity of the trachea and bronchus and allow radiologists to easily visualize any potential pathology, as shown by an example in Fig. 4.18.

General volume VR techniques have more parameters that can be controlled than the MIP and minMIP processes.[7] For example, the color of a light source is often used to produce different effects in addition to its brightness. The resulting VR images can also be presented in color to mimic different anatomies, based on the CT number and other information present in the original CT images. Table 4.1 lists a comparison of control parameters between MIP and VR. Needless to say, the image generation process is more computationally intensive for the VR technique.

Given the complexity of VR, it is impossible to discuss the entire algorithm in depth. Instead, we chose to illustrate the impact of one variable—opacity function—on the VR images. An opacity function is designed to specify the relationship between the pixel intensity of the reconstructed object and the ray transmission characteristics. By adjusting the opacity function (in either a discrete or a continuous form, as illustrated in Fig. 4.19), different material types (e.g., soft tissue, bone, or iodine contrast) can be selectively emphasized or de-emphasized. To demonstrate the utility and impact of the opacity function,

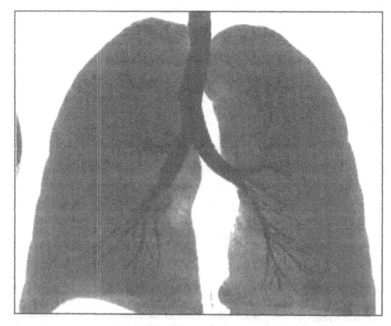

Figure 4.18 Example of minMIP.

Fig. 4.20 depicts two extremely simplified opacity functions that emphasize different CT value zones. Function (a) assigns an opacity value of 1 to all CT numbers less than –400 HU, and a value of 0 to other CT numbers. The original CT image is multiplied by this function prior to performing the line integral calculation. Given the fact that soft tissues, bones, and iodine contrast have CT values significantly higher than –400 HU, this function suppresses most anatomical structures inside the human body and emphasizes only the lung and other less attenuating organs. Indeed, Fig. 4.21(a) depicts a VR image of a patient scan with the lung and the colons clearly depicted. Function (b) of Fig. 4.20, on the other hand, assigns an opacity value of 1 to CT numbers higher than 300 HU, which corresponds to bony structures and organs with significant iodine contrast uptake, and assigns a value of 0 to other CT numbers. The resulting VR image of Fig. 4.21(b) clearly illustrates the effect of the suppressed soft tissue without significant iodine uptake for the same patient scan.

Table 4.1 Comparison of MIP and VR parameters.

Parameters	MIP	VR
light source	brightness	brightness and color
subvolume definition	thickness of the volume traversed by the ray	thickness of the volume or boundaries defined by density thresholds
value accumulation rule	maximum intensity encountered by the ray	model absorption, reflection, other physical and mathematical processes to delineate internal structure

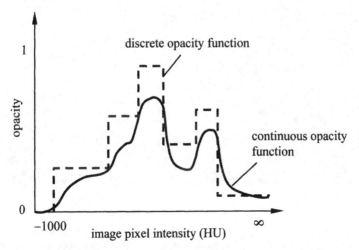

Figure 4.19 Examples of discrete and continuous opacity functions.

As listed in Table 4.1, the VR algorithm uses source color and brightness variations, selection rules of specific subvolumes, functions to govern the transmission or accumulation of the color and intensity of each voxel, and surface reflection properties to produce VR images that emphasize a specific characteristic of the reconstructed CT volume. Two examples of volume-rendered images are shown in Figs. 4.22 and 4.23, which depict a renal CT scan and a head CT angiograph examination, respectively.

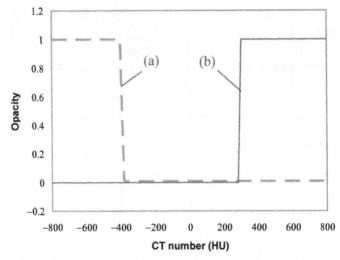

Figure 4.20 Examples of two simple opacity functions (a) and (b).

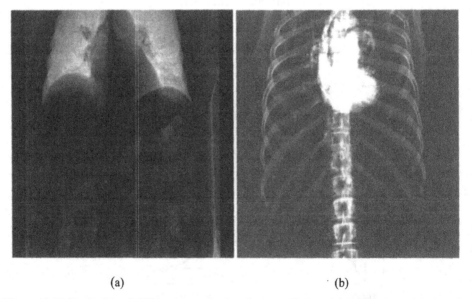

Figure 4.21 Illustration of different opacity functions applied to the same scan. (a) Opacity function in which a value of 1 was assigned to CT numbers less than –400 HU. (b) Opacity function in which a value of 1 was assigned to CT numbers higher than 300 HU.

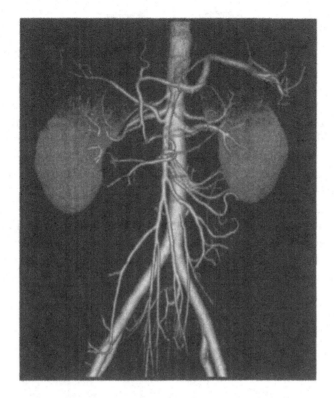

Figure 4.22 VR image of a renal study.

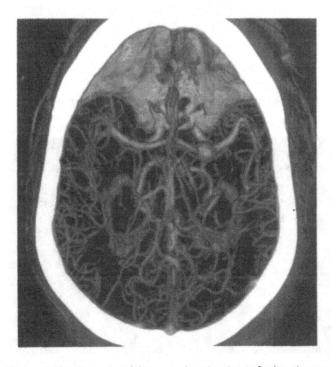

Figure 4.23 VR image of the vascular structure of a head scan.

4.2.3 Surface rendering

Finally, we will briefly discuss the technique of surface rendering, also known as shaded surface display. The first step in surface rendering is to compute a mathematical model for the surface of the object to be rendered. The structure of the object is typically defined by a predetermined threshold. For example, a bony structure can be defined as any pixel with a reconstructed intensity above 225 HU. For a pixel to be considered part of a surface, certain connectivity requirements or criteria must be met. The surface-rendering algorithms then calculate the coordinates and the respective surface norms to describe the orientation of each point on the surface, as shown in Fig. 4.24.[8–11] For a given viewing orientation and light source, a rendered image is created with intensities proportional to the amount of light reflected by the surface toward the observer. Additional shading based on the distance of the reflective surface can also be added. Figure 4.25 shows an example of a surface-rendered ankle produced from a set of CT images.

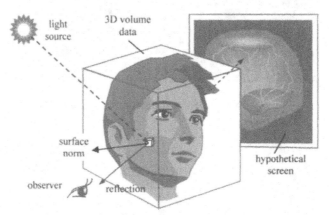

Figure 4.24 Geometric relationships of surface rendering.

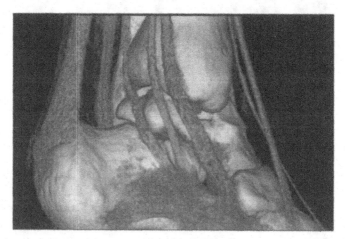

Figure 4.25 Surface-rendered image of a CT scan of an ankle.

4.3 Impact of Visualization Tools

For a long time, the primary CT viewing mode for radiologists was in the axial plane. Radiologists take pride in their ability to formulate 3D objects while viewing 2D images one slice at a time. This skill takes years of training and practice; it is remarkable to observe a radiologist analyzing CT images at a speed that is difficult for many scientists and engineers to comprehend.

With the recent technological advancements of CT, however, radiologists increasingly rely on computer visualization tools such as MIP, MPR, or VR to locate pathologies and make diagnoses. Several forces are at work to drive this change. The first is the large number of images that CT scanners currently produce for each patient. In the single-slice CT era, a typical head or body study consisted of 10 to 30 axial images, each between 5 and 10 mm in thickness. Such a study can be easily presented on a single or a few sheets of film, and radiologists seemingly spent as much time placing the films onto a lightbox as they did on the diagnosis itself. With the introduction of helical and multislice CT, the advancement in x-ray tube power, increased gantry rotation speed, and greater computer processing power, CT scanners routinely produce thin axial slices with much less than a millimeter thickness and with an acquisition time well below a patient's breath-hold. The number of axial images has increased by orders of magnitude to hundreds or even thousands for each study. Reviewing such a large number of images is an insurmountable task and can easily cause fatigue. The advanced visualization tools, on the other hand, offer the radiologist a way of viewing the entire volume in a single glance, and allows him or her to zoom in quickly on suspicious regions, instead of having to process hundreds of slices that are not related to the pathology.

The second factor is the increased ability of computers and algorithms to make the 3D volume-rendering tools fully automated, easy to use, and fast to generate. Not long ago, many institutions had a dedicated 3D laboratory to produce VR images. Radiologists had to send each study to the lab, then received the rendered images a few hours later. The 3D lab employed skilled computer scientists and technicians whose goal was to produce the best-looking 3D images. Nowadays, many of these functions can be completed at a CT console or a workstation in a matter of seconds without the involvement of a dedicated laboratory. The image generation process is quite simple and does not require expert knowledge. Although 3D laboratories still exist in many institutions, the turnaround time has been significantly reduced, and the laboratories are fully connected to CT scanners and image-viewing stations via high-speed networks. Some of these laboratories even offer services to small hospitals or clinics outside the institution.

The third driving force is the need to share information with other medical disciplines outside the radiology department. For example, CT scans are often used by surgeons who need to be informed of any abnormalities in a patient's anatomy to avoid surgical complications, or to predict the outcome of a planned procedure to repair a damaged bony structure in a patient. The 3D-rendered image becomes a communication vehicle that bridges the gap between radiologists and surgeons.

The use of advanced computer visualization tools has had a profound impact on the workflow of radiologists and the radiology department. Radiologists increasingly rely on these tools to serve as their primary reading devices. For example, when examining pathologies in extremities, reading coronal images (see Fig. 4.26) often becomes the preferred way of making a diagnosis, as opposed to reading conventional axial images. The coronal view provides a more direct visualization of the extremity structures and allows radiologist to make quicker diagnoses.

The ability of CT scanners to acquire images with isotropic spatial resolution and advanced visualization tools to generate oblique-plane images has also changed the way data acquisition is conducted. For example, CT scans were often acquired with a gantry tilt for cervical, thoracic, or lumbar vertebrae studies so that the image acquisition plane was perpendicular to the curvature of the spine, as shown by the white lines in Fig. 4.27. Scans acquired without the gantry tilt often produced suboptimal image quality because of the thick slices. Nowadays, isotropic spatial resolution can be routinely obtained, so there is little need to perform data acquisition with a gantry tilt for spine studies. Reformatted images in the oblique plane are of similar image quality compared to the images acquired directly in the oblique plane. The tilting of the CT gantry's large rotating mass during a patient scan was cumbersome and added undesirable delays to the data acquisition.

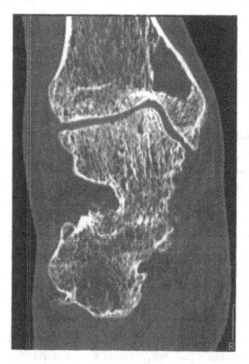

Figure 4.26 Coronal image of an ankle.

One key task of a CT operator is to ensure that acquired images are of good quality. The operator must make certain that regions of suspected pathologies are properly scanned with the desired contrast uptake levels. This requires the operator to review all reconstructed images slice by slice after the data acquisition and before the patient is released. Needless to say, this is a time- and energy-consuming task, so advanced visualization tools can significantly reduce the operator's workload. For example, instead of displaying axially reconstructed images slice by slice during the data acquisition, several 3D images, such as coronal, sagittal, and VR images, can be produced and displayed simultaneously, as illustrated in Fig. 4.28. These 3D images are constantly updated as the newly reconstructed axial images become available. A quick glance at these images allows the operator to rapidly assess the CT scan quality and make decisions prior to releasing the patient.

One major challenge facing x-ray CT is the information explosion. Recent advances in CT technologies allow not only the routine production of thin slice images of isotropic spatial resolution, but also the generation of multiple image volumes over time to monitor the temporal aspect of the study. With the availability of dual-energy scanning capability (to be discussed in Chapter 12), images with functional information can be generated that add yet another dimension to the already enormous amount of data. These images need to be presented to radiologists in a clear, concise, and accurate manner to allow improved efficiency and reduced fatigue. The 3D techniques presented in this chapter are only a first step in achieving such a goal. Research and development activities in this area are progressing rapidly and are urgently needed.

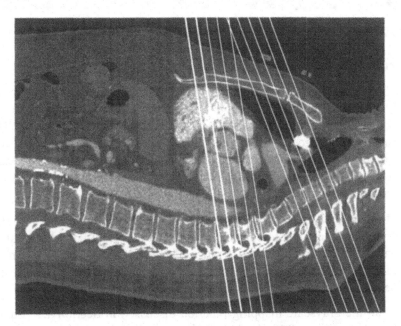

Figure 4.27 Illustration of CT gantry tilt angle for spine scans.

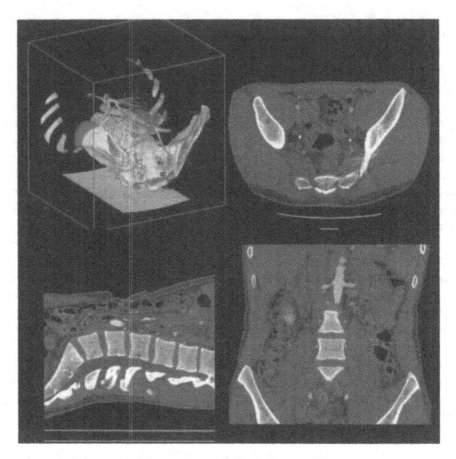

Figure 4.28 Illustration of simultaneous displays of axial, coronal, sagittal, and VR images.

4.4 Problems

4-1   Images of the midbrain section are often displayed with a window width
        of 100 while the corresponding images of the abdomen are often
        displayed with a window width of 400. Assume that a brain image and an
        abdomen image are displayed side by side with their respective custom
        window widths, and that the noise levels in both images are visually
        identical. What is the ratio of the measured standard deviations of the
        original images?

4-2   In Section 4.1, an example was given to demonstrate the use of grayscale
        mapping functions to enhance image contrast [Eq. (4.3)]. For simplicity,
        assume that in the grayscale range of interests, the slope of the mapping
        function increases from 1 to 1.2. Estimate the increase in noise in the
        enhanced image.

4-3   Can the grayscale mapping function discussed in Section 4.1 [Eq. (4.3)]
        be discontinuous? If so, what are the potential issues?

4-4   If one designs a grayscale mapping function [Eq. (4.3)] that has a slope significantly larger than unity, what are the potential issues?

4-5   Show that, like MIP, minMIP exhibits this phenomenon: its value is dependent on the path length of the rendered object. For example, for a uniform cylindrical object, the minMIP intensity is nonuniform across the diameter of the cylinder.

4-6   Predict the amount of MIP change if the path length increases from 100 to 300 samples that are normally distributed with a mean of 0 and a standard deviation of 1. Repeat the exercise when the path length increases from 300 to 500 samples. What observation can you make with the results?

4-7   Repeat problem 4-3 with minMIP. What conclusion can you draw about the amount of change between MIP and minMIP?

4-8   A thin-slab MIP operation is performed on the reconstructed images of a uniform object with normal distributed noise. Assuming the MIP path length through the object is constant over the MIP FOV, will the standard deviation of the MIP image be higher or lower than the original reconstructed images? How will the standard deviation of the MIP image change with an increase in slab thickness?

4-9   As discussed in Section 4.2, one major application of MIP is the visualization of vascular structures in studies with contrast injection. Since bony structures are equally attenuating as the iodine contrast, these structures often obstruct proper viewing of the vessels. Discuss at least two approaches to solve this problem.

4-10   A patient was scanned and images were reconstructed with a 50-cm FOV. To visualize the airway structure of the lung in the A-P direction (parallel to the y axis), a student wrote a program to search the minimum value along the entire image column (y direction). Will the student obtain the desired result? If not, what approaches can be taken to avoid potential issues?

4-11   In a clinical application, four MIP images must be generated that correspond to projection angles of 15 deg, 30 deg, 45 deg, and 60 deg with respect to the x axis. One student wrote a program that rotates the original image by −15 deg in four steps and searches the maximum intensity along each row (x axis) to produce the MIP images. Another student searched the maximum intensities in the original image along straight lines of the specified angle to produce MIP images. Discuss the pros and cons of each approach.

4-12   Section 4.2 discusses issues of MIP value variation with path length due to the influence of noise. Considering the similarity between VR and

MIP in which a ray is traversed through the reconstructed volume, what are the potential noise influences on the quality of the VR images?

4-13 One student wants to use the VR process to generate an image that mimics a conventional x-ray radiograph. The x-ray device consists of an x-ray tube with a single focal spot and a flat-panel detector that is D cm away from the source. The iso-center of the CT acquisition is R cm from the source, and the reconstructed volume is represented by $f(x, y, z)$. Describe the VR generation process.

References

1. W. V. Glenn Jr., R. J. Johnston, P. E. Morton, and S. J. Dwyer, "Image generation and display techniques for CT scan data. Thin transverse and reconstructed coronal and sagittal planes," *Invest. Radiol.* **10**(5), 403–416 (1975).

2. M. Prokop, "CT angiography," *Categorical Course in Diagnostic Radiology Physics: CT and US Cross-Section Imaging*, L. W. Goldman and J. B. Fowlkes, Eds., Radiological Society of North America, Inc., Oak Brook, IL, 143–157 (2000).

3. S. Napel, "3D display for computed tomography," in *Medical CT and Ultrasound: Current Technology and Applications*, L. W. Goldman and J. B. Fowlkes, Eds., Advanced Medical Publishing, Madison, WI (1995).

4. N. J. Mankovich, D. R. Robertson, and A. M. Cheeseman, "Three-dimensional image display in medicine," *J. Digital Imag.* **3**(2), 69–80 (1990).

5. A. B. Strong, S. Lobregt, and F. W. Zonneveld, "Application of three-dimensional display techniques in medical imaging," *J. Biomed Engr.* **12**(3), 233–238 (1990).

6. J. K. Udupa and G. T. Herman, Eds., *3D Imaging in Medicine*, CRC Press, Boca Raton, FL (1991).

7. J. C. Russ, *The Image Processing Handbook*, 5th edition, CRC Press, Boca Raton, FL (2007).

8. H. E. Cline, W. E. Lorensen, S. P. Souza, F.A. Jolesz, R. Kikinis, G. Gerig, and T. E. Kennedy, "3D surface rendered MR images of the brain and its vasculature," *J. Comput. Assist. Tomogr.* **15**(2), 344–351 (1991).

9. M. Magnusson, R. Lenz, and P. E. Danielsson, "Evaluation of methods for shaded surface display of CT volumes," *Comput. Med. Imaging Graph.* **15**(4), 247–256 (1991).

10. J. K. Udupa, "Three-dimensional visualization and analysis methodologies: A current perspective," *Radiographics* **19**, 783–803 (1999).

11. E. Seeram, *Computed Tomography, Physical Principles, Clinical Applications, and Quality Control*, W. B. Saunders Company, Philadelphia (2001).

Chapter 5

Key Performance Parameters of the CT Scanner

The assessment of a CT scanner's performance is a complicated subject. The evaluation of a CT device cannot be simply an extension of the methodologies used to assess film, electronic cameras, and x-ray radiographs.[1] A reconstructed CT image contains quantitative information that requires precision and accuracy; precision describes the reproducibility of a measurement, and accuracy characterizes the closeness of the measurement to the truth. Recent CT technology developments have extended CT images to four dimensions: three dimensions in space and one dimension in time. Clearly, it is impossible to deal with all aspects of these parameters in a short chapter, so we will address in this chapter only the key performance parameters that are not covered in the other chapters of this book. Some parameters are discussed in detail in a separate chapter (such as image artifacts covered in Chapter 7) and will be omitted here to avoid duplication. One parameter that has received increased attention in recent years, energy resolution, will be covered under dual-energy CT in Chapter 12. Because of the importance of the x-ray dose, that topic is also covered in a separate chapter together with a discussion on the biological effects of x-ray radiation and several dose-reduction approaches (Chapter 11). One topic not covered in this book is quality assurance of CT. Although quality control and acceptance testing of CT equipment (which addresses the installed CT system's conformance to the vendor's specification) is an important aspect of CT performance, we will refer interested readers to other literature on the subject.[2-7]

5.1 High-Contrast Spatial Resolution

The high-contrast spatial resolution of a CT scanner describes the scanner's ability to resolve closely placed objects. Spatial resolution is often measured in two orthogonal directions: in-plane (x-y) and cross-plane (z). Historically, the CT scanner was used primarily to generate 2D images, and there was a large discrepancy between the in-plane and cross-plane spatial resolutions. However, the difference between the two is quickly disappearing due to the recent introduction of helical and multislice scanners.

5.1.1 In-plane resolution

The in-plane resolution is typically specified in terms of line pairs per centimeter (lp/cm) or line pairs per millimeter (lp/mm). A line pair is a pair of equally sized black-white bars. Thus, a bar pattern representing 10 lp/cm is a set of uniformly spaced comb-shaped bars with 0.5-mm-wide teeth. Figure 5.1 shows the reconstructed section of a CATPHAN® phantom. Each set of bars represents a certain line pair. By examining a CT scanner's ability to resolve different bar patterns, one obtains an estimate of the system's spatial resolution capability under prescribed conditions. For this particular example, the image was reconstructed with a standard reconstruction kernel (see Chapter 3 for details). Compared to an x-ray radiograph, the spatial resolution of CT is significantly worse. The typical limiting resolution of a screen film is 4 to 20 lp/mm while the limiting resolution of CT is 0.5 to 2 lp/mm[2].

To fully understand high-contrast spatial resolution, we will first discuss the modulation transfer function (MTF). The MTF is defined as the ratio of the output modulation to the input modulation; it measures the response of a system to different frequencies. An "ideal" system has a flat MTF curve such that the system response is independent of the input frequency. For practical systems, however, the input response always degrades in some manner. The MTF of most systems falls rapidly at higher frequencies. The input frequency at which the system response approaches zero is called the limiting frequency. The corresponding spatial resolution is called the limiting resolution.

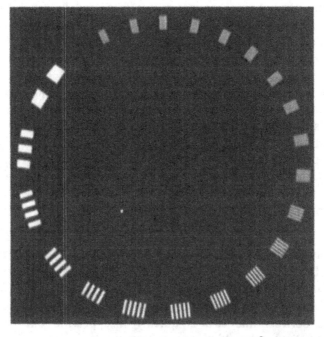

Figure 5.1 Image of bar patterns in a CATPHAN® phantom.

The MTF of a system can be obtained by calculating the magnitude of the Fourier transform of its point spread function (PSF). The PSF is the response of the system to an ideal point object or a Dirac delta function $\delta(x, y)$:

$$\delta(x, y) = 0, \forall (x, y) \neq (0, 0) \tag{5.1}$$

$$\int_{-\infty}^{\infty} \int_{-\infty}^{\infty} \delta(x, y) dx dy = 1 \tag{5.2}$$

and

$$\int_{-\infty}^{\infty} \int_{-\infty}^{\infty} f(x, y) \delta(x - x_0, y - y_0) dx dy = f(x_0, y_0), \tag{5.3}$$

where $f(x, y)$ is continuous at (x_0, y_0). The Dirac delta function has an infinitely small width and an infinitely large magnitude [i.e., $\delta(0, 0) = \infty$]. The area under the function equals unity.

Clearly, ideal point objects do not exist in the real world. In practice, we use the system's response to high-density thin wires to approximate the PSF. As long as the diameter of the wire is significantly smaller than the limiting spatial resolution of the system, its response should accurately model the PSF. For example, the GE Performance™ phantom shown in Fig. 5.2 uses a 0.08-mm-diameter tungsten wire submerged in water. To ensure adequate sampling in the image space, a targeted reconstruction around the wire must be performed. For example, the wire portion of the phantom is reconstructed at a 10-cm FOV so the spatial resolution measurement is not limited by the finite image pixel size

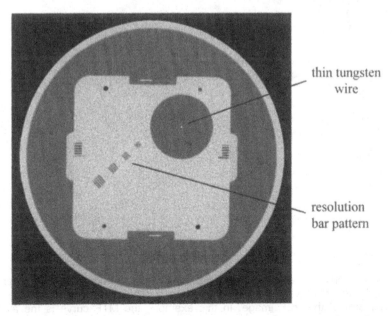

thin tungsten wire

resolution bar pattern

Figure 5.2 Reconstructed image of a GE Performance™ phantom.

(roughly 0.2 mm in this case). For accurate PSF measurement, we must remove background variations in the wire image to avoid a potential bias. Although theoretically the background should be flat, variations in the background do exist due to imperfect preprocessing or calibrations such as beam hardening, off-focal radiation, and other factors (see Chapter 7 for details). The background is removed by fitting a smooth function to the water region near the wire. The resulting image is shown in Fig. 5.3(a). Next, a 2D Fourier transform of the wire image is performed, and the magnitude of the 2D function is obtained. The phase information is not preserved in this process by design. The resulting function is our estimation of the system MTF, as shown in Fig. 5.3(b). For many CT scanners, a single MTF curve is obtained by averaging the MTF function over a 360-deg range. In many cases, MTF values at discrete locations are used to describe the system response instead of the curve itself. For example, CT performance datasheets often use 50%, 10%, or 0% MTF to indicate frequencies (lp/cm) corresponding to points on the MTF curve at which the magnitude is 50%, 10%, or 0% of the dc magnitude.

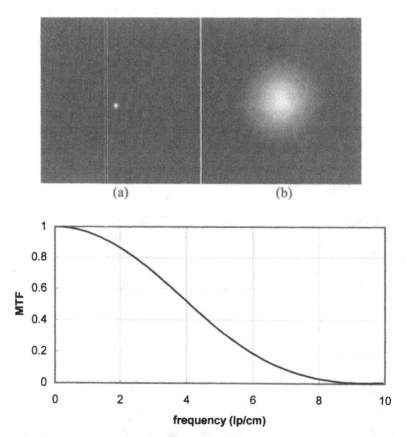

Figure 5.3 Example of a wire scan and MTF measurement. (a) Tungsten wire reconstructed with the standard algorithm to approximate PSF. (b) Magnitude of the Fourier transform of the PSF image. In this example, the MTF curve is the averaged magnitude over 360 deg.

For performance analysis, it is often more convenient to separate the spatial resolution analysis into two orthogonal directions—radial and azimuthal—since the two directions are influenced by different system design parameters. The radial direction is defined along a line that connects the location of the point object to the iso-center of the system, as shown in Fig. 5.4. The azimuthal resolution is defined tangential to the radial direction.

The radial resolution is determined mainly by the x-ray focal spot size and shape, detector aperture, scanner geometry, and reconstruction algorithm. The impact of the x-ray focal spot size, detector aperture, and scanner geometry will be discussed in detail in Chapter 8. Figure 5.5 is an example showing the impact of the reconstruction kernel on spatial resolution. The two images in Fig. 5.5 are reconstructed from the same scan data. The only difference between the two is the selection of the reconstruction kernel. Figure 5.5(a) was reconstructed with a standard kernel and Fig. 5.5(b) with a bone kernel. The two kernels differ in their cutoff frequencies and their window functions (see Chapter 3). It is clear that the bone algorithm produces a much higher spatial resolution as demonstrated by the resolvability of the bar patterns. Note that increased noise often accompanies an improvement in spatial resolution.

One factor that is often overlooked in the pursuit of high spatial resolution in clinical settings is the supportable spatial resolution of the image matrix. That is, the reconstruction pixel size frequently plays an important role in limiting the spatial resolution. Consider an image matrix of 512 × 512 with a reconstruction FOV of 50 cm. To cover such a large area, each pixel needs to be 0.98 mm by 0.98 mm (500/512). Based on the Nyquist sampling criteria, the supportable spatial resolution is roughly 5.1 lp/cm. When we use a high-resolution algorithm

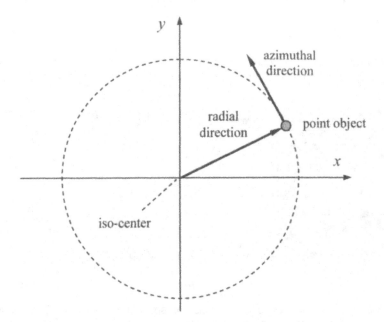

Figure 5.4 Description of the radial and azimuthal direction.

for reconstruction, such as a bone kernel, fine bar patterns in the reconstructed image are not clearly resolved, as shown in Fig. 5.6(a). For better visualization, we interpolated the reconstructed image to a 10-cm FOV (0.20 × 0.20-mm pixel size), and the degraded spatial resolution is evident. When a targeted reconstruction is carried out directly over a 10-cm FOV, the image pixel size is fine enough to support a spatial resolution up to 25.6 lp/cm. The reconstructed image is shown in Fig. 5.6(c). The high-resolution capability of the system is now fully utilized.

The azimuthal resolution, on the other hand, is largely determined by the number of projection views, focal spot size, distance to the iso-center, and detector aperture. Two MTF curves can be obtained separately by averaging the 2D MTF over small angular ranges along the radial and azimuthal directions.

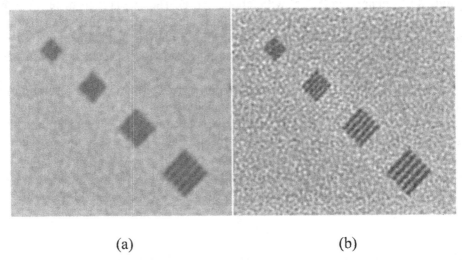

(a)   (b)

Figure 5.5 Impact of the selection of the reconstruction kernel on spatial resolution. (a) Reconstructed with a standard kernel. (b) Reconstructed with a bone kernel.

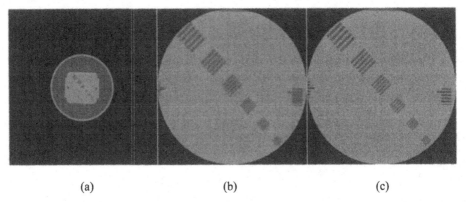

(a)   (b)   (c)

Figure 5.6 Impact of reconstruction pixel size on spatial resolution. (a) Reconstructed to a 50-cm FOV. (b) Interpolation of image (a) to a 10-cm FOV. (c) Target reconstructed to a 10-cm FOV.

The MTF of a CT system is spatially variant. The variation is caused mainly by two factors: the projected focal spot shape change as a function of detector angle γ, and the spatial sampling pattern change as a function of the distance to the iso-center. The first factor contributes to the radial resolution variation, and the second factor contributes to the azimuthal resolution degradation. A detailed discussion on the first factor will be presented in Chapter 8. The second factor can be explained by considering the sampling pattern a projection ray over a single view. Figure 5.7(a) depicts the sampling pattern of an iso-ray of a particular view: two wedge-shaped regions centered at the iso-center. Because the x-ray flux is continuous during the scan and the gantry does not stop between views, each projection sample covers a double wedge-shaped region, not a pencil beam. The increased sampling distance translates to a degraded azimuthal resolution as we move away from the iso-center. One effective way of improving the azimuthal resolution is to increase the view sampling rate of the CT system. Figures 5.7(b) and (c) show the reconstructed images of a resolution phantom at different view sampling rates with an insert position near the edge of the 50-cm FOV to maximize the impact of azimuthal blurring. At this phantom location, the azimuthal direction is parallel to the x axis. It is clear from the figure that an increased view sampling rate substantially improves the spatial resolution.

High-contrast resolution can also be measured with phantoms that contain bar patterns of various spatial frequencies, as illustrated by the CATPHAN® phantom in Fig. 5.1 and the GE Performance™ phantom in Fig. 5.2. By visually inspecting different bar patterns, one can determine the finest bar pattern (and therefore the highest spatial frequency) that is "just resolvable" or "barely separable." The limiting resolution value is generally assumed to represent the 5% point of the MTF.[2] Although this test is useful, the result is somewhat subjective and represents only a single point on the MTF curve. It provides little information about behavior at lower spatial frequencies.[8] In addition, the result is sensitive to the orientation of the bar pattern due to the difference between the radial and azimuthal resolutions away from the iso-center.

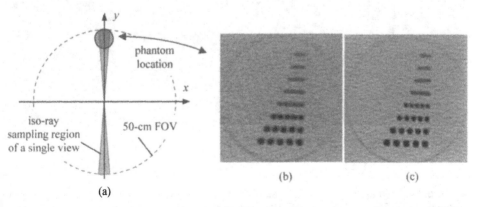

(a)

(b)   (c)

Figure 5.7 Illustration of the impact of view sampling frequency on azimuthal resolution. (a) Schematic diagram to illustrate azimuthal blurring and phantom location, (b) 984 views per rotation, and (c) 2000 views per rotation.

Before concluding the discussion on spatial resolution, we will illustrate the importance of this performance parameter in clinical settings. Consider coronary artery imaging of the heart. One application of cardiac CT imaging is to investigate the integrity of a stent that has been placed in a patient. Stents are typically made of small metal wires that are approximately 200 μm in diameter. The visualization of the shape and position of these wires requires the excellent spatial resolution capability of a CT scanner. Figure 5.8 shows two VR images of a 3-mm stent submerged in water to simulate realistic clinical conditions. The stent was first scanned on a Discovery™ CT750 HD scanner (GE Healthcare, Waukesha, Wisconsin) and reconstructed in a high-resolution mode. The wire mesh is clearly visible in Fig. 5.8(a). The stent was then scanned on an older vintage scanner and VR was performed on the reconstructed images, as depicted in Fig. 5.8(b). The lack of wire mesh details in the latter is evident.

5.1.2 Slice sensitivity profile

Cross-plane spatial resolution is often described by the slice sensitivity profile (SSP). Similar to in-plane resolution, the SSP describes the system response to a Dirac delta function $\delta(z)$ in z. Unlike the specification used for in-plane spatial resolution (lp/cm), the actual system response curve to $\delta(z)$ is commonly used to describe the SSP. In many cases, the curve itself is replaced by two numbers: the FWHM or full width at tenth maximum (FWTM), both of which are illustrated in Fig. 5.9.

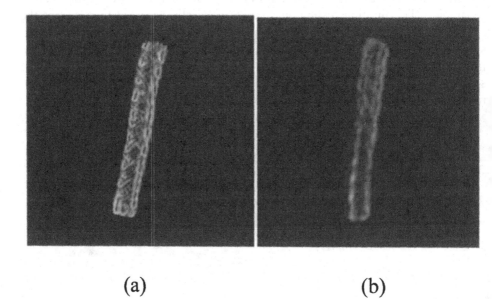

(a)   (b)

Figure 5.8 VR images of a 3-mm stent. (a) Data collected and reconstructed on a Discovery™ CT750 HD. (b) Data collected and reconstructed on an older vintage scanner.

In practice, the Dirac delta function is often approximated by objects whose thickness is significantly smaller than the slice thickness of the data acquisition. For example, researchers use a small bead or a thin disc to perform SSP measurements. Using these types of objects creates difficulties because other factors often impact the measurement accuracy.[9] Thus, if a small bead is used, the table has to be indexed at very fine increments to ensure adequate sampling in z. For example, a 0.1-mm increment is needed to characterize an SSP with a FWHM on the order of 1 mm. A much finer increment (e.g., 0.05 mm) is required for slice profiles that are 0.5 mm or less. This is particularly troublesome for the step-and-shoot mode data collection, since sampling requirements can easily exceed the accuracy and capability of the patient table. In the step-and-shoot mode (also called axial mode), a phantom remains stationary while the gantry rotates about the phantom to collect a complete set of data. Once the data collection is completed, the phantom is indexed to the next location for another scan. The second drawback of this technique is the size limitation on the bead. To approximate an impulse response, very small beads (a small fraction of a millimeter) must be used. When imaging such a small object, it is critical to provide adequate sampling in the reconstruction plane (the x-y plane). If we reconstruct an image with a 10-cm FOV in a 512×512 image matrix (this is the lower limit on many CT scanners), the reconstructed pixel size is roughly 0.2 mm. The pixel size is the same or larger than the diameter of the bead. Based on the Nyquist criteria, the sampling is likely to be inadequate to faithfully represent the bead in the cross-sectional plane.

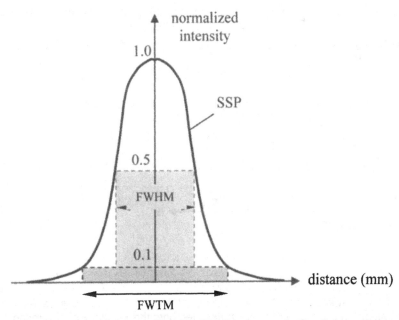

Figure 5.9 Illustration of FWHM and FWTM. FWHM represents the distance between two points on the SSP curve whose intensity is 50% of the peak. FWTM represents the distance between two points on the SSP whose intensity is 10% of the peak.

An alternative SSP measurement can be performed with a highly absorbing thin-disc phantom. The disc is placed perpendicular to the z axis, and scans are collected to construct the SSP, similar to the small bead experiment. The advantage of the thin-disc phantom is improved sampling in the x-y plane because of its large size, which leads to a potential improvement in the accuracy of the attenuation measurement. On the other hand, the alignment of the disc with the x-y plane is critical to the accuracy of the experimental result. A slanted disc may result in a broadened SSP measurement.

A more convenient method of measuring SSP is to use a shallow-angled slice ramp.[10,11] A line or thin strip is placed at a shallow angle with respect to the x-y plane. During the scan, the line or thin strip is projected onto the x-y plane, as shown in Fig. 5.10(a). Based on simple trigonometry, the SSP in z is magnified by a factor $1/\tan(\theta)$, where θ is the angle of the strip formed with the x-y plane. When θ is small, the magnification factor is large. For example, the CATPHAN® phantom uses a 23-deg ramp to produce a 2.4-fold enlargement in the measured SSP. Figure 5.11 shows a GE Performance™ phantom scanned with a 5-mm and a 10-mm acquisition. It is clear from the figure that the image of the slice ramp is much wider in the 10-mm case than in the 5-mm case. This is not surprising, since the image's ramp width scales linearly with the slice thickness. At the same time, the contrast between the ramp and the background is significantly reduced in the 10-mm case, due to the partial volume effect (a detailed discussion on partial volume can be found in Chapter 7). For quantitative results, the intensity profile of the reconstructed ramp along the strip is plotted and scaled by $\tan(\theta)$ to arrive at the SSP.

Using a similar principle, one can drill a series of small airholes parallel to the x-y plane that are spaced at a uniform distance in z to estimate the slice thickness. To ensure adequate airhole size, the airholes are spread over a large distance in the x-y plane, as shown in Fig. 5.10(b). In this approach, the number

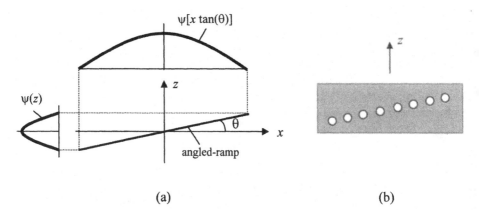

(a)   (b)

Figure 5.10 Illustration of SSP measurement by a shallow-angled slice ramp and an airhole pattern. In (a), the slice ramp is projected onto the x-y plane during reconstruction. A magnified version of the SSP is produced. In (b), the number of airholes included in the slice is proportional to the slice thickness.

of holes included within the reconstructed image is proportional to the slice thickness. By counting the number of holes and knowing the distance (in z) between the holes, one can directly estimate the slice thickness. For example, in the GE Performance™ phantom, the airholes are spaced 1 mm apart in z. Consequently, Fig. 5.11(a) has 5 airholes and Fig. 5.11(b) has 10 airholes.

This section has focused on the step-and-shoot mode of data acquisition. An analysis of SSP for more advanced acquisition modes can be found in Chapters 9 and 10. The use of FWHM and FWTM to define the SSP of a CT system is due solely to historical reasons, not technical reasons. In the past, CT images were generally reconstructed with thicker slices in the range of 5 to 10 mm; FWHM and FWTM were easy to measure and describe for these images. With recent technological advancements that enable the routine generation of submillimeter images, there is no valid reason why MTF cannot be used to fully characterize the system performance in z.

In clinical settings today, the role of SSP is equally important as that of in-plane spatial resolution, largely because of the increased use of 3D visualization tools for primary diagnosis. For example, the primary viewing method of coronary artery imaging of the heart is no longer in the axial plane; instead, it is dominated by curved reformation, MPR, and VR. Chapter 4 demonstrated the impact of SSP on coronal image quality (e.g., see Fig. 4.9). Its impacts on curved reformation, MIP, and VR are equally visible.

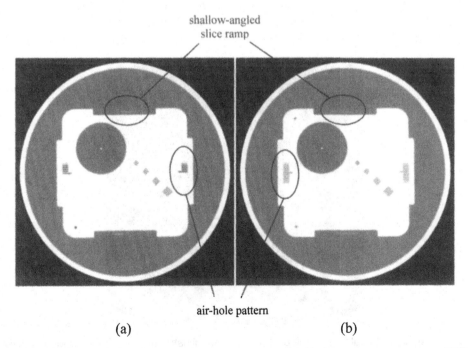

(a)   (b)

Figure 5.11 Illustration of the use of a slice ramp and airhole pattern to measure SSP acquired in step-and-shoot mode with (a) a 5-mm slice thickness and (b) a 10-mm slice thickness.

5.2 Low-Contrast Resolution

The ability of a CT system to differentiate a low-contrast object from its background is an important indication of the scanner's quality. Indeed, low-contrast detectability (LCD) is the key differentiator between CT and the conventional radiograph. This characteristic is one of the main reasons that CT has gained rapid clinical acceptance.

Low-contrast resolution is measured with phantoms that contain low-contrast objects of different sizes. The low-contrast performance of the scanner is typically defined as the smallest object that can be visualized at a given contrast level and dose. In CT, the contrast level is typically specified in terms of the percent linear attenuation coefficient. A 1% contrast means that the mean CT number of the object differs from its background by 10 HU.

The definition of LCD implies that an object's visibility depends not only on its size but also on its contrast (intensity difference) with the background. To illustrate the object-visibility dependency on size and contrast, Fig. 5.12 presents a computer-generated disc pattern in which the size of the disc changes from left to right while the contrast changes from top to bottom. No noise is added to the image. For objects of the same contrast (same row), the difficulty in identifying them increases as their size decreases. Similarly, the visibility of objects of the same size (same column) decreases with the reduction in their intensity contrast to the background. When these two effects are combined, we can conclude that the visibility of an object cannot be determined simply by either its size or contrast. We have to examine both effects simultaneously.

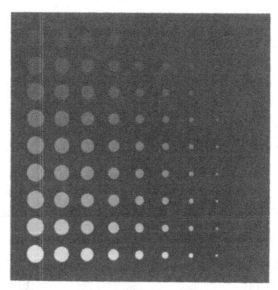

Figure 5.12 Computer-generated test pattern illustrating the size and intensity dependence of object visibility. The diameters of the discs (from left to right) are 33, 29, 25, 21, 17, 13, 9, 5, and 1 pixel. The intensities of the discs (from top to bottom) are 10, 20, 30, 40, 50, 60, 70, 80, and 90 HU.

The LCD definition also implies that an object's visibility is highly influenced by the presence of noise. To demonstrate this effect, a computer-generated test pattern is shown in Fig. 5.13. In this test pattern, the size of the discs is constant, but the contrast of the discs increases from top to bottom. A different amount of Gaussian noise was added to each column, and the standard deviation increases from the left column to the right. It is clear from the figure that the visibility of the disc decreases as the noise increases. Again, neither effect (contrast or noise) can be treated alone, and both effects must be considered simultaneously. Given the discussion on contrast and object size, we can conclude that the three effects interact to determine the LCD.

Many factors influence the noise level in a CT image. For example, noise in the image changes with slice thickness, tube voltage, tube current, scanned object size, and reconstruction algorithms. For illustration, we scanned the low-contrast portion of a GE QA phantom at 120-kV/5-mm aperture and reconstructed it with a standard algorithm. The first image [Fig. 5.14(a)] was acquired with a 200-mA tube current, and the second image [Fig. 5.14(b)] was acquired with a 50-mA tube current. The lower tube current (therefore, lower dose) results in a reduced LCD. The third image [Fig. 5.14(c)] was produced with the same scan data as Fig. 5.14(a) but with a bone algorithm. Because a bone algorithm is a high-resolution kernel that produces a higher level of noise, the visibility of the holes is reduced compared to the image generated with the standard kernel. A more detailed discussion of noise will be presented in Section 5.4.

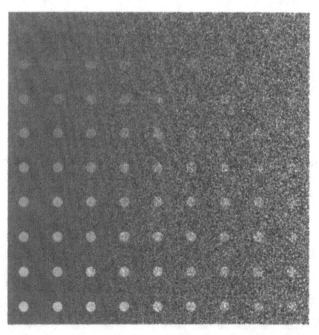

Figure 5.13 Computer-generated test pattern illustrating the noise dependency of object visibility. The diameter of each disc is 15 pixels. The intensities of the discs (from top to bottom) are 10, 20, 30, 40, 50, 60, 70, 80, and 90 HU. The standard deviations of the noise (from left to right) within each sub-block are 0, 11.5, 23, 34.5, 46, 57.5, 69, 81, and 92.5 HU.

Since LCD is determined by human observers, the image display plays a significant role in the results. For example, different display devices have different gamma curves (the mapping of the image intensity to the brightness of the display device). These gamma curves produce different levels of enhancement to the image appearance. Even on the same display device, the selection of display window width and window level significantly influences the visibility of objects. Although a wider display window width reduces the appearance of image noise (which improves LCD), it also reduces the appearance of contrast (which degrades LCD). To illustrate this effect, Figs. 5.15(a) and (b) show the same image displayed at two different window widths. It can be argued that for each low-contrast image, there is an "optimal" display setting that maximizes the performance of a "typical" observer.

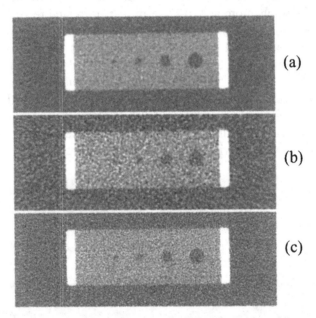

(a)

(b)

(c)

Figure 5.14 Illustration of the impact of noise on LCD for an image acquired with (a) 200 mA and the standard algorithm, (b) 50 mA and the standard algorithm, and (c) 200 mA and the bone algorithm.

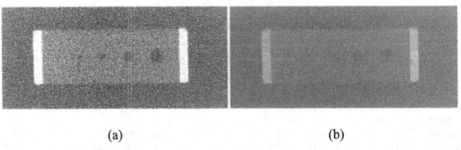

(a)   (b)

Figure 5.15 Illustration of the impact of display WW: (a) WW = 50 HU and (b) WW = 200 HU.

Low-contrast phantoms can be constructed with two different techniques. The first technique is to use materials with slightly different attenuation coefficients. Figure 5.16 shows a reconstructed low-contrast cross section of a CATPHAN® phantom. Since most CT scanners are capable of differentiating less than a 1% contrast, these phantoms must be produced with a tight tolerance on the material composition. Consequently, phantoms constructed with this method are generally more expensive. An alternative technique uses the partial-volume effect. For example, the low-contrast section of the GE QA phantom is made of a thin plastic plate that is 1.59 mm thick. When scanned with thicker slices, the CT number difference in the reconstructed images between the plate and the background (water) becomes a small fraction of the attenuation coefficient difference between the plate and its background due to the partial-volume effect. Therefore, we can use material with a much higher contrast to mimic a low-contrast object. This approach reduces the cost of the phantom, but the object-to-background contrast becomes highly dependent on the nominal thickness of the reconstructed slices. Since slice thickness can vary with the acquisition mode (helical or step-and-shoot) and reconstruction algorithms, special attention must be paid when interpreting the result. Figures 5.17(a)–(d) show four reconstructed images of the low-contrast section of a QA phantom. The scans were acquired with apertures of 2.5, 5, 7.5, and 10 mm. The figure shows that the contrast of the plate is reduced as the slice thickness increases. With reduced contrast, our ability to visualize smaller-sized holes becomes increasingly difficult.

As discussed previously, LCD is determined by scanning a standard test phantom. The phantom is scanned under different techniques (kVp, mAs, slice thickness, etc.) and reconstructed with different algorithms. Images are presented to several observers to identify the smallest object. The LCD is taken as the average of the outcomes from the selected observers. Since the results from this

Figure 5.16 Image of the low-contrast portion of the CATPHAN® phantom (WW = 50 HU).

method are somewhat unpredictable and unverifiable, a statistical method was proposed.[12] The method assumes that when the means (a random variable) of multiple low-contrast objects of the same size (under the same conditions) are measured, they will follow a Gaussian distribution. Similarly, the measured means of the background at multiple ROIs, each with the same size as the low-contrast object, will also follow a Gaussian distribution and have an identical standard deviation, as shown by Fig. 5.18. This assumption is based on the fact that both the low-contrast object and the background are scanned under identical conditions (the same scan), and the difference between their attenuation coefficients is small by definition. The only difference between the two distributions is their expected mean values. If we use the midpoint between the two distributions as the threshold to separate the low-contrast objects from their background, the false-positive rate (the area under the background distribution curve that exceeds the threshold) reaches 5% when the means of the two distributions are separated by $3.29\sigma_\mu$, where σ_μ is the standard deviation of the distribution. Similarly, the false-negative rate (the area under the low-contrast object distribution curve that is below the threshold) is also 5%. Of course, if a higher confidence level is desired, the means of the two classes must be further separated.

Based on the above analysis, the LCD can be determined solely by computer analysis. To begin the analysis, a uniform water phantom is scanned at the desired dose level (by selecting the proper tube voltage, tube current, slice thickness, scan time, etc.). The phantom is then reconstructed, and a center region of the reconstructed image is divided into multiple grids, as shown in the right portion of Fig. 5.18. The grid area is selected to be the same size as the small low-contrast object of interest, and the mean CT numbers inside each grid are calculated (e.g., 49 means in the example shown). The standard deviation

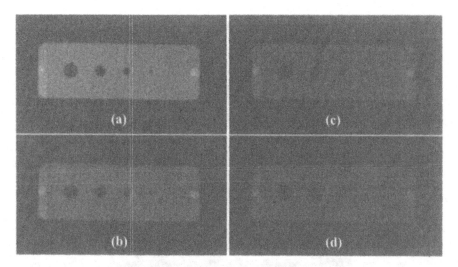

Figure 5.17 Contrast variation as a function of slice thickness for data acquired with the following slice thicknesses: (a) 2.5 mm, (b) 5 mm, (c) 7.5 mm, and (d) 10 mm.

σ_μ of these means is then calculated. Based on our previous discussion, the contrast of these low-contrast objects must be $3.29\sigma_\mu$ in order to be identified from the background with a 95% confidence level. The analysis can be repeated for contrast levels of different object sizes.

When utilizing the statistically based method to measure LCD, the operator should have a thorough understanding of the image-generation process to avoid overestimating the system's capability. For example, consider a case in which a postprocessing filter is applied to the reconstructed images to suppress noise. Although many filters are adaptive and can avoid smoothing real structures, all of them use certain criteria to differentiate between a real structure and noise. When the contrast of the LCD objects falls below the threshold, these objects will be treated as noise and removed by the filtering process. Since the statistically based method presented above uses only a uniform phantom, the LCD always improves when these filters are applied. For illustration, we scanned a 20-cm water phantom [Fig. 5.19(a)] and added computer-simulated, 3-mm disc-shaped LCD objects [Fig. 5.19(b)]. We measured the LCD using the statistical method on the water-only image and obtained a 3-mm LCD value of 0.39%. Both images were then filtered by a median filter; the resulting images are shown in Figs. 5.19(c) and (d). The median filter was purposely selected so that the LCD objects in the filtered image would become nearly invisible [Fig. 5.19(d)]. Clearly, this filter significantly degraded the LCD of the system. However, when we used the statistical LCD method on the filtered water image [Fig. 5.19(c)], we obtained an improved 3-mm LCD value of 0.10%! This simple example illustrates the complexity of LCD measurement. One possible remedy is to use images containing LCD objects and to develop a statistically based method with *a priori* knowledge of the size and location of the objects.[13]

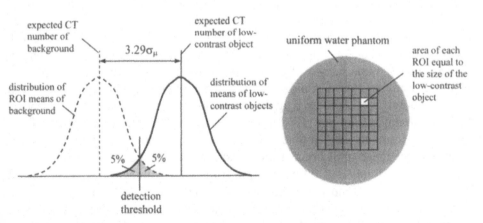

Figure 5.18 Statistical method for LCD. A uniform water phantom is scanned and reconstructed. The center region of the reconstructed image is divided into multiple ROIs with each ROI equal to the size of the low-contrast object. The mean of each ROI is calculated, and the standard deviations of all the ROI means are obtained. The contrast of the object that can be detected with a 95% confidence level equals 3.29 × the standard deviation of the measured means.

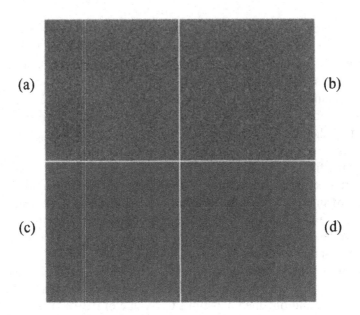

Figure 5.19 Illustration of overestimation of LCD with the statistical method. (a) Original water phantom image, (b) 3-mm LCD objects added, (c) image (a) filtered by a median filter, and (d) image (b) filtered by the same median filter.

5.3 Temporal Resolution

The temporal resolution specification is becoming an increasingly important CT performance parameter for two major clinical applications: CT fluoroscopy and cardiac imaging. CT fluoroscopy is used primarily for interventional procedures, such as CT-guided biopsy. In a typical CT fluoroscopy setup, images are generated at a high rate (5 to 20 frames/sec). The operator relies on the near-real-time feedback from the tableside monitor to adjust the orientation and depth of the interventional instrument. Clearly, a slow temporal response from the CT system is not acceptable. The second major application is cardiac CT scanning, whose primary purpose is to "freeze" cardiac motion. A faster scan speed or a better temporal resolution is highly desirable. Although both applications demand better temporal resolution, each application actually emphasizes different aspects of the temporal resolution. CT fluoroscopy stresses the real-time nature of the temporal resolution, and cardiac CT scanning emphasizes the temporal span. Consequently, these applications must be addressed separately.

First we will look at CT fluoroscopy. Although CT images can be generated in a fraction of a second in the fluoroscopy mode, we cannot conclude that the displayed image represents the condition or the event that took place a fraction of a second ago. It merely indicates that we are observing the CT system's response to the changes in the scanned object during a fraction of a second. The image shown could represent an event that took place some time ago due to the delayed response from the CT system.[14] To characterize the temporal response of the system, we need to define two parameters: time lag and time delay. Time lag specifies the time that the system takes to reveal movement in the interventional instrument after its motion actually takes place. Time delay describes the time interval for the interventional instrument in the CT image to reach its true location after the instrument motion stops.

To understand the impact of time lag and time delay on interventional procedures, consider a scenario in which a needle biopsy is performed, as shown in Fig. 5.20. Our goal is to position the needle tip at the target tumor to collect biopsy samples. After the needle is inserted, the operator watches the needle position displayed on the CT monitor to determine if further adjustment is necessary. Because of the inherent time delay, the image shown on the monitor indicates that the needle tip is still some distance from the target, as shown by the thick black line in Fig. 5.20(a). In reality, however, the needle has already reached the target. If the operator immediately adjusts the needle position, the needle position will overshoot the target, as shown in Fig. 5.20(b). Although the needle can be repositioned after several trials, the biopsy procedure will be significantly prolonged.

Two major factors influence time lag and time delay. The first is the fixed delay due to data transmission, numerical computation, and image display. Fixed delay can be easily calculated or measured. The second factor is the inherent system delay due to the nature of the reconstruction algorithm. A detailed analysis has shown that a delay of this nature can be characterized by the step response of the system.[14] Figure 5.21 shows an example of the calculated step response curve and the corresponding time delay and time lag. Detailed discussions of CT fluoroscopy and its corresponding reconstruction algorithms can be found in Chapter 12.

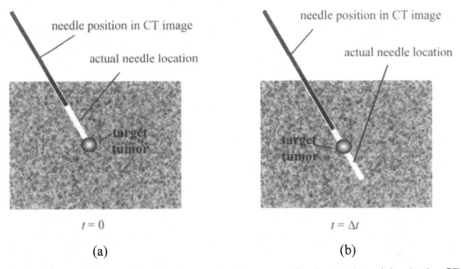

Figure 5.20 Illustration of "overshoot" of the biopsy needle due to time delay in the CT system. (a) Needle position in the CT image versus the actual needle position. Although the needle has reached the target location at $t = 0$ (shown by the white line), the CT image indicates that the needle needs to be inserted farther. (b) At $t = \Delta t$, the needle position is adjusted based on the CT image, which results in an overshoot of the needle position.

For cardiac CT, we are mainly interested in the temporal span of the reconstructed image. Since the heart motion is periodic or near periodic, what is important for cardiac image quality is the width of the temporal window in the cardiac motion cycle during which complete data acquisition takes place. For analysis, consider first the case in which the entire dataset for an image is collected continuously within one cardiac cycle, as illustrated in Fig. 5.22. In this figure, the horizontal axis represents time in seconds, and the vertical axis is the magnitude of the electrocardiogram (EKG) signal. A single cardiac cycle is depicted as the time interval between the R peaks of the EKG. A typical cardiac reconstruction uses half-scan algorithms.[15] These algorithms have the advantage that they use only π plus the fan angle of projections to form an image, so the overall data acquisition time is reduced. To avoid image artifacts caused by redundant data samples, some type of weighting function is used (for detailed discussions, see Chapter 7). The projection data set $p(\gamma, \beta)$ is multiplied by a weighting function $w(\gamma, \beta)$ prior to the FBP operation.

Confusion often arises during the estimation of the system temporal resolution under weighting conditions. In many cases, only the temporal response of the iso-center is quoted and considered. For the iso-center, the temporal response is essentially the weighting function for the iso-ray ($\gamma = 0$) plotted against time (the iso-center is sampled by only iso-rays). Since β scales linearly with time (the gantry rotates at a constant speed), the weighting function $w(\gamma, \beta)$ can be converted to a time function:

$$w'(\gamma, t) = w\left(\gamma, \frac{2\pi t}{s}\right), \qquad (5.4)$$

where s is the time for the gantry to rotate one revolution. The thick black line in Fig. 5.23 shows the temporal response of the iso-center for a 0.5-sec scan speed

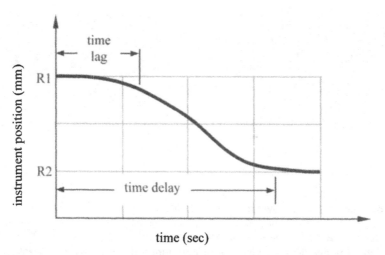

Figure 5.21 Time lag and time delay derived from the system step response curve.

with a half-scan weight. One could conclude that the temporal resolution is 0.25 sec (FWHM). If we use the helical interpolative weight, we arrive at the same 0.25-sec temporal resolution, as shown by the thin gray line in Fig. 5.23 (see Chapter 9 for a detailed description of the helical interpolative weight). Clearly, the two weighting functions produce different kinds of temporal responses, as shown by the different shapes of the curves. Therefore, as in the case of spatial resolution, we should not rely solely on the FWHM to describe the temporal response of a CT system.

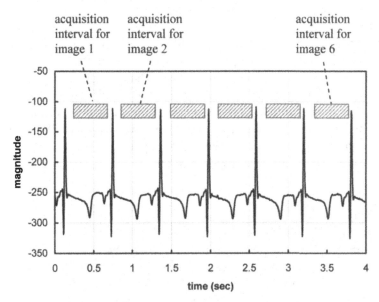

Figure 5.22 Illustration of single-cycle acquisition.

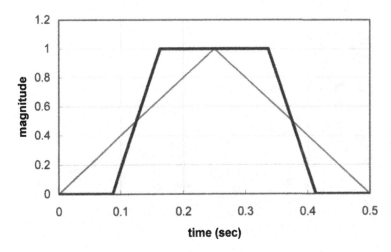

Figure 5.23 Temporal response of a 0.5-sec scan for the iso-channel for different weighting algorithms. Thick black line: half-scan weight; thin gray line: helical interpolation algorithm.

The "temporal response" is location dependent, which means the response changes with the spatial location where the measurement is performed. This is because the locus of a point on the scanned object is a sinusoidal curve in the sinogram space, as shown in Fig. 5.24 (the loci of two points are shown). The background intensity in the figure represents a half-scan weighting function with black representing zero and white representing one. Since the weighting function is not symmetric with respect to the iso-channel line (the dash-dotted vertical line in the middle), the temporal responses of the two points shown in the figure are significantly different. For illustration, the 50% intensity points (corresponding to FWHM) are labeled with captions indicating the best and worst cases. Since a typical cardiac FOV is 25 cm (out of the full 50-cm FOV), the projection loci in the figure represents two edge points inside the cardiac FOV. For quantitative analysis, Fig. 5.25 shows the temporal response curves plotted for four cases: the best case, worst case, iso-channel, and average. The average case is calculated by averaging the weighting function over the center 25-cm regions. A significant variation exists in the temporal response across the cardiac FOV: the percentage difference between the best case (0.213 sec) and worst case (0.286 sec) is more than 34%. On the other hand, the temporal response of the iso-channel (0.250 sec) is quite close to the average response of the system (0.256 sec).

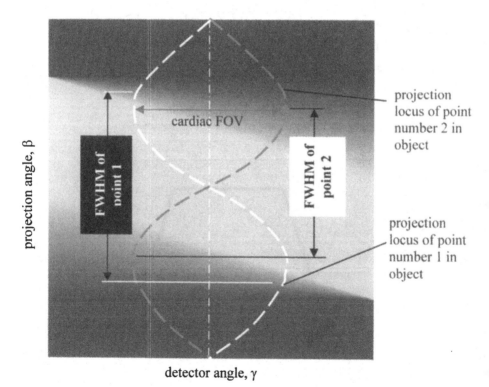

Figure 5.24 Illustration of temporal resolution variation as a function of location. The projection loci of two points (best and worst case) are shown by the dotted lines in the graph. The temporal response (vertical direction) is different for the two points.

Next, consider the case in which reconstructions are performed over more than one cardiac cycle, as shown in Fig. 5.26. In this figure, the entire dataset for a single image is obtained over three cardiac cycles. Because the dataset is divided into multiple segments, the data acquisition window for each sector is significantly reduced compared to the single-cycle acquisition. Besides the spatially dependent temporal response discussed for a single-cycle acquisition, a multisector-based reconstruction is further complicated by several factors.

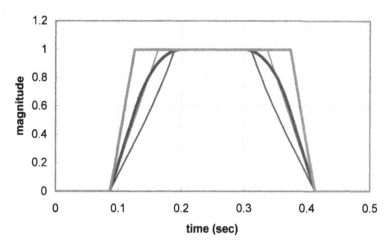

Figure 5.25 Temporal response of a 0.5-sec scan for different channels. Thin black line: best-case channel (FWHM = 0.213 sec); thin gray line: iso-channel (FWHM = 0.250 sec); thick black line: average of all channels for the center 25-cm FOV (FWHM = 0.256 sec); and thick gray line: worst-case channel (FWHM = 0.286 sec).

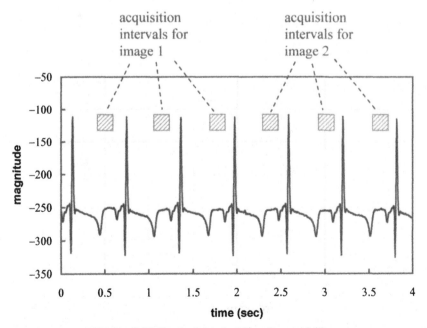

Figure 5.26 Illustration of multicycle acquisition.

First, to avoid abrupt changes in the projections at the boundaries of adjacent sectors (due to misregistration or mistriggering in the data acquisition), an overlapped region is typically present between neighboring sectors so that projections can be blended together to avoid image artifacts. This increases the sector acquisition temporal window beyond Γ/n, where Γ is the temporal span of a single-cycle acquisition, and n is the number of sectors.

The second complication arises from unavoidable misregistrations during the sector data acquisition. If we plot the sector acquisition window against the phase of the cardiac cycle, it is highly likely that the acquisition window will not be centered on exactly the same phase from cycle to cycle. This is a direct result of the nonperiodicity of cardiac motion. Most multisector-based acquisitions rely to some extent on the periodicity of cardiac motion (even algorithms that use several preceding heart motions to predict future motion rely on the stable trending of cardiac motion). When misregistration occurs, the acquisition windows are shifted from one sector to another, as illustrated in Fig. 5.27. The composite temporal response $w_{\text{eff}}(\gamma, t)$ is then

$$w_{\text{eff}}(\gamma,\ t) = \sum_{i=1}^{n} w'_i(\gamma, t + \tau_i - \Delta_i),\qquad(5.5)$$

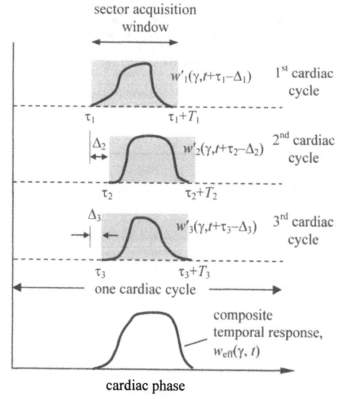

Figure 5.27 Illustration of composite temporal response for multisector reconstruction.

where τ_i is the start time of data acquisition for sector i, and Δ_i is the amount of shift relative to the ideal case for sector i. Under ideal conditions of $\Delta_i = 0$, we arrive at the best temporal response. In general, the acquisition window width for each sector T_i changes from sector to sector. Consequently, the effective temporal response is influenced not only by the amount of phase shift in each sector, but also the widest acquisition window of all sectors.

5.4 CT Number Accuracy and Noise

In many clinical practices, radiologists rely on the value of measured CT numbers to differentiate healthy tissue from disease pathology. Although this practice is not recommended by most CT manufacturers (unless the difference between the healthy and diseased tissues is large), it underlines the importance of producing accurate CT numbers. Two aspects factor into CT number accuracy: CT number consistency and uniformity. CT number consistency dictates that if the same phantom is scanned with different slice thicknesses, at different times, or in the presence of other objects, the CT numbers of the reconstructed phantom should not be affected. CT number uniformity dictates that for a uniform phantom, the CT number measurement should not change with the location of the selected ROI or with the phantom position relative to the iso-center of the scanner. For illustration, Fig. 5.28 shows a reconstructed 20-cm water phantom. The average CT numbers in two ROI locations should be identical. Because of the effects of beam hardening, scatter, CT system stability, and many other factors, both CT number consistency and uniformity can be maintained within

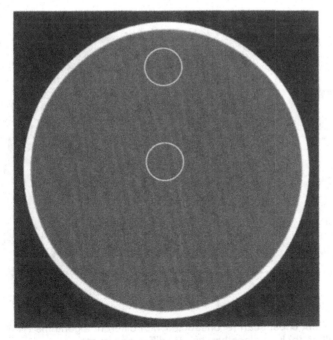

Figure 5.28 Water phantom for CT number uniformity and noise measurement.

only a reasonable range. As long as radiologists understand the system limitations and the factors that influence the performance, the pitfalls of using absolute CT numbers for diagnosis can be avoided.

It is important to point out that the CT number may change significantly with different reconstruction algorithms. Most kernels used for reconstruction are designed for specific clinical applications and should not be used indiscriminately for all applications. For example, the bone or lung algorithm on HiSpeed™ or LightSpeed™ scanners is designed to enhance the visibility of fine structures of bony objects. It was designed with inherent edge enhancement characteristics and not intended to maintain the CT number accuracy of small objects. Consequently, some of the lung nodules reconstructed with the bone or lung algorithm appear to be falsely calcified, since the CT number of the nodules is artificially elevated. Figures 5.29(a)–(c) depict a chest scan reconstructed with, respectively, the standard algorithm, the lung algorithm, and the difference image. The CT number of many lung nodules and even some large vessels is significantly higher for the lung algorithm, as indicated by the bright structures in the difference image.

Another important performance parameter is image noise, which is typically measured on uniform phantoms. To perform noise measurement, the standard deviation σ within an ROI of a reconstructed image $f_{i,j}$ is calculated by

$$\sigma = \sqrt{\frac{\sum\limits_{i,j \in \text{ROI}} \left(f_{i,j} - \bar{f}\right)^2}{N-1}}, \tag{5.6}$$

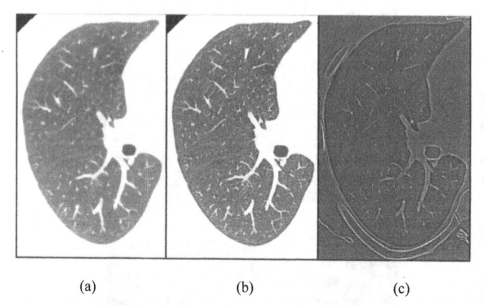

          (a)   (b)   (c)

Figure 5.29 Patient chest scan reconstructed with different algorithms (WW = 1000): (a) standard algorithm, (b) lung algorithm, and (c) difference image.

where i and j are indexes of the 2D image, N is the total number of pixels inside the ROI, and $\bar{f}$ is the average pixel intensity that is calculated by

$$\bar{f} = \frac{1}{N} \sum_{i,j \in \text{ROI}} f_{i,j} \,. \tag{5.7}$$

Note that in both equations, the summation is 2D over the ROI. For reliability, several ROIs are often used, and the average value of the measured standard deviations is reported. Ideally, noise measurements should be performed over an ensemble of reconstructed CT images of the same phantom under identical scan conditions to remove the impact of the correlated nature of the CT image noise and the potential biases due to image artifacts resulting from nonperfect system calibration. In such calculations, the summation over ROI shown in Eqs. (5.6) and (5.7) is replaced by the summation over an ensemble of images of the corresponding pixel.

In general, three major sources contribute to the noise in an image. The first source is quantum noise, determined by the x-ray flux or the number of x-ray photons that are detected. This source is influenced by two main factors: the scanning technique (x-ray tube voltage, tube current, slice thickness, scan speed, helical pitch, etc.), and the scanner efficiency [detector quantum detection efficiency (QDE), detector geometric detection efficiency (GDE), umbra-penumbra ratio, etc.). The scanning technique determines the number of x-ray photons that reach the patient, and the scanner efficiency determines the percentage of the x-ray photons exiting the patient that convert to useful signals. For CT operators, the choices are limited to the scanning protocols. To reduce noise in an image, one can increase the x-ray tube current, x-ray tube voltage, or slice thickness, or reduce the scan speed or helical pitch. The operator must understand the tradeoffs for each choice. For example, although an increased tube voltage helps to reduce noise under an equivalent kW (the product of the tube current and the tube voltage) condition, LCD is generally reduced. Similarly, increased slice thickness may result in a degraded 3D image quality and increased partial-volume effect. A slower scan speed could lead to increased patient motion artifacts and reduced organ coverage. An increased tube current leads to increased patient dose and increased tube loading. As long as the tradeoffs are well understood, these options can be used effectively to combat noise.

The second source that contributes to the noise in an image is the inherent physical limitations of the system. These include the electronic noise in the detector photodiode, electronic noise in the data acquisition system, x-ray translucency of the scanned object, scattered radiation, and many other factors. For CT operators, options to reduce noise in this category are limited.

The third noise-contributing factor is the image generation process. This process can be further divided into different areas: reconstruction algorithms, reconstruction parameters, and effectiveness of calibration. The impact of different classes of reconstruction algorithms on noise was discussed in Chapter

3 (e.g., see Fig. 3.48) and will not be repeated here. For the same reconstruction algorithm, such as FBP, the reconstruction parameters include the selection of different reconstruction filter kernels, reconstruction FOV, image matrix size, and postprocessing techniques. In general, a high-resolution reconstruction kernel produces an increased noise level. This is mainly because these kernels preserve or enhance high-frequency contents in the projection. Unfortunately, most noise presents itself as high-frequency signals. An example of the filter kernel selection on noise is shown in Fig. 5.14.

A few words should be said about postprocessing techniques. Many image-filtering techniques have been developed over the past few decades for noise suppression. To be effective, these techniques need to not only reduce noise and preserve fine structures in the original image, but also to maintain "natural-looking" noise texture. Often, techniques are rejected by radiologists because the filtered images are too "artificial." In recent years, many advances have been made in this area; a more detailed discussion can be found in Chapter 11 (e.g., Fig. 11.24).

As was discussed in Chapter 3, the calibration or preprocessing techniques used in CT to condition the collected data are not perfect. The residual error often manifests itself as artifacts of small magnitude. These artifacts sometimes cannot be visually detected. However, they do influence the standard deviation measurement and consequently should be considered as part of the noise source.

Although standard deviation is the most straightforward way of measuring noise in an image and correlates fairly well with visual observation, it also has many limitations. As an illustration, Fig. 5.30 shows three images with different types of noise. Figure 5.30(a) shows the noise image of a reconstructed water phantom with the standard kernel. In a similar fashion, the noise image with the bone kernel was scaled, as shown in Fig. 5.30(b). The scaling factor was selected so that Figs. 5.30(a) and (b) have an identical standard deviation. Figure 5.30(c) was not produced with a CT reconstruction. Instead, computer-simulated noise with a normal distribution was added to a flat image. Although all three images are identical as measured by the standard deviation, the noise textures are significantly different. Figure 5.30(a) appears to be grainy, which indicates a lack of high-frequency contents. Figure 5.30(c) has the finest grain, and Fig. 5.30(b) falls in between.

To capture the noise texture difference, the noise power spectrum (NPS) is often used.[16-20] NPS can be defined by the following equation:

$$NPS(u, v) = \frac{1}{A} \left\langle \left| \iint_A n(x, y) e^{-2\pi i(xu + yv)} dx dy \right|^2 \right\rangle, \qquad (5.8)$$

where $n(x, y)$ is the reconstructed image containing only noise, u and v are frequency variables, A represents the area over which $n(x, y)$ is defined, and the brackets indicate an ensemble average to be performed over all such images.

Although the generation of the noise image $n(x, y)$ may seem to be straightforward with the reconstruction of a uniform phantom, caution must be taken in the practical implementations. Since the data acquisition and calibration processes are hardly ideal, it is inevitable that errors and artifacts are present in the reconstructed images. These artifacts often appear as low-frequency fluctuations superimposed on otherwise uniform images, and they produce bias in the low-frequency portion of the NPS.

The simplest way to get rid of bias is to scan the same uniform phantom twice with identical protocols. Since both acquisitions are acquired sequentially and close in time, variations due to drifts in the CT system components are negligible. Identical protocols, phantom, and scanning location ensure that image artifacts due to nonideal data acquisition and calibration are identical between two sets of images. The two sets of reconstructed images are then subtracted to produce the final noise image, $n(x, y)$. Because NPS is obtained by squaring the magnitude of the Fourier transform of the noise image, NPS itself is likely to be noisy. The ensemble average, as indicated by the brackets in Eq. (5.8), is used to reduce fluctuations in the final NPS.

The NPS for the three images shown in Fig. 5.30 are depicted in Fig. 5.31. The high-frequency components in the NPS of the standard kernel [Fig. 5.31(a)] are significantly suppressed as a result of the low cutoff frequency of the standard kernel itself. The NPS of the normal distributed noise [Fig. 5.31(c)] is quite flat. This is not surprising since the original image Fig. 5.30(c) was produced by adding computer-generated random numbers to a uniform image with zero mean. The NSP frequency response of the bone kernel [Fig. 5.31(b)] falls between the other two.

The NPS has a profound impact on the LCD. Figures 5.32(a)–(c) show the results when low-contrast discs of various sizes are added to the noise images of Figs. 5.30(a)–(c). Despite the fact that all three images have identical standard deviation, the visibility of the disks is significantly different, with the standard kernel image being the worst and the simulated noise image being the best. The criteria used for the visibility test consist of not only the ability to identify the

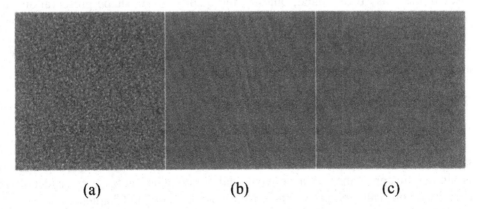

(a)   (b)   (c)

Figure 5.30 Different types of noises with identical standard deviation ($\sigma = 6.8$). (a) Noise in the reconstructed water phantom with the standard algorithm. (b) Noise in the reconstructed water phantom with the bone algorithm. (c) Additive noise with normal distribution.

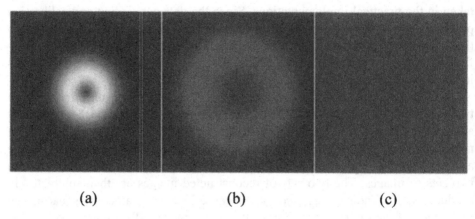

(a)   (b)   (c)

Figure 5.31 NPS for the three types of noise shown in Figs. 5.30(a)–(c).

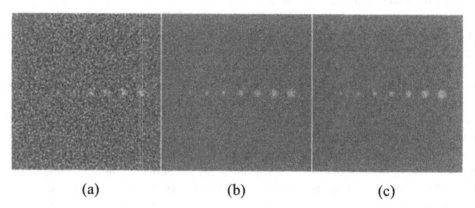

(a)   (b)   (c)

Figure 5.32 Impact of noise spectrum on LCD. Low-contrast discs of different sizes were added to the noise images of Fig. 5.30.

discs in the noisy background, but also the quality of the shape preservation of the original LCD objects. From this illustration, it is clear that standard deviation alone does not provide a complete picture of the noise performance of a system. The impact of the NPS on the detectability of low-contrast objects is complex, and many studies have been devoted to this subject.[16–20]

5.5 Performance of the Scanogram

A scanogram is also known as a scout image or projection radiograph. During a scanogram image acquisition, both the x-ray tube and the detector remain stationary while the patient table travels at a constant speed. The resulting image is similar in appearance to a conventional x-ray radiograph. A more detailed description on scanogram image acquisition can be found in Chapter 6.

Historically, scanogram images were used primarily to localize anatomical regions for CT scan prescription.[21] A scanogram provides anatomical landmarks that enable CT scans to be performed at a desired location. Figure 5.33 shows two examples of typical scanogram images. Figure 5.33(a) was acquired with the x-ray tube in the 12 o'clock position. This type of scan is called an anterior posterior (A-P) scanogram, referring to the fact that x-ray beams travel in the A-P direction. Figure 5.33(b) was taken with the x-ray tube in the 3 o'clock position and is called a lateral scanogram. An operator prescribes the locations of CT scans on the acquired scanogram images, as marked by the horizontal lines.

The most important performance criterion for a scanogram is clear organ definition. Edge-enhancement algorithms are employed to enable better visualization of the internal organs. In the past, little attention has been paid to the specifications for performance parameters with this type of image. Although these images perform satisfactorily in providing anatomical details, they often contain false contours around the boundaries of any high-density object due to the edge-enhancement algorithms, as shown in Fig. 5.34(a).

Clinical studies show that scanogram images can provide additional pathological information that can be complementary to CT.[22] The combined use of CT and scanograms not only reduces the x-ray dose to patients, but also reduces the overhead (time, effort, and resources) needed to transport a patient between different imaging devices. The desire to reduce dose to patient and save time led to the investigation of special scout image acquisition and generation processes.[22] Figure 5.34(b) shows an example of a scanogram image in which edge enhancement is provided without the unwanted artifacts. New clinical applications are likely to demand better scanogram image quality and lead to more detailed specifications on scanogram performance in terms of spatial resolution, contrast visibility, dose, and artifacts.

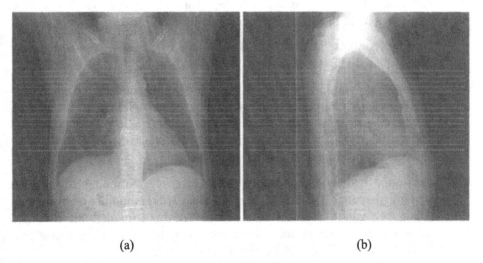

(a)   (b)

Figure 5.33 Examples of scanogram images. (a) A-P scanogram and (b) lateral scanogram.

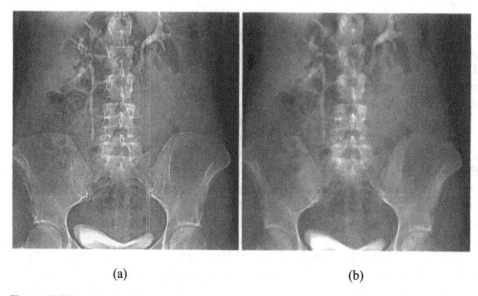

(a)   (b)

Figure 5.34 Images produced with different scout generation processes. (a) Traditional and (b) improved scanogram for uro-radiology application.

5.6 Problems

5-1   When the resolution insert of the CATPHAN® phantom (Fig. 5.1) is positioned at the iso-center, which spatial resolution does it measure: radial or azimuthal? Can it be used to measure both resolutions?

5-2   A engineer who wants to measure the radial and azimuthal resolutions of a CT scanner at the edge of a 30-cm FOV positions a wire phantom 15 cm from the iso-center and 45 deg from the x axis. How does the engineer obtain the radial MTF and azimuthal MTF?

5-3   Can bar patterns, instead of a wire phantom, be used to quantitatively measure the MTF?

5-4   To support high spatial resolution, the reconstructed image pixel size must be sufficiently small to satisfy the Nyquist criteria. This can be accomplished by either reducing the reconstruction FOV (targeted reconstruction) or increasing the image matrix size (e.g., to 1024). Discuss the pros and cons of each approach.

5-5   Section 5.1 demonstrated the impact of view sampling density on azimuthal spatial resolution—that is, when a large number of projection views per gantry rotation are used to reduce azimuthal blurring and improve spatial resolution. What are the practical considerations that limit the constant use of the highest view-sampling rate?

5-6   Develop an analytical expression to perform a first-order estimation of the impact of azimuthal blurring on a CT system. For simplicity, perform the analysis at the edge of a 30-cm FOV in a CT system that samples

1000 views per gantry rotation with a continuous x-ray-on tube. Assume that the PSF of the system with a pulsed x-ray source is $p(x)$, where x is in the azimuthal direction (azimuthal blurring of the pulsed x-ray CT system can be ignored).

5-7   As discussed in Section 5.1, the impact of azimuthal blurring decreases with the distance to the iso-center. A student argues that if one can always place the ROI portion of an object at the iso-center (assuming the ROI is small), the view sampling rate can be extremely low (i.e., a few projections per gantry rotation). Do you agree?

5-8   Using a shallow-angled ramp is a convenient way to measure SSP, as discussed in Section 5.1. If the reconstructed image pixel size is 0.5 mm, what is the maximum ramp angle that ensures accurate estimation of a system with expected FWHM of 0.6 mm, assuming a Gaussian-shaped SSP?

5-9   Standard deviation is still the predominate method of noise measurement. Because it measures CT number variations inside an ROI, the result contains not only noise, but also system bias (e.g., CT number nonuniformity or artifacts due to nonideal calibration). What methods can be used to separate the noise versus other system-related issues, assuming that only the images from a single scan of a uniform phantom are available and the "difference image" method discussed previously cannot be used?

5-10   One method to determine the LCD specification of a CT system is to visually inspect the reconstructed images of an LCD phantom and identify the smallest object with the lowest contrast that is visible. What issues may result from the fact that the observer has full *a priori* knowledge of the size and location of the low-contrast objects? Can you design an experiment to minimize this impact?

5-11   In the statistical LCD method presented in Section 5.2, the image of a uniform phantom is subdivided into small ROIs that match the sizes of the low-contrast objects. How do you calculate statistical LCD when the area of the low-contrast objects is not an integer multiple of the pixel area?

5-12   For the statistical LCD method outlined in Fig. 5.18, the contrast of the object that can be detected with a 95% confidence interval equals $\xi = 3.29 \times$ the standard deviation of the measured mean. Calculate the value of ξ at a 90% confidence interval. Does the result change if the LCD objects have either higher or lower intensities than the background?

5-13   If the ROIs of a uniform phantom image used with the statistical method are slightly larger than the actual low-contrast object size, what is the impact on LCD measurement? How do you adjust the result so it is still accurate?

5-14   The manufacturing process of any LCD phantom contains variations; that is, the contrast level of the low-contrast objects may change from phantom to phantom. How do you minimize the impact of such variability on the LCD measurement?

5-15   In Fig. 5.18, the ROIs are shaped as $N \times N$ pixel squares. What are the pros and cons if we use a rectangular ROI? Can one use $1 \times N$ ROIs?

5-16   Figure 5.24 illustrates the temporal resolution variation as a function of location. One student proposes to always place the patient's heart around the "point 2" location to maximize the temporal resolution. What are the practical limiting factors of such an approach?

5-17   Figure 5.31 shows the NPS of different noise images. Based on these images, one student argues that the NPS is circularly symmetric, and only a 1D profile along the radius is needed to describe the noise property. Do you agree with this argument?

5-18   As illustrated in Fig. 5.32, visibility of the LCD objects with the bone kernel [Fig. 5.32(b)] is better than that of the standard kernel [Fig. 5.32(a)]. One student quickly draws the conclusion that we should always use a bone kernel or other kernels that pass high frequencies to reconstruct images with low-contrast objects. Is this conclusion valid?

References

1. P. F. Judy, "Evaluating computed tomography image quality," in *Medical CT and Ultrasound: Current Technology and Applications*, L. W. Goldman and J. B. Fowlkes, Eds., Advanced Medical Publishing, Madison, WI, 359–377 (1995).

2. C. H. McCollough and F. E. Zink, "Performance evaluation of CT systems," in *Categorical Courses in Diagnostic Radiology Physics: CT and US Cross-Sectional Imaging*, L. W. Goldman and J. B. Fowlkes, Eds., Radiological Society of North America, Inc., Oak Brook, IL, 189–207 (2000).

3. E. C. McCoullough, "Specifying and evaluating the performance of computed tomography (CT) scanners," *Med. Phys.* **7**, 291–296 (1980).

4. American Association of Physicists in Medicine (AAPM), "Specification and acceptance testing of computed tomographic scanners," Report No. 39, AAPM, New York (1993).

5. L. D. Loo, "CT acceptance testing. Specification, acceptance testing," in *Quality Control of Diagnostic X-Ray Imaging Equipment*, Medical Physics Monograph #20, J. A. Seibert, G. T. Barnes, and R. G. Gould, Eds., American Association of Physicists in Medicine, New York (1994).

6. L. N. Rothenberg, "CT dose assessment. Specification, acceptance testing," in *Quality Control of Diagnostic X-Ray Imaging Equipment*, Medical Physics

Monograph #20, J. A. Seibert, G. T. Barnes, and R. G. Gould, Eds., American Association of Physicists in Medicine, New York (1994).

7. National Council on Radiation Protection and Measurement (NCRP), "Quality assurance for diagnostic imaging equipment," NCRP Report No. 9, NCRP, Bethesda, MD (1988).

8. P. M. Joseph and C. D. Stockham, "The influence of modulation transfer function shape on computed tomographic image quality," *Radiology* **145**, 179–185 (1982).

9. J. Hsieh, "Investigation of the slice sensitivity profile for step-and-shoot mode multi-slice CT," *Med. Phys.* **28**(4), 491–500 (2001).

10. D. J. Goodenough, K. E. Weaver, and D. O. Davis, "Development of a phantom for evaluation and assurance of image quality in CT scanning," *Opt. Eng.* **16**, 52–65 (1977).

11. D. J. Goodenough, "Tomographic imaging," in *Handbook of Medical Imaging, Vol. 1, Physics and Psychophysics*, J. Beutel, H. L. Kundel, and R. L. Van Metter, Eds., SPIE Press, Bellingham, WA, 511–554 (2001).

12. E. H. Chao, T. L. Toth, N. B. Bromberg, E. C. Williams, S. H. Fox, and D. A. Carleton, "A statistical method of defining low contrast detectability," Radiological Society of North America, Inc., Annual Meeting, Oak Brook, IL (2000).

13. J. Hsieh and T. Toth, "Low-contrast detectability for x-ray computed tomography," *Med. Phys.* **35**, 2645 (2008).

14. J. Hsieh, "Analysis of the temporal response of computed tomography fluoroscopy," *Med. Phys.* **24**(5), 665–675 (1997).

15. D. L. Parker, "Optimal short scan convolution reconstruction for fan-beam CT," *Med. Phys.* **9**, 254–257 (1982).

16. K. M. Hanson, "Detectability in computed tomographic images," *Med. Phys.* **6**(5), 441–451 (1979).

17. R. F. Wagner, D. G. Brown, and M. S. Pastel, "Application of information theory to the assessment of computed tomography," *Med. Phys.* **6**(2), 83–94 (1979).

18. M. F. Kijewski and P. F. Judy, "Noise power spectrum of CT images," *Phys. Med. Biol.* **32**(5), 565–575 (1987).

19. J. H. Siewerdsen, L. E. Antonuk, Y. El-Mohri, J. Yorkston, W. Huang, and I. A. Cunningham, "Signal, noise power spectrum, and detective quantum efficiency of indirect-detection flat-panel imagers for diagnostic radiology," *Med. Phys.* **25**(5), 614–628 (1998).

20. K. L. Boedeker, V. N. Cooper, and M. F. McNitt-Gray, "Application of noise power spectrum in modern diagnostic MDCT: Part I. Measurement of noise power spectrum and noise equivalent quanta," *Phys. Med. Biol.* **52**, 4027–4046 (2007).

21. L. L. Berland, *Practical CT: Technology and Techniques*, Raven Press, New York (1986).

22. C. H. McCollough, J. Hsieh, T. J. Vrtiska, B. F. King, A. J. LeRoy, and S. H. Fox, "Development of an enhanced CT digital projection radiograph for uroradiologic imaging," *Radiology* **218**, 609 (2000).

days of CT, pulsed x-ray tubes were generally used.[1] In the pulsed mode, x-ray tubes produced x-ray photons in short-duration pulses. The pulse time varied between 1 and 4 ms. The nonoperating period was typically 12 to 15 ms during which no x-ray photons were emitted because x-ray detectors could not take measurements while the signals were sampled. Some benefits of the pulsed x-ray tube include the elimination of a large number of signal integrators, the ability to reset the electronics between pulses, the ability to adjust pulse length (and therefore photon flux) based on patient size, and the potential reduction of azimuthal blurring.

With the advances in electronics and tube technology, the advantages of the pulsed x-ray tube have diminished (although its use in micro-CT continues). In fact, for high-speed CT scanning, pulsed x-ray tubes are disadvantageous since the duty cycle of the tube is less than 100%. The duty cycle is defined as the fraction of the x-ray emitting time over the total operating time. For high-speed scanning, it is necessary to have x-ray tubes constantly producing x-rays to provide sufficient x-ray flux.

Although the size and appearance of the x-ray tube have changed significantly since its invention by Roentgen in 1895, the fundamental principles of x-ray generation have not changed.[2] The basic components of an x-ray tube are a cathode and an anode. The cathode supplies electrons and the anode provides the target. As discussed in Chapter 2, x-ray photons are produced when a target is bombarded with high-speed electrons. The intensity of the produced x rays is proportional to the atomic number of the target material and to the number of electrons bombarding the target. The energy of the generated x-ray photons depends on the electric potential difference between the cathode and the anode.

Most of the x-ray tubes used in CT scanners today employ the heated cathode design, which dates back to 1913 when Coolidge built the first high-voltage tube.[2] Figure 6.2 is a photograph of a glass envelope x-ray tube. The glass frame provides the housing and support to the anode and cathode assemblies and sustains a vacuum of 5×10^{-7} Torr (the vacuum level in the early operating life tends to be in the range of 10^{-6} to 10^{-7} Torr, and the internal pressure decreases over the tube life). The glass frame is a composite of several types of glass. The main section is a borosilicate glass with good thermal and electrical insulation properties. The thickness of the glass is typically between 0.18 and 0.30 mm. The glass seals at both ends of the tube are made with splice rings of various grades of glass to match the thermal expansion coefficients between the metal and the borosilicate glass. In more advanced tube designs, the glass frames are replaced with metal frames, as depicted in Fig. 6.3. The metal frame has the advantage of being able to operate at or near ground potential to improve the efficiency of the motor that drives the anode assembly. Another advantage of the metal frame is the reduced spacing between the frame and anode, which accommodates a larger-sized anode without significantly increasing the size of the tube envelope. The metal case can also collect off-target backscattered electrons.

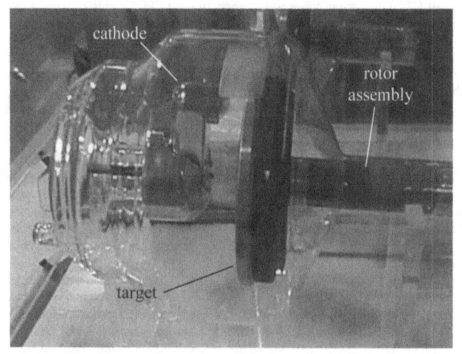

Figure 6.2 Photograph of an early vintage glass envelope x-ray tube.

Figure 6.3 Photograph of a metal frame tube with a cutout to show the anode and cathode assemblies.

Since the intensity of x-ray radiation is proportional to the number of impacting electrons, it is critical to precisely control the tube current. This is accomplished by operating the cathode filament in the thermionic emission mode. In this mode, the tube current becomes strongly dependent on the cathode temperature and independent of the voltage between the cathode and the anode. The process of thermionic emission can be described by the Richardson-Dushman equation:

$$J = AT^2 e^{-W/kT} , \qquad (6.1)$$

where J is the emitted electron current density in amps/cm^2, T is the filament temperature in degrees Kelvin (K), W is the work function or energy required to remove an electron from the cathode (4.5 eV for tungsten), A is a constant and equals 120 amps/(cm K)^2, and k is the Boltzmann's constant and equals 1.38 × 10^{-23} joules/K. A cup in the cathode assembly focuses the electrons emitted from the filament to a small region on the target to ensure a well-defined focal spot size and shape (the focal spot size also depends on the filament length). For many x-ray tubes used in CT scanners today, two different focal spot sizes are provided to handle high-resolution (small spot size) and high-flux (large spot size) applications. This is typically accomplished by placing two different sized filaments either in parallel or in series in the cathode assembly.

Chapter 2 explained that the production of x rays by electron bombardment of a target is highly inefficient. Less than 1% of the input energy to an x-ray tube is converted to x-ray photons; over 99% of the energy becomes heat. The temperature at the impact point can reach between 2600 and 2700°C (the melting temperature of tungsten is 3300°C). To prevent the target from melting, the anode is rotated at a very high speed, typically between 8,000 and 10,000 rpm. This allows the cooler part of the focal track to be brought under the electron beam and the heat to be distributed over a larger area on the target, as shown in Fig. 6.4. To illustrate the impact of heat loading on the anode, Fig. 6.5 shows a schematic diagram of a target's cross-section and the temperature distribution of an older vintage x-ray tube. A high-technique protocol (i.e., the protocol uses a large amount of x-ray flux) was used to enable the temperatures to reach an equilibrium state after 30 min of operation. The location directly underneath the electron bombardment exhibits the highest temperature; the focal track temperature falls between the focal spot and bulk temperatures, due to the fast rotation of the target. Since it is impractical to pass a rotating shaft through a high-vacuum seal, all of the rotating parts of the tube are located and sealed inside the vacuum tube envelope. The rotating parts consist of the anode target connected to a molybdenum shaft that is surrounded by a rotor. A bearing is also located inside the vacuum so that the anode, shaft, and rotor can rotate freely inside the tube envelope. A stator is placed outside the envelope to provide an alternating magnetic field that causes the rotator assembly to rotate.

To further increase the impact area, the focal track is at a shallow angle α with respect to the scanning plane of the CT (see Fig. 6.4). α is often called the target angle. The projected focal spot length h is related to the actual focal spot length L by the following equation:

$$h = L\sin(\alpha).$$ (6.2)

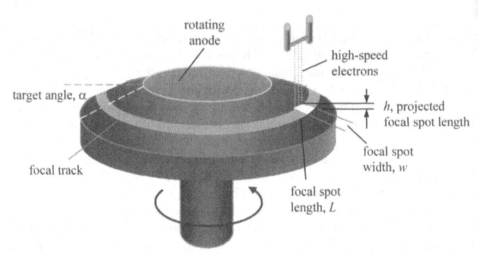

Figure 6.4 Illustration of a target assembly. The target rotates at a very high speed so that the heat generated by the electron bombardment is distributed over a large area (light gray band on the target). In addition, the target surface is at a shallow angle ($\alpha = 7$ deg) with respect to the CT scan plane to increase the exposure area while maintaining a small projected focal spot length.

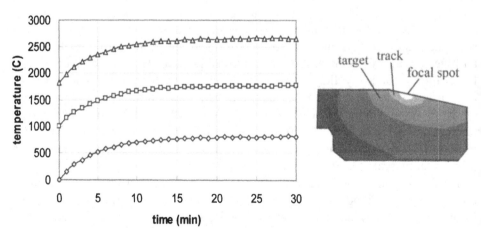

Figure 6.5 Illustration of a target's cross section and anode temperatures of an older vintage tube. Triangles represent focal spot temperature, squares represent focal track temperature, and diamonds represent target bulk temperature.

In a typical x-ray tube design, α is selected roughly at 7 deg. As a result, the actual focal spot length can be more than a factor of 8 larger than the projected focal spot. This is often called the line focus principle. Although this approach has the advantage of increased exposure area, it has two minor problems. The first problem is that the focal spot size and shape become location dependent. Although the focal length described by Eq. (6.2) is valid when viewed at a location perpendicular to the focal line, a significant error is present when the viewing location is farther away. The location-dependent focal spot and shape is one factor that contributes to the spatially variant resolution in CT. This phenomenon will be described in detail in Chapter 8. A second problem is the heel effect, which describes a phenomenon in which the x-ray intensity is not constant along the direction perpendicular to the CT gantry plane. This effect can be explained by the fact that on average, x rays are emitted a certain depth from the anode surface, as shown in Fig. 6.6. Because of the shallow angle of the target surface, the x-ray path length through the tungsten changes significantly. The tungsten target itself serves as an x-ray filter, and a longer path length through tungsten reduces the x-ray intensity, as illustrated in the figure. In general, the x-ray intensity declines as we move from the cathode side of the tube to the anode side. For CT scanners with a small z coverage, this effect can be safely ignored because the coverage along the patient long axis (perpendicular to the CT gantry plane) is relatively smaller. As the volume coverage increases, however, this issue can no longer be ignored and must be properly addressed.

Figure 6.7 shows normalized x-ray output intensities measured by a flat-panel detector on an x-ray radiographic system. As we move toward the anode side, the x-ray intensity drops. Because of the tube target geometry, the intensity drop is nonlinear. For this particular system, the x-ray flux peak-to-valley ratio is slightly over 2. That is, near the edge of the FOV, the x-ray flux is less than half that of the central region. The impact of the heel effect is not limited to the

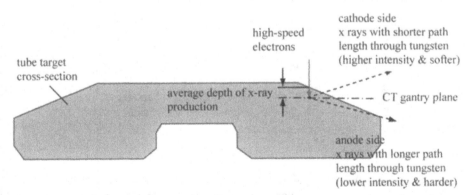

Figure 6.6 Illustration of the heel effect. X-ray photons generated inside the tungsten target travel different path lengths before exiting the anode. The filtration of the tungsten produces variation in the x-ray intensities and spectrums. This effect is negligible for commercial CT scanners with smaller coverage but cannot be ignored when the coverage is large.

intensity drop across the FOV in z; the x-ray spectrum also changes with respect to the imaging location. In general, the x-ray spectrum becomes harder (higher average energy) for x-ray beams toward the anode side of the tube and softer (lower average energy) toward the cathode side. The change in x-ray spectral intensity across the z coverage not only brings complexities to the calibration process, but also introduces variation in the LCD across the z FOV.

One method of reducing the heel effect is to increase the tube target angle. As the target angle increases, the difference between the shortest and the longest path length in Fig. 6.6 is reduced. Since the major cause of the heel effect is differential path length, the overall heel effect should be reduced. On the other hand, a reduced target angle decreases the effectiveness of the line focus principle.

Because of the enormous heat deposited on the target, special attention must be paid to target design. Traditionally, the tube target is made of a molybdenum alloy and the focal track consists of a layer of tungsten-rhenium. The advantage of this design is the quick heat transfer from the focal track to the bulk of the target. With the introduction of helical/spiral CT and faster scan speeds, this design no longer offers the required heat capacity. The newer design uses brazed graphite in which the metal target (similar to the traditional design) is brazed to a graphite body to increase the heat storage per unit weight. This design combines the good focal track characteristics of the metal tube with the heat storage capability of graphite.[2] An alternative design uses chemical vapor deposition to deposit a thin layer of tungsten-rhenium on a high-purity graphite target. The advantages of this approach are the light target weight and high heat storage capacity. The disadvantages are higher cost, increased potential for particles, and limited focal spot loading.

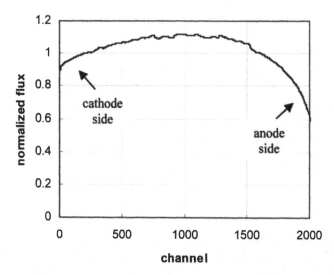

Figure 6.7 Example of the heel effect on an x-ray radiographic tube. The normalized x-ray flux is plotted as a function of the detector channel (from the cathode side to the anode side) of a flat-panel system. The peak-to-valley flux ratio is slightly over 2.

Since one of the most important issues that impacts x-ray tube performance is heat management, x-ray tubes are often rated in terms of their heat capacity. Tube heat management specification is described in heat units (HU):

$$1 \text{ heat unit} = 0.74 \text{ J} \qquad (6.3)$$

For example, a 30-sec clinical protocol operating at 120 kVp and 300 mA deposits a total of 1080 kilojoules of energy (30 × 120 × 300). This corresponds to 1459 KHU. Typical tube specifications include anode heat storage capacity (in MHU), maximum anode cooling rate (in MHU/min), casing heat capacity (MHU), and average casing cooling rate (in KHU/min).

Because heat management is an important and complicated issue, many commercial CT scanners employ computer algorithms (often called tube-cooling algorithms) to estimate the different conditions under which clinical protocols can be used to prevent premature damage to the tube. For example, a protocol may be safely run when the x-ray tube is cold, but the same protocol could potentially cause damage to the tube when it is hot. Under this condition, the algorithm will recommend either a reduced technique (reduced tube current or scan time) or a cooling period before starting a scan. In essence, the tube-cooling algorithms derive a compromised solution between performance and tube life.

With the introduction of multislice CT (see Chapter 10), tube utilization has become highly efficient. Since most CT protocols can be completed in a few seconds, the scanners are capable of executing nearly all clinical protocols without encountering any tube cooling. Therefore, tube heat transfer and the handling of tube cooling have become less important. A more significant performance parameter now relates to the maximum x-ray tube power that can be delivered in a short period of time. As will be discussed in Chapter 12, one of today's most demanding clinical protocols is coronary artery imaging of the heart. To avoid heart motion, scan speeds of less than 0.4 sec are typically used in conjunction with a half-scan (about 60% of the full-scan acquisition time) and thin-slice acquisition. To deliver sufficient x-ray flux to the study and ensure good signal-to-noise ratio, the maximum tube power must be substantially increased. Note that for a given kVp setting, the total x-ray flux received at a detector cell is proportional to the product of the tube current (mA), the detector aperture (mm), and the acquisition time (sec). When the detector aperture and acquisition time are both substantially reduced, a significantly higher tube current is required to maintain the same flux and the x-ray tube target must be able to handle a huge heat load in a short period of time. This is a significantly different requirement from the older vintage scanners in which a moderate amount of heat had to be controlled over an extended period of time.

Other parameters are also important to x-ray tube performance. A good example is the focal spot width and focal spot length shown in Fig. 6.4. These dimensions can significantly impact the in-plane spatial resolution as well as the slice thickness (this topic will be discussed in more detail in Chapter 8). In recent years, more attention has been paid to better controlling tradeoffs between focal spot size and tube power. The majority of existing x-ray tubes employ two focal spot sizes: one for the low tube powers up to a specified limit and the other for the higher tube powers. This design is based on the fact that a larger focal spot allows heat to be distributed over a larger area, which allows for increased tube loading (an increase in x-ray tube current or voltage). The two-focal-spot configuration is accomplished by using a pair of filaments of different lengths in the cathodes, as shown in Fig. 6.8(a). When prescribing a protocol in a clinical practice, the CT operator must consider that a higher technique (a higher tube current) reduces image noise but at the possible expense of spatial resolution. In the new tube design shown in Fig. 6.8(b), the size of the x-ray focal spot can be controlled by applying different electrical fields near the filament to focus the high-speed electrons. In theory, the size of the focal spot can be adjusted dynamically and continuously based on the scanning technique instead of a binary decision. This capability also allows the focal spot position to be tuned during the scan to dynamically compensate for the spot drifts that result from thermal expansion. As will be discussed in Chapter 7, this capability can also be used for focal spot deflection to combat aliasing artifacts.

Other criteria besides image quality considerations are also important in determining tube performance. One such example is the expected tube life. Today, a tube that can scan beyond a half-million slices is not that uncommon. Increased tube life translates to a reduced operational cost and reduced scanner downtime. A different yet related performance parameter is the time duration associated with a tube change. A shorter tube replacement process means less interruption to the clinical operation.

The other important component in the x-ray generation system is the high-voltage generator. To produce and maintain the desired x-ray flux output, the voltage and current to the x-ray tube must be kept at a constant or desired level. Unfortunately, the voltage of the power supply provided by the power line fluctuates sinusoidally from negative to positive relative to the ground. This is undesirable for two reasons. First, as previously discussed, if the cathode voltage becomes positive relative to the anode, the electrons emitted from the hot anode

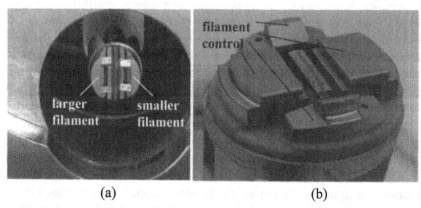

(a)   (b)

Figure 6.8 Examples of focal spot size selection. (a) Older vintage tube cathode with two fixed-size focal spots. (b) Newer tube cathode design with dynamic focal spot control.

will accelerate toward the cathode, causing premature aging of the cathode filament and generating undesired x-ray photons (located significantly off the x-ray focal spot). Second, considerable fluctuation of the voltage causes significant difficulties in the calibration and data conditioning of CT systems. The energy spectrum of the x-ray flux is closely linked to the voltage potential across the cathode and anode, so when a significant variation exists in the power supply voltage, the x-ray spectrum changes constantly from view to view. This makes some of the correction processes, such as beam hardening, nearly impossible. A detailed discussion of beam hardening and other corrections can be found in Chapter 7. To ensure adequate image quality and prevent premature x-ray tube failure, the power supply to the x-ray tube must be rectified and additional efforts made to ensure the voltage is as close to constant as possible (to simulate a dc power supply). The voltage generator from a single-phase rectifier typically has a ripple of 100%. Ripple is defined as the peak-to-valley voltage over the peak voltage (Fig. 6.9). In the early days of CT scanners, three single-phase waveforms were shifted by $2\pi/3$ with respect to each other to reduce the voltage modulation. With a three-phase generator, the ripple was reduced to 13.4% (6 pulses) or 3.4% (12 pulses).[3] This technology was later extended to the constant potential generator (CPG), which uses tetrode tanks and feedback circuits to achieve a waveform with nearly 0% ripple (nearly dc). The x-ray spectrum generated by this technology is slightly higher in average energy. Because x-ray production is more efficient at higher energies, the x-ray tube output (measured by photon fluence or mR per mAs) improves. Another technology introduced in the late 1980s uses digitally controlled high-frequency inverter generators to convert a sinusoidal power supply to dc, chop the dc voltage to a higher frequency with a digital oscillator, input the higher frequency to a high-voltage transformer, and finally to rectify and smooth the power supply for the x-ray tube. Although the ripple produced by such a power supply is about 5%, it has a lower cost and reduced complexity due to the use of digital technology. Interested readers can refer to Ref. 3 for additional information.

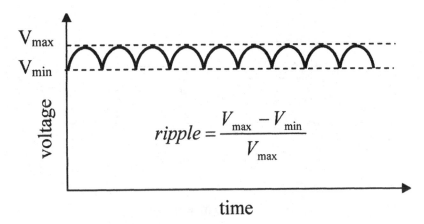

$$ripple = \frac{V_{max} - V_{min}}{V_{max}}$$

Figure 6.9 Illustration of voltage fluctuation and ripple.

A hot topic in recent CT research is dual energy for beam-hardening correction, material separation, and tissue characterization. Although Chapter 12 will discuss this topic in detail, we will mention the impact of one dual-energy CT approach to high-voltage generator design: the fast kVp switching technology used to acquire two x-ray energies nearly simultaneously. During data acquisition, the input x-ray tube voltage is quickly switched between two different energy settings (e.g., 80 and 140 kVp) on a view-by-view basis. This requires the high-voltage generator to be able to provide a voltage waveform that has a short rise and fall time, as shown in Fig. 6.10. In general, the shorter the rise and fall time in the voltage waveform, the less contamination occurs between signals of adjacent views, and the more effective is the energy separation of the collected signals. Because the voltage is switched between views that last less than a millisecond, new technologies must be used in the voltage generator design.

6.3 The X-Ray Detector and Data-Acquisition Electronics

The x-ray detector is as important as the x-ray tube to the performance of a CT scanner. Like x-ray tube technology, detector technology has experienced tremendous growth over the past 30 years. Figure 6.11 shows a collection of CT detectors used on various third-generation scanners, arranged chronologically from top to bottom. The earlier vintage detectors, such as the GE 7800, had smaller FOVs used mainly to perform head scans. The later vintage detectors, such as the GE 9800 HiLight™ detector and GE LightSpeed™ detector, had a much larger FOV to handle entire bodies.

Detectors of the third-generation scanners use either a high-pressure inert gas (usually xenon) or solid state scintillators coupled with photodiodes. The operating principle of the xenon detector is illustrated in Fig. 6.12. The detector is constructed with many thin tungsten plates inside a high-pressure xenon chamber. Every other plate is connected to a high-voltage dc supply (in the GE 8800 detector, the plate is connected to 500 V). The remaining plates are

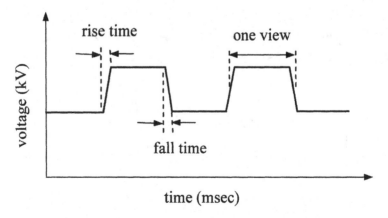

Figure 6.10 Illustration of voltage waveform for fast kV switching.

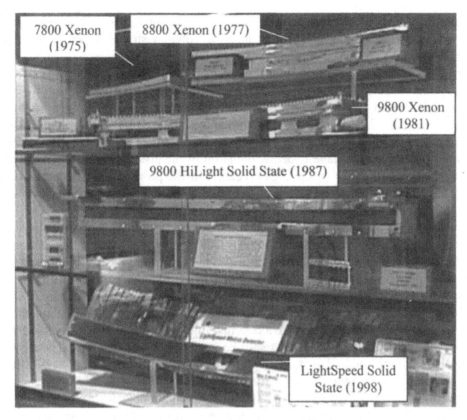

Figure 6.11 A collection of third-generation CT detectors.

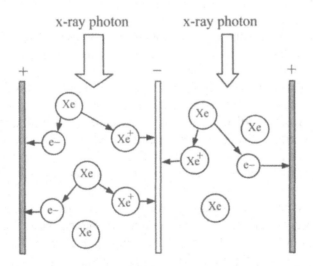

Figure 6.12 Ionization of xenon gas with x-ray photons. The ionization of high-pressure xenon gas cells produces positive xenon ions and negative electrons. The positive ions migrate to the relatively negative collection plates, and the electrons migrate to the positive collection plates. The ionization quantity of xenon gas is linearly proportional to the x-ray intensity.

electrically floating at approximately 0 V. A pair of the high-low voltage plates forms a single detector cell. When an x-ray photon strikes a cell, it causes ionization of the xenon gas in a photoelectric interaction. The interaction releases energetic photoelectrons that ionize more gas ions. The ionized heavy xenon nuclei are collected by the 0 V plates, and the free electrons are collected by the positively biased plates (500 V) to produce a current signal. The bias voltage across the plates is set high enough such that the ions are rapidly collected by the plates with little recombination of positive and negative ions in the gas. At the same time, the bias voltage is below the point of the avalanching effect where a nonlinear magnification of the signal occurs. In the properly biased region, the amount of ionization is linearly proportional to the total energy of the absorbed x-ray photons.

The major disadvantage of the xenon detector is its low quantum detection efficiency (QDE). (The definition of QDE will be presented later in this section.) Because of the relatively low density of xenon gas, some x-ray photons pass through the chamber without causing ionization. Although various efforts have been made over the years to improve the QDE by increasing the pressure of the xenon chamber and making the capture chamber longer, the QDE of the xenon gas detector still falls short compared to that of solid state materials. In addition, it is very difficult to build 2D detector chambers for multislice CT. The major advantage of the xenon detector is its low cost. Many of the low-end single-slice CT scanners manufactured today still use xenon detectors.

Solid state detectors overcome many of the xenon detector's shortcomings. Figure 6.13 shows a schematic diagram of a solid state detector made of small blocks of scintillating material, such as $CdWO_4$, Gd_2O_2S, HiLight[TM], or GEMS Stone[TM] coated with reflective material and coupled to a set of photodiodes. In a solid state detector, an incident x-ray photon undergoes a photoelectric interaction with the scintillator. The photoelectron released from the interaction travels a short distance in the scintillator and excites electrons in other atoms. When the excited electrons return to their ground states, characteristic radiation is emitted in the visible or UV light spectrum. The decay process is exponential with a time constant characteristic of the atom's electronic configuration. This time constant is often called the primary speed of the scintillator. Due to the presence of impurities, a small percentage of the excited electrons are trapped in the excited states for a longer period of time before returning to their ground states, which produces longer decay time constants known as afterglow.

The light photons produced by the scintillation process travel in all directions. The scintillator is coated with a highly reflective material to direct the light toward the photodiodes at the bottom of the detector. Because of the reflection and absorption in the scintillator, only a fraction of the light photons reaches the photodiodes to produce electrical signals.

In recent years, another type of detector, the semiconductor direct-conversion detector, has received much attention.[4,5] This detector is formed with an x-ray photoconductor, such as Se, CdTe, or CdZnTe, with a bias voltage applied across the detector structure. The x-ray photons directly generate electron-hole pairs in the photoconductor layer (Fig. 6.14). However, the attractiveness of such a detector is not its direct conversion of x-ray signals to electrical signals without the intermediate step of the light photons (both the gas detector and x-ray film have this capability). Instead, it is the potential performance of x-ray photon counting: as each x-ray photon interacts with the detector, an electrical signal is generated, allowing the x-ray photon energy to be determined. At the present time, however, these detectors do not have a sufficient count rate capability to handle the high x-ray flux that is required for routine clinical CT applications.

To evaluate the performance of a detector, several important parameters are useful. The QDE of a detector is defined as the incident x-ray photons that are attenuated by the detector.[3,6] Mathematically, the QDE of a detector is described by

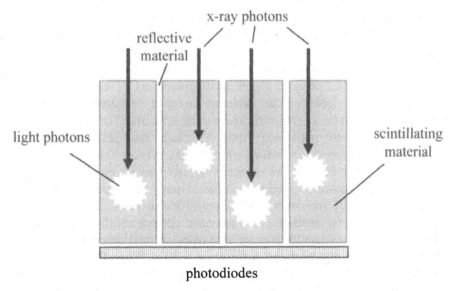

Figure 6.13 Schematic diagram of a solid state detector.

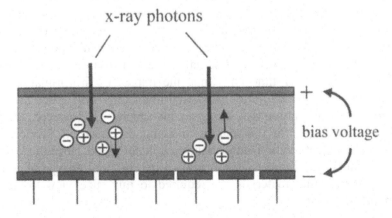

Figure 6.14 Schematic diagram of a semiconductor direct-conversion detector.

$$QDE = \frac{\int_0^{E_{max}} \Phi(E)\left(1 - e^{-\mu(E)\Delta}\right) dE}{\int_0^{E_{max}} \Phi(E) dE}, \tag{6.4}$$

where $\Phi(E)$ is the x-ray spectrum, $\mu(E)$ is the linear attenuation coefficient of the detector, and Δ is the detector thickness. Clearly, $0 \leq QDE \leq 1$. Equation (6.4) states that QDE increases with increased detector thickness Δ and increased attenuation coefficient μ. The detector thickness Δ is often limited due to other design and performance considerations. On the other hand, μ depends solely on the selected detector material. For typical high-pressure xenon detectors, the QDE is between 60% and 70%. This means that, on the average, 30% to 40% of the incident x-ray photons do not contribute to the detector output. In contrast, the QDE of HiLight[TM] detectors is between 98% and 99.5%.

Although QDE provides a good description of the detector's efficiency in absorbing x-ray photons that enter the detector, it does not provide any indication of the percentage of x-ray photons traveling toward the detector that actually enter the detector cells. For example, in third-generation CT scanners, collimator plates are placed in front of the detector to reject scattered radiation. The collimator plates must be a certain thickness to be effective. The typical thickness of a tungsten collimator plate is 0.2 mm. Although these plates stop scattered radiation, they also stop the primary x-ray photons that are intersected by these plates. Since roughly 20% of the detector surface area is covered by tungsten plates (detector cell-to-cell spacing is about 1 mm), the extremely high QDE value does not indicate the overall efficiency of the detector in collecting x-ray photons. Therefore, we need to use geometric detection efficiency (GDE) to describe the ratio of the useful detector area over the total detector area. For the example given above, the geometric efficiency is 80%. When the two factors are combined, we attain a combined efficiency that describes the actual effectiveness of the detector in converting incident x-ray energy into signals: (QDE)(GDE).

Other parameters are also important in describing detector performance. For example, the primary speed and afterglow of a detector describe the time constants in which the excited electrons return to their ground states in the scintillation process. In general, a slower detector primary speed and a larger magnitude of afterglow indicate a higher level of contamination from previous signals to the currently measured signal. Although it is desirable to produce a scintillator with shorter time constants, the impact of the primary speed and afterglow can be reduced or eliminated algorithmically.[7] A detailed discussion of the algorithmic correction will be given in Chapter 7. However, the fast kV switching technology for dual-energy imaging discussed in Section 6.2 puts much more stringent requirements on the scintillator speed, since the x-ray output energies of adjacent views are significantly different. For this type of application, advanced scintillating materials are required to minimize view-to-view signal contamination.

Another important parameter is the hysteresis or radiation damage of a detector. When a scintillator is exposed to x-ray flux, the detector gain changes. The amount of change can vary with exposure history, exposure intensity, recovery period, and many other factors. If no correction is rendered, image artifacts can result. For example, if several head scans are performed prior to a body scan, the gains of the center detector channels are impacted less by the x-ray exposure than by the outer channels, since the x-ray flux to the center channels are partially blocked by the patient's head. For the body scan that follows, the nonuniform channel-to-channel gain variation can lead to an imprint of a centered round object on the reconstructed body image. The best approach to combat this type of image artifact is to ensure that the radiation damage under the worst-case clinical condition is still within the specified artifact-free range. For example, the HiLight^{TM} scinitillator is doped with rare-earth elements to reduce the radiation damage of the detector to a negligible level. For the scintillating materials that cannot be made insensitive to radiation damage (e.g., $CdWO_4$), frequent calibrations or algorithmic corrections are often used to compensate for the ghost image.

Similar to the effect of radiation damage, the thermal stability of a detector can also affect image quality. For solid state detectors, the detector gain generally varies with the ambient temperature. During a long CT scan, the ambient temperature inside the gantry is likely to change. If the detector gain changes significantly and nonuniformly (differing from channel to channel), image artifacts are likely to result. To ensure an artifact-free image, most CT scanners employ temperature-control devices to maintain the ambient temperature within the narrow range required by the detector.

For detectors used in third-generation CT scanners, designers often specify the detector performance parameters (gain, afterglow, radiation damage, etc.) in terms of channel-to-channel uniformity in addition to absolute values. This is done because the relationship between the detector, x-ray source, and iso-center is fixed for third-generation geometry. The line that connects a detector channel to the x-ray source is always at a fixed distance from the iso-center, regardless of the projection angle. If the output of a particular channel consistently deviates from the true measurement, the backprojection step in the image reconstruction maps the erroneous detector signal to a ring concentric to the iso-center. Unfortunately, the human visual system is quite sensitive to the ring pattern; even a low-intensity ring can be easily identified from a complicated human anatomy background. Since ring artifacts are unlikely to result as long as all the detector cells behave in an identical manner, detector channel-to-channel uniformity is critical in performance specifications.

Another important performance parameter for a CT detector is spatial response uniformity. In a sense, this requirement can be derived directly from the channel-to-channel uniformity. If the detector response is not uniform along the z axis (across the slice thickness), the channel-to-channel uniformity cannot be easily maintained, especially when the scanned object is not uniform in z. A detailed analysis and examples can be found in Chapter 7.

So far, our discussion has been limited to the detector performance, but similar performance requirements can be placed on the data acquisition electronics associated with each detector channel. The measured signal is a combination of the detector's analog signal and the digital signal obtained from the data acquisition electronics, which represents the digitized version of the analog signal. Consequently, the quality of the output is also influenced by the quality of the data acquisition system (DAS). For example, the dark current of the DAS must be stable within the operating temperature range. If the dark current drifts significantly during the scan, bias is introduced into the digitized signal. This can lead to image artifacts if proper correction steps are not taken.

Another example of DAS impact is signal crosstalk. Crosstalk is the amount of signal that leaks from the current sample to subsequent samples (in time). Theoretically, one would like to have a zero-crosstalk DAS, indicating that the analog signal from the current sample would have no impact on the next samples. Unlike the photon counting (sampling one photon event at a time) performed on the nuclear medicine equipment, the sampled projection in CT represents the integrated x-ray photon flux over the sampling period. Therefore, the DAS functions as an integrator. At the end of each sampling period, the integrator must be reset so the residual signal from the current sample will not become part of the integration process of the next sample. In practice, however, a small portion of the signal will remain after the reset. This phenomenon is somewhat similar to the detector afterglow discussed previously. If the amount of crosstalk is large, it can potentially impact the spatial resolution or produce image artifacts. Therefore, design specifications are needed to ensure that the level of crosstalk is below an acceptable level. These specifications can be determined either experimentally or with computer simulation.

Another important requirement on the DAS is linearity. Linearity refers to the relationship between the DAS input and output signals. Theoretically, one would like the DAS output to increase or decrease linearly with the input signal—that is, the output of the DAS would be a faithful representation of the input signal without additional correction. DAS linearity specifications are generally derived based on theoretical analysis and phantom experiments. Of course, if the DAS output is nonlinear but can be precisely characterized, a mapping function can be generated with software to correct for the nonlinearity.

As in any measurement system, noise is unavoidable in a CT system. The DAS output contains two major types of noise: x-ray quantum noise and electronic noise. Quantum noise is due mainly to the limited number of x-ray photons collected during the sampling interval. Electronic noise comes from all of the electronic components in the x-ray detection system. This includes the photodiode and the electronic circuits of the DAS. In general, it is desirable to build a system in which the noise is limited or is dominated by the x-ray quantum statistics rather than the electronic noise. The operator must ensure that the electronic noise represents only a small fraction of the total noise in the signal, even at low x-ray photon flux conditions. Again, the requirements can be derived from either theoretical analysis or experimentation.

Other performance requirements, such as the dynamic range (the minimum-to-maximum signal level acceptable by the DAS), thermal stability, quantization accuracy, and many more, also must be specified to ensure the accuracy of measurements. Detailed discussions of these subjects are omitted here.

6.4 The Gantry and Slip Ring

The gantry is the backbone of a CT system, so the amount of mechanical design effort placed on the gantry cannot be taken lightly. The rotating side of the gantry typically contains the x-ray tube, detector, high-voltage device, tube-cooling tank, slip ring, and other supporting devices, as shown in Fig. 6.15 of an older vintage CT gantry. These components weigh well over 1000 pounds. With such a large load, the gantry still needs to maintain angular and position accuracy. Angular accuracy requires the gantry to rotate at highly constant speeds. Position accuracy requires the gantry to be free of significant vibrations in all directions (both in-plane and cross-plane). Consider, for example, clinical applications in which submillimeter slice thickness is required. Since the width of the x-ray beam is less than a millimeter, the position of the x-ray beam should not vary more than a small fraction of the beam width during gantry rotation to ensure true submillimeter imaging (the slice thickness of the reconstructed image is the weighted sum of the x-ray beams at different locations). Consequently, the gantry must be stable within a fraction of a millimeter for all projection angles. At a 500-mm radius (for typical CT scanners), this is not an easy task.

With increasingly demanding scan speeds, the requirements of gantry performance have also increased significantly. These requirements place large design constraints on the components that are mounted on the gantry. Since the centrifugal force increases with the square of the rotation speed, these components must endure a high G-force while conducting routine operations.

Figure 6.15 Gantry of an older vintage third-generation CT scanner.

Figure 6.16 shows the G-force as a function of the gantry rotation frequency for an object located 0.7 m from the iso-center. At a gantry speed of 0.35 sec per revolution, the G-force is already 23 g. Considering the required safety margin of a factor of 8, the mechanical design is not easy!

Another key component of the CT system is the slip ring. Its function is to supply power to the rotating side of the gantry, transmit command signals both ways, and send the CT projection data to the stationary side. An example of a slip ring is shown in Fig. 6.17. Although the slip ring was introduced to CT in the early 1980s to facilitate continuous gantry rotation, it became the de facto standard only after the invention of the helical scan mode. The data signals and the power to the x-ray tube both flow between the continuously rotating gantry and the stationary CT components through the electrical, optical, or RF connections on the slip ring. The amount of data that must be transmitted is large. For a typical multislice scanner, there are roughly 1000 channels per detector row, and each rotation contains nearly 1000 projections or views. To avoid excessive storage on the rotating side of the scanner, the data transfer rate needs to be in sync with the data generation rate. The data transfer rate R for a 0.35-sec-per-revolution 64-slice scanner is:

$$R = \frac{\text{number of samples per rotation}}{\text{time per rotation}} = 1.8 \times 10^8 \text{ samples / sec} . \qquad (6.1)$$

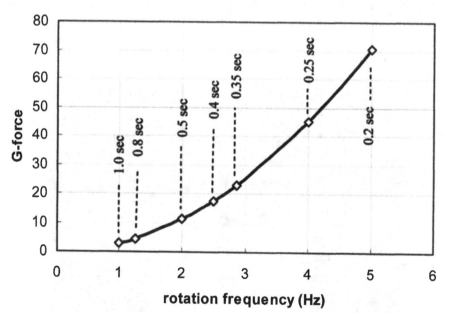

Figure 6.16 Illustration of centrifugal force as a function of rotation frequency on an object 0.7 m from the rotation center.

Figure 6.17 Photograph of a slip ring used for power and data transmission.

If each sample is digitized in a floating-point format (32 bits), the bandwidth of the slip ring must be at least 5.7 Gbauds, without considering other sources of overhead such as error-detection or error-correction bits. The bandwidth requirement of slip rings is certain to rise with the introduction of new CT scanners.

6.5 Collimation and Filtration

Collimation in CT serves two purposes: to reduce unnecessary dose to the patient and to ensure good image quality. Generally speaking, there are two types of collimation: prepatient collimation and postpatient collimation. As its name implies, prepatient collimation is positioned between the x-ray source and the patient. Since x-ray photons emitted from the x-ray tube cover a very wide range in z, prepatient collimation restricts the x-ray flux to the patient to a narrow region, as shown in Fig. 6.18. For single-slice CT, prepatient collimation not only reduces dose to the patient, but also defines the slice thickness of the imaging plane. However, for multislice CT the slice thickness is defined by the detector aperture instead of the collimator (a detailed discussion of multislice CT can be found in Chapter 10). Because over 90% of the x-ray photons emitted from the x-ray tube are blocked by prepatient collimation, x-ray tube utilization for CT is poor.

Due to geometric limitations, the x-ray beam, after passing prepatient collimation, has two regions (in z): umbra and penumbra. The umbra is a region in which the x-ray flux is homogeneous. If one were to look at the x-ray source at any point inside this region, one would see no blockage by the collimation. In other words, the entire x-ray focal spot can be seen at any point inside the region.

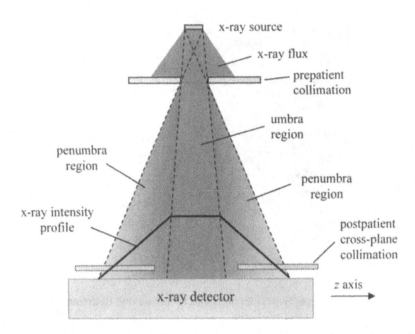

Figure 6.18 Illustration of prepatient collimation and the umbra and penumbra regions.

The penumbra is a nonhomogeneous region. The x-ray focal spot is always partially blocked by the prepatient collimation. For single-slice CT, the slice thickness is defined by the FWHM and FWTM of the entire umbra-penumbra region. Consequently, special attention must be paid to the design of the pre-patient collimation to ensure a satisfactory SSP. For multislice CT, the relative sizes of the umbra and penumbra regions play an important role in the dose utilization of the scanner. In most commercially available multislice CT scanners, only the umbra portion of the x-ray beam is used to form CT images (the active detector cells are located inside the umbra region). The penumbra portion of the x-ray beam represents unused dose to patient. To improve the dose efficiency of the scanner, many attempts are made to limit unused x-ray photons.

The second type of collimation is postpatient collimation. Typically, two kinds of collimators are used, one for in-plane collimation and one for cross-plane collimation. In-plane collimation (a grid) is used by third-generation CT scanners to reject scattered x-ray photons (this technique cannot be employed effectively by fourth-generation scanners due to their wide acceptance angle[8]). The in-plane collimator is made of many thin and highly attenuating plates that are placed in front of the detector to focus on the x-ray source. Since the path of the scattered radiation generally deviates from the original x-ray photon (the primary photon) path, the plates block these photons from entering the detector. Detailed discussions and drawings can be found in Section 7.3.3.

Cross-plane collimation is employed by both third- and fourth-generation scanners. For the third-generation scanner, it serves mainly as additional z-collimation to improve the scanner's SSP. As discussed previously, the slice thickness is determined by the combination of the umbra and penumbra regions. Because of geometric limitations, it is often difficult to design a prepatient collimation that provides very thin slice profiles for the scanner. To achieve this objective, additional collimation may be employed near the detector surface to further restrict the x-ray beam to a narrow slice thickness, as shown in Fig. 6.18. The drawback to this approach, of course, is the dose efficiency penalty of the scanner, since some of the x-ray photons that pass through the patient are not utilized.

A postpatient cross-plane grid is sometimes used with fourth-generation scanners to reject scattered radiation in a manner similar to the in-plane collimation (grid) for the third-generation scanner. Although the role of fourth-generation scanners is fading, the grid is likely to be employed by third-generation scanners. As the volume coverage of the detector increases, the scatter-to-primary ratio (the amount of scattered radiation compared to the primary radiation) will increase. When the ratio cannot be controlled by the in-plane collimation alone, it becomes necessary for scanners to employ cross-plane collimation for scatter rejection.

We will briefly discuss the topic of x-ray beam filtration. The x-ray photons emitted from an x-ray tube represent a wide spectrum. Many soft (low-energy) x-rays are present. The low-energy x rays are primarily absorbed by the patient and contribute little to the detected signal. Therefore, it is necessary to remove these soft x rays to reduce the patient dose. To achieve this objective, most CT manufacturers employ additional x-ray filtration to improve beam quality. The most commonly used filters are the flat filter and the bowtie filter. The flat filter is typically made of copper or aluminum and is placed between the x-ray source and the patient. The flat filter modifies the x-ray spectrum uniformly across the entire FOV. Since the cross section of a patient is mostly oval-shaped, some manufacturers employ a bowtie filter to modify the x-ray beam intensity inside the FOV to further reduce the patient dose (Fig. 6.19). Detailed discussions of both filters can be found in Chapter 11.

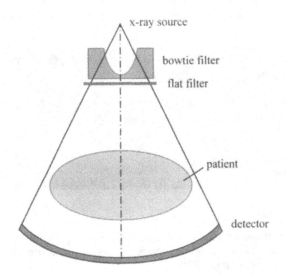

Figure 6.19 Schematic diagram of a bowtie filter and a flat filter.

6.6 The Reconstruction Engine

The reconstruction engine refers to the computer hardware that performs preprocessing (data conditioning and calibration), image reconstruction, and postprocessing (artifact reduction, image filtering, and image reformation). Since details of the reconstruction and artifact suppression algorithms are covered in separate chapters, we will make only a few general comments here regarding the hardware aspect of the reconstruction engine.

Reconstruction speed has experienced tremendous improvement throughout CT history. Figure 6.20 shows some historical images of a GE 7800 console that was the state-of-the-art in the 1970s. Image archiving was performed on either 8-inch floppy disks that held less than 0.5 MB of data, or on 10.5-inch reel 9-track magnetic tapes. The entire computation and storage power shown in the figure is probably less than a regular laptop computer today. Image reconstruction time has decreased from 2½ hours to produce the very first CT image in 1967 to significantly less than 0.1 sec today. In special "proof-of-concept" devices that are not bigger than a typical reconstruction engine used commercially, a speed of over 200 images per second has been demonstrated. One major factor that contributes to the tremendous increase in speed is the rapid advance of computer hardware and architectures. For example, dedicated signal processing and ASIC chips enable highly efficient algorithms, such as FFT and backprojection, to be executed at a very high speed.

Another observation of the reconstruction engine is that more and more tasks are processed in parallel and are distributed to multiple processors. This allows the use of less-powerful computer hardware for high-performance reconstruction engines. Most of the reconstruction algorithms are well suited for parallel processing. For example, since the backprojection of a particular view does not require information from other views, multiple processors can be used to independently backproject different views and sum up the output from all processors to obtain the final image. Similar arrangements can be made for many

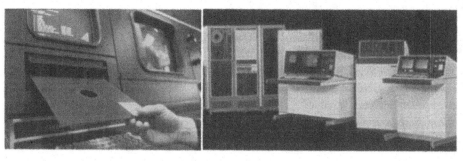

(a)   (b)

Figure 6.20 State-of-the-art equipment from the 1970s. (a) Operator console on a GE 7800 with a floppy disk to store an image. (b) Reconstruction engine, data storage devices, and the console of a GE 7800.

preprocessing and artifact-correction steps. For example, measurements collected on different detector rows can be sent to different processors to perform preprocessing, calibration, and weighting. The resulting projections can then be summed and filtered for backprojection. In recent years, new computing architectures such as the graphics processing unit (GPU) and cell processor have become available; these devices are well suited for the efficient implementation of complicated reconstruction algorithms.[9–11]

Future CT reconstruction engines are unlikely to be limited to producing only CT images. With the development of multimodality scanners such as CT-PET and CT-SPECT, these reconstruction engines must not only handle image reconstruction tasks for other modalities, but also must perform image fusion. The integration of other physiological signals, such as EKG devices and respiratory motion monitors, with CT reconstructions is also becoming a necessity for advanced applications.

6.7 Problems

6-1   Section 6.1 discussed the use of two filaments to produce two different focal spot sizes. Assume that the spacing between two parallel filaments is 5 mm, the distance from the center of the filaments to the target surface is 30 mm, and the two electron beams bombard the anode at the same location with good focusing design in the cathode assembly. Because of thermal expansion, the target surface moves toward the filaments (z) by 2 mm. Calculate the amount of focal spot shift in the lateral direction (x).

6-2   Assume that an x-ray tube has a 7-deg target angle and a rectangular-shaped focal spot. When viewing the focal spot at the iso-center, the focal spot is 1 mm in x and 1 mm in z. Estimate the focal spot size and shape if it is viewed at a detector cell with a 30-deg fan angle (the curved surface of the target can be ignored).

6-3   An engineer is designing a CT system with a source-to-iso distance of 500 mm, and the detector z-coverage of 150 mm is at the iso-center. The engineer has an x-ray tube with a target angle of 7 deg, and the focal spot is positioned on the midplane of the detector in z. Will the tube function properly in the CT system?

6-4   In the rectification process for a high-voltage generator, a 60-Hz input sine waveform is converted to the absolute value of the 60-Hz sine waveform. Calculate the ripple if the final output is the combination of three rectified sine waves shifted by $2\pi/3$ with respect to each other. Repeat the calculation if the output is the combination of six rectified sine waves shifted by $\pi/3$ with respect to each other.

6-5   For simplicity, assume the output energy spectrum of an x-ray tube is shaped as an isosceles triangle from 0 to the maximum keV. For fast kV switching, 80- and 140-kVp settings are used. The view duration is 1 ms, and the rise and fall time is 5% of the view duration (linear ramp up and down). Calculate the amount of mean energy shift for the acquired high- and low-energy data.

6-6   Assume that the linear attenuation coefficient μ of a solid state scintillator is energy independent. A detector made with this material has a QDE of 98%. What is the QDE if the scintillator thickness is reduced by 20%?

6-7   Section 6.3 discussed similarities between DAS crosstalk and detector afterglow in terms of their impact on azimuthal resolution. What are the main differences in their impact on azimuthal resolution?

6-8   With the photon-counting capability enabled by semiconductor detectors, what potential image quality gains can be obtained?

6-9   When the input x-ray flux exceeds the photon counting detector's count rate, multiple x-ray events are combined and treated as a single event. This is called the "pile up" effect. Discuss potential issues resulting from this effect.

6-10   A CT system has a focal spot length of 1 mm in z with a uniform intensity distribution. The source-to-prepatient-collimator distance is 150 mm; the source-to-iso-distance is 500 mm; and the source-to-detector distance is 900 mm. The prepatient collimator is set so that the FWHM of the x-ray beam profile at the iso is 3.3 mm. A student wants to remove all x rays in the penumbra regions with a postpatient collimation to improve the z resolution. Calculate the air-scan dose efficiency (the ratio of the x-ray flux received by the detector over the total x-ray flux passing through the prepatient collimator).

6-11   Calculate the umbra-to-penumbra ratio for the CT system described in problem 6-10 for prepatient collimator settings of 4 mm, 8 mm, and 12 mm FWHM at the iso-center.

6-12   From the viewpoint of minimizing centrifugal force, it seems logical to place the x-ray tube and detector at the same distance from the iso-center. What practical design issues may prevent us from doing so?

6-13   A 500-kg CT gantry is rotating at 0.35 sec per gantry rotation. Calculate the power required to completely stop the gantry in 5 sec with a constant force if the mass is distributed evenly 0.5 m from the iso-center.

6-14   A student argues that there is no advantage in designing a reconstruction engine that can reconstruct images at more than 30 frames per second, since it is unlikely that radiologists can read images at a rate faster than the movie. Do you agree? What applications may require a faster reconstruction speed?

6-15 The use of a flat filter to improve beam quality was discussed in Section 6.5. What parameters must be considered in order to select the appropriate material and thickness for the filter?

References

1. C. L. Morgan, *Basic Principles of Computed Tomography*, University Park Press, Baltimore (1983).

2. S. H. Fox, "CT tube technology," in *Medical CT and Ultrasound: Current Technology and Applications*, L. W. Goldman and J. B. Fowlkes, Eds., Advanced Medical Publishing, Madison, WI, 349–358 (1995).

3. J. M. Boone, "X-ray production, interaction, and detection in diagnostic imaging," in *Handbook of Medical Imaging, Vol. 1, Physics and Psychophysics*, SPIE Press, Bellingham, WA (2000).

4. D. J. Wagenaar, K. Parnham, B. Sundal, et al., "Advantage of semiconductor CZT for medical imaging," *Proc. SPIE* **6707**, 67070I (2007).

5. T. Nakashima, H. Morii, Y. Neo, et al., "Application of CdTe photon-counting x-ray imager to material discriminated x-ray CT," *Proc. SPIE* **6706**, 67060C (2007).

6. H. H. Barrett and W. Swindell, *Radiological Imaging: The Theory of Image Formation, Detection, and Processing*, Academic Press, New York (1981).

7. J. Hsieh, O. E. Gurmen, and K. F. King, "Investigation of a solid-state detector for advanced computed tomography," *IEEE Trans. Med. Imag.* **19**, 930–940 (2000).

8. J. Arenson, "Data collection strategies: gentries and detectors," in *Medical CT and Ultrasound: Current Technology and Applications*, W. Goldman and J. B. Fowlkes, Eds., Advanced Medical Publishing, Madison, WI, 328–347 (1995).

9. P. Despres, M. Sun, and B. H. Hasegawa, "FFT and cone-beam CT reconstruction on graphics hardware," *Proc. SPIE* **6510**, 65105N (2007).

10. M. Churchill, G. Pope, J. Penman, et al., "Hardware-accelerated cone-beam reconstruction on a mobile C-arm," *Proc. SPIE* **6510**, 65105S (2007).

11. O. Bockenbach, M. Knaup, and M. Kachelrieß, "Implementation of a cone-beam backprojection algorithm on the cell broadband engine processor," *Proc. SPIE* **6510**, 651056 (2007).

Chapter 7
Image Artifacts: Appearances, Causes, and Corrections

7.1 What Is an Image Artifact?

Chapter 3 provided a representative flow diagram for the image generation process and stated that the preprocessing and postprocessing steps used to overcome nonideal data collection often outweigh, in terms of the number of operations, "textbook" tomographic reconstruction. In other words, a significant portion of the computation associated with CT image generation is related to the reduction or elimination of image artifacts. This chapter provides a broad overview of the causes of some artifacts and presents solutions to avoid or correct them whenever appropriate.

The definition of an image artifact is not as clearly defined as one might expect. Theoretically, an image artifact can be defined as any discrepancy between the reconstructed values in an image and the true attenuation coefficients of the object. Although this definition is broad enough to cover nearly all types of nonideal images, it has little practical value since nearly every image produced by a CT scanner contains an artifact by this definition. In fact, most pixels in a CT image are "artifacts" in some shape or form. In practice, we have to limit our discussion to the discrepancies that are clinically significant or relevant as judged by the radiologists. We want to examine only the discrepancies that impact the radiologists' performance.

Compared to conventional radiography, CT systems are inherently more prone to artifacts. Recall the discussion in Chapter 3 that explained how a CT image is generated with a larger number of projections (about 1000). In a typical CT system, each projection contains roughly 1000 separate measurements. (In the case of a multislice CT scanner, which will be discussed in Chapter 10, the number of measurements in a single projection can easily be quadrupled.) As a result, nearly 10^6 independent readings or measurements are used to form an image. Because the nature of the backprojection process is to map a point in a projection to a straight line in an image, an error in the projection reading is no longer localized, as is the case for conventional radiography. Since inaccuracies in the measurements usually manifest themselves as errors in the reconstructed images, the probability of producing an image artifact is much higher for CT.

Some errors or artifacts are merely an annoyance to radiologists, but others may cause misdiagnosis.[1-3]

Image artifacts are caused by many things: the nature of the physics, suboptimal system design, limitations of current and new technologies, patient characteristics, and suboptimal or inappropriate use of the scanner. Methods used to combat CT image artifacts can be divided into two major classes: artifact correction and artifact avoidance. Artifact correction or reduction comes mainly from either academic institutions or industries. Many research papers and articles can be found in various journals and conference proceedings, and an even larger portion of the research results can be found in patents. A few summaries and overviews on the subject can be found in several books.[1,2,4-8] However, many of the artifact-combating methods are considered proprietary or trade secrets and are not available in the public domain. Artifact avoidance relies mainly on the combined efforts of CT manufacturers and operators. The best clinical protocols should be recommended by CT designers to optimize image quality. On the other hand, CT operators need to be properly trained to use the scanners in an optimal fashion.

This chapter starts with a general definition of image artifacts and a brief explanation of the appearance of various artifacts. A clear artifact definition will lay the foundation for subsequent discussions and help to avoid potential confusion, while the brief explanations of artifact appearance can be linked to error patterns in the sinogram. These error patterns lead naturally to the physical sources of error. In particular, we want to link artifacts to major components in a CT system. For the purpose of this discussion, the overall system design is considered to be one of the major components (e.g., artifacts caused by inadequate projection or view sampling). Artifacts related to other components are divided into four major sections: x-ray source, x-ray detector, patient, and operator.

Before discussing specific types of artifacts, we want to emphasize the fact that artifacts discussed in this section are general artifacts associated with different data-collection modes. In clinical CT, a patient can be scanned in either step-and-shoot mode or helical mode. In the step-and-shoot mode (also called axial mode), the patient remains stationary while the gantry rotates about the patient to collect a complete set of data. Once the data collection is completed, the patient is indexed to the next location for another scan. This is in contrast to the helical (or spiral) mode in which a patient is translated at a constant speed during the data acquisition (Chapter 9). This chapter covers artifacts that occur in either the step-and-shoot or the helical mode. It does not discuss artifacts that are specific to the helical scan mode or multislice CT (Chapter 10) for two reasons. First, it is difficult to understand the nature and causes of these artifacts without a thorough understanding of the principles of these special configurations. Second, since the artifacts are presented with the discussion of each system, readers will gain a better understanding of the system design tradeoffs that can be made.

7.2 Different Appearances of Image Artifacts

Generally speaking, CT image artifacts can be classified into four major categories: streaking, shading, rings and bands, and miscellaneous. *Streaking artifacts* often appear as intense straight lines (not necessarily parallel) across the image. They can be either bright or dark. In many cases, the bright and dark streaks appear in pairs due to the nature of the reconstruction process. Figure 7.1 shows the parallel projection (upper left), filtered projection (upper middle), and reconstructed image (upper right), of a simulated water phantom. For the projection at the 45-deg angle, errors were added to three detector channels with a magnitude roughly equal to 10% of the projection value, as shown by Fig. 7.1 (lower left). After the filtering process, the errors appear to be significantly enhanced in terms of their relative magnitude, and large undershoots appear next to the erroneous channels, as depicted by Fig. 7.1 (lower middle). This occurs because the ramp filter used in tomographic reconstruction is essentially a derivative operator, so discontinuities in the projection yield overshoots and undershoots at locations of discontinuity after filtering. These patterns are mapped to bright and dark lines by the backprojection process, as illustrated by the reconstructed image shown in Fig. 7.1 (lower right). Under normal conditions, these artifacts are unlikely to cause misdiagnosis, since human pathologies rarely resemble their appearance. However, when they appear in large quantities and large magnitudes, these artifacts can degrade the image quality to such an extent that images become either unreadable or unreliable.

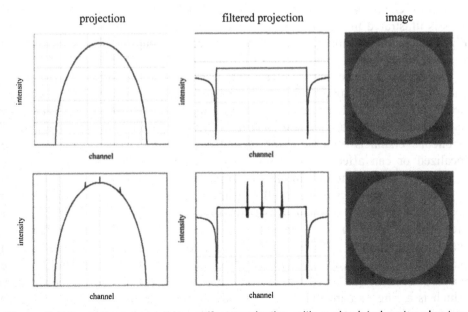

Figure 7.1 Illustration of streaking artifact production with a simulated water phantom. Top: parallel projection, filtered projection, and reconstructed images without error; bottom: about 10% error was added to three channels in a single projection at 45 deg.

As illustrated in the above example, a streak is usually caused by an inconsistency in the isolated measurements (aliasing streaks are a major exception). The inconsistency could be the result of an inherent problem associated with the data collection process (e.g., patient cardiac motion), a mechanical malfunction, or abrupt changes between views. Under normal conditions, the FBP process maps each data point in the projection space onto a straight line in the image domain, the positive and negative contributions among neighboring lines are combined, and no straight lines appear in the final image. When an inconsistency occurs in the projection data set, the reconstruction process is no longer able to properly combine the positive and negative contributions, and lines or streaks result.

Shading artifacts often appear near objects of high contrast. For example, they usually appear in the soft tissue region near bony structures or air pockets. They can be either bright or dark, depending on the nature of the problem. Shading artifacts cause unpredictable CT number shifts in the image and can lead to a misdiagnosis if not correctly identified. Compared to the streaks, shading artifacts are not as easy to identify, because although human anatomies are rarely shaped as straight lines, gradual changes in CT numbers within an organ are not uncommon and do not appear as artificial. As a result, shading artifacts often mimic pathology and lead to misdiagnosis. Thus, shading artifacts are more worrisome and should be handled with greater caution.

The cause of shading artifacts is again an inconsistency in the projection measurement. Unlike streaking artifacts, shading artifacts are generally caused by a group of channels or views that deviate gradually from the true measurements. This is illustrated in Fig. 7.2, which again uses the simulated water phantom in Fig. 7.1. For this example, smoothly varying errors (roughly 10% in magnitude) were added to 40 adjacent channels for a projection at a 45-deg angle, as shown in Fig. 7.2(a). Because there is no sharp discontinuity in the signal, the overshoot and undershoot that appear after the filtering step are small (Fig. 7.2(b)), and the image produced by these errors (excluding the true attenuation measurements) does not contain clear boundaries (Fig. 7.2(c)). Depending on the number of erroneous channels and the error magnitude, shading artifacts can either be localized or can affect larger regions. Sometimes, shading artifacts cover an entire organ and lead to bias in the CT number measurement.

Ring and band artifacts, as their names imply, appear as rings or bands superimposed on the original image structure. They can be either full rings or arcs. The full rings or bands are less dangerous because of their dissimilarity to the human anatomy. Partial rings, on the other hand, can mimic certain pathologies. For example, a dark arc positioned across an aorta could mimic aortic dissection. A special case of the ring/band artifact is the center smudge, which is a ring or band with a small radius. A center smudge is of particular concern since it can mimic pathology and easily lead to misdiagnosis.

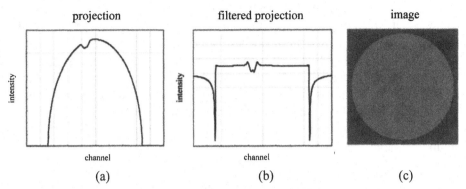

Figure 7.2 Illustration of the shading artifact production with the same water phantom as shown in Fig. 7.1. About 10% error was added to 40 channels in a single projection at 45 deg. (a) Profile of the erroneous projection. (b) Profile after filtering operation. (c) Reconstructed image.

Ring and band artifacts are mainly a third-generation CT phenomenon. They are caused by errors in a single channel or in multiple channels over an extended range of views. Previously we indicated that an error in an isolated view is mapped to a streak (straight line) by the backprojection process. If the same error is persistent over a range of views, the tail portions of the streaks are canceled and an arc or ring is generated. This phenomenon is illustrated with the same simulated water phantom in Fig. 7.3, where an error (roughly 1% in magnitude) was added to a single detector channel over all projections. Given the small error magnitude, it is hardly visible in the projection profile, as illustrated in Fig. 7.3(a). After the filtering step, however, the magnitude of the error is significantly enhanced and can be easily identified in the profile of Fig. 7.3(b). As discussed previously, the backprojection process maps filtered errors to straight lines in the reconstructed image. These lines are at the same fixed distance to the iso-center, since only rotational motion is present in a third-generation detector during data acquisition. Thus, the tail portions of the lines are canceled, and a ring is formed as shown by the reconstructed image in Fig. 7.3(c). Note that although the magnitude of the projection error is small (1%), the intensity of the ring is actually higher than the streaks shown Fig. 7.1 with a 10% projection error. This is because a single projection contributes to only a very small fraction of the final image intensity (roughly $1/N$, where N is the number of projections used for reconstruction). In the ring production process, all projections contain similar errors that reinforce each other, so the error is magnified. Because rings or bands are easily recognized by the human visual system, a projection error as small as a fraction of a percent can lead to a perceivable ring in the image. Of course, the perceptibility of rings is highly influenced by the presence of noise. In general, the higher the noise content in the image, the less likely a ring artifact will be detected by a human observer.

Consider the sensitivity of the ring artifact as a function of the erroneous channel location. To explain the relationship, we will first examine the characteristics of parallel backprojection. Since we are examining the impact of a channel error, we can ignore other channel readings (setting them to zero) and let the projections for the channel of interest be equal to the error ε (in a sinogram, this represents a vertical line with a constant intensity). As discussed previously, the backprojected image of such a pattern forms a ring centered at the iso-center and tangential to the projection rays of the erroneous channel. Consequently, the total intensity integrated along the circumference of the ring Q equals the product of the channel error intensity ε and the number of views n. The intensity of the ring in the image c is

$$ c = \frac{Q}{2\pi r} = \frac{n\varepsilon}{2\pi r}, \tag{7.1} $$

where r is the radius of the ring. Equation (7.1) states that the intensity of the ring artifact produced by a fixed channel error ε is inversely proportional to the radius of the ring. Although this is limited to parallel-beam projections, a somewhat similar relationship exists for fan-beam geometry. Therefore, the sensitivity of the ring artifact is nonuniform across the detector. Thus, if we introduce the same error to projection samples at different locations, we would expect the magnitude of the ring artifacts to be the highest for errors located near the iso-channel (r approaches zero) and the lowest for errors near the periphery channels (large r values). For illustration, we simulated a projection data set of an air scan (under noise-free conditions, all projection samples are zero) and added a fixed amount of error to a few selected detector channels over all projection views. The reconstructed image is presented in Fig. 7.4. The figure shows that although the same error is introduced at different detector channels, the intensities of the resulting ring artifacts are significantly different, with the rings near the iso-center being brighter than those farther away from the iso-center. To illustrate the

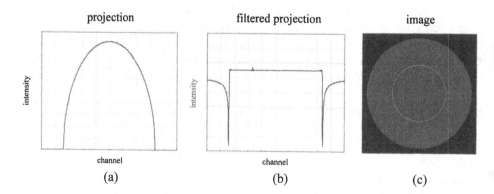

Figure 7.3 Illustration of the production of a ring artifact. About 1% of error was added to a single channel over all projection angles. (a) Projection profile with 1% added error. (b) Profile after filtering operation. (c) Reconstructed image.

difference in the magnitude of the rings, we plotted the intensity of the rings across the image center and presented the result in Fig. 7.5. The graph illustrates that the magnitude of the ring artifact can quickly increase as the error location approaches the iso-channel. Alternatively, we can state that the projection accuracy requirement is much higher for channels located closer to the iso-channels than to the periphery channels. Although our result was obtained with a simulated air scan, the conclusion can be extended to scans of any object, due to the near-linearity of the CT reconstruction process. (Under ideal conditions, an image reconstructed from the sum of two projections equals the sum of two images reconstructed separately from two projections.)

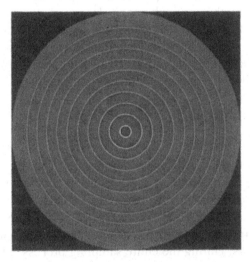

Figure 7.4 Reconstructed image of a simulated noise-free air scan. Errors of identical magnitude were injected into the projection at different channel locations.

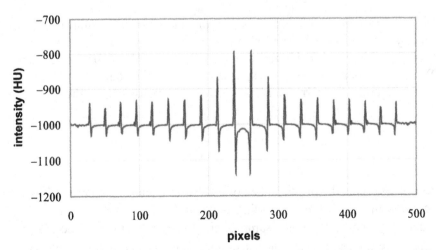

Figure 7.5 Intensity profile of rings produced by introducing errors of the same magnitude at different detector channel locations.

The *miscellaneous artifacts* category covers various less-common artifacts, but by no means does this label de-emphasize their importance. Depending on the application, some artifacts might be as important as those previously discussed. Examples in this category are the basket weave artifact and Moiré pattern. The basket wave artifact appears as a set of periodic horizontal and vertical lines superimposed on the otherwise normal CT images, similar to the weaving pattern of a basket. The artifact can be produced during the image presentation or manipulation process when a CT image needs to be resampled for magnification or demagnification. If a linear interpolation algorithm is used and the magnification factor is close to unity, the new pixel location can periodically overlap precisely with an original pixel location and the noise at this location is, naturally, identical to the original image. However, the noises at other pixel locations are lower than in the original image due to the low-pass nature of the linear interpolation. The periodic sudden change in noise produces the periodic line pattern of the basket weave artifact. However, because of the limited scope of this text, no detailed discussion will be presented on the subject.

7.3 Artifacts Related to System Design

7.3.1 Aliasing

The first step in CT data collection is sampling. The original x-ray intensity distribution (after passing through the patient) is continuous. This continuous waveform is sampled discretely by an array of detectors to produce a set of signals that represents a view or a projection in a third-generation CT scanner. The sampling is further discretized in the temporal domain to divide the projections into views. During the entire sampling process, a set of rules must be followed to avoid the possible production of artifacts.[9]

According to Shannon's sampling theorem, to avoid aliasing, the original data must be sampled at a rate of at least twice the highest spatial frequency contained in the signal. The maximum frequency contained in the CT signal is limited by the focal spot size (as in conventional radiography), the scanner geometry, and the detector cell size (in the design process, a proper balance is achieved so that a reasonably uniform x-ray beam width can be formed across the FOV). If we approximate the response of a detector channel by a rectangular window function, the corresponding frequency response can be shown to be

$$S(f) = \frac{\sin(\pi f \delta)}{\pi f \delta},\tag{7.2}$$

where δ represents the physical size of a detector channel. Equation (7.2) is a sinc function with the first zero-crossing point at $f = 1/\delta$. Therefore, the maximum frequency f_{max} contained in the sampled signal can be assumed to be $1/\delta$.

To fully understand Shannon's sampling theory, let us take a closer look at the sampling process itself. When the detector channels are uniformly spaced at a distance T, the spectrum of the sampled projection consists of the original spectrum plus an infinite number of translated versions of the original spectrum, as shown in Fig. 7.6. The amount of translation is equal to the sampling frequency $(1/T)$ and its harmonics. To faithfully recover the original signal from the sampled projection, none of the shifted spectral components can overlap each other (if portions of any of the shifted functions overlap, it is impossible to separate these components). Figure 7.6 shows the scenario where spectrum overlap occurs. The process of spectral overlap is called *aliasing*. To avoid overlap, the amount of spectrum translation in the sampled projection $(1/T)$ must be at least twice the highest-frequency contents in the original signal $(1/\delta)$. Mathematically, this requires that $T \le \delta/2$, so samples must be collected no farther apart than half of the detector width $(\delta/2)$. This is the well-known Shannon sampling criterion or Nyquist sampling criterion. Unfortunately, third-generation scanners cannot make measurements any closer than the detector width, and seemingly the Shannon sampling criterion cannot be fulfilled. Figure 7.7(a) shows an image reconstructed with simulated projections of a 0.1-mm-diameter wire placed 5 cm off the iso-center. Since the projection is inherently undersampled, aliasing streaks can be clearly observed near the sharp edges. In a later discussion, it will become clear that this characteristic is different from inter-view aliasing, where streaks appear at some distance from the object.

To combat aliasing, the concept of a fourth-generation CT scanner was proposed, as shown in Fig. 7.8(b). In a fourth-generation CT system, the detector is stationary and the x-ray tube rotates around the iso-center [in contrast to the third-generation system, where both the x-ray tube and the detector rotate about the iso-center while remaining stationary with respect to each other, as shown in Fig. 7.8(a)]. A projection view is acquired by sampling a fixed detector channel while the x-ray tube is moving across the fan angle. By adjusting the sampling rate in the temporal domain, the distance between adjacent samples in a projection can be modified. The Shannon sampling criterion can be easily met

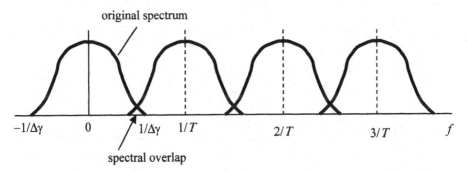

Figure 7.6 Illustration of aliasing in projection sampling. The original spectrum is replicated at an interval $(1/T)$, where T is the spatial sampling distance (equal the detector channel spacing). If $(1/T)$ is too small, overlaps occur between shifted spectrums, and aliasing artifacts result.

with a high sampling rate. This approach, however, brought with it a different set of difficulties. For example, since each detector must view the x-ray source over a very wide angle, no practical and effective collimation can be placed in such systems to reject scattered x-ray photons. In addition, the CT system becomes more sensitive to x-ray tube output fluctuations, since each view is acquired over an extended time period. With the introduction of the multislice CT system (which will be discussed in Chapter 10), the cost of the detector system can become prohibitively expensive. Since nearly all state-of-the-art CT systems developed today are third-generation scanners, we will not discuss in depth the technical issues and possible compensation schemes for fourth-generation CT systems.

For third-generation scanners, various approaches to combat aliasing artifacts have also been proposed. One approach is the detector quarter-offset (also called the quarter-quarter offset). In this arrangement, the detector center is offset by one-quarter of the detector cell width with respect to the iso-center, as shown in Fig. 7.9. Consider two samples that are located on either side of the iso-center. After the gantry rotates 180 deg, one of the samples interleaves the previously acquired samples. This sample provides the required double sampling to avoid aliasing. A similar condition can be found for other channels located on the detector.[10] In fact, it can be shown that the projection sample (γ', β') that provides the interleaved sample between two samples (γ, β) and $(\gamma+\Delta\gamma, \beta)$ also satisfies the following conditions:

$$\begin{cases} \gamma' = -\left(\gamma + \dfrac{\Delta\gamma}{2}\right), \\ \beta' = \beta \pm \pi - 2\gamma - \Delta\gamma. \end{cases} \qquad (7.3)$$

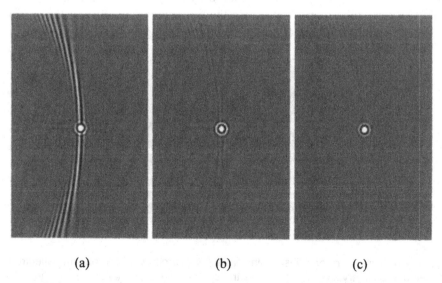

(a)   (b)   (c)

Figure 7.7 Reconstructed image of a 0.1-mm simulated wire. (a) Projections with iso-center and iso-channel in alignment. (b) Projections with detector quarter-offset concept. (c) Projections with focal spot wobble.

Figure 7.7(b) shows the same wire scanned with the detector quarter-offset arrangement. A significant reduction in aliasing artifacts is evident. However, this approach does have a few drawbacks. One disadvantage is that the double sampling relies on two set of samples taken 180 deg apart. Therefore, any patient motion during the sampling interval will compromise the effectiveness of the aliasing reduction. In addition, it can be shown that other than the center rays, the double sampling is only approximately achieved. The amount of displacement of the interleaved samples increases with the distance from the iso-channel. Therefore, the effectiveness of the aliasing artifact reduction is location dependent.

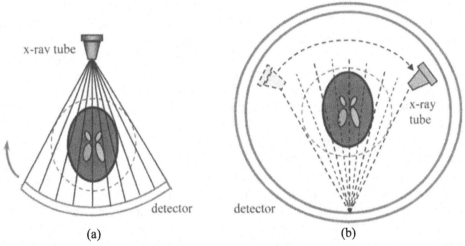

Figure 7.8 Illustrations of projection sampling difference between (a) third- and (b) fourth-generation CT scanners.

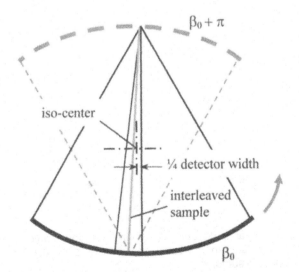

Figure 7.9 Illustration of the quarter detector offset concept.

To overcome these difficulties, an approach to double the samples by wobbling or deflecting the x-ray focal spot has been investigated.[11–13] In this approach, a normal projection is acquired. When the gantry rotates to an angle such that the detector cell positions straddle the positions from the previous view, the x-ray focal spot is deflected, either electrostatically or electromagnetically, to the location of the previous x-ray focal spot, as shown in Fig. 7.10(a). Thus, a double sampling is accomplished [Fig. 7.10(b)]. The two sample sets take place in a very short period of time (typically a fraction of a millisecond), so any patient motion can be ignored. Figure 7.7(c) simulates the same wire scanned and reconstructed with this approach. It is clear that the aliasing artifacts are almost completely eliminated. The detector quarter-offset and the focal spot wobbling can be combined to further reduce aliasing. Note that the maximum-frequency contents in the sampled projection f_{max} were selected as the first zero crossing point in Eq. (7.2). We know that higher-frequency contents do exist in the projection beyond this point (the other side lobes of the sinc function), so the double sampling only approximately satisfies the Nyquist sampling criterion. To further remove higher-frequency aliasing, more than two samples per detector resolution element are needed. The combination of quarter-offset and focal spot wobble can easily satisfy this requirement.

The above discussion is equally applicable for sampling in the view direction. For simplicity, we will consider only the case of a uniform sampling pattern where the angular increment in the view direction is constant. We will examine samples in two directions: radial and azimuthal. The radial direction is defined to coincide with the polar axis in a polar coordinate, and the azimuthal direction is the direction tangential to the radial axis. It can be shown that the x-ray focal spot size and the detector cell size contribute mainly to the radial resolution, while the detector response, the data acquisition system response, and

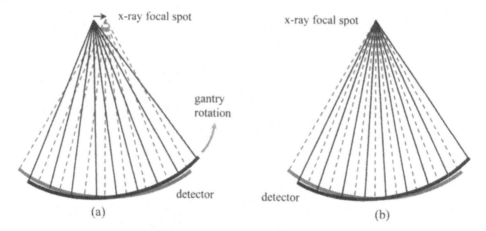

Figure 7.10 Concept of focal spot wobble. (a) First projection location (gray dotted lines) and rotated detector position to straddle previous samples. (b) X-ray focal spot is deflected to achieve double sampling.

the number of views in a scan affect mainly the azimuthal resolution (assuming the x ray is on constantly during the scan). A simple relationship exists between the azimuthal sampling frequency, the frequency contents of the object, and the size of the aliasing artifact-free zone. For the case of parallel rays, this requirement can be described simply by the following equation:[14,15]

$$N_{min} = 2\pi R v_M,$$ (7.4)

where N_{min} represents the minimum number of views over a 180-deg angular span, R is the radius of the artifact-free zone of reconstruction, and v_M is the maximum resolvable spatial frequency. Although this equation was derived with rigorous mathematical analyses, it can be derived somewhat empirically in the following manner. For a circular region centered on any point of a scanned object, the sampling density in the azimuthal direction is highest at the point of interest and lowest away from the point. This can be understood by considering the sampling pattern of all rays that intersect the point. The samples form a star pattern with the point of interest at its center. To avoid aliasing, we must make sure that the Nyquist sampling criterion is satisfied for the worst sampling location (i.e., the periphery of the circular region). Since the circumference of a circle with radius R is $2\pi R$, the distance between adjacent samples at the periphery is simply $\pi R/N$, assuming N views are spaced uniformly over 180 deg. Based on the Shannon sampling theory discussed previously, the maximum supportable frequency v_M is half of the sampling frequency:

$$v_M = \frac{N}{2\pi R}.$$ (7.5)

By multiplying both sides of Eq. (7.5) by $2\pi R$, we obtain Eq. (7.4). The purpose of the empirical derivation is to provide an intuitive understanding of the view-sampling requirement.

For the fan-beam case, this requirement can be described in the following form[16]:

$$N_{min} = \frac{4\pi R v_M}{1 - \sin\left(\dfrac{\psi}{2}\right)},$$ (7.6)

where ψ is the full fan angle. Here, N_{min} represents the minimum number of views required over 360 deg. When $\psi \to 0$, Eq. (7.6) is reduced to parallel-beam sampling. Since the fan-beam algorithm requires projection samples over 360 deg (instead of 180 deg), the degenerated case of Eq. (7.6) is twice that of Eq. (7.4). Figure 7.11 shows an example of the minimum number of views as a function of the maximum spatial frequency. In this example, R = 175 mm and

$\psi = 55$ deg. It is clear from this example that the required number of views is quite high, even for moderate spatial frequencies.

We can also examine the view-sampling requirement from a detector quarter-offset sampling point of view. We illustrated earlier the concept of detector quarter-offset to combat projection aliasing. In this arrangement, the original set of samples is subsidized with additional samples that are 180 deg apart. One of the underlying assumptions is that the projection views are sampled fine enough such that the x-ray source position is able to straddle the two detector cell positions in the original view. This places a constraint on the view-sampling requirement. If we denote the detector cell spacing by δ and the source-to-detector distance by D, the detector angle γ_i formed by detector cell i with the iso-ray can be expressed as follows:

$$\gamma_i = \frac{(i - i_c)\delta}{D}, \tag{7.7}$$

where i_c represents the index of the iso-ray. The set of complementary samples that straddle the samples defined by Eq. (7.7) must then satisfy

$$\gamma_i = \frac{(i - i_c + 0.5)\delta}{D}. \tag{7.8}$$

For convenience, we will examine only the ray samples when the x-ray source is in the 12 o'clock position (for an arbitrary view angle, the entire formulation is merely rotated by a fixed angle). If we define the 0-deg projection angle as the x-ray tube in the 6 o'clock position, the projection angle β_n of the nth view is

$$\beta_n = \frac{2\pi(n - n_c)}{N}, \tag{7.9}$$

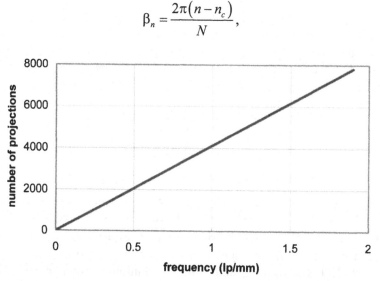

Figure 7.11 Minimum number of views as a function of frequency content in the object.

where N is the number of views per gantry revolution, and n_c is the view index when the projection angle β equals 0 deg. Ideally, we would like to find, for each detector cell i, a β_n that satisfies Eq. (7.8). However, because of the discrete nature of the projection sampling, it is likely that β_n does not match the desired γ_i. Figure 7.12 shows a normalized angular difference between β_n and γ_i for 1000-views-per-gantry rotation. In this example, $D = 949$ mm and $\delta = 1.02$ mm. The normalized angular error is calculated by dividing the actual angular error by $\Delta\gamma$, which is the angular span of a single detector cell. The angular error is periodic, and the angular error can be 1.4 × larger than the angular spacing between adjacent detector channels.

The angular error discussed above represents the amount of "misalignment" for the detector quarter-offset. In other words, the samples that are supposed to straddle the original set of samples are misplaced and become less effective in their ability to combat aliasing. The larger the angular error between the original samples and the complementary samples, the less effective the quarter-offset is at compensating for aliasing.

To reconstruct a point in an image, contributions from samples of different channels are summed (with the exception of a point located at the iso-center). Therefore, the average angular error over the entire detector (since the error pattern is highly periodic) is a good indicator of the residual aliasing present in the image. Figure 7.13 shows the average normalized angular error as a function of the number of projections per rotation. To obtain an effective aliasing artifact reduction, the average error must be a small fraction of the angular spacing between adjacent detector channels. As with the conclusion obtained from Eq. (7.6), several thousand views are required.

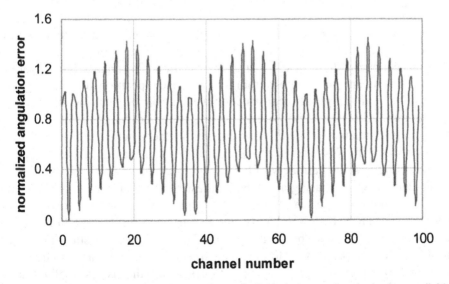

Figure 7.12 Normalized angulation error due to limited view sampling (actual error divided by angular span of each detector cell).

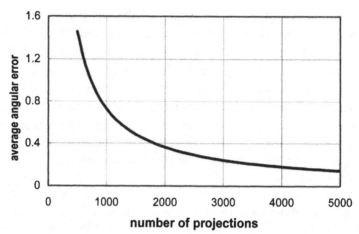

Figure 7.13 Average normalized angular error as a function of number of projections (averaged over all of the detector channels).

In an actual CT design, other factors must be considered in addition to the view aliasing requirement to determine the number of projection views. For example, the noise floor of the data acquisition electronics, the bandwidth of the data transfer, the availability of x-ray photon flux, and other factors must be analyzed to ensure the overall best image quality. Therefore, in practice, the view aliasing guidelines are seldom strictly followed. Instead, careful experiments are conducted (under reasonable clinical conditions) to establish the view-sampling requirement. For reference, a typical commercial CT scanner collects roughly 1000 views per gantry rotation.

View aliasing artifacts typically present themselves as a set of streaks emanating from high-density objects. In the regions near the high-density objects, the sampling frequency may still be high enough to satisfy the Nyquist requirement to create an artifact-free zone. This is in contrast to projection aliasing. To illustrate the patterns of view aliasing, Fig. 7.14(a) shows a chest phantom scanned with 984 views per gantry rotation. To enhance the appearance of view aliasing artifacts, the phantom was purposely positioned at the edge of a 50-cm FOV, and a targeted reconstruction was performed near the outer edge of the phantom. Although minor streaks can be observed near the chest wall, the entire image is relatively free of view aliasing artifacts. When the same phantom was scanned with 704 views per gantry rotation, streaking artifacts became clearly visible near the periphery of the phantom, as shown in Fig. 7.14(b).

To combat view aliasing artifacts, the best remedy, of course, is to increase the view-sampling rate and collect more projection views. However, this approach is often difficult to achieve due to hardware limitations. For example, the number of views can be limited by the bandwidth of the slip ring used to transmit measured projections to the reconstruction engine. It can also be limited by the performance of the data acquisition system (this is especially true for multislice high-speed CT scanners). As a result, alternative algorithmic compensation approaches must be used. One example of a correction scheme is adaptive view synthesis.[17] In this approach, additional views are synthesized based on the measured projections. To minimize the impact on spatial resolution (view synthesis often employs interpolation schemes that suppress high-frequency information), views are synthesized only in locations where view aliasing artifacts are likely to be produced. For example, in a typical patient scan, the shoulder and innominate bones are likely sources of view aliasing streaks. Since most patients are scanned in the prone or supine positions, view aliasing is most likely to be produced by views near the 3 or 9 o'clock angles, and not likely by views near the 6 or 12 o'clock angles. Therefore, view synthesis should be performed only at the selected regions shown in Fig. 7.15. Of course, more advanced techniques can be employed in which projection signal variations in the view direction are constantly monitored. View synthesis is applied only when the signal variation exceeds a predetermined threshold. The contribution amount by these synthesized views to the final image is also dynamically adjusted based on the detected projection variation. Figure 7.14(c) shows a reconstructed

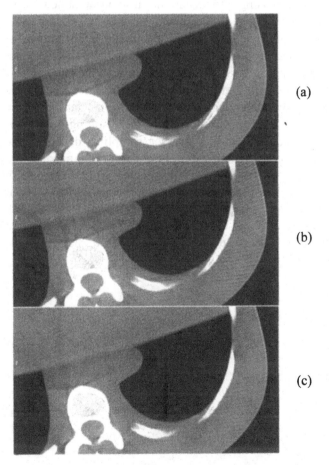

(a)

(b)

(c)

Figure 7.14 Image of a chest phantom (WW = 400). (a) Image reconstructed from a 984-view scan. (b) Image reconstructed from a 704-view scan. (c) Image reconstructed with 704-view scan with adaptive view synthesis.

image using an adaptive view synthesis approach on the same scan data used in
Fig. 7.14(b). It is clear that a significant reduction in view aliasing artifacts is
achieved with minimal impact on the spatial resolution.

   With advances in hardware designs, many of the causes of aliasing artifacts
(either projection or view aliasing) can be reduced or eliminated. As an
illustration, Figs. 7.16(a) and (b) show reconstructed images of a pig head
phantom. To enhance the aliasing artifacts, a copper pipe was inserted inside the
pig skull. One end of the copper pipe was clipped by a tin snipper to create sharp
edges. The phantom was scanned on two types of scanners in a step-and-shoot
mode under identical kV, mA, scan speed, and slice thickness to eliminate other
influencing factors. Figure 7.16(a) was scanned with an older vintage scanner,
and severe aliasing artifacts are clearly visible. Aliasing artifacts are completely
eliminated in Fig. 7.16(b), an image obtained from a newer scanner (the
Discovery™ CT750 HD, which collects more than 2.5 × the projection samples
of the older scanner).

   The sources of aliasing artifacts are not limited to inadequate sampling in the
data acquisition. Other steps in the image generation process can also introduce
such artifacts. Figure 7.17(a) shows an image of a human skull phantom
reconstructed with a fan-to-parallel rebinning algorithm. Linear interpolation was
used in the rebinning process. Closer inspection of the image shows fine streaks
near the edge of the skull highlighted by the oval, since linear interpolation does
not preserve high-frequency contents in the original fan-beam projection. As a
result, the effective sampling frequency of the resulting projection is equivalent
to a projection acquired with a coarser sampling density. When the cutoff
frequency of the reconstruction filter is not reduced to account for such change,
the aliasing artifact will result. When the same scan data were reconstructed with
a high-order interpolation, these artifacts were removed, as shown in Fig. 7.17(b).
Similarly, inadequate sampling in the image space is equally capable of
producing aliasing artifacts, as illustrated by Figs. 7.18(a) and (b) of an x-ray
vascular phantom. The phantom was made of a plastic block with hollow

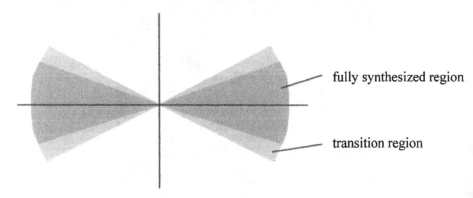

Fig. 7.15 Illustration of adaptive view synthesis.

cylinders of various diameters filled with different iodine contrast agents. The phantom was laid flat onto the patient table so the contrast-filled cylinders were parallel to the z axis. A 16-slice scanner with a 16 × 0.625-mm detector configuration was used to scan the phantom in helical mode with a pitch of 0.5625. Images were first reconstructed at 0.625-mm spacing (0% overlap), and a coronal image was generated as shown in Fig. 7.18(a). The intensity and shape of the rods with smaller diameters appear to be distorted. The same scan was then reconstructed with 0.325-mm spacing (50% overlap) and the resulting coronal image is shown in Fig. 7.18(b). Improvements in image fidelity are evident. This is not surprising since the Nyquist sampling criteria in z were satisfied.

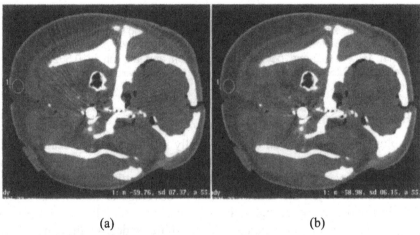

(a)   (b)

Figure 7.16 A pig head phantom scan illustrates both projection and view aliasing artifact reduction in step-and-shoot mode data acquisition (WW = 300). (a) Phantom scanned on an older vintage scanner. (b) Phantom scanned on Discovery™ CT750 HD.

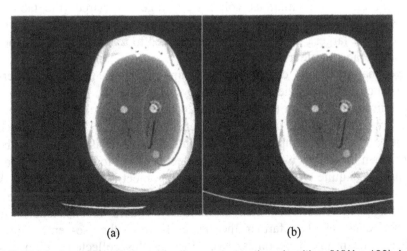

(a)   (b)

Figure 7.17 Aliasing artifact induced by the reconstruction algorithm (WW = 100). Images of a head phantom generated with fan-to-parallel beam reconstruction (a) linear interpolation and (b) high-order interpolation.

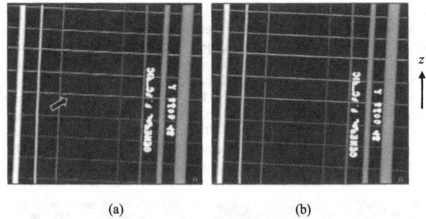

(a)   (b)

Figure 7.18 Aliasing artifacts due to inadequate image spacing sampling. Coronal images of an x-ray vascular phantom of various vessel sizes. Images were acquired on a 16-slice scanner with 16 × 0.625-mm detector configuration and a pitch of 0.5625. (a) Images reconstructed at 0.625-mm spacing (no overlap). (b) Images reconstructed at 0.3125-mm spacing (50% overlap).

7.3.2 Partial volume

Partial volume occurs when an object partially intrudes into the scanning plane. As the slice thickness increases (for example, from 1 to 10 mm), the likelihood of a partial-volume occurrence increases. In a clinical setting, the use of a thicker slice is based mainly on the considerations of coverage or noise. This option is particularly important for a single-slice CT scanner (multislice scanners will be discussed in Chapter 10). For example, to avoid image degradation due to patient motion or to follow a contrast uptake, it is often desirable to complete the scan of an entire organ in a single breath-hold or in a fixed time period. At a given gantry speed, the rate at which the patient table can be translated largely depends on the slice thickness if a continuous volume coverage is desired (the table speed dependency on helical pitch will be discussed in Chapter 9). For large organs (e.g., 20 cm or larger), an operator is often forced to trade off slice thickness for volume coverage on older vintage scanners.

Two major factors cause partial-volume artifacts: projection inter-view inconsistency and intraview inconsistency. Because the x-ray beam profile diverges in the z direction (perpendicular to the scanning plane), the effect of a partially intruded object is angularly dependent. Figure 7.19 illustrates this dependency. In this example, a partially intruded object is located off the iso-center. When the gantry is rotated so the object is closer to the detector, the x-ray beam profile is quite wide, and a portion of the object is within the FOV. When the CT system rotates to the opposite side, the object is completely outside the x-ray beam path. This phenomenon obviously causes inconsistencies in the projection data set. The farther the object is from the iso-center, the more significant the problem becomes. Even for samples collected within a single view, a partially intruded object causes bias in the projection measurement, as shown in Figs. 7.20(a) and (b). The top portion of Fig. 7.20 shows two small detector cells reading the projection measurements independently in z, with a flux of I_1 passing through a high-density object in one detector and a flux of I_2 without the object in the beam in the other detector. The two readings are converted to linear integral measurements with the logarithm operation, and both measurements are accurate. The bottom portion shows a single detector receiving a single flux reading of $I_1 + I_2$. After the logarithmic operation, the resulting line integral measurement does not equal the average of the line integrals obtained previously. This is clearly an inconsistency within the view itself. Mathematically, partial volume can be described by the following equation:[18]

$$P(\beta,\gamma) = P_i(\beta,\gamma)\left(\frac{z}{z_0}\right) - 0.5P_i^2(\beta,\gamma)\left(\frac{z}{z_0}\right)^2\left[\left(\frac{z}{z_0}\right) - 1\right] + \cdots, \qquad (7.10)$$

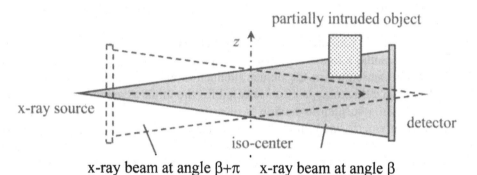

Figure 7.19 Illustration of a cause of the partial volume artifact.

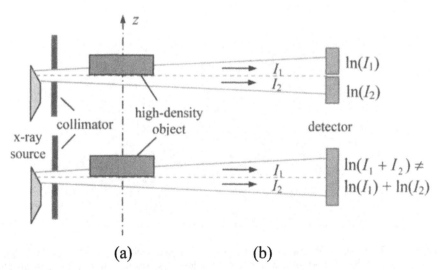

Figure 7.20 Partial-volume effect caused by the nonlinear logarithm operation. (a) Thin-slice and (b) non-thin-slice scanning.

where z_0 and z are the slice thickness and the amount of partial volume into the slice, respectively, and $P_i(\beta,\gamma)$ is the ideal fan-beam projection at fan angle γ and view angle β. The linear component of the error produces a CT number shift in the image, and the nonlinear portion of the error produces streaks. Two phantom experiments demonstrate the two effects. Using the linear term in Eq. (7.10), a thin-plate phantom made of a 1.59-mm (1/16-in.) thick plastic plate was scanned with 2.5-, 5.0-, 7.5-, and 10.0-mm apertures. The thin plate was completely inside the scanning volume at all times. The reconstructed images are shown in Figs. 7.21(a)–(d). The contrast between the plastic plate and the background reduces with the increase in slice thickness. For quantitative analysis, the intensity profiles of the reconstructed plate are plotted in Fig. 7.22. This figure illustrates an inverse linear relationship between the contrast and the slice thickness.

Next, we will illustrate the effect of the nonlinear components on the partial-volume error. For this experiment, we scanned a human skull phantom near the posterior fossa region. This location was selected due to the presence of many fine bony structures that create large variations and discrepancies in the projections. Figure 7.23(a) shows the phantom scanned with a 7-mm aperture. Streaks generated in the cerebellum regions can be clearly observed.

The best method to combat partial-volume artifacts is to use thin slices whenever high variations in the object attenuation are expected. For example, a 5- to 10-mm slice thickness is often selected for the initial screening examination of a lung. When a suspicious nodule is identified, 1-mm or submillimeter scans are subsequently performed for nodule characterization. As another example, a 5- to 10-mm slice thickness is routinely used to scan the midbrain section of the head, while a 1- to 3-mm thickness is recommended for the posterior fossa regions. Figure 7.23(b) shows the same phantom scanned with a 1-mm collimator aperture. Note that streaks in the cerebellum region are almost completely eliminated.

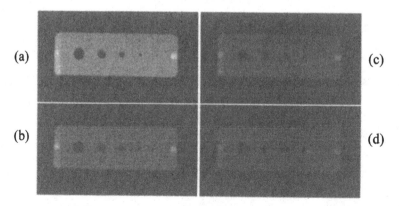

(a)

(b)

(c)

(d)

Figure 7.21 Reconstructed images of a 1.59-mm plastic plate (a) scanned with 2.5-mm slice thickness, (b) scanned with 5.0-mm slice thickness, (c) scanned with 7.5-mm slice thickness, and (d) scanned with 10.0-mm slice thickness.

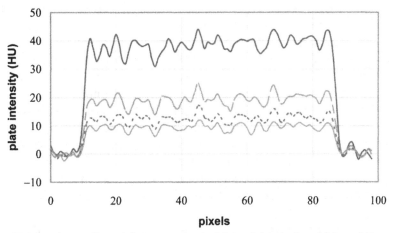

Figure 7.22 Intensity profiles of the reconstructed thin plate. Dark solid line: 2.5-mm scan; gray broken line: 5.0-mm scan; dark dotted line: 7.5-mm scan; gray solid line: 10.0-mm scan.

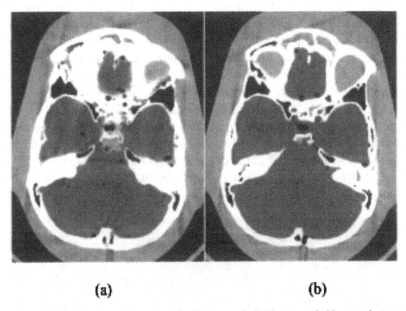

(a)   **(b)**

Figure 7.23 Reconstructed images of a human skull phantom (with equal x-ray flux) (a) scanned with 7-mm collimator aperture and (b) scanned with 1-mm collimator aperture.

In many situations, the reason for scanning a patient with thicker slices is to reduce photon noise. To reduce the partial volume artifact and suppress photon noise, we can simply sum up several thin slices to produce a thicker slice. For a single-slice CT scanner, a penalty must be paid for performing multiple scans instead of only one. However, multislice CT scanners can acquire multiple thin slices simultaneously, and the summation method can be readily applied to the images to produce the final image. The summation process can be performed either in the image space or in the projection space. If the summation is performed in the projection space, it must be carried out after the logarithm step, because the tomographic reconstruction process is nearly linear after the logarithm process (the "textbook" reconstruction described in Chapter 3 is linear). Figures 7.24(a) and (b) compare prelogarithm and postlogarithm summations of a computer-simulated cylindrical phantom. The end of the phantom lines up with the detector row interface, so one of the detector rows measures the full strength of the phantom while the other row misses the phantom completely. The images clearly demonstrate the importance of performing the summation operation after the logarithm step. When the slice thickness of the final image is substantially large, care must be taken when performing projection-space summation to avoid or minimize the effect illustrated in Fig. 7.19.

Computer algorithms can be used to reduce partial-volume artifacts.[19,20] An image space correction scheme has been shown to be quite effective.[20] The algorithm is based on the observation that these artifacts are predominately shadows cast by high-density objects in soft tissue and air regions. Therefore, if the attenuation variation of the high-density objects in z (across the slice thickness) can be estimated using neighboring overlapped images, the artifacts can be "regenerated" by forward projecting the gradient images. Since the reconstruction error caused by the partial-volume effect is roughly proportional to the square of the projection error, the projection of the gradient image is squared prior to the reconstruction. The final image is obtained by subtracting the error image from the original image. Figures 7.25(a) and (b) depict reconstructed images of a CT scan of three cylindrical rods. The rods were carefully positioned so that their ends were located within the width of the x-ray beam to produce the partial-volume effect, as shown in Fig. 7.25(a). Figure 7.25(b) depicts the same image after the proposed correction. The effectiveness of the correction is obvious.

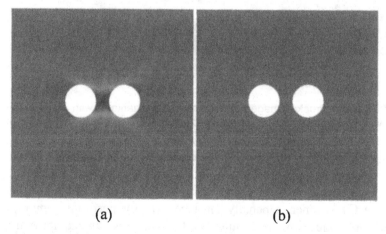

(a)   (b)

Figure 7.24 Simulated partial-volume phantom with projection space summation. (a) Prelogarithm summation and (b) postlogarithm summation.

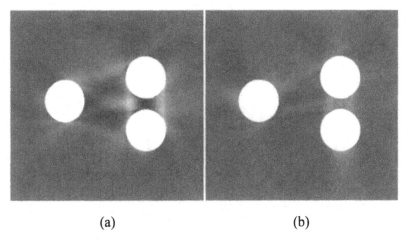

(a)   (b)

Figure 7.25 Three-rod phantom illustrates partial volume and correction. The rods were introduced halfway into the x-ray beam with 7-mm collimation. (a) Original image and (b) image with correction.

7.3.3 Scatter

The most important interaction mechanism between x-ray photons and tissue-like materials is incoherent scattering or Compton scattering, as discussed in Chapter 2. For the convenience of discussion, the process is illustrated schematically in Fig. 7.26. When an x-ray photon collides with an electron, part of the energy is transferred to the electron to free it from the atom, and the rest of the energy is carried away by a photon. Because momentum is conserved in this process, the scattered photon generally deviates from the path of the original photon, as shown by the figure.

Because of Compton scatter, not all of the x-ray photons that reach the detector are primary photons. Depending on the CT system design, a portion of the detected signals is generated from the scatter. These scattered photons make the detected signals deviate from the true measurement of the x-ray intensities and cause either CT number shifts or shading (or streaking) artifacts in the reconstructed images.[21–24] Figure 7.27 is a schematic representation of the Compton scatter effect.

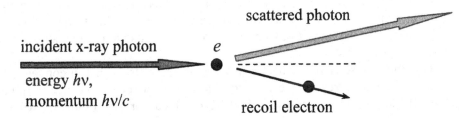

Figure 7.26 Illustration of incoherent scattering or Compton scattering. Part of the energy of the incident photon is given to the recoil electron and the rest is scattered as the scattered photon.

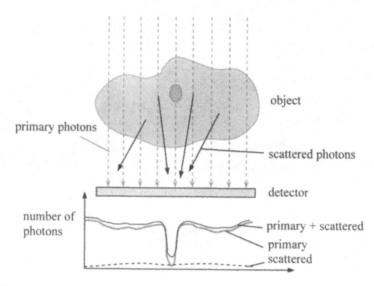

Figure 7.27 Schematic representation of the effect of Compton scattering.

Consider a case in which a highly attenuating structure (represented by a dark shaded oval in Fig. 7.27) is located inside the scanned object (represented by the lightly shaded irregular shape). The light gray dashed lines represent the primary x-ray photons, and a thick gray line at the bottom of the figure plots their intensities at the detector. A significant reduction in the detected number of photons is depicted that corresponds to the location of the highly attenuating structure. Next, consider the impact of Compton scatter. Since the deflection angle of the scattered photon is random, the intensity distribution of the scattered photons received by the detector is a background signal with low frequencies, represented by the dark dotted line in the figure. When combined with the primary photon, the composite signal is a projection with reduced contrast, as shown by the dark solid line. To quantify the scatter, the ratio of the scattered photons' intensity over the primary photons' intensity, called the scatter-to-primary ratio, is often used. The valley of the composite signal (corresponding to the highly attenuating structure) is significantly higher than the scatter-free case. Because of its low-frequency nature, the scatter signal contains little information about the original object. On the other hand, it does contribute to the projection noise. Therefore, the impact of Compton scatter is a reduction in the contrast and the signal-to-noise ratio.

The effect of scattered radiation in CT is different from that in conventional radiography, where scatter merely reduces image contrast since it contributes additively to the film exposure.[25] In CT, the logarithm operation makes this phenomenon significantly more nonlinear (the logarithm of the sum of two numbers does not equal the sum of the logarithms of two numbers). Because the distribution of the scattered photons is object dependent (although it is usually low frequency in nature), it adds a position-dependent structure to the true x-ray intensities.

Given our understanding of the impact of Compton scatter, it is natural to ask the following question: Can we somehow separate the scattered photons from the primary photons once they reach the detector? The answer, unfortunately, is "no" for medical CT scanners, for the following two reasons. First, input x-ray photons have a wide and continuous spectrum, and the x-ray photons emitting from the x-ray tube have different energies. Second, most CT detectors do not have energy discrimination capability. The signal measured by the detector represents simply the integrated x-ray flux over the sampling period. However, with continuing advancement of CT tube and detector technology, it is almost certain that these obstacles can be overcome in the foreseeable future.

One way to characterize scatter is to use the scatter-to-primary ratio. Clearly, the lower the scatter-to-primary ratio, the better the system performs. As discussed previously, it is difficult to measure the scattered signal directly. One experimental technique often used to estimate the scatter-to-primary ratio in a real system is to use different prepatient collimation settings. During the experiment, the desired phantom is first scanned with the prepatient collimator set at a very narrow beam width (e.g., 1 mm in z) but still wide enough to allow full exposure of the detector cell (e.g., the detector cell width in z is set at 0.625 mm). Thus, only 1 mm of the phantom in z is exposed to the x-ray radiation. Since the amount of scatter reduces with the exposed volume, the measured signal can be safely assumed to contain only the primary signal. The phantom is then scanned again at the desired configuration (e.g., 40 mm in z) with identical scanning parameters (e.g., tube voltage, tube current, scan speed, etc). The difference between the two readings should be the contribution from the scattered radiation. Averaging over multiple samples is often used to reduce noise impact.

Most CT designers combat scatter by preventing the scattered radiation from reaching the detector. Much of the scattered radiation can be effectively eliminated by a postpatient collimator, as illustrated in Fig. 7.28, because most of the scattered photons travel along paths that are significantly different from the fan-beam paths. Therefore, by limiting each detector cell's FOV to the vicinity of the x-ray focal spot, a majority of the scattered x-ray photons can be rejected. This can be easily accomplished in third-generation CT scanners by placing a collimator near the detector surface that focuses on the x-ray focal spot (this is very difficult to accomplish in fourth-generation systems). Figure 7.29 shows a computer simulation study to demonstrate the effectiveness of scatter rejection with a postpatient collimator. Figure 7.29(a) was obtained on a LightSpeed^{TM} CT scanner (GE Medical Systems, Milwaukee, Wisconsin) with a postpatient collimator. Figure 7.29(b) was reconstructed by first injecting scattered signals into the same scan data using a Monte Carlo simulation (details of the Monte Carlo simulation are discussed in Chapter 8). Shading artifacts can be clearly observed.

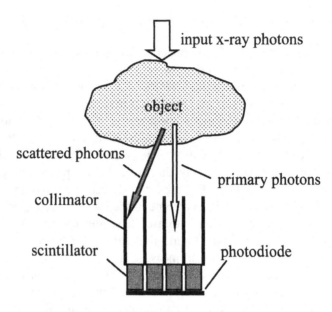

Figure 7.28 Illustration of scattered radiation and the effect of the collimator.

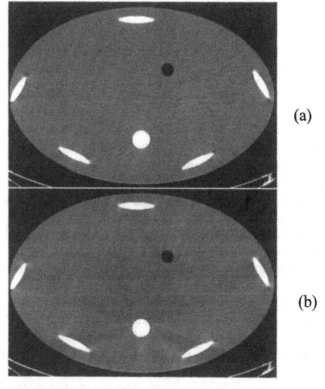

(a)

(b)

Figure 7.29 Reconstructed images of an elliptical phantom (a) with postpatient collimator. (b) Simulation of increased scattered radiation.

When the width of the detector in z (along the patient long axis) is small, the 1D postpatient collimator is quite effective. For example, the scatter-to-primary ratio for the LightSpeed™ 16 scanner is less than 4%. With the increased z coverage (more detector rows), the relative contribution of the scatter is likely to increase, and the aspect ratio of the collimator must increase. At some point, the 1D scatter-rejection collimator may become insufficient. To ensure adequate image quality, a 2D collimator or another scatter rejection scheme must be used. Experiments have shown that many software correction approaches can be effective in combating scatter. Since the scattered signal is typically low frequency in nature, it can be estimated quite accurately with a few sparsely located detector cells outside the primary x-ray beam or with the measured projection signal itself, using signal decomposition or a Monte Carlo simulation. Figures 7.30(a) and (b) show an image without and with an algorithmic scatter correction that seems to sufficiently remove the shading artifacts caused by scatter. Many scatter correction algorithms for flat-panel CT are now available in the public domain.[26–28]

7.3.4 Noise-induced streaks

Chapter 5 stated that one major source of measurement noise is x-ray photon flux. Conventionally, we think of x-ray flux in terms of the photons emanating from an x-ray tube. However, for noise analysis, it is the x-ray photons that reach the detector that determine the image noise (after passing through the patient, many of the original photons are absorbed or scattered). Since the variation in

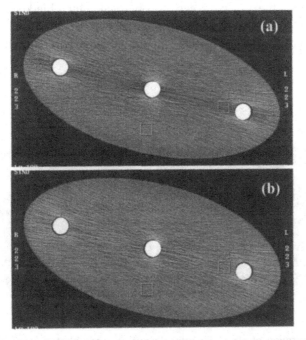

Figure 7.30 Example of an image (a) without scatter correction and (b) with scatter correction.

photon flux follows approximately a compound Poisson distribution, a diminished flux recorded at the detector means a larger signal variation.

In many clinical situations, excessive photon noise can cause severe streaking artifacts.[29] This often occurs as the result of inadequate patient handling (e.g., the patient arm is inside the scanning plane), or improper selection of scanning parameters (e.g., low tube current). In many cases, however, it is simply the result of CT scanner limitations. For example, when a large patient needs to be scanned with thin slices, even the highest x-ray tube voltage and current setting on the scanner cannot deliver sufficient x-ray flux. Sometimes, the artifacts are so severe that the images are rendered unusable, as shown in Fig. 7.31(a). Note that the entire area below the spine is corrupted by severe streaks.

The root cause of streaking artifacts can be understood by analyzing CT noise behavior. In general, the variance of a detected signal σ^2 can be expressed by

$$\sigma^2 = \sigma_e^2 + \sigma_q^2 \cong \sigma_e^2 + \omega x. \tag{7.11}$$

In this equation, σ_e represents the electronic noise floor for the data acquisition system, σ_q represents the x-ray photon quantum noise, ω is the system gain factor that relates the measured signal to the number of x-ray photons, and x is the measured output at the detector. The second part of Eq. (7.11) is based on the fact that x-ray photons follow approximately a Poisson distribution. One key step in the preprocessing of CT projections (as described in Chapter 3) is to take the logarithm of the signal so that the integral of the linear attenuation coefficients along the ray path can be obtained. Our objective is to derive a relationship between the measured signal x and the variance of the random variable y, where $y = g(x) = \log(x)$. To a first-order approximation, σ_y^2 can be shown to be

$$\sigma_y^2 \cong \left| g'(x) \right|^2 \sigma^2 = \frac{\sigma_e^2 + \omega x}{x^2}. \tag{7.12}$$

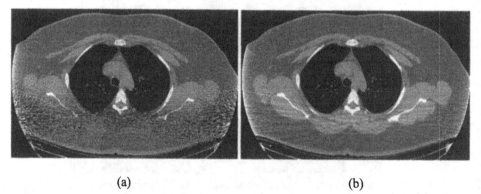

(a)   (b)

Figure 7.31 Illustration of streaking artifacts due to x-ray photon starvation. (a) Original scan and (b) with adaptive noise reduction.

Equation (7.12) is accurate when the function $g(x)$ is approximately linear over the range of statistical variation of the argument x. This condition can be satisfied for the logarithm function when the independent variable is large. When x approaches zero, the approximation is no longer valid, since the value of $\log(x)$ changes quickly with x. Extensive computer simulations have shown, however, that Eq. (7.12) does provide a reasonable estimation of the noise even at small signals. Analyses indicate that the projection noise after the logarithmic operation increases quickly with the decrease in measured photon signal. Since the tomographic filtering operation is essentially a derivative operator, the projection noise is further magnified after filtering. The backprojection process maps these highly fluctuating samples to bright and dark straight lines in the image to form severe streaks.

Careful planning can be quite useful in combating artifacts. For example, care can be taken to instruct the patient to keep his or her arms outside the scan field of view. Various scanning parameters (e.g., kV, mA, scan speed, and aperture) also can be optimized based on the age, weight, and size of the patient. When the problem is caused mainly by scanner limitations, however, computer algorithms must be used to suppress artifacts.

Some approaches use signal-processing techniques to combat artifacts.[29–31] To minimize resolution loss, these techniques are typically nonlinear and adaptive. For example, the characteristics of the employed filters can be made to change with the input x-ray flux based on Eq. (7.12). Minimal or no filtering is applied to the channels that received a high x-ray flux, and the amount of filtering increases with the decrease in x-ray flux. Figure 7.31(b) depicts the same patient scan corrected with the adaptive filtering algorithm described in Ref. 29. Its effectiveness is clearly demonstrated. Because of the filter's adaptive nature, we expect it to be effective not only for severe cases, but for moderate streaking artifacts cases as well. Indeed, a study performed with a shoulder phantom scanned over a wide range of techniques (from 40 to 200 mA) showed that image quality can be maintained nearly constant over a wide signal range by applying the adaptive filtering approach, as illustrated in Fig. 7.32.

Other noise sources besides limited x-ray photon statistics impact the image quality of a CT scan. For example, special attention must be paid to avoid electromagnetic interference between different components of the scanner. Figure 7.33 shows an experiment conducted to demonstrate the importance of properly shielding the subsystem from its environment. To establish the baseline, a 48-in. polyphantom was scanned to ensure that the bay was in good condition, as shown by a reconstructed image in Fig. 7.33(a). During the experiment, a copper shield was removed and the phantom scan was repeated under otherwise identical conditions. Rings and streaking artifacts are clearly visible in Fig. 7.33(b).

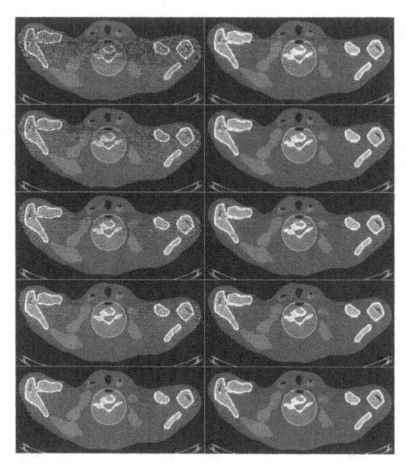

Figure 7.32 Reconstructed images of a shoulder phantom (display WW = 600). Images in rows 1–5 were acquired with 40, 60, 80, 100, and 200 mA, respectively. Images in the left column were reconstructed without streak artifact compensation. Images in the right column were reconstructed with adaptive filtering.

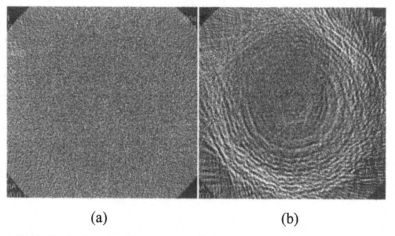

(a)   (b)

Figure 7.33 Artifacts due to electromagnetic interference of DAS signals. (a) Proper shielding of DAS signals and (b) with the shielding removed.

7.4 Artifacts Related to X-ray Tubes

7.4.1 Off-focal radiation

In previous discussions, we treated the x-ray source as a single point known as the x-ray focal spot. All photons emitted from the x-ray tube were assumed to come from that location. In reality, however, photons are emitted from a larger area on the target. This is referred to as off-focal radiation, or more precisely, extra-focal radiation. The right-hand side of Fig. 7.34 shows a cutout view of a cathode-anode assembly. The off-focal radiation is depicted by a shaded region centered on the x-ray focal spot. Off-focal radiation is caused mainly by two effects: secondary electrons and field-emission electrons.[32] The secondary electrons are usually the most dominant source. When a high-speed electron beam strikes a target, electrons are emitted from the impact area. Most of these high-velocity secondary electronics (backscattered electrons) return to the target at points outside the focal spot and produce x-ray photons at their point of impact, as depicted by the bottom-left portion of Fig. 7.34.[33] Therefore, from a CT imaging point of view, the x-ray source can be modeled as a high-intensity center spot surrounded by a low-intensity halo.

Although the halo is low in intensity, it can cause degradation in low-contrast detectability and shading artifacts that either mimic pathology or produce images of unacceptable quality. To illustrate the impact of off-focal radiation, we performed the following experiment. A water phantom with a Plexiglas® casing was scanned and reconstructed, as shown in Fig. 7.35(a). Note the presence of a bright rim near the water-Plexiglas® interface. Although the magnitude of the rim is small, it can lead to misdiagnosis in a clinical setting. For example, if a

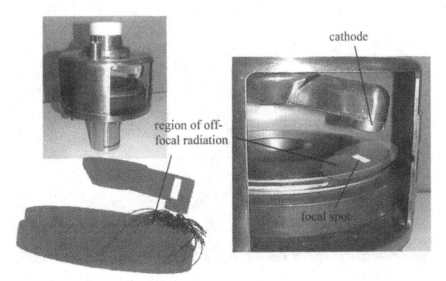

Figure 7.34 Illustration of the off-focal radiation region. Top left: photograph of an x-ray tube. Bottom left: schematic diagram of secondary electrons emitted from the target. Right: cutout view of the cathode-anode assembly and off-focal radiation region.

patient's head is scanned at the midbrain section with excessive off-focal radiation, the bright rim near the bone-brain interface can mimic the presence of fresh blood. In body scans, the appearance of off-focal radiation can be reverted to dark shadows, as shown in Fig. 7.36(a). Note that the lung appears to "bleed" into the tissue regions between the ribs, as shown by the arrows.

Off-focal radiation can be partially controlled by placing collimators outside the x-ray tube. For example, a lead diaphragm with a small port can be placed outside the tube glass envelope to stop some of the off-focus beam spread, as illustrated in Fig. 7.37(a). Although this approach can limit the amount of off-focal radiation, it cannot completely eliminate it. The schematic diagram in Fig. 7.37(b) clearly illustrates this effect by showing the relationship between the x-ray fan beam emitted from the x-ray focal spot and the position of the off-focal collimation. The smallest port opening is determined by the scanner FOV and the distance from the collimator to the focal spot. If the opening is too small, the primary x-ray beam is blocked and the effective FOV of the scanner is reduced. As illustrated in the figure, the detector is still exposed to a significant portion of the off-focal radiation (e.g., the x rays shown by the dashed lines). In general, the closer the port is to the target, the better performance it will produce.

An alternative approach employs a metal center section that is an integral part of the tube envelope. A beryllium window is provided for the primary x-ray beam.[5] Although this approach is an improvement over the stand-alone collimator, it also fails to totally eliminate the off-focal radiation for reasons similar to those outlined above.

Artifacts caused by off-focus radiation can be reduced or eliminated by software correction, though this is by no means simple or straightforward. The difficulty arises from the fact that the point-spread function (or the system's impulse response due to off-focal radiation) is not stationary. It changes from

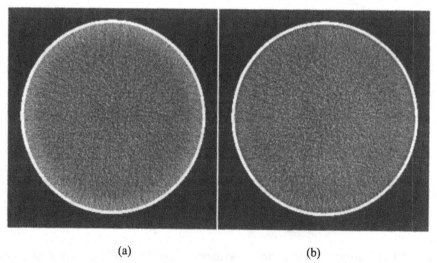

(a)   (b)

Figure 7.35 Reconstructed images of a water phantom encased in Plexiglas. (a) Original scan and (b) with off-focal radiation compensation.

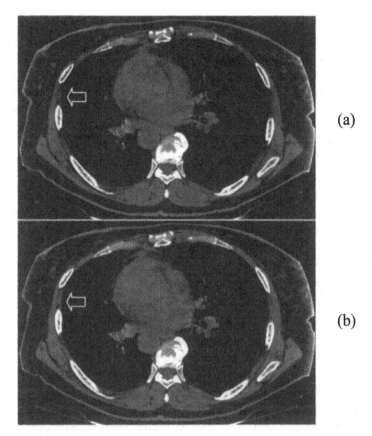

Figure 7.36 Example of off-focal radiation impact on patient chest scan (a) without proper off-focal compensation and (b) with off-focal compensation.

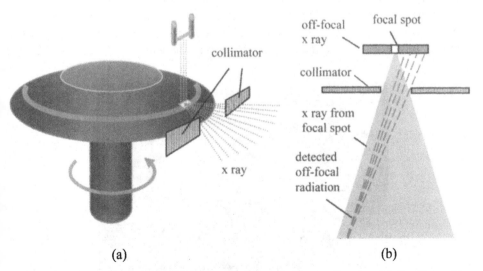

Figure 7.37 Illustration of the off-focal radiation reduction collimator. (a) Position of the collimator. (b) Limitations on the collimator opening.

location to location due to the change in magnification. The situation can be further complicated by the presence of a bowtie filter, which is intended to reduce the surface dose to the patient (see Chapters 5 and 6 for details). Figures 7.35(b) and 7.36(b) show the same scans corrected by a software algorithm. It is quite clear that the water-Plexiglas® interface was restored to the desired CT number for water. Similarly, the intensity of the soft tissue between the ribs was fully recovered.

In recent years, new technologies have been introduced into the x-ray tube design to control off-focal radiation. One such approach uses an electron collection cup, as shown in Fig. 7.38. The collection cup is biased to a higher electrical potential to absorb the backscattered electrons and prevent them from re-entering the target. This approach controls off-focal radiation with little need for additional compensation techniques.

7.4.2 Tube arcing

When impurities are present in the x-ray tube, a temporary short-circuit can result that is often called tube arcing or tube spit. To prevent continued arcing of the tube, a tube-spit detection mechanism is typically built into the power-supply unit of the CT system. For example, one could constantly monitor the output voltage and current of the power supply. When an x-ray tube-spit event occurs, a significant increase in current and a significant drop in voltage often occur. Once a tube-spit event is detected, the power supply to the x-ray tube is temporarily turned off to prevent further arcing. After a short period of time (typically in the millisecond range), the power-supply unit returns to its normal working condition. Therefore, the output x-ray photons are either significantly reduced or eliminated during the tube-spit event. When tube spit is an isolated event, the data acquisition and reconstruction continues, since sophisticated compensation schemes are built into the scanner to ensure image quality. If repeated tube-spit events occur (e.g., due to aging of the tube), the system should automatically abort the scan to prevent a degraded image quality and unnecessary dose to the patient.

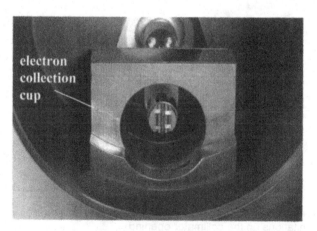

Figure 7.38 Off-focal radiation reduction using an electron collection cup.

Depending on the intensity and duration of a tube-spit event, one or more projection views can be affected. For these views, the phenomenon is similar to an extremely low x-ray dose scan. In the worst case, we must essentially measure the noise in the data acquisition system, since few x-ray photons are delivered to the patient. When the noise is magnified by the filtering process as part of the tomographic reconstruction, streaking artifacts can often be observed in conjunction with the tube-spit event. Figure 7.39 shows reconstructed images of a human skull phantom to illustrate the impact of tube spit. Figure 7.39(a) was obtained under normal tube operating conditions. Figure 7.39(b), on the other hand, was acquired with a severe tube spit. An array of streaks can be clearly observed in the figure. Based on the orientation of the streaks, we can conclude that the arcing occurred when the x-ray tube was at roughly the 2 o'clock position (if we trace all the straight lines in the image, these lines eventually intersect at the x-ray source location).

Since tube spit can be easily detected, algorithmic correction schemes can be applied to eliminate the streaking artifacts by either replacing the corrupted projection with a synthesized projection using neighboring uncorrupted views, or by the adaptive filtering scheme discussed in Section 7.3.4. Figure 7.39(c) shows an image reconstructed from the same scan as in Fig. 7.39(b) but with a tube-spit correction. The image is visually indistinguishable from the one reconstructed with the spit-free data shown in Fig. 7.39(a).

Frequent x-ray tube-spit events are usually a major indication that the x-ray tube is near the end of its life and needs to be replaced. Therefore, by monitoring the frequency of tube spit (together with other tube performance parameters), the life expectancy of the tube can be predicted and a tube replacement schedule arranged before tube failure occurs.

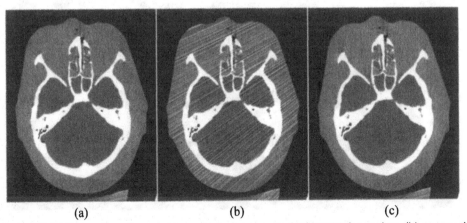

(a)   (b)   (c)

Figure 7.39 Examples of x-ray tube arcing: (a) scanned without tube arcing, (b) scanned with tube arcing and without compensation, and (c) scanned with tube arcing and with compensation.

7.4.3 Tube rotor wobble

In some cases, streaking artifacts are caused by mechanical failure or imperfection. This can be the result of a lack of rigidity in the gantry, a mechanical misalignment, or x-ray tube rotor wobble.[1,2] In all cases, the actual x-ray beam position deviates from the ideal position assumed by the reconstruction algorithm. As discussed in Chapter 6, a typical x-ray anode assembly consists of bearings, a rotor hub, a rotor stud, a rotor, and an anode. To dispense the intense heat generated by the bombardment of a large number of electrons, the rotor shaft spins up to 10,000 revolutions per minute (RPM). Such a high rotating speed in conjunction with the high temperature (several hundred degrees) produces significant wear and tear on a mechanical device. Thus, after an extended period of use, the tube rotor is unlikely to maintain the same stability and accuracy. Because of the fast rotating speed, x-ray rotor wobble typically occurs at a high frequency. Figure 7.40(a) shows a patient scan with severe x-ray tube rotor wobble. The same patient was later scanned after the problem was resolved, as shown in Fig. 7.40(b). The best approach to fix this type of problem is to replace the faulty component.

7.5 Detector-induced Artifacts

7.5.1 Offset, gain, nonlinearity, and radiation damage

Dark currents are present in all electronic devices, and CT detectors (including DAS) are no exception. This phenomenon is often called detector offset. The

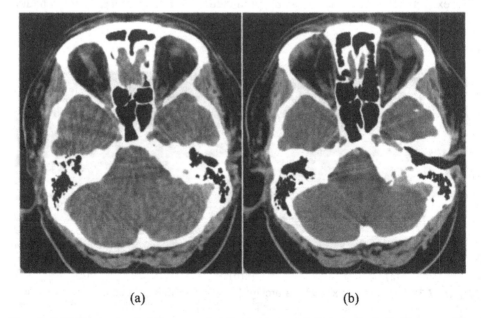

(a)   (b)

Figure 7.40 Reconstructed images of patient scans. (a) Data collected with x-ray tube rotor wobble. (b) Data collected without x-ray tube rotor wobble.

magnitude of offset not only changes with the ambient temperature and time, but also varies from channel to channel. Since the electronic dark current of a particular channel is a random process, the projection samples that merely fluctuate about the mean do not lead to significant image artifacts (this, of course, assumes the magnitude of the offset variation is smaller than the detected x-ray signal). The channel-to-channel variation, on the other hand, can lead to ring or band artifacts because the discrepancy between adjacent channels is consistent over all views. Based on our discussion at the beginning of the chapter, a consistent error at a fixed channel produces a ring (due to the nature of the FBP reconstruction).

To compensate for detector offset, the average value of the detector offset is measured just before each scan. This can be accomplished by calculating the average detector readings over a short period of time (e.g., one second) prior to the x-ray exposure. This process must be carried out on a channel-by-channel basis. Figure 7.41 shows an example of the measured offset values of 100 detector channels. Because the average offset of any channel can drift over time (for example, due to changes in the ambient temperature), it needs to be measured frequently to ensure the accuracy of the offset estimation. The average offset value is then subtracted from the projection samples prior to other preprocessing and reconstruction steps. To illustrate the accuracy requirement of offset, we introduced a 20-count error to the measured offset values at several channels. Compared to the typical value of an offset (over 2000 counts), the error is less than 1%! Images reconstructed with an accurate offset vector and an erroneous offset vector are shown in Figs. 7.42(a) and (b). Ring artifacts can be clearly observed.

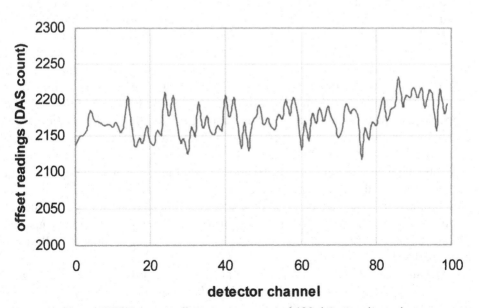

Figure 7.41 Average offset measurement of 100 detector channels.

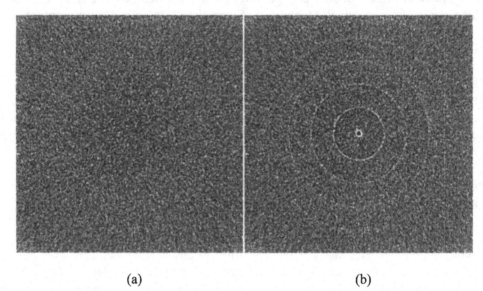

(a)   (b)

Figure 7.42 Reconstructed images of a 35-cm water phantom to illustrate the impact of an offset vector on the image artifact. (a) Properly measured offset and (b) offset with 1% error.

In the detector manufacturing process, it is impossible to produce absolutely identical detector cells. For example, the size of the detector cells can vary slightly when they are diced in different batches. The roughness of the detector cell surface may change from cell to cell because of the slight variation in the treatment process. The reflective material placed between detector cells cannot be guaranteed to be identical. The photodiode connected to each scintillator cell can exhibit a slightly different spectral response and conversion efficiency. The data acquisition electronics are inherently different from channel to channel. And the list can go on and on. The net effect is that the gains of individual detector cells cannot be made equal. In addition, detector gains are likely to change over time. Similar to the case with detector offset, ring or band artifacts will result if a proper correction is not rendered. To compensate for the gain variation from channel to channel and over time, most CT manufacturers use a calibration technique called air scans. In this process, a set of scans is acquired without any object inside the scanning FOV. If we assume that the x-ray flux impinging on all detector cells is the same, the detector cell gain should be proportional to its measurement in the air scans. To reduce statistical fluctuations, a large number of samples are averaged to produce a gain vector. The gain vectors must be produced for different scanning conditions (different tube voltage, collimator aperture, etc.).

The subsequently collected patient scans are divided by the appropriate gain vector to produce gain-normalized scan data. In the early days of CT development, air scans were often performed after each patient examination, due to the instability of the detector components. Thanks to the tremendous improvements in the design and manufacturing process, air scans need to be performed only once per day for current CT scanners.

Throughout our discussion, we have assumed that the detector output signal was proportional to the x-ray flux impinging on the detector. In reality, however, this condition is not strictly observed. The linearity of the detector and the DAS is approximately accurate over the dynamic range of the input signal. With aging of the detector and the DAS, the linearity can be degraded outside the range of specification. Like offset drift, linearity degradation leads to rings and bands.

Another cause of detector signal degradation is radiation damage or hysteresis. For many solid state scintillation materials, the detector gain varies with the radiation exposure history and recovery time. One such example is $CdWO_4$. The net effect of the radiation damage is a sustained bias on the measured signal. If sufficient correction is not rendered, rings and bands can result. In more severe cases, footprints of previously scanned objects can be superimposed on the current CT image.

Offset, nonlinearity, and radiation damage of the detector all lead to ring or band artifacts. Several correction approaches can be made to combat these artifacts. One approach is to locate and correct the error source.[6] For example, the radiation damage of some solid state scintillators can be suppressed significantly by material doping. For errors that are predictable and can be characterized, methods can be derived to eliminate the errors during detector calibration.[34,35]

Other techniques for reducing or eliminating rings are also interesting from a research point of view. These techniques often involve intensive digital signal and image processing techniques, and are listed here for the purpose of illustration.

Based on our previous discussion, a ring in the image space corresponds to a vertical line in the sinogram space. Therefore, one could search through the sinogram for vertical lines and remove them to eliminate rings in the image. The advantage of this approach is its simplicity, since it is much easier to identify a vertical line than a curved line (ring). The difficulty with this approach is the lower contrast between the lines and their background. The added noise and human anatomy make the search process difficult and less reliable.

Ring identification and correction in the image space faces different challenges.[36] The advantage of dealing with these artifacts in the image space is the better contrast between the rings and their background compared to the projection approach. The drawback is that we must search for errors that form an arc or annuli concentric to the iso-center, and since rings are detected by edge enhancement, it is often difficult to separate a ring artifact from the real anatomical boundary. To overcome this deficiency, a "supervised" or "guided" ring detection scheme has been proposed.[37] In this scheme, detector or DAS information is incorporated into the ring detection process. Once a ring is detected in an image, the ring location is mapped to the corresponding detector channel. The ring-to-channel mapping is straightforward: By calculating the radius of the ring (its distance to iso-center in the image), we can identify two channel locations (located on either side of the iso-channel) that form the same distance to the system iso-center. If either detector channel is active (based on *a priori* information), the likelihood of this ring being an artifact is significantly increased. *A priori* information may include radiation damage, linearity, thermal sensitivity, and other parameters related to the detector and DAS. Once an artifact ring is identified, the average intensity of the ring is determined for each annulus or arc. The error is then subtracted from the original image.

7.5.2 Primary speed and afterglow

Nearly all solid state scintillation materials exhibit certain levels of signal decay and afterglow. This behavior is similar to the phenomenon of turning off a television set in a dark room. The television screen does not immediately revert to black. Instead, the intensity of the screen gradually fades away. Similar characteristics exist for solid state detectors. If we expose a solid state detector to x-ray photons for a period of time and swiftly shut off the input x ray, the detector output does not reach zero immediately. The residual signal is sustained from a few microseconds to a few hundred milliseconds, depending on the scintillating material.

For illustration, two decay curves of a HiLight™ (GE Medical Systems, Milwaukee, Wisconsin) detector built more than 15 years ago are depicted in Fig. 7.43. This figure illustrates various time constants in the primary speed and afterglow of the detector. In more recent HiLight™ detectors, afterglow is eliminated by doping the scintillators with rare-earth materials. A casual examination of the curves in the figure indicates that there are multiple decay components in decay curves (different decay components correspond to different decay time constants). Some of the time constants are short (in the range of several milliseconds), and others are long (several hundreds of milliseconds). Their impact on image quality depends largely on the magnitude of the time constant. For the convenience of future discussion, we will call the decay components with short time constants the primary speed of the detector, and the decay components with long time constants the afterglow of the detector.

The primary speed of the detector impacts mainly the spatial resolution of the reconstructed images, while the afterglow components primarily affect image artifacts.[34,35] To illustrate the impact of the primary speed, we performed the following experiment. A GE Performance™ phantom was scanned twice on a LightSpeed™ CT scanner with a HiLight™ detector. To reduce other factors that could potentially bias our measurements, we made sure that these two scans had the same total x-ray photon flux (200 mA for 0.5 sec and 50 mA for 2 sec). Images from both scans were reconstructed with a high-resolution reconstruction kernel (with the bone algorithm) at a 12-cm reconstruction FOV, as shown in Figs. 7.44(a) and (b). For better presentation, only a small portion of each reconstructed image around the wire is displayed. It is clear from the figure that a significant degradation in spatial resolution is present for the 0.5-sec scan. In particular, a comet-like appearance can be observed around the wire along the azimuthal direction in Fig. 7.44(b).

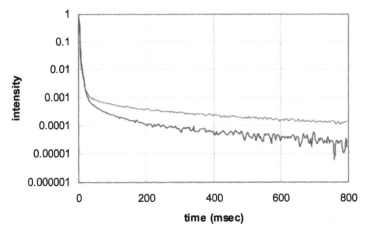

Figure 7.43 Illustration of solid state scintillating detector decay. Gray and black lines represent the step response of two detector cells built more than 15 years ago.

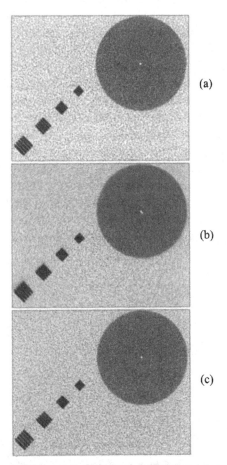

(a)

(b)

(c)

Figure 7.44 Reconstructed images of a GE Performance phantom (a) scanned at 2.0-sec scan speed, (b) scanned at 0.5-sec scan speed without detector decay correction, and (c) scanned at 0.5-sec scan speed with detector decay correction.

To further quantify the resolution impact, we calculated the MTF of the wire; the results are shown in Fig. 7.45. In the MTF calculation, a 2D Fourier transform of the reconstructed wire was performed, and its magnitude (averaged over all orientations) was plotted as a function of spatial frequency. The MTF for the 2-sec scan (without correction) is shown by the thin dotted line, and the 0.5-sec MTF (without correction) by the thick gray dotted line. The loss of resolution is evident.

In addition to the resolution loss, the noise pattern is no longer homogeneous for the 0.5-sec scan, as a result of the detector primary speed. A closer examination of the background in Fig. 7.44 shows that the noise follows the azimuthal orientation in Fig. 7.44(b), while the noise has no preferred orientation in Fig. 7.44(a).

Another impact of the detector primary speed is the distortion introduced in the reconstructed images. The bars in the phantom are physically rectangular in shape. The shape of the larger bars in Fig. 7.44(b) is significantly distorted compared to Fig. 7.44(a), as a result of residual signal contamination from previous views.

The impact of detector afterglow on image quality is somewhat different. Note that afterglow signals constitute a few percent of the overall signal. Therefore, their impact on the spatial resolution is relatively small. Their effects are revealed mainly as image artifacts, such as shading and rings, caused by a channel-to-channel variation in the afterglow characteristics.

As discussed previously, the afterglow contains multiple time constants ranging from a few tens of milliseconds to a few hundred milliseconds. The impact of the relatively short time-constant component is limited to the surface of the scanned object where the contrast is the highest (object versus air). To illustrate this phenomenon, Fig. 7.46(a) presents a clinical example in which a patient was scanned at a 1-sec scan speed. The skin in the lower left-hand side of the image disappears, as indicated by an arrow. Although this type of artifact does not present any risk of misdiagnosis, it degrades image quality and is annoying to radiologists.

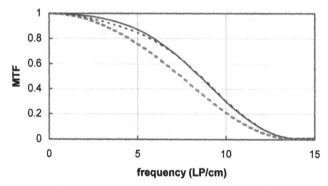

Figure 7.45 MTF comparison of a high-resolution reconstruction (bone algorithm). Solid line: 0.5-sec scan with correction; thin dotted line: 2.0-sec scan without correction; thick gray dotted line: 0.5-sec scan without correction.

The more serious image artifacts are caused by afterglow components with long time constants (in the range of a few hundred milliseconds). A detector cell-to-cell nonuniformity at an afterglow level can cause rings and arcs deep inside the reconstructed object. To demonstrate this artifact, we scanned a large oval phantom with its center positioned roughly 10 cm below the iso-center. The phantom was scanned at a 1-sec scan speed with a vintage detector that exhibited a significant amount of afterglow. The resulting image is shown in Fig. 7.47(a). Multiple rings and bands are clearly visible deep inside the phantom.

Several methods are available to combat image artifacts related to scintillator decay. The best method, of course, is to remove the root cause of the artifacts. Thus, if we can sufficiently suppress the afterglow and primary speed of the scintillating material itself, the projection samples become free of contamination from previous samples. Studies have shown that detector decays can be sufficiently suppressed by doping the scintillators with rare-earth materials.

Alternatively, the detector decay signals can be suppressed with algorithmic corrections.[34,35,38,39] One compensation algorithm has a correction scheme that makes use of the multiple exponential decay model of the HiLight^{TM} scintillator and the stability of the decay parameters to radiation dose, detector aging, x-ray photon energy, and temperature. Based on the theory of the solid state detector afterglow mechanism and extensive experiments, it was found that the detector impulse response can be modeled by the following multi-exponential:

$$h(t) = \sum_{n=1}^{N} \frac{\alpha_n}{\tau_n} U(t) e^{\frac{-t}{\tau_n}}, \qquad (7.13)$$

where

$$U(t) = \begin{cases} 1 & t \geq 0, \\ 0 & t < 0. \end{cases}$$

α_n represents the relative strength of the decay component with time constant τ_n, and N is determined from measurements of the detector decay curves.

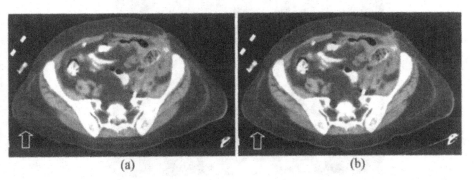

(a)   (b)

Figure 7.46 Example of shading artifact caused by detector decay. (a) Original image and (b) image corrected for detector decay.

Based on linear system theory and the fact that the line integral of the attenuation coefficient is nearly constant within the data sampling interval Δt, a recursive relationship can be derived between the sampled measurement $y(k\Delta t)$ and the true sampled line integral $x(k\Delta t)$:

$$x(k\Delta t) = \frac{y(k\Delta t) - \sum_{n=1}^{N} \beta_n e^{\frac{-\Delta t}{\tau_n}} S_{nk}}{\sum_{n=1}^{N} \beta_n}, \qquad (7.14)$$

where

$$S_{nk} = x\big[(k-1)\Delta t\big] + e^{\frac{-\Delta t}{\tau_n}} S_{n(k-1)} \text{ and } \beta_n = \alpha_n \left(1 - e^{\frac{-\Delta t}{\tau_n}}\right).$$

Note that β_n is a constant that depends on the detector afterglow characteristics and the view-sampling rate. Figure 7.44(c) depicts a reconstructed image of the same phantom scan shown in Fig. 7.44(b), but using the recursive algorithm described above. The image quality is visually indistinguishable from the scan acquired with the 2-sec scan speed [Fig. 7.44(a)]. A quantitative analysis confirmed that the spatial resolution (as measured by the MTF) of the 0.5-sec scan was essentially the same as that of the 2-sec scan, as shown in Fig. 7.45. In addition, the recursive correction effectively eliminates the shading and ring artifacts shown in Figs. 7.46(b) and 7.47(b).

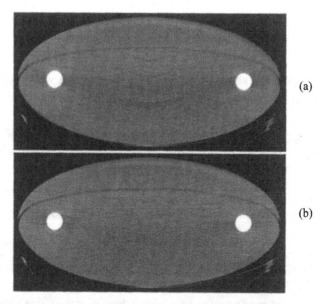

(a)

(b)

Figure 7.47 Illustration of the effect of nonuniform detector afterglow. (a) Original image and (b) image corrected for detector decay.

7.5.3 Detector response uniformity

In all previous discussions, we have assumed that the detector response to x-ray photons was spatially uniform. That is, the output of a detector did not change with the location where the x-ray photon impinged on the detector. If the response of the detector is plotted as a function of location, it should be a rectangular-shaped function, as shown by the thin solid line in Fig. 7.48. In this figure, the horizontal axis represents the location along the detector z axis (across the slice thickness), and the vertical axis represents the normalized detector response. For a single-slice scanner, the largest slice thickness is 10 mm at the iso-center. Considering the magnification of a CT scanner and the extra spacing at both ends of the detector cell used to avoid edge effect, the length of a detector cell is over 30 mm, as shown in the figure. In reality, the detector uniform response assumption is often invalid. For example, the response of a newly manufactured detector cell approximates the "ideal" response, but as the detector ages, the detector response is likely to change. For instance, the reflectivity of the material placed between the detector cells (to channel light photons toward the photon diodes) can change over time due to radiation exposure, mechanical stress between the detector cells, or the aging process. Another possible cause of a nonuniform detector response is the scintillating material property change due to radiation damage. The net effect is that some detector responses are significantly different from the responses of their neighboring cells. The thick line in Fig. 7.48 depicts a degraded detector cell response. In this particular example, the detector cell underwent an excessive number of accelerated tests to investigate the worst-case scenario. These types of tests provide guidelines and feedback on the detector manufacturing process to ensure that the degradation in actual CT detectors is within specification.

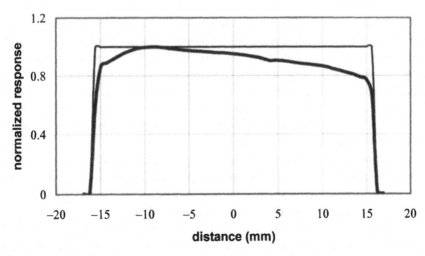

Figure 7.48 Normalized detector response. Thin line: the ideal detector response; thick line: a degraded detector response.

Let us examine the image quality impact of the detector response degradation. The impact of the detector response in x (across detector channels) is different from that in z (along the patient long axis) due to the small physical dimension of the detector cells in x (roughly 1 mm). Light photons that reach the photodiodes are reflected from both sides of the reflective material. Consequently, the impact of a degraded reflector will affect the response across the entire 1-mm region (rather than in a localized region). Now consider the detector response degradation in z of a single-slice scanner (the multislice case will be discussed later). As mentioned previously, a typical detector cell dimension in z is over 30 mm. As a result, the likelihood of a detector response variation in z is much higher. If a portion of a reflector becomes defective, it will likely affect the detector response in its immediate neighborhood but not affect the response several millimeters away. This results in variation in the detector response profile, as shown in Fig. 7.48.

We must ask ourselves the following question: What impact does a degraded detector response in z have on image quality? To estimate the impact, we need to understand why a degraded response cannot be calibrated with a gain vector obtained from air scans, as discussed in Section 7.5.1. Consider the following simple example. Figure 7.49 shows three hypothetical detector cell responses (in z). All three cells have the same average detector gain but with different variations as a function of location. The first cell represents an "ideal" detector in which the detector response is independent of the location. The second and third cells have different gain slopes as a function of the location (z). When an air scan is performed, all three cells (A, B, and C) are exposed to a uniform x-ray flux (in the z direction) of the same intensity. Because the average gains of the three cells are the same, the air scan produces identical gain calibration values for all three cells. The outputs of the calibrated readings are identical, since the input x-ray flux is divided by the same average gain value derived from the air calibration, as shown by the top row on the right-hand side of Fig. 7.49. Next, consider the case in which the input x-ray flux has a gradient in the z direction as depicted by the middle row in Fig. 7.49. In this case, the output obtained from each cell (without calibration) is the integration of the product of two functions: the detector response function and the x-ray flux function. For a detector with a slope similar to that of the x-ray flux (cell B), the resulting signal is large. On the other hand, a detector with a dissimilar slope (cell C) results in a small signal. Since the three cells have identical gain calibration values, the relative difference between the output signals of the three cells (after calibration) remains the same. This shows that the output signal depends on the slope of the detector response, even after calibration. In other words, the detector response cannot be calibrated by the simple air-calibration method. A similar situation exists for x-ray flux with the opposite slope, as shown by the bottom row in Fig. 7.49.

The input x-ray flux gradient is induced by the scanned object. If a uniform x-ray flux is passed through a sloped object (e.g., scanning near the top of a human head), the x-ray flux is no longer uniform. The x-ray intensities toward the top of the head are higher than they are toward the bottom, since the radius of the head reduces quickly. Most human anatomies do not change drastically in shape along the long axis (z axis). As a result, a variation in the x-ray flux becomes significant only over a reasonably large distance (e.g., 5 to 10 mm).

Now we can examine the impact of a degraded detector response on image quality. For illustration, we scanned a human skull phantom (a human skull filled with plastic materials) with a degraded detector at two locations: the mid-brain section and the tip of the skull. The phantom was scanned with a 7-mm slice thickness to ensure sufficient variation in z over the thickness of the slice. At the midbrain section, the shape of the skull is roughly a uniform cylinder with little variation in z. Based on our previous analysis, we do not expect any artifacts associated with the degraded z profile. Indeed, Fig. 7.50(a) shows the reconstructed image with no visible image artifacts. For the scan taken near the tip of the skull, image artifacts are clearly visible as concentric rings or bands, as shown in Fig. 7.50(b).

Several approaches can be employed to combat detector z-slope-related image artifacts. For example, the detector design and manufacturing process can be improved to ensure stable gain over the life of the detector. This includes improvements in both the reflector and the scintillating material. By eliminating the root cause of the image artifact, sloped objects can be scanned as safely as the nonsloped objects.

Another approach is to avoid using thick slices when scanning sloped anatomies. For example, when scanning near the top of the head, the slice thickness should be limited to 3 mm or less. Because of the gradual change in the shape of the skull, we do not expect significant x-ray flux variation (after passing through the head) over a 3-mm distance. To ensure an adequate signal-to-noise ratio, images from multiple scans can be added together (in either the image space or projection space after the logarithmic operation). The drawbacks, of course, are longer scan time and increased tube loading for single-slice CT.

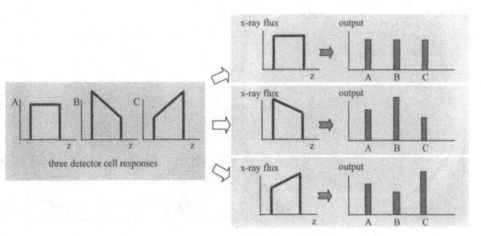

Figure 7.49 Examples of detector responses to different input x-ray flux. Top: input x-ray flux is uniform and calibrated detector outputs are identical. Middle: input x-ray flux is sloped and calibrated detector outputs are different. Bottom: input x-ray flux is sloped and calibrated detector outputs are different.

An alternative approach to eliminating slope-related image artifacts is to use algorithmic corrections. Since we know *a priori* the shape of the detector response as a function of distance in *z*, the amount of projection error can be reliably estimated once the slope of the object is estimated. This approach, of course, relies on knowledge of the neighboring reconstructed images or projections of neighboring slices. A better approach relies on the "signature" of the projection error pattern. For each degraded detector, a unique set of "signatures" is produced when scanning a sloped object. These signatures can be easily calculated based on the measured detector response. When an object is scanned, the projections are filtered to produce an "error candidate." The error candidate is then matched against the detector signature using cross correlation. The amount of slope error present in the projection is determined by the correlation coefficient of the two vectors. Once the projection error is known, we can modify the original projections to arrive at an artifact-free image reconstruction. These approaches have been shown to be quite effective. Interested readers can refer to Refs. 40 and 41 for details. Figure 7.43 shows an example of the correction. Figure 7.51(a) is identical to Fig. 7.50(b) with a dark-band *z*-slope artifact. Figure 7.51(b) was reconstructed from the same scan using one of the software correction methods described above. It is clear that image artifacts are nearly completely eliminated. A similar technique can be applied to multislice scanner *z*-slope corrections.[42] When many small detector cells are grouped together to form thicker slices (for example, using four 1.25-mm cells to form a 5-mm slice thickness detector), the detector *z* slope is unavoidable (due to the cell-to-cell gain variations in *z*). Because of similarities between single-slice and multislice scanners, the multislice discussion is omitted here.

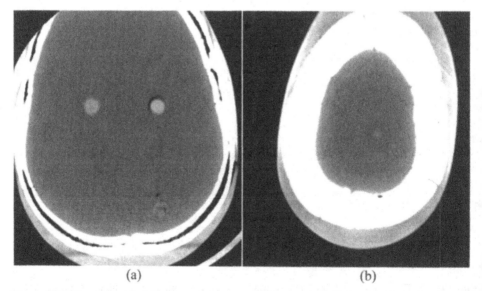

(a)   (b)

Figure 7.50 Reconstructed images of a human skull phantom (WW = 150) (a) scanned at midbrain section (nonsloped anatomy) with a degraded detector, and (b) scanned near the top (sloped anatomy) with the same detector.

Testing of the detector z-slope-related performance can be carried out quite easily with simple objects. For example, instead of building a specialized phantom, we can fill a two-liter soda pop bottle with water and scan it near its neck where a significant size variation is present. The long axis of the soda pop bottle must be placed parallel to the patient table. Figures 7.52(a) and (b) show the soda pop bottle (scanned with a 7-mm slice thickness) reconstructed without and with the z-slope correction. Since the same detector was used for both experiments (a soda pop bottle and a skull phantom), the image artifact patterns are nearly identical [Figs. 7.51(a) and 7.52(a)]. This demonstrates that the soda pop bottle test can be used effectively to predict the detector z-slope performance under clinical conditions.

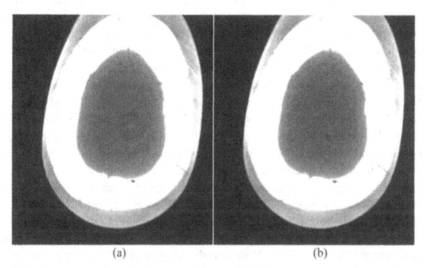

(a)   (b)

Figure 7.51 Reconstructed images of a human skull phantom with a degraded detector (WW = 150). (a) Scanned near the top (sloped) without correction. (b) Same scan (sloped) with correction.

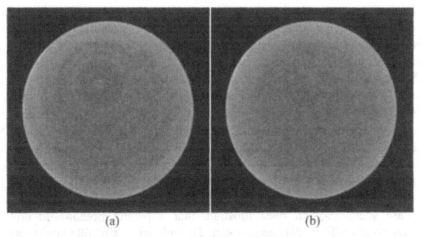

(a)   (b)

Figure 7.52 Reconstructed images of a soda pop bottle near the neck (WW = 80). (a) Original image without correction and (b) reconstructed image with correction.

7.6 Patient-induced Artifacts

7.6.1 Patient motion

Previous sections have described data inconsistencies caused by detector defects. A similar situation can occur when a patient moves during the scan. The motion can be either voluntary (respiratory motion) or involuntary (peristalsis and cardiac motion), and is usually 3D. The object of interest can move, expand, or shrink within the scanning plane. The entire object can also move in and out of the plane. Based on our previous discussions on projection data inconsistency, the link between patient motion and image artifacts is apparent.[43,44] Because these artifacts can sometimes mimic diseases and cause misdiagnosis,[3] particular attention must be paid to recognize, avoid, or correct them.

First we will examine the image artifacts caused by respiratory motion. Contrary to common belief, respiratory motion is hardly periodic. We conducted some experiments to examine the position of the chest wall under "normal" breathing conditions. For the experiment, we asked volunteers to lie in a supine position and rest comfortably for a short period of time. An infrared device was mounted some distance from the subject to monitor the chest wall motion. The subject was instructed to breathe normally while motion curves were recorded. One could argue that the subject under examination could not breathe "normally" since he or she was aware of being monitored. However, this situation was not much different from a clinical setting where a patient is under CT examination. Of all the subjects we examined, none of the collected respiratory motion curves was periodic, either based on peak-to-peak time intervals or based on the location of the chest wall at the end of inspiration or expiration. For illustration, one of the curves is presented in Fig. 7.53. Note that the magnitude of the chest wall motion from peak to valley (corresponding to the end of inspiration and the end of expiration) varies more than a factor of 4 from cycle to cycle! A closer examination of the time intervals between the peaks also indicates a significant variation. In fact, in the worst case, the variation exceeds 40%.

Another surprising finding from this experiment is that a significant amount of chest wall motion existed even when the person was instructed to hold his or her breath. This finding indicates that even under favorable clinical conditions (a cooperative patient), a patient cannot hold his or her breath perfectly. A significant amount of air redistribution takes place during a breath-hold. An example of the chest wall location under the breath-hold condition is shown in Fig. 7.54. In this experiment, the subject was asked to take a deep breath (at the beginning of the motion curve) and hold his breath for a period of 8 sec. Note that the chest wall reached its peak position fairly quickly and gradually retreated to a lower position. The amount of chest wall motion during the entire breath-hold period was roughly 25% of the distance that the chest wall travels from end-of-inspiration to end-of-expiration. This particular example represents one of the worst-case scenarios, but does illustrate that a patient breath-hold does not guarantee "motion-free" data acquisition. The subject was healthy and was asked to hold his breath for only 8 sec. In routine clinical CT examinations, many patients are asked to hold their breath for well over 20 sec. Obviously, respiratory motion-induced variability cannot be ignored.

Now we will examine the impact of motion on image artifacts. We scanned a dry-lung phantom under no motion and under different motion conditions. The phantom was placed on a device whose up-and-down motion was controlled by a computer with measured motion curves. The phantom was first scanned in a stationary condition to provide the "gold standard" for image quality evaluation. The reconstructed image is shown in Fig. 7.55(a). Note that all of the nodule and vessel structures can be clearly observed. Next, we scanned the phantom under three different conditions. In the first experiment, data acquisition was centered at the end of the expiration (the valley portion of Fig. 7.53). Because the data acquisition was relatively short (1 sec) compared to the duration of the valley, the amount of motion in the projection is small. As a result, only minor motion-

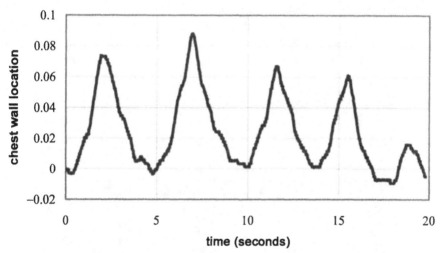

Figure 7.53 Example of a respiratory motion curve measured at the chest wall under "normal" breathing condition.

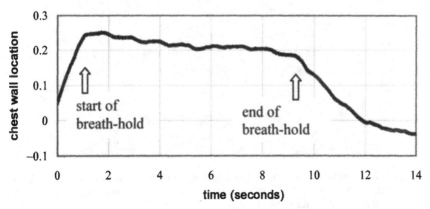

Figure 7.54 Example of a chest wall motion under "breath-hold" condition.

induced image artifacts can be observed, as shown in Fig. 7.55(b). In the second experiment, data acquisition was centered either on the ascending portion or the descending portion of the motion curve. In both cases, the position difference between the beginning and end of the scan was at its maximum; therefore, the projection discontinuity between the beginning and ending of the experiment is large. The purpose of this experiment was to illustrate the worst-case condition. The reconstructed images are shown in Figs. 7.55(c) and (d). As expected, the appearance of the image artifacts is more pronounced. Streaking artifacts are present near the high-density vessel structures, and a close examination of the vessel structures shows ill-defined boundaries; many of the vessel boundaries are blurry.

To illustrate respiratory motion artifacts under clinical conditions, Figs. 7.56(a) and (b) show chest CT scans of a patient taken at two different time instants. The first scan was taken under relatively motion-free conditions, and the second scan was taken during significant patient respiratory motion. Figure 7.56(b) clearly shows that dark and bright shadings are present throughout the lung region as a result of the respiratory motion. A closer examination of the figure indicates that the shapes of the vessels and nodules are also distorted.

(a)   (b)

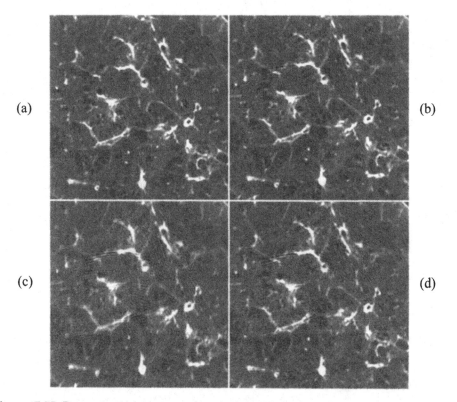

(c)   (d)

Figure 7.55 Reconstructed images of a dry-lung phantom. (a) No motion; (b) image centered at the end of expiration; (c) image centered at the inspiration cycle; and (d) image centered at the expiration cycle.

Various methods can be used to reduce motion artifacts. For respiratory motion, one could instruct the patient to hold his or her breath during the scan. However, in many cases a patient may be unable to hold his or her breath or is simply unconscious or uncooperative, so this approach becomes ineffective. Alternat-ively, the patient can be immobilized by sedation, but this is obviously not comfortable for the patient and may be prohibitive. Another approach is to shorten the scan time. For example, scan time on the order of 40 ms can practically freeze any type of patient motion. However, a very short scan time may lead to a lower signal-to-noise ratio than is provided by a conventional scan time.[45] The tradeoffs among motion artifacts, system design complexity, x-ray photon noise, and many other parameters must be carefully analyzed.

If we assume that patient motion is continuous, we can conclude that the maximum discrepancy caused by the motion occurs between the beginning and end of the scan (the longest time span between two views). Therefore, the views near both ends of the scan should be treated with the least confidence. To ensure artifact-free reconstruction, the contributions of the projection samples must be properly normalized. If the contributions of some projection samples are arbitrarily reduced, shading artifacts will result. To explain the compensation techniques, we will again examine the sampling geometry of a fan-beam CT scanner.

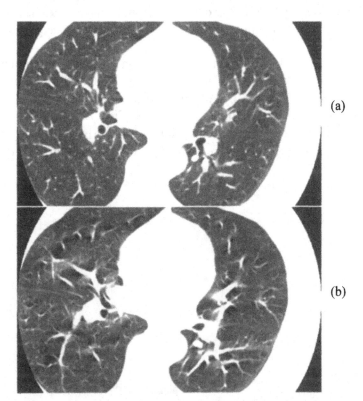

(a)

(b)

Figure 7.56 Illustration of a respiratory motion artifact. (a) Chest scan relatively free of respiratory motion and (b) same patient scanned while breathing.

Figure 7.57 shows the sampling geometry of a third-generation CT scanner. For clarity, only five projection samples per view are shown with the sample of interest depicted as a thick line for the projection with view angle β. The entire projection is shown with a solid black color. The sample of interest forms a detector angle γ with respect to the iso-ray, as shown in the figure. When the CT gantry is rotated to the view angle of $\beta + \pi + 2\gamma$, the original sample of interest is then sampled by the detector cell that forms an angle $-\gamma$ with respect to the iso-ray, as shown by the dotted gray line. When two samples satisfy this condition, they are often called a conjugate sampling pair. This simple illustration demonstrates that a fan-beam scan taken with a full rotation (2π) contains two samples for each ray path. Consequently, each 2π scan can be divided into two complete subsets of projections, and each projection is capable of reconstructing an image. Each subset is commonly called a half-scan.

The property discussed above points to two approaches by which motion-induced image artifacts can be suppressed. The first approach is to reduce the data acquisition time used to form an image. Instead of using the entire set of 2π projections to reconstruct an image, only one subset is used for the reconstruction. The minimum projection angle of a subset is $\pi + 2\gamma_m$, where γ_m is the fan angle of the projection. For a typical CT scanner geometry, this corresponds to roughly 225 to 235 deg of rotation. Compared to the original 360-deg data acquisition, this approach effectively reduces data acquisition time by 35% to 38%.

The formulation of a half-scan reconstruction algorithm is problematic and deserves careful examination. The problem is related to the fact that the projection samples in the $\pi + 2\gamma_m$ angular span contain duplicate samples. This can be explained by the sinogram shown in Fig. 7.58. In this figure, the horizontal axis represents the detector angle γ formed by each sample with the iso-ray, and the vertical axis represents the projection angle β. The conjugate

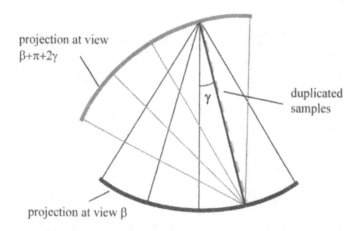

Figure 7.57 Illustration of duplicated samples obtained in a fan-beam sampling geometry.

samples to the line $\beta = \beta_0$ fall along the line $\beta = \beta_0 + \pi - 2\gamma$. Therefore, the two shaded triangles shown in Fig. 7.58 represent two redundant sampling sets. To ensure artifact-free image reconstruction, the total contribution from each conjugate sampling pair must sum up to unity. This can be accomplished by using a widely cited algorithm proposed several decades ago.[46] This algorithm calls for the normalization of projections by a weighting function $w(\gamma, \beta)$ prior to the FBP:

$$w(\gamma, \beta) = 3\theta^2(\gamma, \beta) - 2\theta^3(\gamma, \beta), \tag{7.15}$$

where

$$\theta(\gamma, \beta) = \begin{cases} \dfrac{\beta}{2\gamma_m - 2\gamma}, & 0 \le \beta < 2\gamma_m - 2\gamma, \\[2mm] 1, & 2\gamma_m - 2\gamma \le \beta < \pi - 2\gamma, \\[2mm] \dfrac{\pi + 2\gamma_m - \beta}{2\gamma_m + 2\gamma} & \pi - 2\gamma \le \beta < \pi + 2\gamma_m. \end{cases} \tag{7.16}$$

The above weighting function satisfies two essential conditions. First, the total contribution from conjugate samples equals unity. Second, the weighting function is continuous and differentiable in γ, because the tomographic filtering operation is essentially a derivative operator. Any discontinuity in the projection will likely cause streaks in the reconstructed image. The advantage of the half-scan approach is improved temporal resolution, since the data acquisition time is significantly reduced. The disadvantage is an increased image noise level, since only a fraction of the available samples is used for reconstruction. Another potential disadvantage is that this approach excludes high-resolution applications unless dynamic focal spot wobbling is used. The detector quarter-offset concept discussed in Chapter 3 is no longer valid. This concept relies on conjugate samples that are 180 deg apart. A compromise solution was later proposed that uses more than the minimum amount of data to form an image.[47]

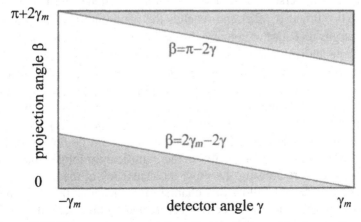

Figure 7.58 Sinogram representation of redundant samples in a half-scan.

The second approach to suppress motion artifacts minimizes the contributions of the most inconsistent projections in the image. This can be accomplished by applying small weights to the projections near the start and end of a scan. The entire 2π projection data are used to reconstruct an image. In general, the weighting function varies in both β (view angle) and γ (fan angle), and must satisfy the following condition:

$$w(\gamma, \beta) + w(-\gamma, \beta+\pi+2\gamma) = 2. \qquad (7.17)$$

In addition, $w(\beta, \gamma)$ must be continuous and differentiable to avoid streaking artifacts. An example of this type of weighting scheme is the underscan weight[48]:

$$w(\gamma, \beta) = \begin{cases} 3\theta^2 - 2\theta^3, & \text{where } \theta = \dfrac{\beta}{\beta_u}, & 0 \leq \beta < \beta_u, \\[2ex] 2 - \left(3\theta^2 - 2\theta^3\right), & \text{where } \theta = \dfrac{|\beta - \pi + 2\gamma|}{\beta_u}, & \pi - \beta_u - 2\gamma \leq \beta < \pi + \beta_u - 2\gamma, \\[2ex] 3\theta^2 - 2\theta^3, & \text{where } \theta = \dfrac{2\pi - \beta}{\beta_u}, & 2\pi - \beta_u \leq \beta < 2\pi, \\[2ex] 1, & & \text{otherwise,} \end{cases} \qquad (7.18)$$

where β_u is a parameter that controls the size of the transition region. In general, the larger the value of β_u, the better the performance is in terms of motion suppression. However, a larger β_u produces a worse noise performance in the final image. In addition, the effectiveness of the detector quarter-offset is compromised, since the conjugate samples no longer contribute equally to the final image. As a result, we must expect either a reduction in spatial resolution or an increase in aliasing artifacts in the final image. Clearly, parameter selection must be carefully considered in terms of motion artifact suppression, noise, and spatial resolution.

Note that the underscan weighting function approaches zero at both ends of the data set ($\beta = 0$ and $\beta = 2\pi$) and reaches the maximum near the center of the dataset. For scans that start at nonzero projection angles, we can simply replace β by $\beta - \beta_0$ in the above equation, where β_0 is the starting angle of the dataset. To demonstrate the effectiveness of the underscan correction, a patient abdominal scan with significant motion artifacts is shown in Fig. 7.59(a), and a patient head scan is shown in Fig. 7.60(a). The same scan was then weighted by underscan weighting prior to the FBP. The resulting images are shown in Figs. 7.59(b) and 7.60(b). A significant reduction in streaking artifacts can be observed. Since the underscan relies on the motion period being significantly longer than the gantry rotation time, therefore causing the most inconsistency at the start and end of a scan, motions that do not follow this characteristic, such as the swallow motion shown in Fig. 7.61, will not be effectively corrected by this method.

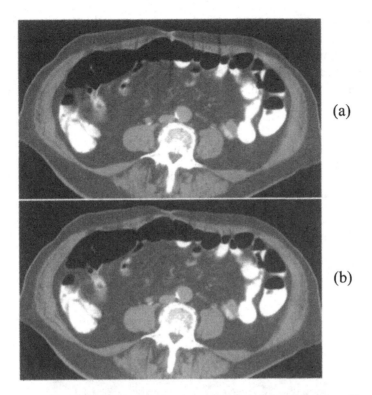

Figure 7.59 Reconstructed images of a patient scan. (a) Original data without motion correction. (b) Reconstruction with underscan weighting.

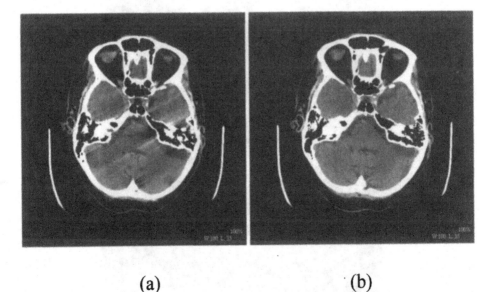

(a)   (b)

Figure 7.60 Illustration of patient's involuntary head motion (a) without compensation and (b) with correction.

Another type of weighting scheme suppresses patient motion artifacts by using more than 2π projection data. Like the previous scheme, this type of weighting scheme tries to reduce the contributions of the projections at the start and end of a data set. The redundant or duplicate samples, in this case, differ by exactly one gantry rotation. Therefore, the weighting function satisfies the following equation:

$$w(\gamma, \beta) + w(\gamma, \beta+2\pi) = 1. \tag{7.19}$$

An example of this type of weighting function is overscan weighting, which can be expressed mathematically by the following equation:

$$w(\gamma, \beta) = \begin{cases} 3\theta^2 - 2\theta^3, & \text{where } \theta = \dfrac{\beta}{\beta_u}, & 0 \le \beta < \beta_u, \\ 1 - 3\theta^2 + 2\theta^3 & \text{where } \theta = \dfrac{\beta - 2\pi}{\beta_u}, & 2\pi \le \beta < 2\pi + \beta_u, \\ 1 & & \text{otherwise.} \end{cases} \tag{7.20}$$

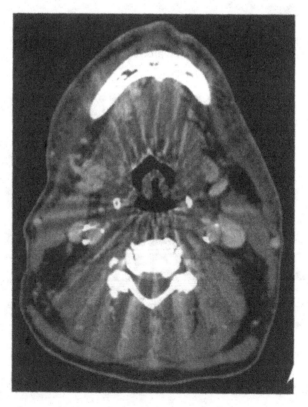

Figure 7.61 Artifacts due to swallow motion.

Similar to the underscan weight, β_u is a parameter that represents the size of the transition region. The disadvantage of the overscan is that it needs more than one rotation of projections to reconstruct an image. This translates to a longer scan time. On the other hand, the noise distribution in the reconstructed image is more uniform than the underscan type of weighting.

The severity of a motion artifact depends on the orientation of the motion. In general, motion artifacts can be reduced by aligning the initial position of the x-ray source with the primary direction of motion. Intuitively, this can be explained by considering a point in the object moving along a line. If the motion is parallel to the direction of the x-ray beam, the effect on the projection is negligible, since the integration along the path has not changed. When the motion is perpendicular to the x-ray beam, however, the change is noticed immediately because when the point moves outside the original ray path, the integration along the ray path changes. Alternatively, we can state that when a point moves along the direction of the x-ray beam, the shadow generated on the detector does not change. On the other hand, a movement of the point perpendicular to the direction of the x-ray beam produces a motion of the shadow on the detector. Therefore, one can devise a scheme to minimize motion artifacts by forcing the start angle of a scan to be approximately aligned with the direction of motion. For example, when scanning a chest or abdomen, the motion induced by breathing most likely moves along the y axis (up and down) rather than sideways. Therefore, the starting angle of the scan should be close to the 12 or 6 o'clock positions instead of the 3 or 9 o'clock positions.

In the past, various attempts have been made to compensate for patient motion by performing physiological gating.[50–53] The basic idea is that the time it takes to acquire a complete set of projections is usually shorter than a respiratory motion cycle, and in a typical respiratory cycle, the amount of patient motion is relatively small near the peaks of inspiration or expiration. This is known as the quiescent period of respiration. The objective in physiological gating is to acquire the scan during the time window when the patient motion is at its minimum. This can be accomplished by employing an adaptive prediction scheme.[53] The adaptive approach is necessary since the breath pattern changes frequently during a scan. Figures 7.55(c) and (d) show a phantom study performed without the predictive algorithm, and Fig. 7.55(b) shows the phantom study performed with the predictive algorithm. The reduction in streaks is obvious, and the blurring caused by the motion has been significantly reduced. Physiological gating devices are often limited in the types of motion that they can detect. For example, a recent study has shown that the motion of air bubbles in the bowel or other liquid-containing structures can cause image artifacts.[54] This type of artifact has a peculiar appearance, as illustrated by the computer simulation in Fig. 7.62. For this simulation, a water cylinder with an air bubble was placed at the iso-center of the system. During the data acquisition, the air bubble moved vertically upward toward the periphery of the water cylinder at a constant speed. At the present time, an effective gating device does not exist that can both accurately monitor the motion of air bubbles and be easily integrated with CT. Although an ultrasound device may be able to handle the task, it would require that an operator be by the patient's side, and the ultrasound probe itself may cause other artifacts.

Today, patient breath-holding is still the most widely used clinical practice to combat respiratory motion. Although this approach has its limitations as described earlier, it has been shown to be adequate for most CT studies. To further improve its effectiveness, other motion-detecting devices can be used to monitor a patient's breath-holding status. When excessive motion is detected, a warning message can be sent to the scanner to abort the data acquisition. Alternatively, the motion monitoring result can be incorporated into the image reconstruction process to guide the motion-correction algorithms. The motion-monitoring device can be a position-sensing device to examine the chest wall location, or a pressure-sensing device to monitor the chest volume change or the collected CT projections. For projection-based monitoring, the inherent set of consistency conditions in the dataset must be used to determine the presence of motion.

So far, our discussion has focused on the motion artifacts that occur within a scan (intrascan motion), but *inter*scan motion is equally important. If we examine images slice by slice, we may not observe any abnormalities or artifacts. However, when these images are used to form 3D rendered images, discontinuities between adjacent scans become visible, as shown by the arrows in Fig. 7.63. For this particular example, the best approach is to stabilize the patient's head with straps that are attached to the patient head-holder prior to the data acquisition. For interscan motion related to physiologic activities such as cardiac motion, EKG gating is often used to reduce interscan artifacts [Figs.

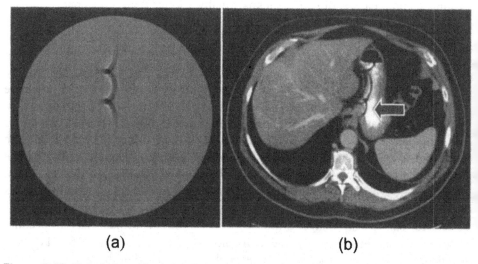

(a)   (b)

Figure 7.62 Air bubble artifact. (a) Computer simulation of an air bubble inside a water cylinder moving up from the iso-center to the periphery of the water cylinder at a constant speed. (b) Clinical example of an air bubble artifact. (Clinical image courtesy of Dr. P. E. Kinahan. Reprinted from Ref. 53, Copyright 2008, American Roentgen Ray Society.)

7.64(a) and (b)]. Additional techniques that have been investigated and developed in cardiac scanning to combat high-frequency motion are presented in Chapter 12.

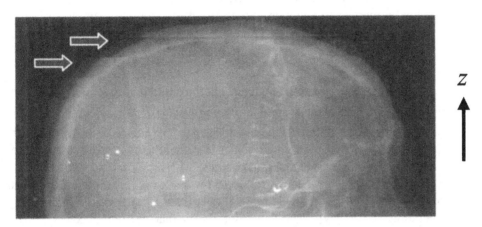

Figure 7.63 Illustration of interscan motion artifact with a step-and-shoot mode data collection on a four-slice scanner with 4×2.5-mm detector configuration. Reconstructed images are forward projected in the lateral direction to produce volume-rendered views of the patient's head. Discontinuities in z correlates well with the axial scan boundaries.

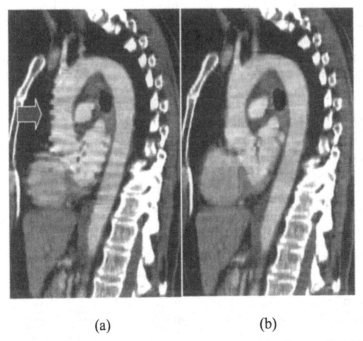

(a)   (b)

Figure 7.64 Sagittal images illustrate interscan motion artifacts. Data collected on a four-slice scanner with 4×3.75-mm detector configuration, 0.75 helical pitch, 1.0-sec gantry rotation, and 31-sec total acquisition. (a) Without EKG gating and (b) with EKG gating.

7.6.2 Beam hardening

Beam hardening is caused by the polychromatic x-ray beam spectrum and the energy-dependent attenuation coefficients. Most materials absorb low-energy x rays better than they absorb high-energy x-ray photons, mainly due to photoelectric absorption. To emphasize the energy-dependent nature of attenuation, the Beer–Lambert law for monoenergetic x ray can be rewritten in the following form:[54]

$$I_E = I_{0,E} e^{-\int \mu_{E,s} ds},\tag{7.21}$$

where $I_{0,E}$ and I_E represent the incident and transmitted x-ray intensities with energy E, and $\mu_{E,S}$ is the linear attenuation coefficient of the object at the same energy. The subscript E indicates the energy-dependent nature of these variables. The attenuation produced by a given object is defined as the negative logarithm of the ratio of transmitted to incident intensities.

In practice, the x-ray beam produced by an x-ray tube covers a broad spectrum. Figure 7.65 shows an example of the output x-ray spectrum of an x-ray tube operating at 120 kVp. The spectrum shows that the output x-ray photon energy ranges from 20 to 120 keV. In this particular example, the extremely low-energy photons (less than 20 keV) are absorbed by the oil, glass, and other materials in the x-ray tube. In general, the energy spectrum varies with the tube design.

Figure 7.66 shows the linear attenuation coefficients of bone, muscle, and water as a function of x-ray energy. Two immediate observations can be made. First, the attenuation characteristics of muscle and water are nearly identical. This is the reason many CT manufacturers use water phantoms for calibration and quality control to approximate soft tissue in the human body. Second, it is

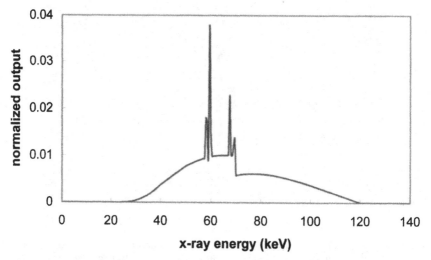

Figure 7.65 Example of an x-ray tube output energy spectrum.

evident from the figure that the linear attenuation coefficient changes drastically with x-ray energy (note the logarithmic scale used in the graph). At lower energies, the linear attenuation coefficient for nearly all materials is higher than its value at higher energies. For example, photons with energies below 10 keV are almost totally absorbed by skin, thus administering a high dose to the patient with no detected signal. As a result, the low-energy x-ray photons are preferentially attenuated, and the x-ray beam spectrum becomes harder (averages a higher energy) as it passes through an object. Figures 7.67(a)–(c) show, respectively, the original x-ray spectrum of a 120-kVp setting, the x-ray spectrum after the beam passes through 15 cm of water, and the spectrum after the beam passes through 30 cm of water. For comparison, Fig. 7.68 shows three spectrums that are normalized for the same total x-ray flux, which clearly shows a spectrum shift to the higher-energy x-rays as the beam passes through an increasing amount of water. Thus, a heterogeneous x-ray beam becomes proportionately richer in high-energy photons (more penetrating or "harder"), which is why most people call this phenomenon "beam hardening."

Although the amount of attenuation by a material for a single energy follows the exponential relationship, the transmitted intensity is the summation of the intensities over all energies, each of which experiences a different attenuation because of the energy dependency of the attenuation coefficients. Mathematically, the relationship can be described by the following relationship:

$$ I = I_0 \int \Omega(E) e^{-\int \mu_{E,s} ds} dE , \tag{7.22} $$

where $\Omega(E)$ represents the incident x-ray spectrum, the area under $\Omega(E)$ equals unity, and I_0 and I represent the total incident and transmitted intensities, respectively.

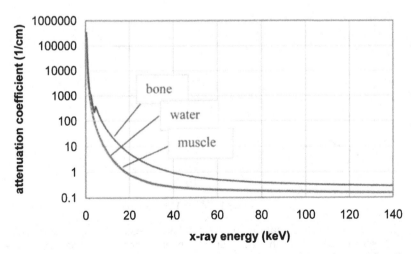

Figure 7.66 Linear attenuation coefficient for bone, water, and muscle as a function of x-ray energy.

Chapter 2 explained that CT projection samples are obtained by taking the logarithm of the transmitted to incident intensity ratio. Mathematically, this is the following operation:

$$p = -\log\left(\frac{I}{I_0}\right) = -\log\left(\int \Omega(E)e^{-\int \mu_{E,s}ds}\,dE\right). \qquad (7.23)$$

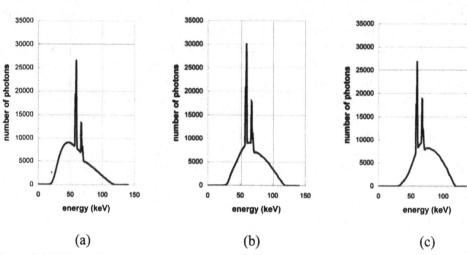

(a)   (b)   (c)

Figure 7.67 Effect of beam hardening as the x-ray beam traverses different object sizes. (a) Original spectrum, (b) after traversing 15 of cm water, and (c) after traversing 30 of cm water.

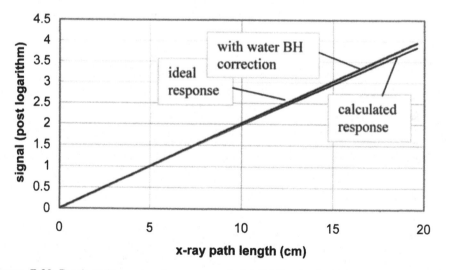

Figure 7.68 Postlogarithm signal versus material thickness for water. Dark solid line: calculated response based on the x-ray spectrum and attenuation coefficient of water. Gray solid line: ideal signal based on linear relationship. Dark dotted line: calculated response after fourth-order water beam hardening correction.

Equation (7.23) indicates that the relationship between the measured projection p and the path length of the material is no longer linear. For illustration, the projection measurement of water is plotted as a function of the path length, as shown by the dark solid line in Fig. 7.68. For reference, a linear function is plotted in the same figure by a solid gray line. The nonlinear nature of the curve is obvious.

Because of the beam-hardening-induced inaccuracy, intensity cupping can result in the reconstructed images, as shown in Fig. 7.69(a).[54] This artifact can be explained by the nonlinear nature of beam hardening. Figure 7.68 shows that the measured projection is always less than the "true" projection, regardless of the path length. In addition, the difference between the two increases with the path length. For a cylindrically shaped object, the path length is the longest at the object center and approaches zero at the object boundaries. As a result, the measured projection is significantly smaller than the true projection at the phantom center, and the error diminishes near the phantom boundaries.

Cupping artifacts can mimic certain pathologies and lead to misdiagnosis. They also severely limit our ability to perform quantitative analysis on CT images. In addition, the optimal display window level becomes difficult to select due to the intensity variation from the center to the peripheries of the object.

Beam-hardening artifacts can be reduced by adequate beam filtration.[55,56] For example, a thin layer of aluminum, copper, or brass can be placed between the x-ray target and the patient to shape the spectrum of the x-ray beam. The flat filter will preferentially filter out the soft x-ray photons and, in effect, harden the beams before they reach the patient. The selection of the filter material and the filter thickness is based on tradeoff analyses of contrast, artifact reduction, noise, CT number uniformity, and patient dose. Therefore, the design process requires not only theoretical calculations, but also extensive experiments and testing.

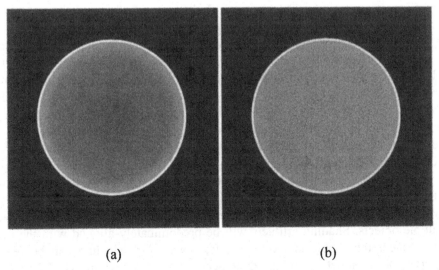

(a)   (b)

Figure 7.69 Reconstructed images of a 35-cm water phantom (a) without beam-hardening correction and (b) with water beam-hardening correction.

For CT, the flat filter alone cannot meet all clinical requirements. As a result, additional software corrections are needed.[1,2,57] Since over 80% of a human body is made of water, an operator can compensate for the beam-hardening error by remapping the projection samples based on known water attenuation characteristics. Again, examine Fig. 7.68, which indicates that the relationship between the measured signal and the path length is a smooth, monotonically increasing concave function. Theoretically, any such curve P_c can be mapped onto a straight line P by the following operation:

$$P = \sum_i \alpha_i P_c^i , \tag{7.24}$$

where α_i is the coefficient for the ith polynomial term. These coefficients can be determined, for example, by a minimum least square fit using the nonlinear curve as the independent variable and the ideal straight line as the dependent variable. The labeled line in Fig. 7.68 shows the result of the measured signal after a fourth-order polynomial remapping. A nice match between the remapped function and the ideal linear function is obtained. Figure 7.69(b) depicts the same water phantom scan corrected by this method. This type of beam-hardening correction is generally called water correction, since uniform water phantoms are used in the parameter calculation.

The polynomial coefficients depend highly on the output spectrum of the x-ray tube. For different x-ray tube spectrums, these coefficients must be recalculated to ensure the accuracy of the correction. For example, two different sets of coefficients are needed for tube settings of 120 and 140 kVp. Also, new sets of coefficients are probably needed after an old x-ray tube is replaced.

When a scanned object is highly heterogeneous and its attenuation characteristics deviate significantly from those of water, the water beam-hardening correction is inadequate. This is particularly apparent for a head scan, which normally has many bony structures (e.g., the skull) in addition to soft tissues.

The attenuation characteristics of bone are significantly different from those of water. To illustrate this phenomenon, Fig. 7.70 plots the calculated projection value of bone as a function of path length. An ideal straight line is also plotted for reference. Compared to the beam-hardening characteristics of water (Fig. 7.68), the bone attenuation curve deviates more significantly from the linear curve. For this type of CT examination, one can expect residual errors in the projection, as well as artifacts in the reconstructed images, if only the water beam-hardening correction is applied.

In general, there are two types of image artifacts. One is dark banding between dense objects.[57] This is caused by discrepancies between the projection rays that pass through only one of the objects and the rays that pass through multiple objects. Shading artifacts will be predominately aligned with the paths that connect these objects. For example, severe dark shading can be readily observed between petrous pyramids in head scans, as shown in Fig. 7.71(a). The

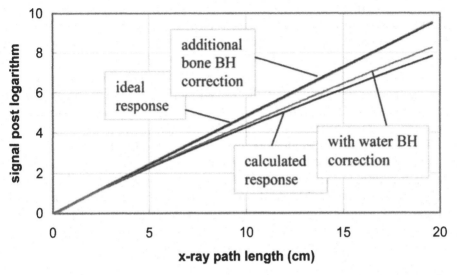

Figure 7.70 Postlogarithm signal versus path length for bone. Dark solid line: calculated response based on the x-ray spectrum and attenuation coefficient of bone. Gray solid line: ideal signal based on linear relationship. Gray dotted line: calculated response after fourth-order water beam-hardening correction. Dark dotted line: calculated response after additional second-order bone beam-hardening correction.

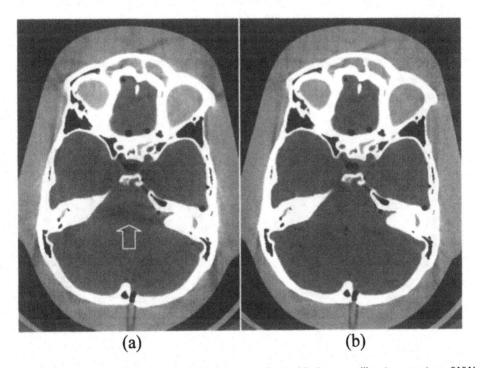

Figure 7.71 Images of a human skull phantom taken with 1-mm collimator aperture (WW = 200) (a) reconstructed with only water beam-hardening correction and (b) reconstructed with bone beam-hardening correction.

second type of artifact is the degraded bone-brain interface. The image intensity of the soft-tissue portion of the interface is elevated to produce a fuzzy boundary. Figure 7.72(a) shows an example of a human skull phantom scanned at the mid-brain level and displayed with a window width of 80 HU.

To combat these artifacts, various approaches have been proposed. For example, the shading artifact at the bone-brain interface can be reduced by modifying the line spread function (LSF) of the reconstruction algorithm to cancel the cupping artifact.[58] Because the method assumes a "typical" skull thickness and density, it often fails to eliminate the artifacts. In some cases, it even introduces artifacts due to overcorrection. More sophisticated algorithms are iterative in nature.[59,60] These corrections can be derived as follows. Let us denote the incident x-ray intensity by I_0, the transmitted beam intensity after attenuation by water only as I_w, and the line integral of the attenuation coefficient after the water beam-hardening correction by P_w. Based on our discussion of water correction in Eq. (7.24), the following relationship exists:

$$P_w \approx \sum_{i=1}^{N} \alpha_i \ln^i \left(\frac{I_0}{I_w} \right). \tag{7.25}$$

In this equation, N represents the order of the polynomial used for the water beam-hardening correction. In the majority of cases, a third- or fourth-order polynomial is sufficient to accurately compensate for the x-ray attenuation behavior.

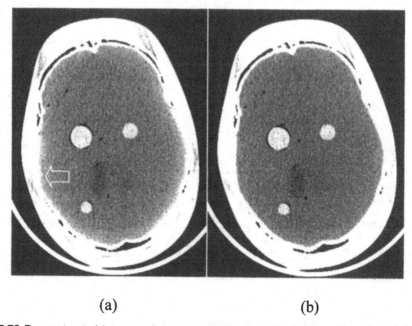

(a)   (b)

Fig. 7.72 Reconstructed images of a human skull phantom (WW = 80) (a) reconstructed with water beam-hardening correction only and (b) reconstructed with additional bone beam-hardening correction.

When two different materials are present (bone and water), the above relationship is no longer valid. However, if we place the two materials in sequence (the output x-ray intensity from one material is the input to the other material), the x-ray beam attenuation relationship can be approximated by a cascade application of Eq. (7.25). If we denote the transmitted beam intensity after the first material (water) by I_w, the line integral estimation can be represented by

$$P_L \approx \sum_{i=1}^{N} \alpha_i \ln^i \left(\frac{I_0}{I_w} \right) + \sum_{i=1}^{N} \beta_i \ln^i \left(\frac{I_w}{I} \right). \tag{7.26}$$

Similar to α_i, β_i depends on the attenuation characteristics of the second material and the x-ray beam spectrum. If we treat the entire object as a single material in the beam-hardening correction, the error in the projection ΔP_L is simply the difference between Eq. (7.25) and Eq. (7.26). For the same I_0 (input x-ray intensity) and I (measured output intensity), this is

$$\Delta P_L \approx \sum_{i=1}^{N} \alpha_i \ln^i \left(\frac{I_0}{I} \right) - \left[\sum_{i=1}^{N} \alpha_i \ln^i \left(\frac{I_0}{I_w} \right) + \sum_{i=1}^{N} \beta_i \ln^i \left(\frac{I_w}{I} \right) \right]$$

$$\approx \sum_{i=1}^{N} (\alpha_i - \beta_i) \ln^i \left(\frac{I_w}{I} \right) = (\alpha_1 - \beta_1)\xi + (\alpha_2 - \beta_2)\xi^2 + \cdots, \tag{7.27}$$

where $\xi = \ln(I_w / I)$. The quantity ξ is the logarithm of the ratio of the intensity incident on bone (the material that has not been properly accounted for) to the measured output intensity. The nonlinear terms in Eq. (7.27) are responsible for producing the streaking and shading artifacts. The linear term in Eq. (7.27) causes a low-frequency CT number shift in the reconstructed images.

To examine the accuracy of Eq. (7.27), we performed the following calculation. Using the attenuation coefficient of bone (Fig. 7.66), the energy spectrum of the x-ray tube (Fig. 7.65), and Eq. (7.23), we calculated the measured projection signal (after the minus logarithm) as a function of the bone thickness. The result is plotted in Fig. 7.70. The dark solid line in the figure represents the projection value without any beam-hardening compensation. The curve is significantly nonlinear. Next, we applied the water beam-hardening correction to the curve; the result is shown by the gray dotted line. An additional second-order correction for the bone beam-hardening effects maps the gray line to the ideal straight line, as shown by the dark dotted line. A good match is obtained.

Therefore, to correct for bone beam-hardening artifacts, we need to estimate the quantity ξ. Based on Eq. (7.27), this is the projection value of the bony structures only. Since the measured projection contains both soft tissue and bone (it is nearly impossible to separate them directly), we turn to the reconstructed image for estimation. To a first-order approximation, the densities of the reconstructed bony structures are accurate. Therefore, the quantity ξ can be estimated by calculating line integrals (forward projecting) through the reconstructed bony structures. Figure 7.73 shows a flowchart of the bone beam-hardening correction process.

Figures 7.71(b) and 7.72(b) show images corrected by the above method. An improvement in image quality is apparent. This technique is quite effective in reducing dark banding, shading, or streaking artifacts in the posterior regions. The bone-brain interface also becomes sharp and well defined.

Although the above two-pass correction is effective, often we need to pay special attention to the interactions of such a correction with other artifact-producing sources. For example, the effectiveness of the correction can be affected by the partial-volume effect, since the algorithm relies on an accurate assessment of the bone content in the water-corrected images. When the partial-volume effect is present, the amount of bone in the image can be either overestimated or underestimated. Consequently, either an overcorrection or an undercorrection occurs in the bone-corrected images. Figure 7.74 shows an example of overcorrection. To produce Fig. 7.74, a patient was scanned in a step-and-shoot mode with a 1.25-mm detector aperture on a four-slice CT scanner. Four images were reconstructed. Figure 7.74(a) was produced by first averaging

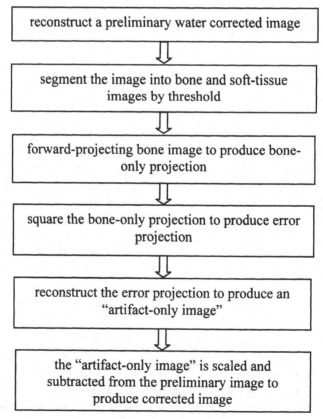

Figure 7.73 Flowchart of the bone beam-hardening correction process.

the four images together to produce a single water-corrected image. The bone beam-hardening correction was then applied. Figure 7.74(b) was obtained by applying the bone beam-hardening correction to four images individually, prior to performing image averaging. The resulting overcorrection in Fig. 7.74(a) is shown by the arrow.

Beam-hardening-induced shading artifacts are not limited to bony structures. In many cases, they are caused by the high-density contrast agent injected into the patient. Figure 7.75(a) shows a scan of a rat lung with an injected contrast agent. Severe streaking artifacts are obvious. Several methods can be used to combat image artifacts under these conditions. One method is to optimize the time interval between the contrast intake and the scanning. Another method is to select a contrast agent with a lower beam-hardening effect. The software correction schemes outlined previously can be equally effective,[60] as shown by Fig. 7.75(b).

Beam artifacts can also be corrected with a dual-energy approach. In this scheme, the same object is scanned with two different x-ray tube settings (e.g., 80 and 140 kVp) or with specially designed detectors that can differentiate low- and high-energy photons. Two independent measurements allow the operator to better characterize the material and perform a more robust correction. Details of this approach will be discussed in Chapter 12.

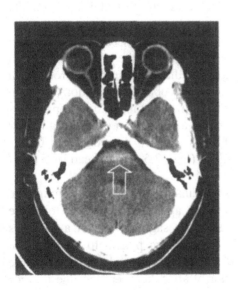 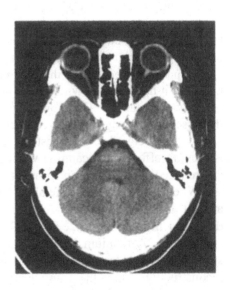

(a)   (b)

Figure 7.74 Illustration of the interaction between partial-volume effect and beam-hardening correction. (a) Four adjacent 1.25-mm slices are summed and bone correction is applied to the summed image. (b) Four adjacent 1.25-mm slices are bone corrected prior to the summation.

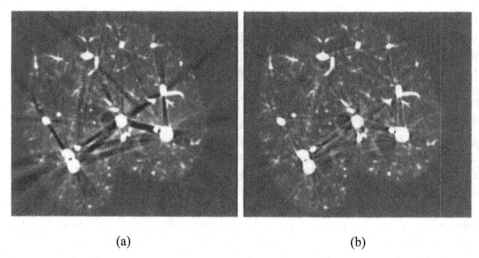

(a)   (b)

Figure 7.75 Scan of a rat lung illustrates contrast-induced artifacts. (a) Original image and (b) image with beam-hardening correction.

7.6.3 Metal artifacts

The causes of metal artifacts are quite complicated. Depending on the shape and density of the metal objects, the appearance of this type of artifact can vary significantly. In medical applications, metal objects can be metallic orthopedic hardware inside the patient (e.g., surgical pins and clips) or equipment attached to the patient's body (e.g., biopsy needles). A metal object can produce beam hardening, partial volume, aliasing, under-range in the data acquisition electronics, or overflow of the dynamic range in the reconstruction process. In addition, it has been shown that the motion of metal objects is a major culprit in producing artifacts.[61] Therefore, many of the artifact compensation approaches discussed previously can be used to combat metal artifacts. For example, a higher kVp setting (e.g., 140 kVp instead of 120 kVp) can help to reduce the beam-hardening effect, and scanning with thin slices (e.g., 2.5 mm instead of 5 mm) can reduce the partial-volume impact.

When the cause of a metal artifact is dominated by beam hardening, the artifact can be corrected algorithmically with approaches similar to those discussed in Section 7.6.2. Figure 7.76(a) shows a sagittal view of a patient's spinal cord. A titanium implant was placed inside the patient to stabilize the spine. Only water beam-hardening correction was used in the image reconstruction. Since the scan was taken shortly after the operation, the interior of the implant device was hollow, and the CT image intensity should be low (closer to air than bone). Because the attenuation characteristics of titanium are significantly different from those of water, the intensity inside the implant is elevated by the beam-hardening artifact. Since we know that the implant is made of titanium and its attenuation characteristics can be measured or calculated, we can use correction algorithms similar to the bone correction to compensate for the image artifacts. Figure 7.76(b) shows the same scan corrected for titanium beam hardening. The intensity inside the implant is properly restored.

When metal objects are highly attenuating, under-range in the data acquisition electronics often results. When this occurs, image artifacts similar to photon starvation appear. As discussed in detail in Section 7.3.4, this type of artifact is dominated by high-intensity streaks. When combined with the beam-hardening artifacts caused by metal, both shading and streaking artifacts appear in the reconstructed image. Figure 7.77(a) shows an example of patient hip implants made of titanium. After beam-hardening correction, streaks caused by under-range in the DAS still appear in the image, as shown in Fig. 7.77(b). Of course, a combined beam-hardening correction and adaptive noise-filtering approach should further improve the image.

When many highly attenuating metal implants are present in a scanned object (e.g., hip implants made of stainless steel), the magnitude of the projection signal is so high that the dynamic range of the DSP chips can be exceeded after the filtering step. Chapter 3 mentioned that the tomographic filtering operation is essentially a derivative operator. Since metal implants create a significant discontinuity in the projection data, the magnitude of its derivative can be very large. For reconstruction speed, DSP chips [e.g., to perform FFT and inverse Fourier transform (IFFT)] are often employed in the image reconstruction engine. These chips typically have a limited dynamic range. As a result, the filtered projection data are truncated and cause additional image artifacts.

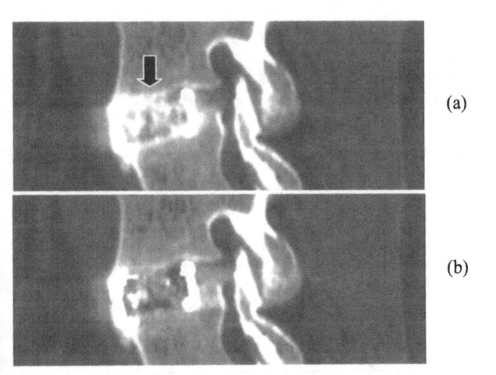

(a)

(b)

Figure 7.76 Reconstructed images of a patient scan with Ti implant. (a) Original image and (b) image with Ti correction.

Many studies have been conducted to overcome metal-induced image artifacts.[61–65] Some approaches replace the projection signal produced by the metal object with a synthesized projection signal based on neighboring projection samples that do not contain metal implants, as shown in Fig. 7.78. These approaches are quite effective at combating artifacts in regions that surround the metal object (projection signals that do not contain the metal object are well preserved). On the other hand, any information on the metal object itself is completely lost. In addition, the region that is immediately adjacent to the metal object is also destroyed. For many clinical applications, the interface between the implant and its neighboring bone and soft tissue is of great interest to the physician. To the best of our knowledge, effective and affordable metal artifact correction schemes have yet to be found.

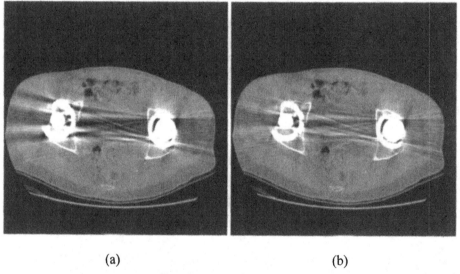

(a)   (b)

Figure 7.77 Reconstructed images of a patient with Ti implants. (a) Original image and (b) image with Ti correction.

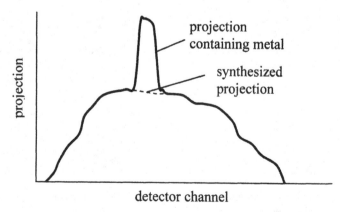

Figure 7.78 Illustration of metal artifact suppression with synthesized projection samples.

Often, metal artifact correction is complicated by patient motion, which creates additional projection inconsistencies and aggravates the streaking artifacts. To illustrate the problem, Fig. 7.79 shows the scan of a patient's heart. A pacemaker was implanted in the patient. Unlike respiratory motion where a patient can be instructed to hold his or her breath for a brief period of time, cardiac motion is unavoidable. As a result, the metal leads (wires) of the pacemaker produce significant streaking artifacts. This scan was acquired with ECG gating, and the data were collected in the diastole phase of the cardiac motion (while the motion of the heart was minimal). Metal artifacts will be more pronounced when the data are collected during the systolic phase of cardiac motion.

7.6.4 Incomplete projections

An incomplete projection occurs when a portion of a projection is not available for reconstruction. Incomplete projections occur more often than one would expect. For example, a projection can be truncated when the scanned object is partially outside the scan FOV. For most commercially available CT scanners, the size of the gantry opening and the size of the scan FOV are different, as shown in Fig. 7.80. In most cases, the gantry opening is significantly larger than the scan FOV. For example, while the average gantry opening of a CT scanner is about 70 cm, the scan FOV is limited to 50 cm. The purpose of this design is

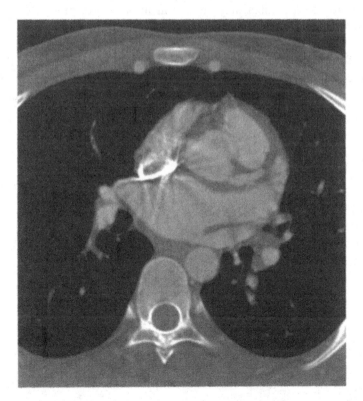

Figure 7.79 Illustration of a metal artifact enhanced by patient motion.

mainly ease of patient handling and placement of accessories. Although the operator's manual often recommends centering the patient in the scanning FOV, these recommendations are not strictly followed for various reasons. As a result, a portion of the patient projection is truncated. Truncated projections produce bright shading artifacts near the edge of the truncation.

Because the magnitude of the reconstruction filter kernel falls off quickly from the center point (in the spatial domain), shading is typically limited to a small region near the location of truncation. However, the severity of shading artifacts depends on the amount of projection that is truncated. Usually, the larger the truncation, the more pronounced is the truncation artifact. For illustration, we performed the following experiment. A shoulder phantom was scanned in a step-and-shoot mode inside a 50-cm scan FOV. Since the projection was complete, no truncation artifact was produced, as shown in Fig. 7.81(a). Next, we artificially set the projection readings outside a 40-cm FOV to zero to simulate a scanner with a smaller FOV. The reconstructed image is shown in Fig. 7.81(b). Note that a bright rim can be observed near the tips of the shoulder where projection truncation takes place. The extent of the artifact, however, is limited to the regions immediately adjacent to the edge of the FOV. Then we set the projection readings outside a 35-cm FOV to zero to simulate a more severe truncation (so a bigger portion of the projection would be truncated). The reconstructed image in Fig. 7.81(c) shows a significant increase in the size and intensity of the truncation artifact.

One approach to combat projection truncation is prevention. If a patient can be carefully centered in the scanning FOV, most truncation artifacts can be avoided. Occasionally, however, projection truncation is unavoidable. For example, if the largest dimension of a patient cross-section exceeds the limiting FOV of the scanner, projection truncation occurs regardless of patient centering. For these cases, the only remedy is to use algorithmic correction. Much research

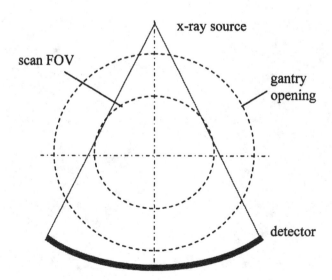

Figure 7.80 Schematic diagram of scan field-of-view and gantry opening.

has been conducted over the years in this area.[66–71] One approach, for example, tries to fit a water cylinder at the location of truncation based on the magnitude and slope of the projection profile at that location.[71] Since the fitting is performed on a view-by-view basis and constrained by a projection consistency condition, the correction can perform reasonably well even if the shape of the truncated object is not similar to a cylinder, as illustrated in Figs. 7.82(a) and (b).

Other issues arise when a scanned object extends beyond the scanning FOV. In some CT scanners, a few detector channels are used to normalize the output flux fluctuation of the x-ray tube. When the scanned object is extended into the x-ray path of these channels, the measured signal no longer represents the x-ray tube fluctuation. If proper compensation is not applied, the overall intensities of the projection views affected by the extruded object can be significantly reduced, as shown by the dark horizontal bands in the sinogram of Fig. 7.83(a). The biased projection intensities can result in biased CT numbers in the reconstructed images, as shown by the horizontal bands in the sagittal image of Fig. 7.84(a). When compensation is made for these signals, the horizontal bands are removed from both the sinogram data [Fig. 7.83(b)] and the sagittal image [Fig. 7.84(b)].

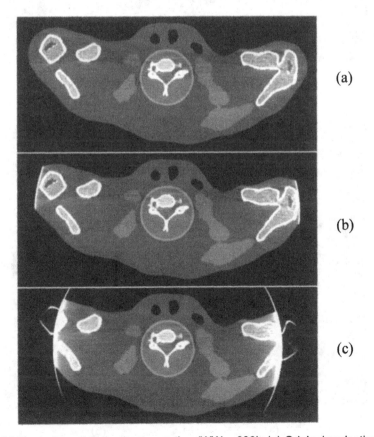

(a)

(b)

(c)

Figure 7.81 Illustration of projection truncation (WW = 600). (a) Original projection without truncation. (b) Simulated moderate level of projection truncation. (c) Simulated severe level of projection truncation.

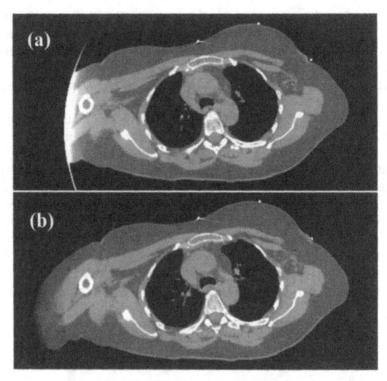

Figure 7.82 Correction for truncated projections.

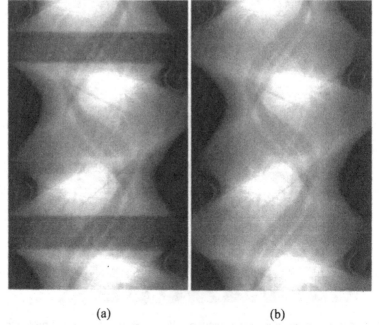

(a)   (b)

Figure 7.83 Sinograms of a patient scan with a blocked reference (a) without proper reference normalization and (b) with proper reference normalization.

In some cases, projections may be corrupted or truncated by over-range in the DAS. This can occur when the scanned object is off-center and a high-scanning technique (using a high dose) is selected for scanning. For projection samples that undergo little or no attenuation from the object, the input x-ray flux can exceed the dynamic range of the data acquisition electronics. Therefore, the measured projections deviate from the true line integrals of the object's attenuation coefficients. Because over-range often occurs in a large number of detector channels, the impact is significant. The impact of DAS over-range is illustrated by the patient head scans in Fig. 7.85. Figure 7.85(a) was acquired without DAS over-range, and Fig. 7.85(b) was acquired with DAS over-range. It is clear that anatomical information is nearly completely obliterated by the over-range artifact.

The best approach to combating over-range artifacts is always prevention so that the DAS over-range does not occur in the first place, even under the worst conditions (i.e., when the detector channels are fully exposed to x-ray flux without any attenuation). This can be accomplished by clever system design. In fact, the original LightSpeed™ 4-slice scanner introduced in 1998 (GE

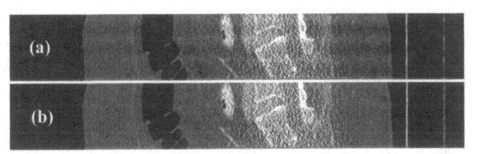

Figure 7.84 Sagittal images of a patient scan with blocked reference (a) without proper compensation and (b) with proper compensation.

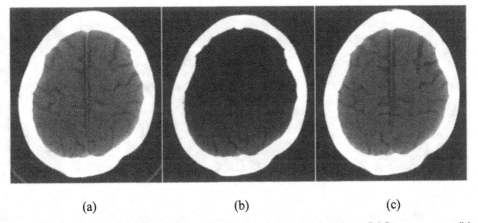

(a)   (b)   (c)

Figure 7.85 Reconstructed images of a patient head scan (a) without DAS over-range, (b) with DAS over-range, and (c) with DAS over-range and correction.

Healthcare, Waukesha, Wisconsin) employed DAS with multiple gains. When a protocol with a higher technique is used (higher tube voltage, tube current, slower gantry speed, or thicker slices), the DAS gain reduces automatically to ensure that the highest signal received by a detector cell does not cause over-range. For scanners that do not have this capability, another means of prevention can be used. In addition to proper patient centering, the proper scanning technique can be automatically selected to avoid DAS over-range. Many methods for automatically selecting the optimal scanning technique (x-ray tube current) have been proposed.[72-74] Although the original objective of these methods was to reduce unnecessary dose to the patient, a side benefit is the elimination or reduction of DAS over-range, since the output x-ray flux is automatically adjusted based on the measured signal at the detector.

An alternative approach to combating over-range artifact is to perform a software correction.[75,76] The objective of this correction is to estimate the missing projection samples corrupted by the DAS over-range. This correction scheme is based on the fact that the cross-sections of most scanned objects are oval in shape. Therefore, the slope of the missing projections can be estimated based on the remaining projection samples. By extrapolating the projection profiles based on the gradient information, the missing projection can be reliably synthesized. Figure 7.85(c) shows the reconstructed image of the same head scan with the proposed correction. The original object is nearly completely restored. The same technique also can be applied to DAS over-range cases in body scans, as shown in Figs. 7.86(a) and (b).

7.7 Operator-induced Artifacts

CT operators are the most important and most direct link to patients. They provide the answers to many questions that cannot be easily obtained by CT manufacturers, academic researchers, or even radiologists. The knowledge and skill of the operators makes a significant difference in the final image quality.

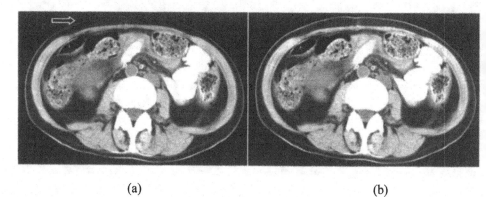

(a)   (b)

Figure 7.86 Patient abdominal scans (a) with DAS over-range and (b) with DAS over-range and correction.

As explained at the beginning of this chapter, there are two aspects to combating image artifacts: correction and avoidance. So far, this chapter has covered a significant amount of information on artifact correction and reduction. This section will illustrate the important role that the CT operator plays in artifact prevention and avoidance.

The previous section presented image artifacts associated with patient motion. Although involuntary motion is difficult to control, voluntary motion can often be reduced or minimized. For example, proper training and instruction can be given to a patient prior to the CT scan to hold his or her breath during the data acquisition. If the patient initiates the breath-hold too late or terminates the breath-hold prematurely, image artifacts will likely occur. Motion restraint accessories can be used during a head scan to avoid motion. Many head-holders have elastic Velcro straps that can be placed on the patient's forehead and attached to the sides of the restraint to prevent any slight rotating motion by the patient. Even during a phantom experiment, it is a good idea to use masking tape and padding to ensure that the phantom does not move during the table translation [compare Figs. 7.87(a) and (b)]. For illustration, Figs. 7.88(a) and (b) compare patient head scans without and with the application of a Velcro strip.

Another example of the operator's impact on image quality is proper protocol selection. CT scanning protocols need to be not only organ- or disease-specific, but also patient-dependent. For example, when scanning a large patient, higher kVp settings, a higher tube current—and if acceptable, a slower scan speed, a slower helical pitch, or thicker reconstructed slices—need to be selected. These parameters can help to avoid image artifacts associated with photon starvation and excessive image noise.

One factor that is often overlooked during the scan is the patient's position. For example, the patient needs to be properly centered (in both the x and y

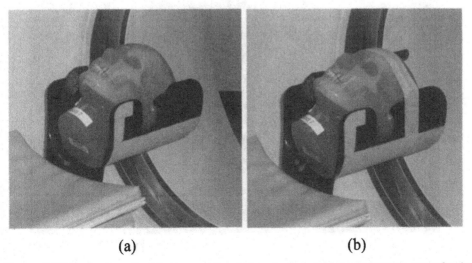

(a)   (b)

Figure 7.87 Illustration of motion prevention in head scan. (a) Head phantom rests freely in a head-holder. (b) Masking tape is placed on head phantom to prevent motion.

directions) to allow optimal performance of the bowtie filter. This will be discussed in more detail in Chapter 11. Another example is the positioning of the patient's arms. Because of the large and dense humerus, radius, and ulna of human arms, placing the arms on the patient's side while performing chest or abdomen scans can lead to truncation, photon starvation, and beam-hardening artifacts, as shown in Figs. 7.89(a) and (b). Whenever possible, the patient should be instructed to hold his or her arms over his or her head during body scans.

Since the introduction of 64-slice scanners, many people have thought that the partial-volume artifact was a thing of the past because submillimeter data acquisition is routinely performed on every patient. Unfortunately, the sources of partial-volume artifacts can be introduced by the CT operator. An example of such a case is shown in Fig. 7.90, in which a special "homemade" head-holder was used to stabilize the patient. The portion of the head-holder that connects to the patient table is made of metal, and its end terminates abruptly with a flat surface parallel to the x-ray plane. When the patient table is positioned such that the transition region is within the imaging slice, partial-volume effects still occur despite the submillimeter slice thickness [Fig. 7.90(a)]. Severe shading artifacts can be clearly observed. When the table is indexed 1 mm away from the transition region, shading artifacts are significantly reduced, as shown in Fig. 7.90(b).

Before concluding this discussion on image artifacts, it should be reiterated that there are many causes for artifacts, including the nature of the physics, technology limitations, shortcomings in new technologies, characteristics of the patients, and suboptimal use of the scanner. Given the complexity of these causes, it is clear that a combined effort is needed from CT manufacturers, academic researchers, patients, radiologists, and CT operators to reduce or eliminate artifacts and minimize their impact.

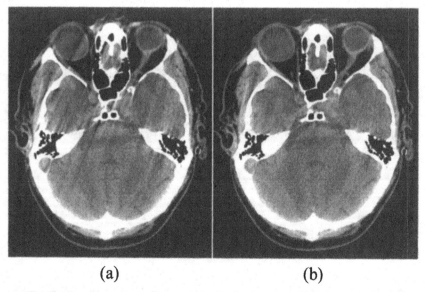

(a)   (b)

Figure 7.88 Patient head scans (a) without and (b) with the use of motion-constraint accessories.

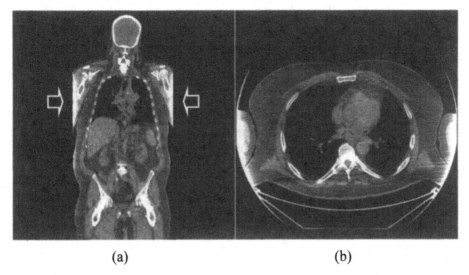

(a)   (b)

Figure 7.89 Example of image artifacts caused by the patient's arms. (a) Coronal image and (b) axial image.

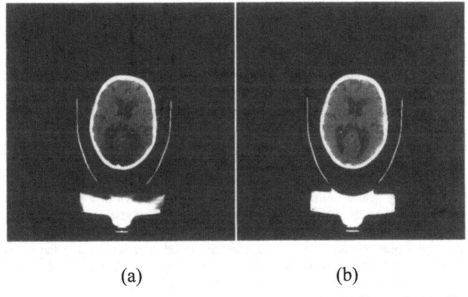

(a)   (b)

Figure 7.90 Partial-volume artifacts induced by foreign objects. (a) Slice at the transition between thicker and thinner part of the holder. (b) Slice at 1-mm distance from the transition region of the holder.

7.8 Problems

7-1   Equation (7.1) shows the relationship between ring artifact intensity and erroneous channel location. What is the ring artifact intensity for an error ε in the detector iso-channel?

7-2   Equation (7.1) shows that the intensity of the ring artifact falls off as $1/r$, where r is the distance to the iso-center. To combat detector-error-induced ring artifacts, some people have suggested shifting the detector along its arc back and forth during the scan so that the iso-channel of the detector changes constantly, as shown in Fig. 7.91. Since a one-to-one relationship would no longer exist between the detector channel and its distance to the iso-ray, the intensity of the ring artifact induced by the detector should be reduced. In one configuration, a detector is shifted by three channels in either direction for each view. Estimate the reduction in ring intensity.

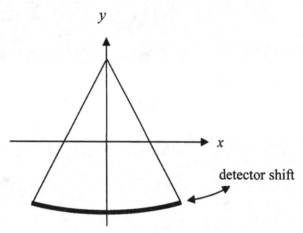

Figure 7.91 Schematic diagram of a shifting-detector system.

7-3   For the system presented in problem 7-2, discuss the potential issues of this approach.

7-4   In Fig. 7.8, a fourth-generation CT system is shown in which the x-ray tube is rotating about the iso-center and a ring of stationary detectors is positioned outside the tube trajectory. If we assume that each detector samples the projections at a constant speed, can we use the fan-beam reconstruction algorithms presented in Chapter 3 for reconstruction? If not, how do you reconstruct an image?

7-5   A design engineer wants to use the constant-angle fan-beam reconstruction algorithm directly for the fourth-generation system shown in Fig. 7.8. If changes to the reconstruction algorithm are not allowed, what modifications to the CT system are needed?

7-6   Figure 7.10 illustrates the achievement of double sampling with focal spot wobble. Do all projection samples shown by the solid line in Fig. 7.10(b) straddle the dotted lines exactly at the half-way locations? If not, what is the potential impact on the effectiveness of the aliasing artifact reduction?

7-7   To ensure that Nyquist sampling criteria are fully satisfied, one engineer wants to obtain a triple sampling with focal spot wobble. That is, the focal spot is deflected when the detector is rotated one-third and two-thirds of a detector width (instead of one-half as described in Section 7.3). Discuss the potential issues of this approach.

7-8   Figure 7.22 shows the intensity profile of a reconstructed thin plate with 2.5-, 5.0-, 7.5-, and 10.0-mm slice acquisitions. A student claims that the μ values of the thin plate calculated based on the images in Figs. 7.21(b), (c), and (d) should be one-half, one-third, and one-fourth of the μ value calculated from Fig. 7.21(a). Is the student correct? If this is an approximation, what factors influence the accuracy of this approximation?

7-9   A postpatient collimator is commonly used in CT for scatter rejection. List the pros and cons of using a 1D collimator (a set of parallel plates) versus a 2D collimator grid (two sets of parallel plates placed orthogonally to each other).

7-10 The scatter-to-primary ratio of a centered 20-cm water phantom measured at a collimatorless detector of 1-mm width in x is 25%. If you need to design a 1D postpatient collimator to reduce the scatter-to-primary ratio to 5% (with the same phantom and at the same z exposure), what is the minimum height of the collimator blade, assuming the scatter is uniformly distributed?

7-11 The height of a postpatient collimation relates closely to its scatter rejection capability. One student argues that we should always use a collimator as high as the space allows, in order to attain the best scatter-to-primary ratio possible. What other potential issues must be considered?

7-12 Assume that the mean energy of an x-ray tube output at a 100-kVp setting is 60 keV, and at a 140-kVp setting is 80 keV. For a constant tube power (the product of kVp and mA), what is the noise ratio of the 140-kVp over 100-Vp settings after passing through 40 cm of water [$\mu(60$ keV$) = 0.206$ cm^{-1} and $\mu(80$ keV$) = 0.184$ cm^{-1}]?

7-13 Based on the calculation of problem 7-12, what advice do you give to a CT operator when scanning a large patient to avoid the photon starvation problem?

7-14 To combat the photon starvation problem at a fixed kVp setting, one student suggests using a higher tube current, a second student suggests using a thicker slice, and a third student suggests using a slower gantry speed. Discuss the pros and cons of each approach.

7-15 Figure 7.35(a) shows the impact of off-focal radiation on a 20-cm water phantom. Explain why the artifact appears as a bright rim near the water-Plexiglas® interface instead of appearing as a dark rim.

7-16   Explain why the dark shadow in Fig. 7.36(a) that resulted from off-focal radiation is different from that in Fig. 7.35(a).

7-17   Derive the recursive relationship described by Eq. (7.14) for the detector afterglow correction.

7-18   Figure 7.47(a) illustrates the appearance of detector afterglow artifacts. For this scan, is the iso-center near the top, middle, or bottom of the oval phantom? Explain.

7-19   For the example shown in Fig. 7.47(a), why do artifacts appear as short arcs (roughly 45 deg) rather than as long arcs that cover the entire phantom (over 180 deg)?

7-20   Derive the half-scan weighting function shown in Eqs. (7.15) and (7.16).

7-21   A CT operator accidently collected a set of projection data over the angular range $\pi + 2\gamma_m + \Gamma < 2\pi$. Develop a weighting function that utilizes all the collected data.

7-22   Which algorithm has better motion suppression capability: the half-scan or the underscan? Justify your answer.

7-23   In the early days of CT, water bags surrounding the patient head were used for beam-hardening compensation. Describe a beam-hardening correction scheme with this approach.

7-24   A ring artifact is found on a third-generation CT image reconstructed with a 40-cm FOV on a 512×512-image matrix. The diameter of the ring is 310 pixels. If the detector iso-channel is 400.75, the detector channels are spaced 1 mm apart, the source-to-iso distance is 500 mm, and the source-to-detector distance is 950 mm, what detector channel numbers need to be examined as the potential source of artifacts?

7-25   Horizontal streaking and shading artifacts are found on a CT image reconstructed with a 30-cm FOV. What are the potential culprits of these artifacts? Assuming that the original projection data are available, what steps do you recommend for isolating the source(s) of the artifacts?

7-26   A patient was scanned on a third-generation CT scanner in a step-and-shoot full-scan mode (360 deg) at a 1-sec scan speed. Image artifacts appear in the reconstructed image; you suspect they are related to a patient's jerky motion. How do you convince the radiologist that they are indeed motion artifacts (assuming the original projection data are available and re-scanning the patient is not an option)?

7-27   For the situation in problem 7-26, the radiologist wants to know whether the patient motion occurred near the beginning, middle, or end of the scan. What method can you use to isolate the timing?

7-28 How do you measure and characterize the off-focal radiation of an x-ray tube?

7-29 Figure 7.60(a) shows a patient head scan with motion artifacts. What was the x-ray tube's probable angular position when the patient motion occurred? Justify your answer.

References

1. J. Hsieh, "Image artifacts, cause and correction," in *Medical CT and Ultrasound: Current Technology and Applications*, L. W. Goldman and J. B. Fowlkes, Eds., Advanced Medical Publishing, Madison, WI (1995).

2. J. Hsieh, "Image artifact in CT," in *Categorical Course in Diagnostic Radiology Physics: CT and US Cross-sectional Imaging*, L. W. Goldman and J. B. Fowlkes, Eds., Radiological Society of North America, Oak Brook, IL, 97–115 (2000).

3. R. D. Tarver, D. J. Conces, Jr., and J. D. Godwin, "Motion artifacts on CT simulate bronchiectasis," *Am. J. Roentgen.* **151**(6), 1117–1119 (1988).

4. L. L. Berland, *Practical CT, Technology and Techniques,* Raven Press, New York (1986).

5. T. H. Newton and D. G. Potts, *Radiology of the Skull and Brain, Technical Aspects of Computed Tomography*, The C. V. Mosby Company, St. Louis (1981).

6. C. M. Coulam, J. J. Erickson, F. D. Rollo, and A. E. James, *The Physical Basis of Medical Imaging,* Appleton-Century-Crofts, New York (1981).

7. A. B. Wolbarst, *Physics of Radiology,* Appleton & Lange, East Norwalk, CT (1980).

8. S. Takahashi, *Illustrated Computed Tomography,* Springer-Verlag, Berlin (1983).

9. C. R. Crawford and A. C. Kak, "Aliasing artifacts in computed tomography," *Appl. Optic.* **18**(21), 3704–3711 (1979).

10. T. M. Peters and R. M. Lewitt, "Computed tomography with fan beam geometry," *J. Compt. Assist. Tomogr.* **1**, 429–436 (1977).

11. A. R. Sohval and D. Freundlich, "Plural source computed tomography device with improved resolution," U.S. Patent No. 4637040 (1986).

12. A. H. Lonn, "Computed tomography system with translatable focal spot," U.S. Patent No. 5173852 (1990).

13. J. Hsieh, M. F. Gard, and S. Gravelle, "A reconstruction technique for focal spot wobbling," *Proc. SPIE* **1652**, 175–182 (1992).

14. B. F. Logan, "The uncertainty principle in reconstructing functions from projection," *Duke Math. J.* **42**, 661–706 (1975).

15. D. L. Snyder and J. R. Cox, Jr., "An Overview of Reconstructive Tomography and Limitations Imposed by a Finite Number of Projections," in

Reconstruction Tomography in Diagnostic Radiology and Nuclear Medicine, M. M. Ter-Pogossian, et al., Eds., University Park Press, Baltimore (1977).

16. P. M. Joseph and R. A. Schulz, "View sampling requirements in fan beam computed tomography," *Med. Phys.* **7**(6), 692–702 (1980).

17. J. Hsieh, C. Slack, S. Dutta, C. L. Gordon III, J. Li, and E. Chao, "Adaptive view synthesis for aliasing artifact reduction," *Proc. SPIE* **4320**, 673–680 (2001).

18. G. H. Glover and N. J. Pelc, "Nonlinear partial volume artifacts in x-ray computed tomography," *Med. Phys.* **7**(3), 238–248 (1980).

19. G. Henrich, "A simple computational method for reducing streak artifacts in CT images," *Comput. Tomogr.* **4**, 67–71 (1980).

20. J. Hsieh, "Nonlinear partial volume artifact correction in helical CT," *IEEE Trans. Nucl. Sci.* **46**(3), 743–747 (1999).

21. M. Endo, T. Tsunoo, N. Nakamori, and K. Yoshida, "Effect of scattered radiation on image noise in cone beam CT," *Med. Phys.* **28**(4), 469–474 (2001).

22. R. Ning, B. Chen, R. Yu, D. Conover, X. Tang, and Y. Ning, "Flat panel detector-based cone-beam CT angiography imaging: system evaluation," *IEEE Trans. Med. Imag.* **19**, 949–963 (2000).

23. J. H. Siewerdsen and D. A. Jaffray, "Cone-beam computed tomography with a flat-panel imager: magnitude and effects of x-ray scatter," *Med. Phys.* **28**(2), 220–231 (2001).

24. G. H. Glover, "Compton scatter effects in CT reconstructions," *Med. Phys.* **9**, 860–867 (1982).

25. P. M. Joseph, "Artifacts in computed tomography" in *Radiology of the Skull and Brain, Technical Aspect of Computed Tomography*, T. H. Newton and D. G. Poffs, Eds., The C. V. Mosby Company, St. Louis (1981).

26. R. Ning, X. Tang, and D. Conover, "X-ray scatter correction algorithm for cone beam CT imaging," *Med. Phys.* **31**, 1195–1202 (2004).

27. J. Siewerdsen, M. J. Daly, B. Bakhtiar, D. J. Moseley, S. Richard, H. Keller, and D. A. Jaffray, "A simple, direct method for x-ray scatter estimation and correction in digital radiography and cone-beam CT," *Med. Phys.* **33**, 187–197 (2006).

28. M. Kachelrieß, K. Sourbelle, and W. A. Kalender, "Empirical cupping correction: a first-order raw data precorrection for cone-beam computed tomography," *Med. Phys.* **33**, 1269–1274 (2006).

29. J. Hsieh, "Adaptive streak artifact reduction in computed tomography resulting from excessive x-ray photon noise," *Med. Phys.* **25**(11), 2134–2147 (1998).

30. M. Kachelrieß, O. Watzke, and W. A. Kalender, "Generalized multi-dimensional adaptive filtering for conventional and spiral single-slice, multi-slice, and cone-beam CT," *Med. Phys.* **28**(4), 475–490 (2001).

31. P. J. La Riviere, "Penalized-likelihood sonogram smoothing for low-dose CT," *Med. Phys.* **32**, 1676–1683 (2005).

32. J. A. Randmer, T. J. Koller, and W. P. Holland, "X-ray sources and control," in *Radiology of the Skull and Brain, Technical Aspect of Computed Tomography*, T. H. Newton and P. G. Potts, Eds., The C. V. Mosby Company, St. Louis (1981).

33. T. S. Curry III, J. E. Dowdey, and R. C. Murry, *Christense's Physics of Diagnostic Radiology*, Lea & Febiger, Philadelphia (1990).

34. J. Hsieh, "Radiation detector offset and afterglow compensation technique," U.S. Patent No. 5331682 (1994).

35. J. Hsieh, O. E. Gurmen, and K. F. King, "Investigation of a solid-state detector for advanced computed tomography," *IEEE Trans. Med. Imag.* **19**, 930–940 (2000).

36. D. A. Freundlich, "Ring artifact correction for computerized tomography," U.S. Patent No. 4670840 (1987).

37. J. Hsieh, "Guided ringfix algorithm for image reconstruction," U.S. Patent No. 5533081 (1996).

38. J. Hsieh, O. E. Gurmen, and K. F. King, "Optimization of solid state detector design for 0.5-s multislice CT," *Radiology* **213**, 318 (1999).

39. J. Hsieh, "Compensation of computed tomography data for detector afterglow," U.S. Patent No. 5265142 (1993).

40. J. Hsieh and D. R. Thayer, "Computed tomography system with correction for z-axis detector non-uniformity," U.S. Patent No. 5473656 (1995).

41. J. Hsieh, "Computed tomography system with z-axis correction," U.S. Patent No. 5301108 (1994).

42. G. Besson, H. Hu, M. Xie, D. He, G. Seidenschnur, and N. Bromberg, "A z gain nonuniformity correction for multislice volumetric CT scanners," *Med. Phys.* **27**(5), 873–884 (2000).

43. L. A. Shepp, S. K. Hillal, and R. A. Schultz, "The tuning fork artifact in computerized tomography," *Comput. Graph. Imag. Process.* **10**, 246–255 (1979).

44. J. R. Mayo, N. L. Muller, and R. M. Henkelman, "The double-fissure sign: a motion artifact on thin-section CT scans," *Radiology* **165**, 580–581 (1987).

45. R. G. Gould, "Principles of ultrafast computed tomography: historical aspects, mechanisms of action, and scanner characteristics," in *Ultrafast Computed Tomography in Cardiac Imaging: Principles and Practice*, W. Stanford and J. A. Rumberger, Eds., Futura Publishing Co., Mount Kisco, NY, 1–15 (1992).

46. D. L. Parker, "Optimal short scan convolution reconstruction for fan-beam CT," *Med. Phys.* **9**, 254–257 (1982).

47. M. D. Sliver, "A method for including redundant data in computed tomography," *Med. Phys.* **27**(4), 773–774 (2000).

48. N. J. Pelc and G. H. Glover, "Method for reducing image artifacts due to projection measurement inconsistencies," U.S. Patent No. 4580219 (1986).

49. C. R. Crawford, J. D. Godwin, and N. J. Pelc, "Reduction of motion artifacts in computed tomography," *Proc. IEEE Engr. Med. Biol. Soc.* **11**, 485–486 (1989).

50. S. C. Moore, P. F. Judy, J. D. Garnic, G. X. Kambic, F. Bonk, and G. Cochran, "Prospectively gated cardiac computed tomography," *Med. Phys.* **10**, 846–855 (1983).

51. W. Kalender, H. Fichie, W. Bautz, and M. Skalej, "Semiautomatic evaluation procedures for quantitative CT of the lung," *J. Comput. Assist. Tomogr.* **15**, 248–255 (1991).

52. C. J. Ritchie, J. Hsieh, M. F. Gard, J. D. Godwin, Y. Kim, and C. R. Crawford, "Predictive respiratory gating: a new method to reduce motion artifacts on CT scans," *Radiology* **190**, 847–852 (1994).

53. F. Liu, C. Cuevas, A. A. Moss, O. Kolokythas, T. J. Dubinsky, and P. E. Kinahan, "Gas bubble motion artifact in MDCT," *Am. J. Roentgen.* **190**(2), 294–299 (2008).

54. H. E. Johns and J. R. Cunningham, *The Physics of Radiology*, Charles C. Thomas Publisher Ltd., Springfield, IL (1983).

55. R. A. Brooks and G. DiChiro, "Beam hardening in reconstructive tomography," *Phys. Med. Biol.* **21**(3), 390–398 (1976).

56. R. J. Jennings, "A method for comparing beam-hardening filter materials for diagnostic radiology," *Med. Phys.* **15**(4), 588–599 (1988).

57. P. K. Kijewski and B. E. Bjarngard, "Correction for beam hardening in computed tomography," *Med. Phys.* **5**, 209–214 (1978).

58. R. C. Chase and J. A. Stein, "An improved image algorithm for CT scanners," *Med. Phys.* **5**, 497–499 (1978).

59. P. M. Joseph, and R. D. Spital, "A method for correcting bone induced artifacts in CT scanning," *J. Comput. Assist. Tomogr.* **12**, 100–108 (1978).

60. J. Hsieh, R. C. Molthen, C. A. Dawson, and R. H. Johnson, "An iterative approach to the beam hardening correction in cone beam CT," *Med. Phys.* **27**(1), 23–29 (2000).

61. G. H. Glover and N. J. Pelc, "An algorithm for the reduction of metal clip artifacts in CT reconstructions," *Med. Phys.* **8**(6), 799–807 (1981).

62. D. D. Robertson, P. J. Weiss, E. K. Fishman, D. Magid, and P. S. Walker, "Evaluation of CT techniques for reducing artifacts in the presence of

metallic orthopedic implants," *J. Comput. Assist. Tomogr.* **12**(2), 236–241 (1988).

63. J. S. Choi, K. Ogawa, M. Nakajima, and S. Yuta, "A reconstruction algorithm of body sections with opaque obstructions," *IEEE Trans. Sonic. Ultrason.* **SU-29**(3), 143–150 (1982).

64. J. Hsieh, "Artifact correction for highly attenuating objects," U.S. Patent No. 6035012 (2000).

65. R. P. Tolakanahalli, J. E. Medow, J. Hsieh, G.-H. Chen, and C. A. Mistretta, "Reduction of metal artifacts using subtracted CT projection data," *SPIE Proc.* **5535**, 636–643 (2004).

66. G. H. Glover and N. J. Pelc, "Method and apparatus for compensating CT images for truncated projections," U.S. Patent No. 4550371 (1985).

67. K. F. King, "Compensation of computed tomography data for objects positioned outside the field of view of the reconstructed image," U.S. Patent No. 5043890 (1991).

68. R. M. Lewitt, "Processing of incomplete measurement data in computed tomography," *Med. Phys.* **6**(5), 412–417 (1979).

69. G. T. Herman and R. M. Lewitt, "Evaluation of a preprocessing algorithm for truncated CT projections," *J. Comput. Assist. Tomogr.* **5**(1), 127–135 (1981).

70. B. Ohnesorge, T. Flohr, K. Schwarz, J. P. Heiken, and K. T. Bae, "Efficient correction for CT image artifacts caused by objects extending outside the scan field of view," *Med. Phys.* **27**(1), 39–46 (2000).

71. J. Hsieh, E. Chao, J. Thibault, B. Grekowicz, A. Horst, S. McOlash, and T. J. Myers, "A novel reconstruction algorithm to extend the CT scan field-of-view," *Med. Phys.* **31**(9), 2385–2391 (2004).

72. J. Hsieh, "Methods and apparatus for modulating x-ray tube current," U.S. Patent No. 5696807 (1997).

73. J. Hsieh, J. R. Mayo, K. P. Whittall, K. Kim, M. J. Brown, and O. Dake, "An algorithm for automatic prediction of optimal x-ray tube current," *Radiology* **205**(p), 352 (1997).

74. W. A. Kalender, H. Wolf, C. Sues, M. Gies, D. Hentschel, and W. A. Bautz, "Dose reduction in CT by automatically adapted tube current modulation: experimental results and first patient studies," *Radiology* **205**(p), 471 (1997).

75. J. Hsieh, W. K. Braymer, K. F. King, and P. D. Kosinski, "Over-range image artifact reduction in tomographic imaging," U.S. Patent No. 5,165,100 (1992).

76. J. Hsieh and W. K. Braymer, "Reduction of image artifacts from support structures in tomographic imaging," U.S. Patent No. 5,225,980 (1993).

Chapter 8
Computer Simulation and Analysis

8.1 What Is Computer Simulation?

Computer simulation uses theoretical models to predict the performance of a real system. Computer simulation can be divided into two general categories: analytical and statistical. Analytically based simulation generates system models using known analytical equations. A good example of this category is the generation of projections of mathematical phantoms. This type of projection is calculated based on the line integrals of attenuation coefficients of objects whose shapes can be described by closed-form equations (the phantom is formed with cylinders, spheres, ellipsoids, or bars). For a given source and detector cell position, the line integral can be calculated precisely for each object. The final result is the weighted summation of these integrals. Statistically based simulation, on the other hand, uses random number generators and the physical property of the interaction process to predict the system performance. A good example is the Monte Carlo simulation used to predict the scatter distribution.

Computer simulation is applicable in two major areas: system optics and system physics. System optics mainly addresses issues related to the geometric factors of the system. For example, computer simulation is helpful in understanding the impact of x-ray focal spot size, detector size, source-to-detector distance, and source-to-iso distance to the system spatial resolution. In our analyses, we focus mainly on the geometric factors that impact the system performance. System physics focuses on the physics properties of each process. For example, we can use a known input x-ray spectrum, the attenuation characteristics of different materials as a function of the x-ray energy, and the relative concentration of each material to understand the beam-hardening phenomenon and its impact on image quality. Physics-based simulation can also help to explain the impact of component characteristics on system performance parameters, such as LCD. For each application category (optics or physics), either analytical or statistical simulation tools can be used.

Computer simulation is a useful tool for designing a new CT system. During the design process, the designer is often faced with the problem of determining various system parameters. The parameter selection process is generally iterative and trial-and-error in nature. If one must build different CT components and iterate the design process, it can be very time-consuming and expensive. When the design is limited by the availability of real system components, it is likely to be suboptimal. This is the situation in which computer simulation can play a significant role. Once a simulation model is built, the computer program can produce results for different configurations in a short period of time. The optimal selection of these parameters can then be made based on tradeoff analyses of the simulation results.

Computer simulation can also help to isolate problems and develop specifications for complicated systems. For example, one important performance parameter of a CT system design is the stability of its mechanical structure. In today's state-of-the-art CT system, the gantry rotates at least 0.4 sec per revolution. The x-ray tube and detector typically weigh several hundred kilograms. Spinning a large mass at such a high speed unavoidably introduces mechanical vibrations. Our goal is to build a system with a level of vibration that does not produce image artifacts and impact the system performance. To develop a vibration specification, additional weights can be physically placed on an existing mechanical structure to induce imbalance to the system. By scanning different phantoms with different levels of imbalance, one could establish a relationship between the amount of mechanical vibration and image artifacts. This approach is obviously expensive and time consuming. An alternative approach is to mimic mechanical vibrations by computer simulation. For example, the iso-center of the backprojection can be dynamically changed as a function of the projection angle to simulate mechanical in-plane vibrations. Figure 8.1(a) and (b) depict reconstructed images of a Teflon rod. In Fig. 8.1(a), the phantom was scanned with an imbalanced CT system (mechanical vibration is present). The bright and dark shading near the rod can be clearly observed.

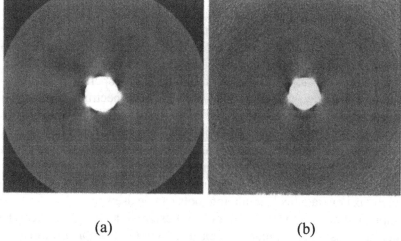

(a)   (b)

Figure 8.1 Use of computer simulation for mechanical vibration study. (a) Actual scan taken with an unbalanced gantry with mechanical vibration. (b) Computer simulation of vibration using vibration-free scan data.

Figure 8.1(b) was produced by computer simulation. The scan data were collected on a well-balanced system with little mechanical vibration. An angular-dependent bias was introduced in the reconstruction process to mimic mechanical vibration. Note the similarity of the artifacts in both images. Since vibration is simulated during the reconstruction process, different vibration magnitudes and characteristics (phase and frequency) can be easily analyzed.

8.2 Simulation Overview

Figure 8.2 shows a simplified view of a CT system. The entire CT system is divided roughly into five major components: x-ray generation, scanned object, x-ray photon detection and signal generation, mechanical structure, and image reconstruction. For the simulation discussion in this book, we focus mainly on the first four components (the process of image generation itself can be considered a computer simulation).

The first step in the CT imaging chain is x-ray production. As mentioned previously, the x-ray photons are generated by bombarding a target with high-speed electrons. The produced x-ray photons pass through additional filtrations, such as oil, glass, beryllium, aluminum, copper, or other materials before reaching the patient. The x-ray energy spectrum has a significant impact on the quality of the produced CT images. Consequently, understanding this linkage has direct impact on the optimal selection of clinical protocols. Different clinical applications emphasize different aspects of image quality. For example, in CT examination of the midbrain, the gray-white matter differentiation is important and requires the separation of materials that differ by a few tens of CT numbers. Low-contrast differentiation, in turn, relies on the availability of low-energy photons. For this type of examination, a narrow 80- to 100-HU display window width is often used. Consequently, images must have a low noise level. For inner auditory canal (IAC) studies, the bony structures are the primary objects

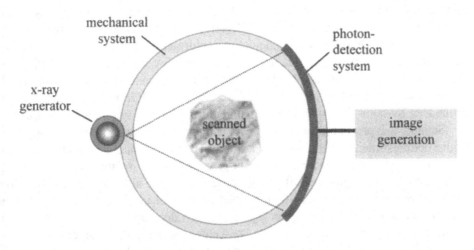

Figure 8.2 Major components for simulation in a CT system.

of interest. Because the display window width is quite large (more than 1000 HU), the scanner's low-contrast capability is not of great concern. In addition, the wide display window offers better tolerance to image noise. Computer simulation can help a designer to achieve the proper balance between x-ray flux and x-ray spectrum. This includes the selection of proper tube operating voltage, x-ray beam filtration, and patient shape-dependent filters (e.g., the bowtie filter).

The second part of the CT system deals mainly with the interaction between the x-ray photons and the scanned object. This includes photon energy-dependent attenuation characteristics, scattered radiation, partial volume, and many other effects. Although many of these issues can be addressed with real experiments, it can be difficult to separate the impacts of different factors. In addition, since excessive exposure to x-rays is known to be harmful to humans, many experiments are faced with regulatory and ethical constraints. For example, to understand the relationship between patient motion (either due to the patient's physiological motions or to special scanning protocols) and image quality, it is impossible to subject a patient to multiple scans to experiment with different motion conditions. Computer simulations, on the other hand, are free from these constraints. Another example is the study of beam-hardening artifacts. Our clinical experience has shown that various contrast agents can produce different levels of beam-hardening artifacts. If these interactions must be studied by real experiments, the number of required experiments can quickly reach an unmanageable level.

The third major component of a CT system is the x-ray detection system. This includes both the detector and the data acquisition electronics. As discussed in Section 8.1, it is important during the early design phase to predict the required detector geometries, quantum detection efficiency, and electronic noise prior to producing an actual system. Detector manufacturing processes are expensive and time consuming. If any one of the geometric parameters must be determined by trial and error using actual detector modules, the cycle time becomes prohibitively long. Another aspect of detection system simulation is to understand the impact of nonideal component characteristics on system performance. A good example is the detector channel-to-channel variation. To ensure artifact-free images, the detector performance characteristics must be maintained at a certain specification level over the life of the detector. Although these specifications often can be determined with an accelerated life test of a real system, this is both time consuming and expensive. In addition, the results can be influenced by other factors that also change during the test (e.g., the x-ray tube).

Section 8.1 presented an example of a mechanical system simulation. The figure demonstrated that the reconstruction process could be modified to simulate different mechanical vibrations. Similarly, phenomena related to mechanical misalignment or angular positioning inaccuracy can also be simulated. For example, the detector quarter-quarter offset (discussed in Chapter 7) requires the detector iso-channel to be offset by a quarter detector width from the line connecting the x-ray source and the iso-center. Since a detector cell width is roughly 1 mm and the source-to-detector distance is roughly 1 m, a detector is unlikely to be positioned exactly a quarter-cell from the iso-line. Computer simulation can help to determine the alignment tolerance level.

Although our simulation discussion is limited to four of the five major components of the imaging chain, it should be pointed out that computer simulations are often crucial in the development of preprocessing and reconstruction algorithms (the fifth major system component). For example, many reconstruction algorithms for helical and multislice CT were developed based on simulated projections. If reconstruction algorithms had to be developed after a real system was built, a prolonged product development cycle could be expected. In addition, it becomes difficult to separate the impact of the inexact nature of the reconstruction algorithm from the nonideal behavior of the system.

Computer simulation is a complicated topic, so it is impossible to cover all aspects of the subject in a short chapter. For each of the five major CT system components, simulations can be performed for the system optics or physics. For each category, either analytical or statistical methodology can be used. Since each major component is composed of many subcomponents, numerous combinations are possible. The following sections offer a few examples to illustrate the general methodologies of simulation.

8.3 Simulation of Optics

In the CT system design process, the answers to many seemingly simple questions are, in fact, quite complex. For example, consider the geometry of a scanner. The geometry includes the detector cell size, the x-ray focal spot size, the distance between the detector and the source, and the distance between the source and the iso-center (for the time being, only the geometrical aspect of the system will be discussed and physics-related parameters will be ignored). We want to make sure that the selected geometric parameters can produce a CT system with the proper spatial resolution, noise performance, and speed.

Consider first the selection of the detector cell size to satisfy the above performance requirements. From a spatial resolution point of view, we want to use a small detector cell. On the other hand, the detector cell size needs to be sufficiently large to ensure the dose efficiency of system (the effective detector area diminishes with smaller cell size due to the postpatient collimator and cell gaps). We also need to achieve a proper balance between detector size and x-ray focal spot size, and we want to avoid the scenario in which one of them dominates spatial resolution. If the spatial resolution was totally dominated by the x-ray focal spot size, the system spatial resolution would become insensitive to the change in detector cell size. This would be an indication that the detector cell size was too small, and the penalty would be paid in dose efficiency. However, if the spatial resolution was dominated by the detector, it would imply that the x-ray focal spot size was probably too small. Since the amount of x-ray flux the tube can deliver is closely related to the focal spot size (the smaller the focal spot size, the less x-ray flux that can be delivered), the system's noise performance is likely to suffer.

A similar discussion can be rendered on the selection of source-to-detector and source-to-iso distance. By changing the magnification of the system (the magnification is defined as the ratio of the source-to-detector distance over the source-to-iso-center distance), the use of the x-ray tube can be modified. A larger magnification generally improves tube utilization. On the other hand, a larger magnification can lead to higher noise nonuniformity, which will be discussed in Chapter 9.

Often, system performance can be estimated without a full-fledged simulation. The ability to perform a "back-of-the-envelope" estimation can be quite useful in setting the parameter ranges and reducing the number of full-fledged simulations. Even with advanced computing hardware, the amount of time needed to produce a complete simulation is usually not negligible.

The second utility of the back-of-the-envelope estimation is to check the simulation results. Since most simulation codes are large and prone to programming errors, the simulation results must be checked independently. If the answer provided by the simulation deviates significantly from the theoretical prediction, it may be necessary to check the accuracy of the code. Finally, a back-of-the-envelope prediction can provide a good estimation of the system performance when a full-fledged simulation tool is not available. A quick calculation based on these parameters can provide good insight into a scanner's capability.

Let us discuss the technique to estimate the spatial resolution of a CT system near the iso-center for a given set of detector cell size, x-ray focal spot size, source-to-detector distance, and source-to-iso distance. For convenience, we will introduce the concept of "projected response" at the iso-center. A "projected detector-cell response" at the iso-center is derived by examining the system response at the iso-center when the x-ray focal spot is an ideal point source. Similarly, the "projected focal spot response" is the system response to the x-ray source when the detector cell is assumed to be a point. These projected responses, depicted in Figs. 8.3(a) and (b), are the scaled-down versions of the real responses. If we denote the detector response measured at the surface of the detector cell by $d(x)$, the projected detector response at the iso-center by $d_p(x)$, the source response measured at the x-ray focal spot by $s(x)$, the projected source response at the iso by $s_p(x)$, the source-to-detector distance by L, and the source-to-iso distance by D, then the following relationships exist:

$$d_p(x) = \lambda \cdot d\left(\frac{xL}{D}\right) \tag{8.1}$$

and

$$s_p(x) = \eta \cdot s\left(\frac{xL}{L-D}\right). \tag{8.2}$$

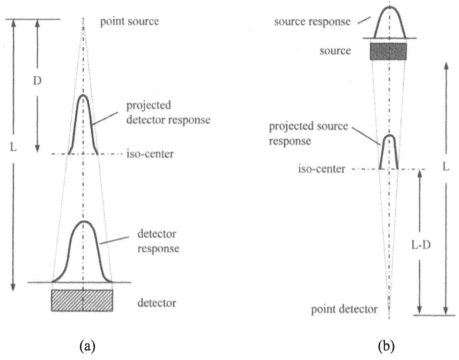

Figure 8.3 Illustration of the "projected response." (a) Projected detector response at iso-center is obtained by treating the x-ray source as a point. (b) Projected source response at iso-center is derived with a point detector cell.

The parameters λ and η are scaling factors to ensure that the areas under the projected curves equal unity. The combined system response at the iso-center is then simply the convolution of the projected detector response with the projected source response. We can justify the convolution process by linear system theory. From a linear system point of view, $d_p(x)$ or $s_p(x)$ can be viewed as the impulse response (or point-spread function) of the system, since either the x-ray focal spot or the detector cell is assumed to be an impulse function. The linear system theory states that the output of a system to an input function is the convolution of the input function with the impulse response of the system.

To illustrate the analysis, consider an example of a CT system with the following parameters. The source-to-detector distance is 950 mm, and the source-to-iso-center distance is 540 mm. The detector cells are formed on an arc concentric to the x-ray focal spot. The spot size is 0.7-mm wide and the detector cell spacing is 1.0 mm. Because of the postpatient collimator, 80% of the cell surface is exposed to the x-ray photons. To estimate the MTF of such a CT system, we assume that the reconstruction filter is an ideal ramp function multiplied by a cosine-square window with its cutoff frequency at the Nyquist frequency of the CT system.

Given the above set of conditions, the analysis can be carried out as follows. Assume ideal responses for both the detector and the x-ray source; that is, their responses can be represented by a rectangular function. The projected focal spot width s_w at the iso-center, calculated based on Eq. (8.2), equals 0.3 mm. The width of the rectangular function for the detector is $0.8 \times 1 = 0.8$ mm. The projected width of the detector at the iso-center d_w is then 0.45 mm. The combined system response is the convolution of the two rectangular functions. The resulting function $g(x)$ is a trapezoidal function:

$$g(x) = \begin{cases} (d_w s_w)^{-1}\left(x + \dfrac{s_w + d_w}{2}\right), & -\dfrac{s_w + d_w}{2} \le x < \dfrac{s_w - d_w}{2}, \\[2mm] d_w^{-1}, & \dfrac{s_w - d_w}{2} \le x < \dfrac{d_w - s_w}{2}, \\[2mm] (d_w s_w)^{-1}\left(\dfrac{d_w + s_w}{2} - x\right), & \dfrac{d_w - s_w}{2} \le x < \dfrac{d_w + s_w}{2}, \\[2mm] 0, & \text{otherwise.} \end{cases} \tag{8.3}$$

Next, we calculate the Fourier transform of the combined system response $g(x)$ to arrive at the function $G(f)$. The cosine-square window $H(f)$ can be expressed by the following equation:

$$H(f) = \begin{cases} \cos^2\left(\dfrac{\pi \cdot f}{2 f_c}\right), & -f_c \le f \le f_c, \\[2mm] 0, & \text{otherwise.} \end{cases} \tag{8.4}$$

The MTF of the system is then simply the magnitude of $H(f)\,G(f)$, shown by a solid line in Fig. 8.4. To assess its accuracy, we performed a full computer simulation (to be discussed shortly) with the same set of parameters. The

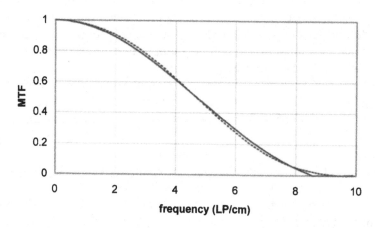

Figure 8.4 Example of the estimated MTF at iso-center. Solid line: MTF calculated based on the analytical model; dotted line: MTF based on computer simulation.

resulting MTF curve is plotted by a dotted line in Fig. 8.4. It is clear from the figure that the back-of-the-envelope prediction is quite accurate and therefore provides insight into the factors that impact the system spatial resolution. However, for a more in-depth investigation of system performance (for example, spatial resolution away from the iso-center), we must rely on more elaborate computer simulation approaches. To illustrate the complexity, consider the following complication.

In the previous analysis, we assumed that the shape of the focal spot could be modeled by a stationary function $s(x)$ for all detector channels. In reality, this condition is rarely satisfied. For illustration, examine the formation of the x-ray focal spot. As discussed in Chapter 6, the majority of the input energy to the x-ray tube is converted to heat. Less than 1% of the total energy is converted to useful x-ray photons. The biggest challenge in tube design is to reduce the impact that temperature has on the target. One solution is to distribute the high-speed electrons over a larger area. To maintain the small focal spot size viewed by the detector (for spatial resolution), the target surface is positioned at a shallow angle with respect to the scanning plane, as shown in Figs. 8.5(a) and (b). This is the so-called *line focus principle*. Some tube designs set the target angle ξ at 7 deg. When viewed directly from the top (a top-down view of the target), the focal spot is shaped roughly as a rectangle with a width w and a height h. When the same focal spot is viewed from the iso-center of the scanning plane, the focal spot becomes w in width and $h \times \sin(\xi)$ in height. The shape of the focal spot is still a rectangle, as shown in Fig. 8.6. When the x-ray source is viewed at a location away from the iso-line, however, the focal spot shape is no longer a rectangle; it is a trapezoid. The width of the trapezoid is influenced not only by the width of the focal spot w, but also by the height of the focal spot $h \times \sin(\xi)$ and the fan angle γ. To quantitatively demonstrate this effect, Fig. 8.7 plots the x-ray focal spot responses to a 0.1-mm wire with an "infinitely" fine detector sampling, viewed at the iso-center and 20 deg off the iso-ray (we normalized the peak response at the iso-center to unity for easy comparison). The response was obtained by a simulation method that will be described in the next paragraph. The figure clearly illustrates the degradation of the focal spot shape due to the increased detector angle. Although the detector-angle-dependent focal spot shape can be incorporated into our analytical model, the model's complexity increases quickly when the x-ray intensity is spatially nonhomogeneous. As a result, different computer simulation approaches need to be explored.

One of the most commonly used techniques to simulate the optics of a CT system is to divide the complicated component geometries into smaller elements. When the size of each element is small enough, we can approximate its response by a point. For example, in Fig. 8.8 the x-ray focal spot is divided into an $\mathbf{M} \times \mathbf{N}$ matrix (in width and height). For ease of discussion, we will call each element in the matrix a source-let. Similarly, each detector cell is divided into a $\mathbf{K} \times \mathbf{L}$ matrix (in x and z), and each element is called a detector-let. For each source-let

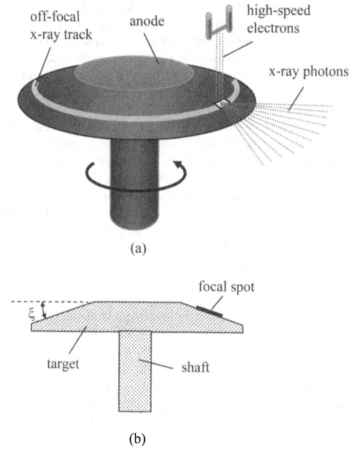

Figure 8.5 Illustration of the x-ray tube target. (a) For heat dissipation, the x-ray tube target rotates at a high speed and is at a shallow angle with respect to the scan plane. The focal spot is shaped like a rectangle viewed from the top. (b) Cross-sectional view of the x-ray tube target. The angle ξ is typically a few degrees.

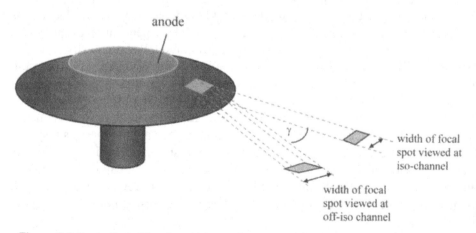

Figure 8.6 Illustration of focal spot shape change as a function of the detector angle.

and detector-let pair, the x-ray path is approximated by a pencil beam with an infinitesimal width. We call each pencil beam a path-let. Since each path-let is a pencil beam, its line integral along a mathematical phantom can be calculated analytically (this will be described in detail later). The total signal received by a detector cell is then the weighted summation of all path-lets that intersect the detector cell. The weights are determined by the product of the source and detector response functions (assuming these responses are nonuniform). If a

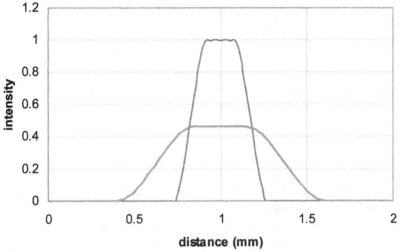

Figure 8.7 X-ray focal spot shape viewed by the detector changes with detector angle. The x-ray focal spot is 0.9 mm (in x) and 0.7 mm (in z). The target angle is 7 deg. Each detector is 0.02 mm, so its impact can be ignored. The wire radius is 0.1 mm. Thin black line: response to the wire located at iso-center. Thick gray line: response to the wire located at 20 deg off iso-ray. The distance between this wire and the source is the same as the distance between the iso-center wire and the source.

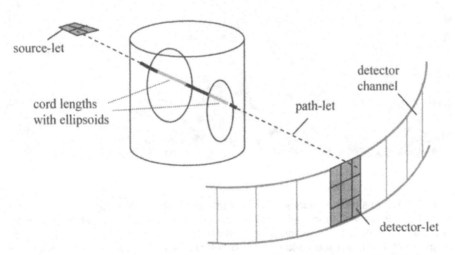

Figure 8.8 Illustration of source-let, detector-let, path-let, and cord lengths.

particular source-let contributes only 5% of the overall x-ray flux and a certain detector-let contributes 7% of the overall detector signal (to a uniform incoming flux), the weight for this particular path-let is 0.35%:

$$p = \sum_{i=1} w_i \left[\sum_{k=1} \mu_k l_{i,k} \right], \tag{8.5}$$

where $l_{i,k}$ is the cord length for object k intersecting path-let i, and w_i is the weighting factor for path-let i. For ideal focal spot and detector responses (flat responses independent of location), a simple average can be used. Since we are focusing only on the optics of the system, we do not take into consideration the energy-dependent nature of the linear attenuation coefficient. When energy dependency is considered, the summation process must be carried out prior to the minus logarithm operation. This will be discussed in more detail in Section 8.4.

In the above operation, it is necessary to divide the source and detector cells into finer "lets." Although a single or fewer lets are more desirable from a computational complexity point of view, this often leads to inaccurate results. The inaccuracy occurs at the projection samples with a high-intensity gradient (for example, near the edge of a high-density object). For illustration, we performed a computer simulation of a phantom consisting of a high-density wire, a high-density rod, a medium-density rod, and an air pocket located inside a water cylinder. Seven source-lets and seven detector-lets were used to generate a set of projections. The reconstructed image is shown in Fig. 8.9(a). Streaks near

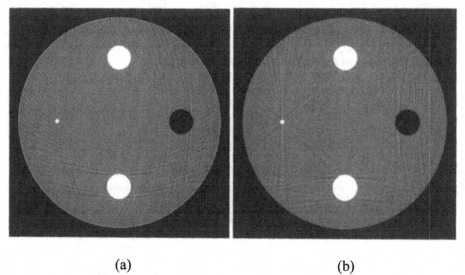

(a)   (b)

Figure 8.9 Reconstructed images of a simulated phantom (WW = 20). (a) Projection was generated with seven source-lets and seven detector-lets. The object boundaries are sharp and well defined. Projection aliasing artifacts can be well observed. (b) Projection was generated with two source-lets and two detector-lets. The object boundaries are less defined. Other undersampling artifacts are superimposed on the aliasing artifacts.

the edge of the smaller objects resemble the aliasing artifacts observed in images produced on real scanners. The wire is also well defined with sharp boundaries. The simulation was repeated with two source-lets and two detector-lets, and the suboptimal result is shown in Fig. 8.9(b). Streaking artifacts near the wire and rod do not correlate well with real scanner results. In addition, the sharpness of the object boundaries is degraded. These boundaries are clearly outlined in the difference image shown in Fig. 8.10, indicating a reduced spatial resolution. Indeed, the intensity profiles of the reconstructed wire (shown in Fig. 8.11) confirm our assessment.

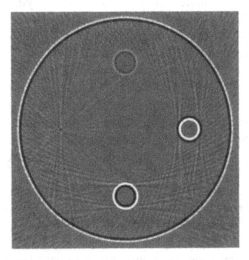

Figure 8.10 Difference image between images produced by seven and two source-lets and detector-lets (WW = 20).

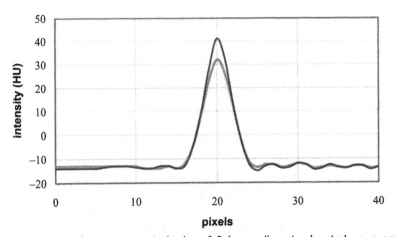

Figure 8.11 Profile of a reconstructed wire of 0.1-mm diameter located on a soft-tissue background. This figure is used to illustrate the impact of the number of source-lets and detector-lets on the accuracy of simulation. Thin black line: seven source-lets and seven detector-lets were used in the simulation. Thick gray line: two source-lets and two detector-lets were used in the simulation. A reduced peak intensity results from inadequate sampling.

Now consider the mathematical phantom itself. Any complicated object can be formed with a combination of spheres, ellipsoids, cylinders, or rectangles (we will call each component an object-let). The line integral along a path-let for each object-let can be calculated analytically by solving for the cord length of the intersection between the ray and each object-let. The final result is the weighted summation of the signals from all object-lets, the weight being the linear attenuation coefficients. To cover the scenario in which the object-lets located inside a larger object-let may have lower attenuation coefficients, the attenuation coefficient for each object-let can be assigned either positive or negative values. For example, we will simulate a human chest with a cylinder (to simulate the chest wall) and two low-density ellipsoids (to simulate the lung lobes). The attenuation coefficient of the chest wall is 0.2 cm^{-1} (similar to water) and the attenuation coefficient of the lung lobes is 0.02 cm^{-1}. For the computer simulation, we set the attenuation coefficient of the cylinder at 0.2 cm^{-1} and the two ellipsoids at –0.18 cm^{-1} (0.02 – 0.2). The line integral of any path-let is simply the weighted sum of the cord length of the cylinder (ignoring the presence of the ellipsoids) and the cord length of the ellipsoids. Alternatively, one could use the true attenuation coefficients and keep track of the geometric relationship among the object-lets. The latter approach is often preferred when simulating partially overlapped object-lets.

The above calculations must be repeated for all detector channels and all projection views. For the example shown in Fig. 8.8, the x-ray source is divided into 2 × 2 source-lets and a detector cell into 3 × 3 detector-lets. For each source-let, we calculate 9 line integrals (path-lets). For four source-lets, there are a total of 36 line integral calculations (but only one such calculation is depicted in the figure). In Fig. 8.8, we select a simple object composed of a cylinder and two ellipsoids. Therefore, each line integral calculation consists of the calculation of three cord lengths. For the example given, a total of 108 (36 × 3) cord-length calculations are involved. For a detector that has 900 detector channels, the total amount of the cord-length calculation is 108 × 900 = 97,200.

Next, we introduce the concept of a view-let. For most commercially available CT scanners, the x-ray source is constantly energized during the scan; that is, the x-ray source is not pulsed between adjacent views. For a scanner that samples at N projection views per gantry rotation, each view covers an angular range of 360 deg/N. Therefore, an averaging or blurring effect takes place. To accurately simulate this effect, the angular span of each view is further divided into multiple subangular regions, and each region is defined as a view-let. Within each view-let, a pulsed x-ray source is assumed. Each projection reading is then the average of all view-let samples. If we assume that a scanner collects 1000 views, and three view-lets are used for each projection, a total of 3000 projection sets are needed for the simulation of a single gantry rotation. When this number is combined with the 97,200 cord-length calculations per view-let, the total number of cord-length calculations for this extremely simple example is nearly 292 million! Although the amount of calculation is huge, the computational time is manageable, even using the inexpensive computer hardware available today.

The methodology presented here is applicable not only to address questions related to spatial resolution, but also to other problems linked to the system's geometric factors. For example, if we want to determine the impact of system misalignment on the final image quality, we can perform a computer simulation in the following manner. We generate sets of projections by assuming different iso-channel locations. In the reconstruction process, we ignore the simulated iso-channel locations and use only the default location value. It is important to select appropriate objects for analysis, because if a set of large low-density objects is used, it is likely that little impact on image quality will be observed until the magnitude of the misalignment becomes large. For illustration, we simulated two cylinders inside a larger cylinder. The linear attenuation coefficients of the two cylinders and the background cylinder were 0.21 cm^{-1} and 0.20 cm^{-1}, respectively. The iso-channel was shifted by two channels in the projection generation process and was ignored in the reconstruction. The reconstructed image is shown in Fig. 8.12(a). No visible image artifact can be observed. Clearly, the absence of artifacts is due to the improper selection of simulated objects. For this type of study, we need to use high-density objects (high contrast to the background). For example, we can use a thin tungsten wire (e.g., 0.1 mm in diameter) located in air as our phantom. By comparing the MTF measurements of the reconstructed wires as a function of the magnitude of misalignment, we establish a quantitative relationship between the magnitude of misalignment and the spatial resolution impact. We can also simulate high-density rods and observe streaking and shading artifacts in the background. Figure 8.12(b) depicts a reconstructed image of two dense rods ($\mu = 0.37$cm^{-1}) inside a larger cylinder ($\mu = 0.20$cm^{-1}). The iso-channel of the simulated projection was again shifted by two channels. Streaks can be clearly observed in the image. This example demonstrates the importance of selecting appropriate objects for computer simulation.

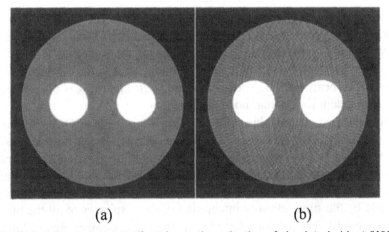

(a)   (b)

Figure 8.12 Variation in image artifact due to the selection of simulated object (WW = 20). The attenuation coefficient for the background cylinder is 0.2 cm^{-1} for both images. The attenuation coefficient of the two cylinders is (a) 0.21 cm^{-1} and (b) 0.37 cm^{-1}. A two-channel shift was introduced in the simulated projection.

8.4 Computer Simulation of Physics-related Performance

Often, the need arises to investigate system performance related to the fundamental physics of the system instead of its optics. For example, we may want to select an optimal x-ray spectrum to produce the required LCD while maintaining an acceptable image quality and patient dose. Although the geometric factors of the CT system do impact the available photon flux and patient dose to a certain extent, the low-contrast differentiation is influenced mainly by the presence of lower-energy x-ray photons. This is due to the fact that x-ray attenuation coefficients for most materials are a function of energy. Therefore, computer simulation tools are needed that take into account the physical properties of the interaction between the x-ray photons and the material.

Physics-related simulation can be classified into two categories: analytical and statistical. The advantage of the analytical approach is its computational efficiency. For physics properties that can be described or approximated by analytical forms on a macroscopic level, simple equations are sufficient to gain insight into the nature of the problem. Some physical properties, however, are not easily describable in a concise form (e.g., scattered radiation). Often it is easier to understand these phenomena by tracking a large ensemble of interaction events. For these cases, we often rely on statistical approaches (e.g., Monte Carlo simulation) to perform computer simulation. Clearly, Monte Carlo simulation is computationally more expensive. Both methodologies are described in the remaining sections.

We start with an example that investigates the beam-hardening effect. Chapter 7 stated that under monoenergetic x-ray beam conditions, the attenuation characteristics of an object can be described by a simple exponential relationship (the Beer–Lambert law):

$$I_E = I_{0,E} e^{-\int \mu_{E,S} dS}, \qquad (8.6)$$

where $I_{0,E}$ and I_E represent the incident and transmitted x-ray intensities, $\mu_{E,S}$ is the linear attenuation coefficient of the object, and subscript E indicates the energy dependency of these variables. The integration in the exponential term is along the x-ray beam path S. The attenuation coefficient $\mu_{E,S}$ is both energy- and location-dependent for an inhomogeneous object.

A polychromatic x-ray flux with an energy spectrum of $\Omega(E)$ can be placed into a large number of bins based on their energies. The x-ray photons within each bin should follow approximately the attenuation relationship specified in Eq. (8.6), since all of the photons inside a specific bin have similar x-ray energies. Therefore, the final response of the material to the polychromatic x-ray flux should be the summation or integration of the responses of all the bins. If the attenuation coefficient $\mu_{E,S}$ as a function of the input x-ray energy is known, the relationship between the incident and the transmitted x-ray fluxes is simply

$$I = I_0 \int \Omega(E) e^{-\int \mu_{E,S} dS} dE , \qquad (8.7)$$

where I and I_0 represent the total transmitted and incident x-ray intensities, respectively. The measured projection p is the negative logarithm of the intensity ratio:

$$p = -\log\left(\frac{I}{I_0}\right) = -\log\left(\int \Omega(E) e^{-\int \mu_{E,S} dS} dE \right). \qquad (8.8)$$

To reduce the amount of computation, the number of summations over energy can be minimized with larger energy bin sizes.[1] As an example, we generated a set of phantom projections to investigate bone beam-hardening artifacts using the simulation process described above. We used an ellipsoidal bone ring to simulate a human skull and two ellipsoids to simulate two petrous bones. The x-ray spectrum used for simulation is shown in Fig. 8.13. The attenuation coefficient $\mu_{E,S}$ for bone and soft tissue as a function of energy is depicted in Fig. 8.14 by the black line and the gray line, respectively. The reconstructed image is shown in Fig. 8.15. The dark bands between the two petrous bones resemble typical artifacts observed when a patient is scanned in the posterior fossa region without proper correction for beam hardening. Chapter 7 contains representative human skull phantom images.

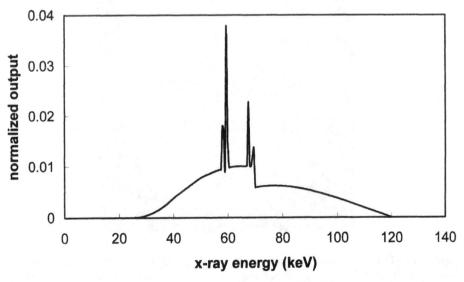

Figure 8.13 X-ray tube spectrum for 120 kVp.

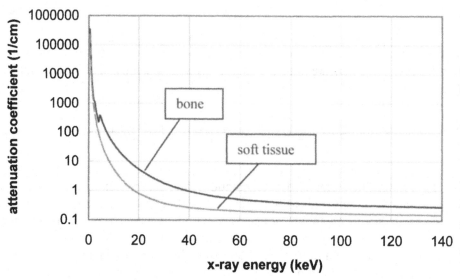

Figure 8.14 Attenuation coefficient of bone as a function of x-ray photon energy.

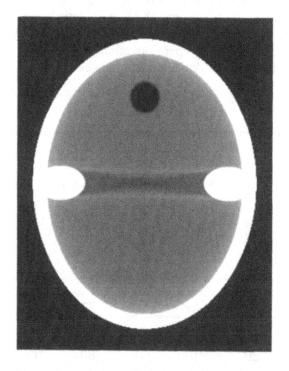

Figure 8.15 Computer simulation to investigate the beam-hardening effect in the posterior fossa region (WW = 80). The outer rings and the two ellipsoids near the center are simulated bone material. The internal material is simulated brain tissue. Simulation assumes the 120-kVp x-ray tube spectrum shown in Fig. 8.13. Projection signals for bone and soft tissue were produced using the graph shown in Fig. 8.14. The dark band between the two bone ellipsoids is a typical appearance of bone-induced artifact.

To illustrate the utility of computer simulation for contrast analysis, we will return to the problem discussed at the beginning of this section. Specifically, we want to demonstrate the impact of the x-ray energy spectrum on the contrast differentiation ability of a CT scanner. We selected two different x-ray kVp settings: 120 and 200 kVp. The phantom contains two cylinders (muscle and diluted iodine) inside a water container. As in the previous example, we generated an x-ray tube spectrum for both kVp settings. Projections as a function of photon energy were calculated for water, muscle, and diluted iodine using Eq. (8.8). Figure 8.16 shows the intensity profiles of the reconstructed cylinders, with the 120-kVp case depicted by the gray line and the 200-kVp case by the black line. In the reconstruction, only the water beam-hardening correction was applied. Quantitative measurements show that the muscle-water contrast was reduced by less than 2%, while the iodine-water contrast was reduced by nearly 18% as a result of the change in the x-ray spectrum.

The investigation of scattered radiation is a little more complicated. In the x-ray photon energy range of CT (40 to 150 keV), the dominant interaction in soft tissue is Compton scattering. It is usually reasonable to assume that the scattering direction is nearly isotropic. If we assume that the object is a slab extending infinitely across the direction of the x-ray beam and that none of the detected photons has undergone more than one scattering event, we can arrive at an approximate PSF of scatter as a function of distance from the pencil beam location.[2] Because of its complexity, the equation is omitted, but general observations will be made here. First, the scatter PSF is low frequency in nature and shows broad structureless wings around the pencil beam. The magnitude of the scattered flux at the detector is proportional to the area of the incident x-ray beam. Therefore, as the exposed volume increases (such as in the case of volumetric CT), the scatter-to-primary ratio increases. Second, the scattered

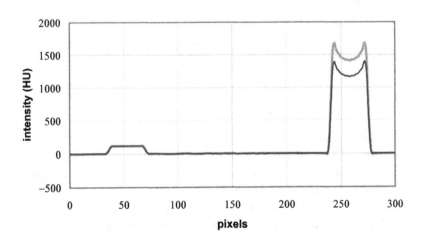

Figure 8.16 Intensity profiles of muscle (on left side of graph) and diluted iodine contrast (on right side of graph) in the reconstructed images to illustrate the impact of kVp on contrast. Thick gray line:120 kVp; thin black line: 200 kVp.

radiation can be effectively controlled by reducing the solid angle at which the detector elements are exposed to the scattering media. The magnitude of the wings derived from the PSF of a pencil beam is low (a small percentage of the primary beam). The resulting magnitude of the scatter for a large area of the x-ray beam is the integration of the pencil beam PSF over the solid angle of the detector element. Although these results provide good insight into x-ray scattering properties, they are insufficient to provide quantitative measurements when the object is complex and multiple scattering events need to be considered.

One methodology to prevail over these shortcomings is Monte Carlo simulation. For this approach, we start with a large collection of x-ray photons of different energies. The energy distribution of these photons follows the x-ray tube spectrum (for example, the one shown in Fig. 8.13). Next, we build a model for the scanned object, such as a human chest modeled with an elliptical cylinder and two large air pockets. Ribs can be modeled by a set of small bony cylinders. For most CT applications, only two types of interactions are considered: photoelectric absorption and Compton scatter. For each photon, we follow its path into the scanned object (in nuclear medicine, the direction of incident photons is *not* well defined, but the direction is well defined in CT). We randomly determine (with a random number generator) the distance d that the x-ray photon travels before an interaction takes place. The probability density function $p(d)$ for the random variable is

$$p(d) = \mu e^{-d\mu}, \qquad (8.9)$$

where μ is the linear attenuation coefficient of the material. Then we randomly determine (again with a random number generator) whether the interaction is Compton or photoelectric, based on the known probability of photoelectric p_{photo}, and the Compton interaction $p_{Compton}$:

$$\begin{cases} p_{photo} = \dfrac{\sigma^p}{\sigma^p + Z\sigma^c} \\[2mm] p_{Compton} = \dfrac{Z\sigma^c}{\sigma^p + Z\sigma^c} \end{cases}, \qquad (8.10)$$

where σ^p is the collision cross-section of the photoelectric absorption, Z is the atomic number of the material, and σ^c is the collision cross-section of the Compton interaction. If it is determined to be a Compton interaction, a new photon traveling direction is determined randomly based on the angular distribution (for simplicity, we assume an isotropic distribution). The energy of the scattered photons can be determined by the following formula:

$$\frac{1}{hv'} = \frac{1}{hv_0} + \frac{1}{m_0 c^2}(1 - \cos\theta), \qquad (8.11)$$

where hv_0 is the energy of the incoming photon, hv' is the energy of the scattered photon, θ is the scattering angle, and $m_0c^2 = 0.511$ MeV. This process continues until the photon either reaches the detector, dissipates all of its energy inside the object, or escapes from the system (scatters outside the detectable area). At this point, the fate of a new x-ray photon will be traced in the same fashion. The entire procedure is repeated numerous times to ensure that an adequate statistical sampling is obtained.

Obviously Monte Carlo simulation is computationally expensive, but this obstacle can be partially overcome by the use of powerful computers and improved algorithms. The accuracy of the Monte Carlo simulation depends on the quality of the random number generator, so it is important to make sure that the numbers produced by the random number generator are indeed random. To ensure adequate performance, a series of useful tests can be performed.[3] Many research papers have been published on the subject of using Monte Carlo simulation for medical imaging.[4–10] Details of these studies are omitted here.

We will briefly discuss the simulation of x-ray photon noise under quantum-limited conditions (i.e., when noise due to the x-ray photon flux is the dominating factor and other noise sources, such as electronic noise, can be ignored). The noise property of CT is more complicated than that of nuclear medicine for several reasons. First, although the monoenergetic x-ray photon statistics follow a Poisson distribution, postpatient polychromatic x-rays follow a compound Poisson distribution.[11] Second, the measured signals undergo many preprocessing or calibration steps to remove the imperfectness of the CT system. These operations are often nonlinear and recursive in nature. Consequently, the noise property after these operations are performed no longer follows a Poisson distribution. The third factor is the logarithm operation needed to convert the measured signals to line integrals of the attenuation coefficients. The logarithm operation is highly nonlinear. The matter is further complicated by the fact that some preprocessing and calibration steps are carried out after the logarithm step. To illustrate the complexity, we performed a simple experiment. A GE Performance™ phantom was scanned in a stationary mode (the x-ray tube and detector were stationary throughout the scan) and 1000 projection samples were collected. Since the gantry was stationary, all of the projections measured a constant x-ray flux through the same object paths. Consequently, if we examine the measurements of a single detector channel, they represent an ensemble of measurements of the same underlying signal. Their distribution is a good representation of the statistical properties of the system. Figure 8.17 shows the distribution of the measurements before the calibration process (except offset removal). Figure 8.18 depicts the same measurements after the calibration steps. The function that maps from the noise distribution before the calibration steps to the distribution after the calibration steps must be thoroughly researched.

Another important issue is the accuracy of the simulation itself. Most computer simulation codes are large and contain many functions or subroutines, so some programming errors in the myriad lines of computer code are

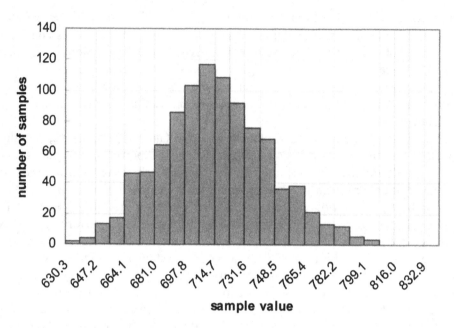

Figure 8.17 Distribution of measured samples of a single channel in a stationary scan. Only offset correction was performed on the data.

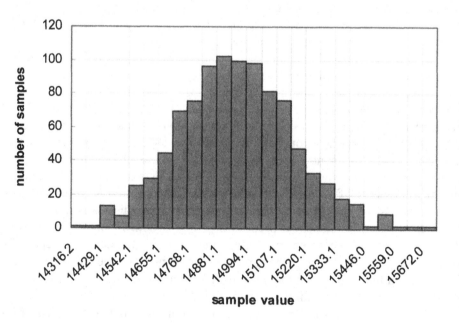

Figure 8.18 Distribution of measured samples of a single channel in a stationary scan. All calibration steps were performed on the data.

unavoidable. Therefore, any computer simulation should be checked against theoretical calculations, back-of-the-envelope calculations, or real experiments. Only after these tools have been fully checked against known results can they be used to analyze problems and predict the performance of future systems.

8.5 Problems

8-1 A student who was asked to write a simulation program for a CT system divided the source into 3 × 3 source-lets and each detector cell into 3 × 3 detector-lets. In the projection generation process, the student calculated the line integral for the corresponding source-let–detector-let pairs so that the upper-left source-let paired with the upper-left detector-let, the upper-middle source-let paired with the upper-middle detector-let, and so on. Do you think the student's simulation will work properly? If not, under what condition would this simulator not fail?

8-2 Ellipsoids are often used in a simulation to approximate an object. Develop an expression to calculate the path length of a pencil beam that passes through (x_1, y_1, z_1) and (x_2, y_2, z_2) and an ellipsoid located at the iso-center with equatorial radii of a and b and a polar radius of c. Can you use this formula to calculate the path length when the ellipsoid is not located at the iso-center and its axes are not parallel to the x, y, and z axes?

8-3 The source-to-iso distance of a CT system is 550 mm and the source-to-detector system is 950 mm. If both the source and detector cell can be modeled by rectangular functions with a 1-mm width, analytically estimate the iso-center MTF of the system in the x-y plane.

8-4 Given the source and detector sizes in problem 8-3, how do you change the CT geometry (the source-to-iso-center and source-to-detector distances) to achieve a balanced design so that the system MTF is not dominated by either the focal spot size or the detector cell size?

8-5 Figure 8.7 shows the change in an x-ray focal spot shape viewed by the detector cell at two different locations. If the focal spot size, target angle, detector cell size, and iso-center-to-detector distance remain the same, and the source-to-iso-center distance increases, how will the focal spot function be impacted? What other system performance parameters will be negatively impacted by the increased source-to-iso-center distance?

8-6 Describe the steps involved in performing computer simulation of the detector afterglow. Derive a recursive formula based on Eq. (7.14) for the afterglow simulation.

8-7 Describe the steps involved in simulating patient head motion in CT. Describe in detail the characteristics of head motion, noting that most head scans are performed by placing the patient's head inside a head holder that is shaped like a horseshoe and is slightly larger than the head.

Head motion typically occurs when head straps are not used to secure the head.

8-8 In many commercial CT scanners, a bowtie filter is used to equalize the signal and reduce the skin dose to the patient. Describe in detail the steps involved in simulating the beam-hardening effect when scanning an object in such a scanner. The path length of a Teflon bowtie changes as a function of the fan angle γ, $b(\gamma)$. The input x-ray spectrum is $\Omega(E)$.

8-9 A student simulates the partial-volume effect by placing two ellipsoids, A and B, inside a background cylinder, C. Discuss the parameter selection considerations that will ensure a successful simulation. Give an exemplary parameter set.

8-10 A system engineer uses a simulator to predict the performance of a CT system. The engineer uses the proper number of source-lets, detector-lets, and view-lets, inputs the correct system optics parameters, and uses a thin wire parallel to the z axis as a test object. When the engineer compares the simulated MTF against the MTF measured on a real system, the engineer discovers that the real system has a 1.0 lp/cm loss at a 50% MTF. What steps can the engineer take to isolate the root cause of the resolution loss?

References

1. B. DeMan, "Iterative Reconstruction for Reduction of Metal Artifacts in Computed Tomography," Ph.D. dissertation, Katholieke University Leuven, Leuven, Belgium (2001).

2. H. H. Barrett and W. Swindell, *Radiological Imaging: The Theory of Image Formation, Detection, and Processing*, Academic Press, New York (1981).

3. R. L. Morin, D. E. Raeside, J. E. Goin, and J. C. Widman, "Monte Carlo advice," *Med. Phys.* **6**, 305–306 (1979).

4. J. M. Boon, "Monte Carlo simulation of the scattered radiation distribution in diagnostic radiology," *Med. Phys.* **15**, 713–720 (1988).

5. H. Chan and K. Doi, "Studies of x-ray energy absorption and quantum noise properties of x-ray screens by use of Monte Carlo simulation," *Med. Phys.* **11**, 38–46 (1984).

6. H. Chan and K. Doi, "Physical characteristics of scattered radiation in diagnostic radiology: Monte Carlo simulation studies," *Med. Phys.* **12**, 152–156 (1985).

7. H. Chan and K. Doi, "The validity of Monte Carlo simulation in studies of scattered radiation in diagnostic radiology," *Phys. Med. Biol.* **27**, 109–129 (1983).

8. M. Sandborg, J. O. Christoffersson, G. A. Carlsson, T. Almen, and D. R. Dance, "The physical performance of different x-ray contrast agents:

calculations using a Monte Carlo model of the imaging chain," *Phys. Med. Biol.* **40**, 1209–1224 (1995).

9. D. R. Dance and G. J. Day, "The computation of scatter in mammography by Monte Carlo methods," *Phys. Med. Biol.* **29**, 237–247 (1984).

10. J. M. Boone, "X-ray production, interaction, and detection in diagnostic imaging," in *Handbook of Medical Imaging, Volume 1: Physics and Psychophysics*, J. Beutel, H. L. Kundel, and R. L. VanMeter, Eds., SPIE Press, Bellingham, WA (2000).

11. B. R. Whiting, "Signal statistics in x-ray computed tomography," *Proc. SPIE* **4682**, 53–60 (2002).

Chapter 9
Helical or Spiral CT

9.1 Introduction

Helical CT (also called spiral CT) was introduced commercially in the late 1980s and early 1990s. Helical CT has expanded the traditional CT capability by enabling the scan of an entire organ in a single breath-hold. It is safe to state that helical CT is one of the key steps that moved CT from a slice-oriented imaging modality to an organ-oriented modality.

The difference in the naming convention between helical and spiral CT is due mainly to different CT manufacturers. For all practical and technical purposes, there is no difference between the two. To avoid confusion, we will use the term "helical" throughout this chapter.

9.1.1 Clinical needs

All previous chapters have focused on a single scanning protocol: the step-and-shoot mode. This scanning protocol contains both a data acquisition period and non-data-acquisition period. During the data acquisition period, the patient remains stationary while the x-ray tube and detector rotates about the patient at a constant speed. Once a complete projection dataset is acquired for the slice, the non-data-acquisition period starts. The x-ray tube is turned off and the patient is indexed to the next scanning location. For typical CT scanners, the minimum non-data-acquisition period is on the order of seconds as a result of both mechanical and patient constraints. The mechanical constraint is due to the fact that a typical patient weighs over 45 kg, and the patient table requires a certain amount of time to move a large mass from one location to another. The cause of the patient constraint may not be as obvious. From the law of physics we know that to move a resting object over a short distance, first we must accelerate the object up to a certain speed and decelerate the object when it is near the target location. Since the distance between adjacent scanning locations is typically a few millimeters, the amount of acceleration and deceleration is fairly large. A human body is not rigid (the internal organs can move and deform), so the acceleration and deceleration will likely induce motion in the patient.[1,2] As a result, a certain amount of time must elapse to minimize motion artifacts.

In the late 1980s, the CT scan speed approached one second per revolution. Compared to the non-data-acquisition period of at least one second, the duty cycle (or the scanning time efficiency) of a step-and-shoot CT is 50% at best. Although the scan time is not as critical from the perspective of patient throughput (which is dominated by patient handling time instead of scanning), a longer scan time does impact patient coverage and contrast utilization. The contrast uptake and washout time in human bodies is relatively short (in seconds or tens of seconds), so once a contrast agent is injected, the entire scan should be completed during the same phase of contrast uptake (e.g., the arterial phase or venous phase). A prolonged scan time translates to suboptimal contrast-enhanced images (since half of the precise time has been spent on patient indexing). In addition, the volume that can be covered in a single breath-hold is reduced due to the less-efficient duty cycle. If an organ must be scanned in multiple breath-holds, slice-to-slice misregistration can occur or portions of the organ can be omitted, because it is nearly impossible to train patients to repeat the same breath-hold level in multiple breath-holds. Clearly, the ability to scan an entire organ in a single breath-hold is important.

To overcome these difficulties, the helical scan mode was proposed.[3-6] In helical CT, projections are continuously acquired while the patient is translated at a constant speed. Although theoretically a patient can be translated with a variable speed, nearly all commercial scanners use a constant speed for simplicity and other advantages. Since there is no acceleration or deceleration of patients during the scan, the non-data-acquisition period is eliminated. Consequently, a nearly 100% duty cycle can be achieved. The scanning mode is called a helical (or spiral) scan because the traveling path of a point on the gantry relative to a fixed point on the patient is shaped like a helix, as shown in Fig. 9.1.

Helical CT has several advantages. First, because the interscan delay (non-data-acquisition period) is essentially eliminated, a complete organ coverage in a single breath-hold becomes possible. For illustration, consider a scenario in which a 20-cm-long organ needs to be covered in 25 sec (most patients can hold their breath for a 20- to 30-sec period). This goal requires the patient table to

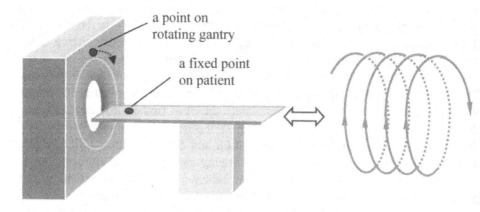

Figure 9.1 Illustration of helical scan. In a helical scan mode, the traveling path of a point on the rotating gantry viewed by a fixed point on the patient is a helix.

move at a speed of 8 mm/sec. If a 5-mm collimation is selected, the 8-mm/sec speed is well within the helical CT capability. Of course, tradeoffs must be made between table speed, collimator aperture, slice thickness, and image artifacts. We will address these issues in Sections 9.3 and 9.4.

The second advantage of helical CT comes from its method of sampling. If we examine the sampling pattern along the patient axis (z axis), we conclude that the sampling density of helical CT is uniform in z since the patient table is translated at a constant speed. This is in contrast to the step-and-shoot mode, in which slices are acquired at discrete locations. In the step-and-shoot case, the reconstruction plane is identical to the acquisition plane. However, helical CT does not have a "preferred" slice location for reconstruction because all locations are sampled in an identical pattern. Consequently, images can be reconstructed at arbitrary locations. This feature is particularly important when a small pathology must be isolated. Figure 9.2 illustrates a scenario in which the slice thickness of a CT image is roughly equal to or larger than the size of the object of interest. When the object is partially intruded into the plane of reconstruction, its intensity (and therefore, its contrast with the background) is reduced due to the partial-volume effect, as shown in Fig. 9.2(a). In this particular example, the contrast of the object is roughly 50% of what the true value should be. If the imaging plane can be centered to the object (by adjusting the location of the reconstruction plane), the true intensity of the object is recovered, as shown in Fig. 9.2(b).

Another important feature of the helical scan is its ability to produce "overlapped" images without overlapped scans. An overlapped image results when the spacing between reconstructed images is smaller than the slice thickness, as shown in Fig. 9.3. The degree of overlapping is often described by the percentage of an image that is shared by adjacent images. For example, a 50% overlap means that half of the current image slab is covered by the preceding image and the other half by the trailing image, as shown in Fig. 9.3(b). Each point in the scanned volume is contained in exactly two reconstructed

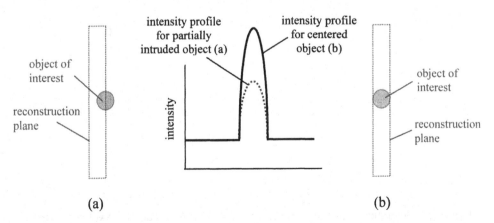

Figure 9.2 Illustration of the impact of object centering on the contrast of the reconstructed image. (a) The object of interest is partially intruded into the reconstruction plane. (b) The object of interest is fully centered on the reconstruction plane.

images. A 0% overlap means that no overlapping occurred in the covered volume. For the single-slice step-and-shoot mode, overlapped images require data collection to be overlapped due to the one-to-one relationship between the acquisition location and reconstruction location. Therefore, multiple x-ray exposures must be rendered at the overlapped regions. This is undesirable from both a patient-dose and scan-time point of view. For helical scans, overlapped images can be produced independent of the data acquisition and without an additional x-ray dose to the patient.

The advantage of overlapped image reconstruction is mainly in the production of reformatted or 3D images. This can be understood from sampling theory. For the time being, assume that the shape [or slice sensitivity profile (SSP)] of each reconstructed image is a rectangular function, as shown in Fig. 9.4(a). The SSP of a helical reconstruction is shaped more like a Gaussian than a rectangle, which will be explained in Section 9.3. However, this does not affect the general conclusion. The Fourier transform of a rectangular function is simply a sinc function, shown in Fig. 9.4(b). For a faithful representation of the signal in z, it is important that we obey the Nyquist sampling criterion discussed in Chapter 7. If we assume that frequency information beyond the first zero-crossing point $(1/\Delta)$ can be ignored, Nyquist sampling requires images to be produced at an interval of $\Delta/2$ (two samples per image slice thickness). In terms of reconstruction, this indicates that images need to be produced at a 50% overlap.

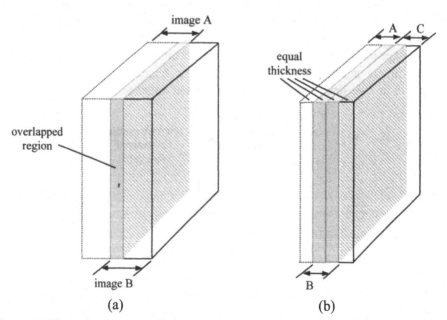

Figure 9.3 Illustration of overlapped reconstruction. (a) A portion of the volume covered by image A is covered by image B. (b) 50% overlap between volume A and volume B coverage.

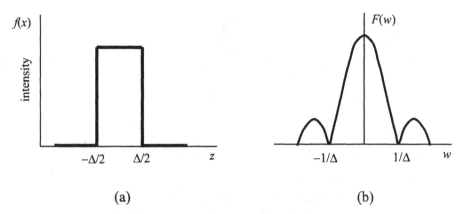

Figure 9.4 Sampling requirement in z for reformatted images. (a) Slice sensitivity profile of a reconstructed image. (b) Fourier transform of the slice sensitivity profile. Note that the first zero-crossing point is one over the slice thickness of the reconstructed image.

9.1.2 Enabling technology

There are two key enabling technologies for helical CT: the slip-ring and the advanced reconstruction algorithm. Since the gantry rotates continuously during data acquisition, both the x-ray tube input power and the detector output signals need to be transmitted through the slip-ring, as shown in Fig. 9.5. Although this task is seemingly simple, the slip-ring design is a complicated technology. To appreciate this complexity, consider the bandwidth of a typical data transmission. In typical single-slice scanners, roughly 1000 views are collected during a 0.5-sec time interval. For each view, there are roughly 1000 detector channels. The configurations of multiple detector rows (to be discussed in Chapter 10) at least quadruples the number of output signals, which requires the transmission of roughly 8 million samples per second through the slip-ring. To ensure artifact-free image reconstruction, additional error-correction and error-detection signals must be transmitted with the original signal. The task of transmitting over 50 kW of electricity over the slip-ring is equally challenging, considering that the power supply to most residential households is only 24 kW. The second enabling technology for helical scanning is the reconstruction algorithm. Detailed analyses of reconstruction techniques will be presented in Section 9.2.

Although only two key enabling technologies have been mentioned for helical CT, the demand of helical scanning on other scanner components also increases significantly. For example, the duty cycle for a step-and-shoot mode x-ray tube is 50% at most. This automatically provides a tube-cooling period between scans. In helical scan mode, the output of the x-ray tube must be maintained at a constant level throughout an entire examination (instead of during a single scan). This requires that the tube provide not only the necessary instant peak power, but also sustained power. As discussed in Chapter 6, this makes tube heat management a much more difficult task. The increase in data storage needs is also not trivial. To compensate for the slow scan speed and the

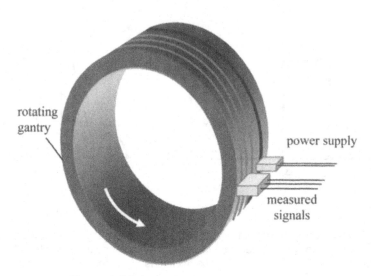

Figure 9.5 Schemetic diagram of a slip-ring.

inefficiency of the step-and-shoot mode for data acquisition, thicker slices were often used to cover an organ. With the faster scan speed of helical acquisition, thinner slices and overlapped images are routinely used. As a result, the size of the scan data set and the number of images to be stored increase substantially.

9.2 Terminology and Reconstruction

9.2.1 Helical pitch

For single-slice CT, helical pitch is defined as the ratio of the table's traveling distance in one gantry rotation over the collimator aperture at the iso-center. Mathematically, if we denote the collimator aperture in millimeters by s and the distance the patient table moves over a 360-deg gantry rotation by d, the helical pitch h is defined by the following formula:

$$h = \frac{d}{s} = \frac{vt}{s},\qquad(9.1)$$

where v is the patient table speed in mm/sec, and t is the gantry rotation time per revolution in seconds. For example, if a patient is scanned with a 3-mm collimation at a 9-mm/sec patient table speed and a 0.5-sec gantry rotation speed, the corresponding helical pitch is 1.5:1. Note that we use the collimator aperture, instead of slice thickness, to define a helical pitch, because unlike in the step-and-shoot case, the slice thickness of a helical image depends not only on the collimator aperture, but also on the helical pitch and the reconstruction algorithm.

It is clear from the above definition that a higher helical pitch translates to a faster volume coverage (assuming that other parameters remain constant). On the other hand, a higher helical pitch also implies fewer data samples per volume.

When helical pitch becomes sufficiently large, the gaps between the samples become significant. A common misconception is that a 1:1 helical pitch will result in uniform volume coverage (no gaps or overlaps). The argument is that for a 1:1 helical pitch, the table travels at exactly one slice thickness in one rotation. In reality, however, this is true only for the samples near the iso-center. For most commercial CT scanners, the size of the x-ray focal spot (in z) is less than 1 mm, while the collimated x-ray beam at the iso-center is larger than 1 mm (clinically, a slice thickness from 1 to 10 mm is commonly used). Consequently, the shape of the x-ray beam (and therefore the sampled volume) is trapezoidal in z. If we examine two projection views that are exactly one gantry rotation apart, part of the volume near the detector is sampled by both views, while part of the volume near the focal spot is sampled by neither view, as shown in Fig. 9.6. Only the volume near the iso-center is uniformly covered.

9.2.2 Basic reconstruction approaches

One basic assumption in CT image reconstruction is that the projection measurements are taken at a fixed slice location. The Fourier slice theorem assumes the object inside the scanning plane remains unchanged during the entire data acquisition. This assumption is obviously not true for helical CT. For example, examine the source trajectory in a helical scan shown in Fig. 9.7. In this figure, the source trajectory is a spiral curve that uses a fixed point on the patient as a reference. Because the source trajectory has a single intersection point with

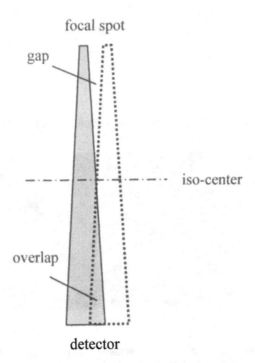

Figure 9.6 Gap and overlapped regions in 1:1 helical pitch sampling pattern.

any plane that is perpendicular to the z axis, only one projection is acquired properly for any reconstruction plane. All other samples represent projections collected adjacent to the reconstruction plane. If we ignore the inherent inconsistency in the data collection and reconstruct helically acquired data in the same fashion as in a step-and-shoot mode, image artifacts will result. Figure 9.8(a) shows a reconstructed body phantom scanned with a 2.5-mm collimation at a pitch of 1:1. Shading artifacts are obvious. Based on the artifact pattern, we can identify the starting angle of the projection dataset from which the image was reconstructed. If we draw lines along the directions of the shading and streaking artifacts, these lines intersect at the two o'clock position. This is the starting x-ray source location, since the largest inconsistency (or difference in z) between two adjacent views in the dataset is between the first and last projections.

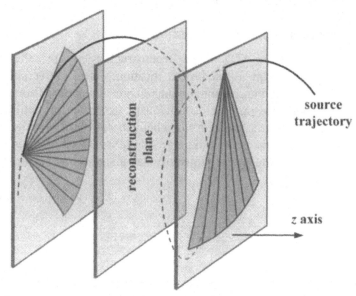

Figure 9.7 Illustration of helical scan mode and reconstruction plane.

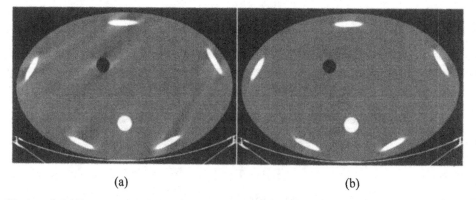

(a)   (b)

Figure 9.8 Reconstructed images of helically acquired data (a) without helical compensation and (b) with helical correction.

In the early phase of helical CT investigation, fairly simplistic approaches were proposed to deal with the projection inconsistency caused by helical motion. For example, a reduced helical pitch scanning protocol was suggested.[3] The projection data corresponding to consecutive 360-deg projections are averaged and weighted. The end result is equivalent to using a large slice thickness with a small helical pitch. Although this method is effective at reducing shading or streaking artifacts due to data inconsistency, the reduction is obtained at the cost of a reduced volume coverage.

A slightly more elaborate scheme uses 4π interpolation to reduce the projection inconsistency.[4] The method uses projections from two 360-deg sets to estimate one set of projections at a constant location. The reconstruction plane is selected exactly halfway between two rotations and perpendicular to the axis of rotation. Linear interpolation is used; the interpolation takes place between views of the same projection angle, as illustrated in Fig. 9.9. If we denote the two projections to be interpolated by $p(\gamma, \beta)$ and $p(\gamma, \beta + 2\pi)$, the distance from $p(\gamma, \beta)$ to the reconstruction plane by x, the distance the patient table moves over a 360-deg gantry rotation by d, and the interpolated projection by $p'(\gamma, \beta)$, we have

$$p'(\gamma,\beta) = \left(\frac{d-x}{d}\right)p(\gamma,\beta) + \left(\frac{x}{d}\right)p(\gamma,\beta+2\pi). \qquad (9.2)$$

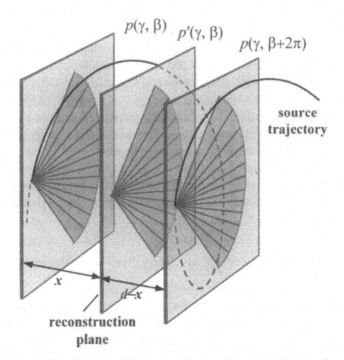

Figure 9.9 Interpolation of two projections differ by 360 deg. Weights for this type of interpolation are detector channel independent.

Note that the interpolation weights change only as a function of the projection angle (z is linearly proportional to β) and are independent of the detector angle γ. The advantage of this type of interpolation is its simplicity. Its drawback is the significantly degraded slice profile.[7] Since the longitudinal resolution is an important issue in high-resolution CT imaging, helical reconstruction algorithms that offer minimal slice broadening are desirable.[8]

An alternative approach minimizes helical artifacts by minimizing the contributions from the most inconsistent projections in the dataset. Examples of this approach are the underscan and overscan algorithms. These algorithms (discussed in Chapter 7) were originally designed to reduce motion artifacts in the step-and-shoot mode for data acquisition. In helical data acquisition, the largest discontinuity exists between projections at angles β and $\beta + 2\pi$. This discontinuity can be minimized by reducing the data contribution near both ends of the scan to the reconstructed image. Since both algorithms use less than 4π views of data, the slice profile is improved.

Most helical reconstruction algorithms used on commercial CT scanners today are based on the philosophy of estimating or interpolating a set of projections at a single reconstruction plane to mimic the step-and-shoot data acquisition. Interpolation takes place between two samples that are either 180 deg or 360 deg apart. An example of 360-deg interpolation was given in Eq. (9.2). Interpolation between samples that are 180 deg apart is slightly more complicated, because sample locations depend on both γ and β. If we use the concept of conjugate samples introduced in Chapter 7, two samples that satisfy (γ, β) and $(-\gamma, \beta + \pi + 2\gamma)$ form conjugate samples, and they represent the same ray path in a step-and-shoot scan. For a helical scan, the same samples represent two parallel ray paths with different z locations due to the constant patient table motion. Consequently, the interpolation weights change with detector angle γ and projection angle β. This is illustrated in Fig. 9.10, where the two conjugate samples are denoted by dotted lines and the interpolated sample by a solid line. Only one conjugate sampling pair exists in the two selected projections. To form a single projection, a number of projections are needed.

Since the backprojection process is essentially a summation operator, projection interpolation can be achieved by multiplying projections with weights prior to the FBP.[7,9] Therefore, the reconstruction formula for fan-beam helical scanning becomes

$$f(r, \varphi) = \int_{\beta_0}^{\beta_0 + \Pi} L^{-2} d\beta \int_{-\gamma_m}^{\gamma_m} w(\gamma, \beta) q(\gamma, \beta) h''(\gamma' - \gamma) D \cos\gamma \, d\gamma . \qquad (9.3)$$

In this equation, D is the distance of the x-ray source to the iso-center, L is the distance from the x-ray source to the point of reconstruction (r, ϕ), γ is the detector angle, β is the view angle, β_0 is the starting angle of the helical dataset, Π is the angular span of the dataset, $w(\gamma, \beta)$ is the weighting function, $q(\gamma, \beta)$ is the fan-beam projection, and $h''(\gamma)$ is the filter function (see Chapter 3 for the equiangular fan-beam reconstruction formula). The advantage of this approach is its computational efficiency and ability to minimize spatial resolution impact.

An example of a weighting function is the helical extrapolative algorithm.[7,10] This algorithm makes use of the fact that each 2π projection set can be divided into two sets of half-scans. Figure 9.11 shows a sinogram representation of the projections in which the horizontal axis represents detector angle γ, and the vertical axis represents the projection angle β. The two half-scans are represented by regions AEA'E' and A'E'A''E''. For each ray sample in AEA'E', we can find a parallel ray sample in A'E'A''E''. These two parallel samples are at different z locations. Our goal is to estimate a projection sample at the plane of reconstruction. For the helical extrapolative algorithm, the reconstruction plane is selected as a flat plane perpendicular to the patient axis (z axis). The plane is located halfway in the dataset, and in the sinogram space, is represented by the line DC'. If all of the conjugate samples are located on opposite sides of DC', a simple linear interpolation can be used. Unfortunately, some of the conjugate samples are located on the same side of DC'. For example, the conjugate samples to region ABC are located in region A'B'C' (both regions are below DC'). Similarly, the conjugate regions DB'E' and B''D'E'' are all located above DC'. For these regions, extrapolation must be used. A graphical representation of interpolation and extrapolation is illustrated in Figs. 9.12(a) and (b). The helical weighting function can be shown to be

$$
w(\gamma, \beta) = \begin{cases} \dfrac{\beta + 2\gamma}{\pi + 2\gamma}, & 0 \le \beta < \pi - 2\gamma, \\[2mm] \dfrac{2\pi - \beta - 2\gamma}{\pi - 2\gamma}, & \pi - 2\gamma \le \beta < 2\pi. \end{cases} \tag{9.4}
$$

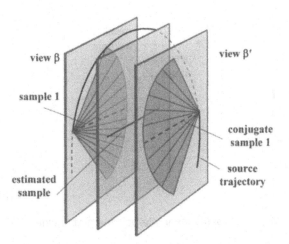

Figure 9.10 Interpolation of conjugate samples that are 180 deg apart. Weights for this type of interpolation are detector-channel dependent.

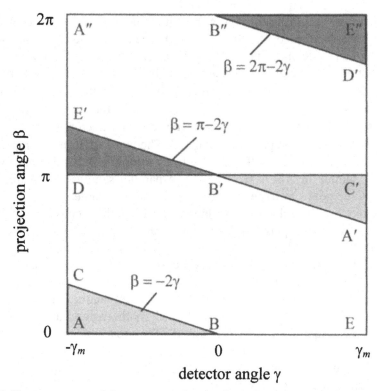

Fig. 9.11 Sinogram space data representation for the helical extrapolative algorithm. The 2π projection set is divided into two half-scans (AEA′E′ and E′A′E″A″). Interpolation takes place between the unshaded regions and extrapolation takes place between the shaded triangular regions.

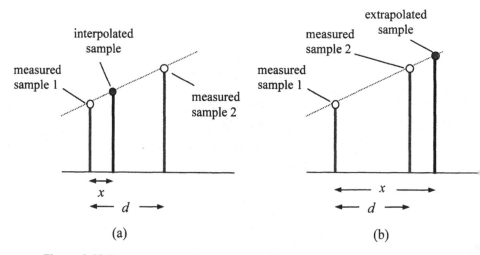

Figure 9.12 Illustration of (a) linear interpolation and (b) linear extrapolation.

Note that the function $w(\gamma, \beta)$ is discontinuous in γ along the line $\beta = \pi - 2\gamma$. Since the convolution filter used in the reconstruction process is essentially a derivative operator (see Chapter 3), discontinuities in the projection are likely to cause streaking artifacts. To eliminate the discontinuity, $w(\gamma, \beta)$ is feathered across the line $\beta = \pi - 2\gamma$. Figure 9.8(b) shows the reconstructed image using the helical extrapolative algorithm. It is clear that helical-acquisition-related shading artifacts are nearly eliminated. Since the distance between conjugate samples is smaller than the distance between two consecutive rotations, the slice profile is significantly improved.

The selection of conjugate regions is flexible. Different conjugate region selections often result in different reconstruction algorithms.[7–13] Despite their differences, all of the algorithms need to resolve two major issues: the selection of the interpolation algorithm, and the selection of the reconstruction plane. These topics will be addressed in the following sections.

9.2.3 Selection of the interpolation algorithm and reconstruction plane

The interpolation of a signal from samples is an active area of research. Over the past few decades, many interpolation algorithms have been proposed and investigated. These algorithms differ in their performance and tradeoffs. Although they have been tested extensively on 1D or 2D signals, their performance in CT cannot be deduced based solely on their performance elsewhere. Each will need to be carefully investigated and examined.

The most commonly used interpolation approach for helical CT is linear interpolation. Its popularity arises from two properties of the algorithm: simplicity and stability. The first property is illustrated in Eq. (9.2). The interpolation coefficients are inversely proportional to the distance between the sample and the interpolation location. Once the coefficient is obtained, the final result requires only two multiplications and one addition. The stability of the interpolation is due to the fact that the interpolated value always lies on the straight line that connects the values of the two samples used for interpolation, as shown in Fig. 9.12. If the interpolation location is situated between the two measured samples, the interpolation coefficients are always less than unity. Therefore, the interpolated value always falls between the measured sample values, as shown in Fig. 9.12(a). On the other hand, when the interpolation point is located outside two samples, one of the interpolation coefficients is larger than unity and the other is less than zero. Although the interpolated value in this case does not fall between the measured values, it still falls on the straight line that connects the two samples, as shown in Fig. 9.12(b). This process is often called extrapolation. Because of the linear constraints, the interpolated value does not deviate from the neighboring measured samples as long as these samples are located somewhere within the vicinity of interpolation.

Although linear interpolation has many advantages, it is unable to preserve high-frequency information. For certain CT applications, it is desirable to obtain a high resolution in z. For example, Fourier interpolation and spline interpolation have been proposed.[14] The key to the success of any algorithm is still its overall performance in spatial resolution, noise, and image artifacts.

Now we will discuss the selection of the reconstruction plane. In the initial helical reconstruction algorithm investigation, the reconstruction plane was always selected as a flat plane located exactly halfway in the dataset and perpendicular to the patient translation axis (z axis), as shown in Fig. 9.13. The selection was based partially on historical reasons and partially on scientific merits. For step-and-shoot CT, the reconstruction plane has always been flat and perpendicular to the z axis (ignoring the case of the gantry tilt). Therefore, it is not surprising that the reconstruction plane for helical scans was selected in a similar fashion. However, there is some scientific merit in placing the reconstruction plane exactly halfway between the start and the end of the dataset. If we assume that any error introduced in the projection measurement is proportional to its distance from the plane of reconstruction, the selection of a middle plane minimizes the overall inconsistency in the dataset.

Although the above selection of the plane of reconstruction is convenient, it is somewhat restrictive. To provide some flexibility in the reconstruction algorithm derivation, the concept of a region of reconstruction was proposed.[9] The region of reconstruction can be a warped plane or even an irregularly shaped volume, as illustrated in Fig. 9.14. By eliminating the constraint that the projection estimation be performed with respect to a single, flat plane, perpendicular to the z axis, we gain the flexibility of being able to optimize the weighting function with respect to other image-quality parameters. The optimization criteria can be noise uniformity, continuity, smoothness of the weighting function, or other parameters deemed important to image quality.

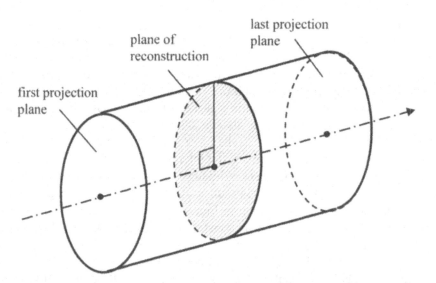

Figure 9.13 Example of the plane of reconstruction for a helical scan.

To illustrate the impact of the plane or region of reconstruction on the helical weighting function, we will re-examine the helical extrapolative algorithm. A detailed look at Eq. (9.4) indicates that the range of the weighting function is not limited to [0, 1] due to the nature of extrapolation. Note that $w(\gamma, \beta)$ is less than zero in the triangular regions ABC and B″D′E″. For regions A′B′C′ and DB′E′, $w(\gamma, \beta)$ is larger than unity. In addition to the issue of stability, the expanded range of $w(\gamma, \beta)$ has noise implications. From the basic property of statistics, if a sample p is obtained from the weighted sum of two samples p_1 and p_2, then

$$p = w_1 p_1 + w_2 p_2 . \tag{9.5}$$

The standard deviation of p, σ_p is related to the standard deviations of the original samples, σ_{p1} and σ_{p2}, by the following equation:

$$\sigma_p = \sqrt{w_1^2 \sigma_{p1}^2 + w_2^2 \sigma_{p2}^2} . \tag{9.6}$$

Consequently, the expanded range of the weighting function results in a greater noise level than that of the weights in the [0, 1] range.

Consider an example in which the region of reconstruction is defined as a straight line, $\beta = \pi - 2\gamma$, in the sinogram space.[9] In real space, the region of reconstruction is a double-cone-shaped plane, shown in Fig. 9.15. For detector channels located in the region $0 < \gamma < \gamma_m$, the reference plane is the bottom portion of the double cone. For channels in the region $-\gamma_m < \gamma < 0$, the top portion of the double cone is used as the reference plane.

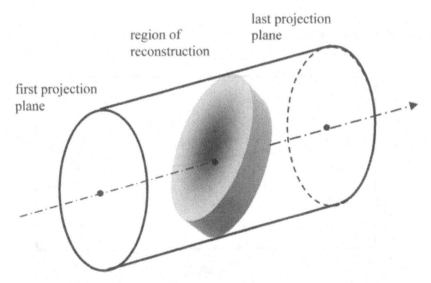

first projection plane

region of reconstruction

last projection plane

Figure 9.14 Illustration of the region of reconstruction.

To derive a weighting function, we denote by p_1 and p_2 a complimentary sampling pair with z coordinates of z_1 and z_2, respectively, as shown in Fig. 9.16. We further denote the intersections of the region of reconstruction with two lines that are parallel to z and pass through p_1 and p_2 by z_{r1} and z_{r2}. Following the rules of linear interpolation, the weighting factor w_1 for point p_1 is given by

$$w_1 = \frac{z_2 - z_{r2}}{(z_2 - z_{r2}) + (z_{r1} - z_1)}. \tag{9.7}$$

Similarly, the weighting factor w_2 for point p_2 is

$$w_2 = \frac{z_{r1} - z_1}{(z_2 - z_{r2}) + (z_{r1} - z_1)}. \tag{9.8}$$

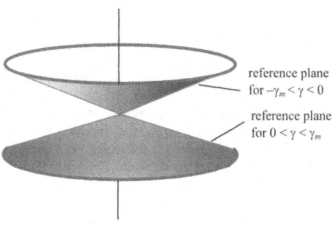

Figure 9.15 A region of reconstruction corresponds to a straight line in the sinogram space.

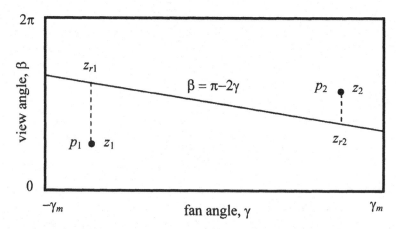

Figure 9.16 Description of a conjugate sampling pair for a helical interpolative algorithm.

When the patient table and gantry move at constant speeds, the table position z is proportional to the projection angle β. Therefore, the ratios of the z locations can be replaced by the corresponding ratios of the projection angle β:

$$w_1 = \frac{\beta_2 - \beta_{r2}}{\beta_2 - \beta_{r2} + \beta_{r1} - \beta_1} \tag{9.9}$$

and

$$w_2 = \frac{\beta_{r1} - \beta_1}{(\beta_2 - \beta_{r2}) + (\beta_{r1} - \beta_1)}. \tag{9.10}$$

For two conjugate samples, the following relationship must exist:

$$\beta_2 = \beta_1 + \pi + 2\gamma_1. \tag{9.11}$$

Since the region of reconstruction is defined by $\beta = \pi - 2\gamma$ for β_{r1} and β_{r2}, we obtain

$$w(\gamma, \beta) = \begin{cases} \dfrac{\beta}{\pi - 2\gamma}, & 0 \le \beta < \pi - 2\gamma, \\[2mm] \dfrac{2\pi - \beta}{\pi + 2\gamma}, & \pi - 2\gamma \le \beta < 2\pi. \end{cases} \tag{9.12}$$

At the region of reconstruction ($\beta = \pi - 2\gamma$), both equations equal unity. This indicates that the weighting function is continuous everywhere. Note also that $w(\gamma, \beta)$ approaches zero when $\beta \to 0$ and $\beta \to 2\pi$. This is quite desirable since the projection inconsistency is the worst at both locations. In addition, it can be verified that the range of $w(\gamma, \beta)$ is [0, 1]. This method is called the *helical interpolative algorithm*.

To demonstrate the importance of the weighting function continuity, we reconstructed a patient scan with the helical extrapolative algorithm described in Eq. (9.4) and the helical interpolative algorithm described in Eq. (9.12). For the helical extrapolative algorithm, the width of the feathering operation is 20 channels. Sets of parallel streaks (shown by arrows) can be observed in the image reconstructed with the helical extrapolative algorithm shown in Fig. 9.17(a). These streaks are absent in the image reconstructed with the helical interpolative algorithm shown in Fig. 9.17(b).

9.2.4 Helical fan-to-parallel rebinning

Similar to the case of the step-and-shoot mode of data acquisition, the fan-to-parallel rebinning process can be applied to helical scanning as well. The rebinning process itself is nearly identical to the step-and-shoot case, since when looking down at the z axis, the relationship between (γ, β) and (t, θ) is identical between the two acquisitions:

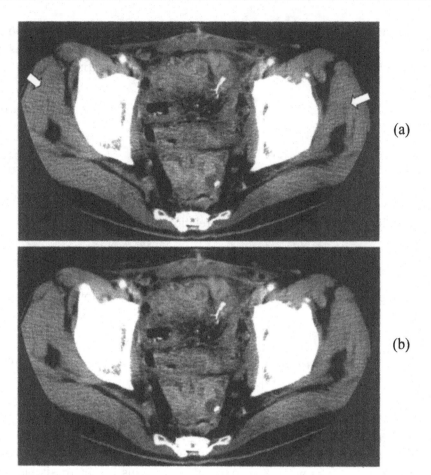

Figure 9.17 Comparison of images generated with (a) helical extrapolation algorithm and (b) helical interpolation algorithm. Reconstruction FOV = 32 cm, standard algorithm (WW = 200).

$$\begin{cases} \theta = \beta + \gamma, \\ t = D \sin \gamma. \end{cases} \tag{9.13}$$

where γ and β are the fan angle and the projection angle for the fan-beam sample, and t and θ are the ray distance and projection angle for the corresponding parallel beam, respectively. The two-step rebinning process is described by Eqs. (3.61) and (3.62) and will not be repeated here.

There is one major difference between the two cases. In the step-and-shoot mode acquisition, the projection data after the rebinning process are truly a set of parallel samples, as if they were generated by parallel x-ray beams. The rebinned parallel-beam dataset in the helical case, however, is not the same as the dataset generated by parallel x-ray beams in a helical scan. To understand the difference, examine Fig. 9.18. When a real parallel x-ray source is used to produce a projection, the x-ray plane formed by all projection samples is parallel to the x-y plane, as shown by Fig. 9.18(a). For the ease of illustration, the projection view angle is selected to be parallel to the x axis. The x-ray plane rotates about the z axis and moves up or down during helical data acquisition. For this configuration, the reconstruction process is simple, since all rays within a single projection have the same z value. The projection interpolation that forms a virtual set of projections at the plane of reconstruction is straightforward and can be performed on a view-by-view basis. The weighting function can be obtained by setting $\gamma = 0$ in the fan-beam helical weighting functions. For the rebinned parallel helical projection, however, the situation is quite different. Since the x-ray source is always located on the helical trajectory, any rebinned projection sample must intersect the helical trajectory, as shown in Fig. 9.18(b). When the patient table moves at a constant speed during data acquisition, the rebinned projection samples no longer form a plane that is parallel to the x-y plane. As a matter of fact, the plane formed by a rebinned parallel projection forms a curved plane. Since the filtering operation takes place after the rebinning process, the filtering is actually performed along a curved line tangential to the helix trajectory.

Now consider the approaches to reconstructing an image based on the rebinned projection dataset. Three different strategies can be deployed:

(1) Weight the fan-beam helical data with the weighting function described in the previous section, rebin the data to a parallel projection dataset, filter the projection, and backproject the data to form an image.

(2) Rebin the fan-beam helical data to a set of parallel-beam data, weight the parallel projections with a helical weighting function, filter the weighted projection, and backproject the dataset to form an image.

(3) Rebin the fan-beam helical data to a set of parallel-beam data, filter the rebinned projections, weight the filtered projection with a helical weighting function, and backproject the dataset to form an image.

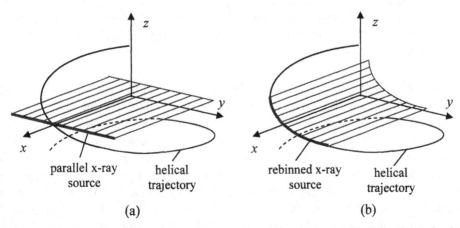

Figure 9.18 True parallel and rebinned parallel helical sampling. (a) True parallel projection in which the x-ray plane is parallel to x-y plane. (b) Rebinned parallel projection in which the x-ray plane is a curved plane.

Flowcharts of these three approaches are shown in Fig. 9.19. The approach depicted in Fig. 9.19(a) is the most straightforward to implement, since both the weighting function and the filter function are readily available. For the approaches described by Figs. 9.19(b) and (c), a helical weighting function must be derived for the rebinned parallel geometry. To accomplish this, we need to rewrite Eq. (9.13) to the following form:

$$
\begin{cases}
\gamma = \sin^{-1}\!\left(\dfrac{t}{D}\right), \\[2mm]
\beta = \theta - \sin^{-1}\!\left(\dfrac{t}{D}\right).
\end{cases}
\tag{9.14}
$$

Now we can substitute this equation into the fan-beam helical weighting functions to derive the corresponding parallel-beam helical algorithm. As an example, when Eq. (9.14) is substituted into Eq. (9.12), we obtain:

$$
w(t,\theta) =
\begin{cases}
\dfrac{\theta - \sin^{-1}\!\left(\dfrac{t}{D}\right)}{\pi - 2\sin^{-1}\!\left(\dfrac{t}{D}\right)}, & \sin^{-1}\!\left(\dfrac{t}{D}\right) \le \theta < \pi - \sin^{-1}\!\left(\dfrac{t}{D}\right), \\[6mm]
\dfrac{2\pi - \theta + \sin^{-1}\!\left(\dfrac{t}{D}\right)}{\pi + 2\sin^{-1}\!\left(\dfrac{t}{D}\right)}, & \pi - \sin^{-1}\!\left(\dfrac{t}{D}\right) \le \theta < 2\pi + \sin^{-1}\!\left(\dfrac{t}{D}\right).
\end{cases}
\tag{9.15}
$$

For illustration, Figs. 9.20(a) and (b) show the grayscale images of the two helical weighting functions described by Eqs. (9.12) and (9.15), respectively. Because of the fan-beam parallel rebinning process, the number of views from the rebinned parallel projections is larger than the original fan-beam projections. This can be explained by Eq. (9.15), in which the range of the variable θ is $(-\gamma_m, 2\pi + \gamma_m)$.

The weighting function described by Eq. (9.15) can be used either before or after the filtering operation. Other than the noise characteristics, the quality of the images reconstructed by these three approaches should not show appreciable differences. For illustration, a shoulder phantom was scanned in a helical mode on an older vintage CT scanner, and the same scan data were used to reconstruct three images corresponding to the above three approaches, as shown by the images in Fig. 9.21. The top image was reconstructed with a fan-beam helical interpolative algorithm; the middle image was reconstructed with a parallel-beam helical interpolative algorithm using the weighting function applied prior to the filtering operation; and the bottom image was reconstructed with a parallel-beam helical interpolative algorithm in which the weighting and filtering operations were reversed. Visually, the three images are quite similar.

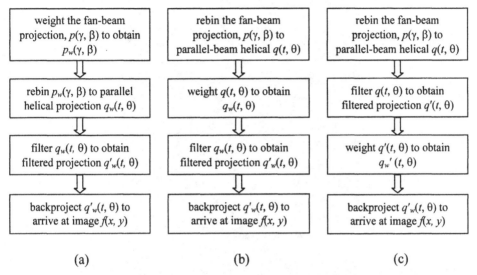

weight the fan-beam projection, $p(\gamma, \beta)$ to obtain $p_w(\gamma, \beta)$	rebin the fan-beam projection, $p(\gamma, \beta)$ to parallel-beam helical $q(t, \theta)$	rebin the fan-beam projection, $p(\gamma, \beta)$ to parallel-beam helical $q(t, \theta)$
rebin $p_w(\gamma, \beta)$ to parallel helical projection $q_w(t, \theta)$	weight $q(t, \theta)$ to obtain $q_w(t, \theta)$	filter $q(t, \theta)$ to obtain filtered projection $q'(t, \theta)$
filter $q_w(t, \theta)$ to obtain filtered projection $q'_w(t, \theta)$	filter $q_w(t, \theta)$ to obtain filtered projection $q'_w(t, \theta)$	weight $q'(t, \theta)$ to obtain $q_w'(t, \theta)$
backproject $q'_w(t, \theta)$ to arrive at image $f(x, y)$	backproject $q'_w(t, \theta)$ to arrive at image $f(x, y)$	backproject $q'_w(t, \theta)$ to arrive at image $f(x, y)$
(a)	(b)	(c)

Figure 9.19 Flowcharts for three helical reconstruction strategies with fan-to-parallel rebinning.

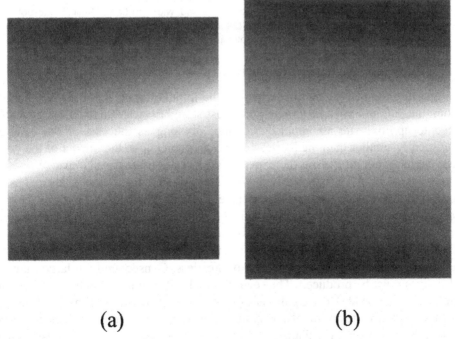

(a)   (b)

Figure 9.20 Examples of helical weighting functions. (a) Fan-beam helical interpolation weighting. (b) Rebinned parallel-beam helical interpolation weighting.

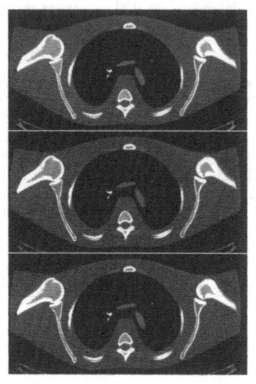

Figure 9.21 Chest phantom images scanned on an older vintage scanner at 120 kV, 80 mA, 1 sec, 0.75:1 pitch (WW = 600). Top: reconstructed with fan-beam helical interpolative algorithm; middle: reconstructed with parallel-beam helical interpolative weight prior to filter; bottom: reconstructed with parallel-beam helical interpolative weight post filter.

9.3 Slice Sensitivity Profile and Noise

To measure the SSP, which is defined as the impulse response of the CT system along the patient long axis (z axis), the most straightforward method uses a small bead to approximate an impulse function. As long as the size of the bead is sufficiently smaller than the FWHM of the slice thickness, the impact of the bead size can be ignored. During the experiment, a series of uniformly spaced images are collected in z. The intensity of the bead is plotted across multiple images to form the SSP. This method does have several drawbacks. First, to ensure adequate sampling in z, the images must be reconstructed at very fine increments. For example, a 0.1-mm increment is needed to characterize an SSP with a FWHM on the order of 1 mm. A much finer increment (e.g., 0.05 mm) is required for slice profiles that are 0.5 mm or less. Consequently, a large number of images must be produced. The second drawback of this technique is related to the size of the bead. Very small beads (a small fraction of a millimeter) must be used to approximate an impulse response. For example, to accurately measure the system SSP of 1 mm, the diameter of the bead needs to be 0.1 mm or less. When imaging such a small object, it is critical to ensure that adequate sampling is provided in the reconstruction (x-y) plane. If we reconstruct an image with a 10-

cm FOV in a 512×512-image matrix (which is the lower limit on many CT scanners), the reconstructed pixel size is roughly 0.2 mm, which makes it larger than the bead diameter. Based on the Nyquist criteria, the sampling is unlikely to faithfully represent the bead in the cross-sectional plane.

Alternatively, SSP measurement can be performed with a highly absorbing thin-disc phantom placed perpendicular to the z axis to approximate an impulse response in z. A helical scan is collected, and the SSP is constructed as it was constructed in the small-bead experiment. Because of its large size, the thin-disc phantom improves the sampling in the cross-sectional plane. The improved sampling leads to a potential improvement in the accuracy of the attenuation measurement. On the other hand, this method suffers from the same drawbacks found in the small-bead experiment, because it still requires a large amount of data processing. In addition, the alignment of the disc with the x-y plane is critical to the accuracy of the experimental result.

To overcome these drawbacks, a different experimental approach was proposed.[15,16] Although the methodology was originally proposed for the step-and-shoot mode, the same principle can be applied to helical CT. Instead of measuring the SSP directly along the z axis, an indirect measurement of the SSP is taken in the x direction. The experiment can be arranged as follows. A thin strip of aluminum or used x-ray film (or a piece of thin tungsten wire) is placed at a shallow angle with respect to the x-ray plane. The short axis of the film is parallel to the CT y (vertical) axis, and the long axis of the film forms a small angle α with respect to the x (horizontal) axis. The film is sandwiched between two foam blocks (very low x-ray attenuation) to ensure its flatness. The experimental setup is illustrated in Fig. 9.22. If we assume that the thickness of the x-ray film is negligible compared to the SSP and that the CT reconstruction is an exact mapping between the measurement and the reconstructed intensity, the thin film simply maps the detector sensitivity profile in z to the reconstructed image intensity in x, as shown in Fig. 9.23. The measured intensity of the reconstructed image is merely an expanded or magnified version of the SSP. The

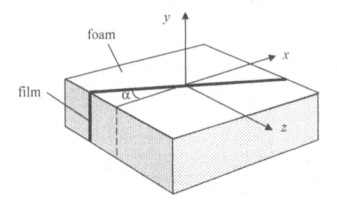

Figure 9.22 Illustration of the experimental setup. The thin film is slanted at an angle α with respect to the x-y plane (scan plane).

expansion factor χ is $[\tan \alpha]^{-1}$. For the small angle α, the expansion factor is large. In practice, the impact of the reconstruction MTF cannot be totally ignored, so a de-convolution approach is often used to obtain a more accurate SSP.

Now consider the problem of how to analytically predict the SSP. Many studies have been conducted on this topic.[12,17] To fully understand the helical SSP, we will start with an SSP model for the step-and-shoot mode CT. Contrary to common belief, the step-and-shoot mode SSP is not a rectangular function, even assuming ideal detector and focal spot responses. Figure 9.24 depicts a typical configuration of a single-slice scanner. The x-ray flux distribution (and therefore the slice thickness) of a single-slice CT is defined primarily by the prepatient collimator. If we denote the source-to-collimator and source-to-iso distances by d and D, respectively, the width of the penumbra at the iso-center, s, is simply

$$s = \frac{(D-d)}{d}\Delta,\tag{9.16}$$

where Δ is the size of the focal spot in z. Consider an example where $d = 250$ mm, $D = 630$ mm, and $\Delta = 0.7$ mm. The width of the penumbra s is 1.06 mm. When the slice thickness is small, the size of s is a significant factor that influences the shape of the SSP. The influence of s decreases with the increase in slice thickness. Figure 9.25 shows an example of a calculated step-and-shoot SSP for a 5-mm slice thickness (thick solid line). Although the FWHM is 5 mm, the FWTM is 5.8 mm (Table 9.1 at pitch = 0) as a result of the penumbrae. (Recall that FWHM is defined as the distance between two points on the profile curve whose intensities are equal to 50% of the peak value; FWTM measures the distance between two points that are 10% of the peak intensity.)

To estimate the SSP for helical scans, both the data acquisition and reconstruction processes must be considered. In the helical mode, the detector (or the patient table) travels a distance r over one gantry rotation, where r equals the

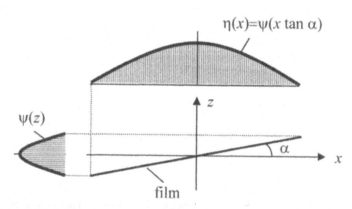

Figure 9.23 Slice sensitivity profile in z can be measured by profile in x.

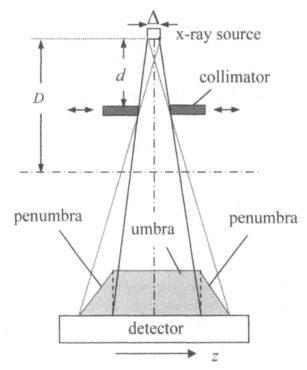

Figure 9.24 Slice thickness selection for single-slice CT.

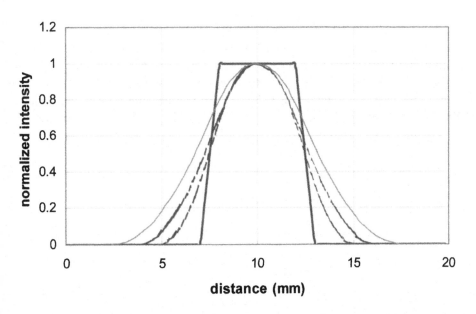

Figure 9.25 SSP for different scanning modes with 5-mm collimation. Thick solid line: step-and-shoot; dashed line: helical mode at 1:1 helical pitch; dash dotted line: helical mode at 1.5:1 helical pitch; thin gray line: helical mode at 2:1 helical pitch.

product of the helical pitch and collimation. If we use helical reconstruction algorithms that use $2k\pi$ of the projection data, the reconstructed image is then the weighted average of projection samples over a distance kr (for the helical interpolative and helical extrapolative algorithms described earlier, $k = 1$). The weight is determined by the helical weighting function used in the reconstruction. This effect is illustrated graphically in Fig. 9.26. The original step-and-shoot mode SSP is shifted (due to the helical table translation), weighted by the helical weighting function, and summed to formulate the final result. This is exactly the convolution process described in Chapter 2. Mathematically, this can be described by the following equation:

$$\psi(z) = \int_0^{kr} w(\gamma,z')\Omega(z-z')dz' , \qquad (9.17)$$

where $\psi(z)$ is the SSP of the helical scan, $\Omega(z)$ is the step-and-shoot mode SSP, and $w(\gamma, z)$ is the helical weighting function used for image reconstruction. Previously, we denoted the helical weighting function by $w(\gamma, \beta)$, where γ is the detector angle and β is the projection angle. In helical mode, the patient table travels at a constant speed and the gantry rotates at a constant velocity, so β is linearly proportional to z:

$$z = \left(\frac{hs}{2\pi}\right)\beta . \qquad (9.18)$$

γ is always selected as the ray that passes the location of interest for SSP evaluation. To calculate the SSP at the iso-center, $\gamma = 0$. It is clear from this equation that $\psi(z)$ changes with the helical pitch, since helical pitch scales with the integration distance r. For illustration, we calculated the SSP at the iso-center for different helical pitches with a 5-mm collimation (the helical interpolative algorithm was assumed). The results are shown in Fig. 9.25. It is clear that SSP

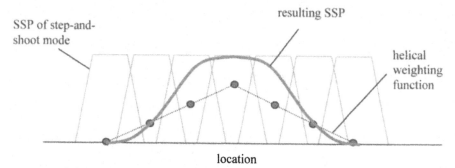

Figure 9.26 SSP for helical scan is the weighted sum of the shifted profiles of step-and-shoot mode. Therefore, helical SSP can be derived by convolving step-and-shoot SSP with the helical weighting function.

degrades with increased helical pitch. The quantitative calculations of FWHM and FWTM are shown in Table 9.1.

One of the more convenient ways to analyze the impact of the data acquisition and reconstruction on SSP is to examine the FWHM and FWTM ratios. The ratios are obtained by dividing the measured widths (in mm) by the collimation (in mm). For the example given in Table 9.1, the measured values are divided by 5. The ratios give a clear picture of the percentage increase in SSP due to the helical scans. For convenience, both the FWHM ratio and FWTM ratio are tabulated in Table 9.1.

An examination of Eq. (9.17) also indicates that SSP changes with the selection of helical reconstruction algorithms. When a different helical reconstruction algorithm is used, the weighting function $w(\gamma, z')$ changes. To demonstrate the impact of the helical algorithm on SSP, Table 9.2 tabulates the FWHM and FWTM ratios for four different helical reconstruction algorithms. Again, these ratios are calculated at the iso-center. It is clear that SSP changes significantly with the selection of a helical algorithm.

The SSP is nonstationary (in the x-y plane), and it changes with the measurement location relative to the iso-center. Two major factors influence the location-dependent characteristics. The first factor is the weighting function used in the helical reconstruction. As mentioned previously, the weight is generally a

Table 9.1 FWHM and FWTM for different helical pitches with 5-mm collimation. FWHM ratio and FWTM ratio are calculated by dividing the actual widths by the collimation (5 mm).

Pitch	0:1	1:1	1.5:1	2:1
FWHM (mm)	5.00	5.03	5.43	6.35
FWHM ratio	1.00	1.01	1.08	1.27
FWTM (mm)	5.80	7.86	9.35	11.14
FWTM ratio	1.16	1.57	1.87	2.23

Table 9.2 FWHM, FWTM, and noise ratios for different helical reconstruction algorithms (1:1 helical pitch). For overscan and underscan, a 45-deg angle is assumed for the transition region. FWHM ratio is the calculated FWHM over collimation. FWTM ratio is the calculated FWTM over collimation.

Algorithm	4π linear	Overscan	Underscan	2π HI/HE
FWHM ratio	1.28	1.06	1.01	1.01
FWTM ratio	2.24	1.82	1.68	1.57
Noise ratio	0.82	0.98	1.09	1.16

function of both γ and β. The different location impacts the γ value, which contributes to the reconstructed region. The second factor is the trapezoidal shape of the x-ray beam profile shown in Fig. 9.25. Theoretically, one can replace the stationary step-and-shoot mode SSP, $\Omega(z)$, in Eq. (9.17) with a nonstationary SSP model, $\Omega(x, y, z)$. In practice, however, most people ignore the location dependency of SSP and use the iso-center SSP to describe a system.

As in SSP modeling, the noise performance of a helical scan can be calculated based on its weighting function. For simplicity, we consider only the noise performance at the iso-center. Section 9.4 will discuss how noise behavior can be highly inhomogeneous due to the interaction between helical weights and backprojection weights.

At the iso-center, the scaling factor for the fan-beam backprojection is unity. If we ignore the filtering operation (since it affects both the helical and step-and-shoot mode in the same manner, and the magnitude of the kernel concentrates on the center few terms), the reconstructed image value at the iso-center f_0 is the weighted summation of all projection samples:

$$f_0 = \sum_{n=1}^{N} w(0, \beta_n) p(0, \beta_n),$$
(9.19)

where N is the number of projections used for reconstruction. When a centered uniform phantom is used, the standard deviation of $p(0, \beta_n)$ becomes independent of the projection angle and can be denoted by σ_p. The standard deviation of f_0, σ, is then simply

$$\sigma^2 = \sigma_p^2 \sum_{n=1}^{N} w^2(0, \beta_n).$$
(9.20)

Since what we are actually seeking is the noise ratio r_n of helical scans over step-and-shoot scans, the noise ratio can be calculated by:

$$r_n = \sqrt{\frac{\sum_{n=1}^{N} w^2(0, \beta_n)}{\sum_{k=1}^{K} q^2(0, \beta_k)}},$$
(9.21)

where $q(0, \beta_n)$ is the weighting function used for the step-and-shoot mode reconstruction, and K is the number of views used in the step-and-shoot mode. For most cases, $q(0, \beta_n)$ is constant. The noise ratios for four different helical reconstruction algorithms are calculated and tabulated in Table 9.2, and they assume a constant weighting function for the step-and-shoot scan. Table 9.2 indicates a clear tradeoff between SSP and noise. In general, an algorithm that optimizes the SSP is likely to perform poorer in terms of noise.

A careful examination of Eq. (9.21) indicates that the noise ratio is independent of the helical pitch. This may be somewhat surprising. If we examine the data acquisition of a uniform water phantom that does not vary in z, the collected projections remain unchanged regardless of the helical pitch. Consequently, the noise in the reconstructed images becomes pitch independent.

9.4 Helically Related Image Artifacts

9.4.1 High-pitch helical artifacts

For single-slice helical CT, the level of helical-related image artifacts increases with the helical pitch; that is, a more degraded image quality is expected for data collected with a higher helical pitch. A higher helical pitch means that the patient table travels at a greater distance in one gantry rotation. For the same object, a larger distance between the start and end of scanning locations generally results in a larger discrepancy or inconsistency in the collected data (assuming the object is heterogeneous in z). A large inconsistency translates to a higher level of image artifacts. When the helical pitch approaches zero, we can revert to the step-and-shoot mode of data collection and no helically induced image artifacts will be present.

To illustrate the monotonic relationship between the helical pitch and image artifacts for single-slice CT, we performed the following experiment. We scanned an oval body phantom with four different helical pitches ($h = 0$, 1, 1.5, and 2). The phantom was made with a uniform elliptical cylinder to simulate soft tissue, and high-density objects near the periphery were Teflon rods inserted at different angles with respect to the x-y plane, to simulate ribs. The void near the center was a cone-shaped air pocket. Since the shape of the ribs and air pocket changed significantly in z, the phantom accentuated helically related image artifacts. For this study, the helical interpolation algorithm described by Eq. (9.12) was used for all helical reconstructions. The results are shown in Figs. 9.27(a)–(d). The image that corresponds to $h = 0$ is used as the "gold standard." We can conclude from these figures that image artifacts increase with increase in h. For example, the image corresponding to $h = 1$ is nearly identical to the step-and-shoot mode except for slight shading artifacts near the rib tips. When the helical pitch increases to $h = 1.5$, shading artifacts become more pronounced and more widely spread. A small amount of geometric distortion can be observed near the air pocket. For the case of $h = 2$, shading artifacts are more severe, and significant geometric distortion can be observed around the air pocket.

Although the experiment was conducted with a phantom, the same conclusion can be made on clinical scans. In typical clinical applications, helical pitches between 1 and 1.5 are routinely used. The selection of helical pitch depends largely on the tradeoffs between coverage, slice thickness, and image artifacts.

The tradeoff between slice thickness and image artifacts can also be performed by a simple algorithmic manipulation.[18] For example, if we denote by p_{z0} an image reconstructed at location z_0 using one of the helical reconstruction algorithms described above, the composite image p_c, calculated from the weighted summation of a series of images located in the vicinity of z_0, should contain fewer helical artifacts. Mathematically, p_c is described by the following equation:

$$p_c = \sum_{n=-N}^{N} s_{n\Delta z} p_{z0+n\Delta z} \, ,\qquad (9.22)$$

where Δz is the spacing between adjacent images, $s_{n\Delta z}$ is the weighting factor, and N defines the range of the averaging operation. Because this operation, illustrated in Fig. 9.28(a), takes place along the z axis (no in-plane smoothing is performed), this technique is often called z filtering. Figures 9.28(b) and (c) compare the reconstructed images of an oval body phantom without and with z filtering. The phantom was scanned with 2.5-mm collimation at a 1.5:1 helical pitch. Figure 9.28(c) was obtained by averaging four neighboring images to produce an image with a resulting slice thickness that is roughly twice the slice thickness of the original image. Shading artifacts are nearly completely eliminated. Clearly, we must trade a degraded slice thickness for a reduction in image artifacts.

Since the FBP is a linear operator, the z-filtering operation performed in the image space can be carried out in the projection space. In other words, we can incorporate the z filtering into the helical weighting functions by convolving the original weighting function with the smoothing function in β. This is based on

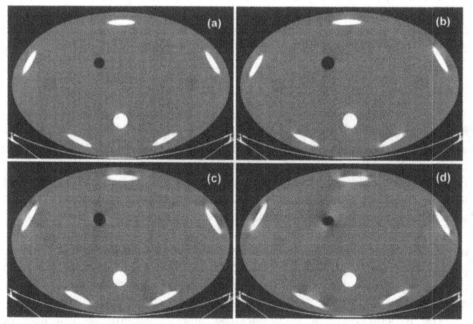

Figure 9.27 Relationship between helical pitch and image artifacts. (a) Pitch = 0.0, (b) pitch = 1.0, (c) pitch = 1.5, and (d) pitch = 2.0.

the linear relationship between z and β for a constant patient table speed. Mathematically, the composite helical weighting function $w'(\gamma, \beta)$ is calculated from the original helical weighting function $w(\gamma, \beta)$:

$$w'(\gamma,\beta) = \sum_{n=-N}^{N} s_{n\Delta\beta} w(\gamma,\beta+n\Delta\beta). \qquad (9.23)$$

The analyses and equations used to calculate SSP and noise ratio can be readily applied by noting that helical reconstruction with z filtering is nothing but a utilization of a different helical weighting function. All we need to do is to

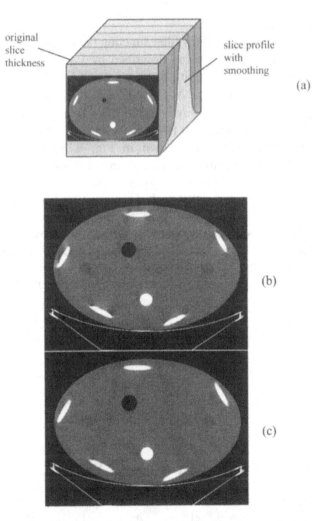

Figure 9.28 Illustration of the effect of z filtering. (a) Original image is smoothed to create thicker slices. (b) Phantom scanned at 1.5:1 helical pitch with 2.5-mm collimation. (c) Z-direction filtered image such that the slice thickness is twice that of the original image (reconstructed from the same scan data).

replace the weighting functions used in Eqs. (9.17) and (9.21) with the composite weighting function of Eq. (9.23). The SSP and noise ratio equations become

$$\psi(z) = \int_0^{kr} w'(\gamma, z')\Omega(z - z')dz' \tag{9.24}$$

and

$$r_n = \sqrt{\frac{\sum\limits_{n=1}^{N} w'^2(0, \beta_n)}{\sum\limits_{k=1}^{K} q^2(0, \beta_k)}}. \tag{9.25}$$

For illustration, we chose a uniform filter function ($s_{n\Delta\beta}$ = constant) and applied it to the helical interpolation algorithm described by Eq. (9.12) at a 1:1 pitch to arrive at composite helical weighting functions using Eq. (9.23). For five of the filter functions, we set parameters $N = 1$ and $\Delta\beta = 0.2\pi$, 0.4π, 0.6π, 0.74π, and 0.82π. The other two filters were $N = 2$ and $\Delta\beta = 0.6\pi$ and 0.9π. The seventh filter was slightly nonuniform, with two edge values equal to 70% of the center values. The calculated SSPs are shown in Fig. 9.29; the innermost curve represents the SSP without z filtering. The seven z filters are ranked in ascending order of smoothing, so the corresponding SSP curves are positioned from the narrowest to the widest. A curve that plots the noise ratio versus the FWHM ratio is shown in Fig. 9.30 (noise ratios were calculated). These calculated values match the experimental values described in Ref. 18, shown by the triangles. Similar calculations were performed for the case of a 1.5:1 helical pitch using a different set of filter kernels. The results are shown in Figs. 9.31 and 9.32. A similar tradeoff behavior between SSP and noise can be observed.

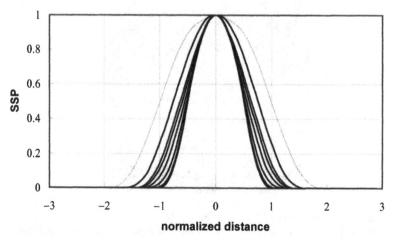

Figure 9.29 Normalized SSP with respect to collimation for pitch 1:1. Normalized distance is the real distance divided by the collimation.

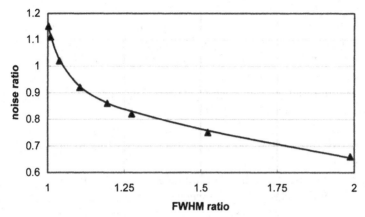

Figure 9.30 FWHM ratio versus noise ratio for helical pitch 1:1.

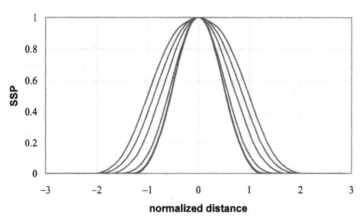

Figure 9.31 Normalized SSP with respect to collimation for pitch 1.5:1. Normalized distance is the real distance divided by the collimation.

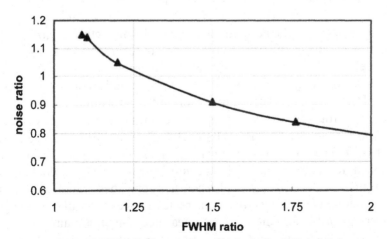

Figure 9.32 FWHM versus noise ratio for helical pitch 1.5:1.

9.4.2 Noise-induced artifacts

Section 9.2 discussed the use of weighting functions for helical interpolation. When a fan-beam reconstruction algorithm is applied, the interaction between the helical weights and the scaling function employed in the backprojection can produce modulated noise patterns. These noise patterns are hardly noticeable when we examine images one slice at a time. They are clearly visible when the images are displayed in a fast cine-paging mode. These artifacts can also produce periodical patterns on the MIP or VR images.[19] Figures 9.33(a) and (b) show examples of MIP and VR images produced with a patient scan. Bright and dark horizontal strips are clearly visible.

To explain the impact of the weighting function on noise distribution, we will investigate the noise relationship between a helically reconstructed image and a reconstructed image without weighting. As in the approach used previously, we define a noise ratio, in terms of the standard deviation, of the helically reconstructed images to the reconstructed images with uniform weights.[19] The inner integral represents the filtering process used in the tomographic reconstruction [see Eq. (9.3)]. For noise analysis, this operation can be ignored because the magnitude of the filter kernel coefficients in the spatial domain falls off as a quadratic function of the distance from the center. The noise impact of the filter on the projection becomes localized and, to a first order of approximation, should be cancelled out in the noise ratio calculation. Therefore, the variance can be estimated as the integration of the projection variance weighted by the square of the scaling factor in Eq. (9.3). The standard deviation ratio at location (r, ϕ) of the reconstructed images $\xi(r, \phi)$ can then be described by the following equation:

$$\xi(r,\varphi) = \left[\int_{\beta_0}^{\beta_0 + \Pi} w^2(\gamma, \beta - \beta_0) \sigma^2(\gamma, \beta) L'^{-4} \, d\beta \right]^{1/2} \left[\int_{\beta_0}^{\beta_0 + 2\pi} \sigma^2(\gamma, \beta) L'^{-4} \, d\beta \right]^{-1/2}, \quad (9.26)$$

where $\sigma(\gamma, \beta)$ is the standard deviation of the projection, and $L' = L\cos\gamma$. It is clear from Eq. (9.26) that $\xi(r, \phi)$ is dependent on position (r, ϕ) and starting angle β_0. $\xi(r, \phi)$ also depends on the noise distribution of the projection $\sigma(\gamma, \beta)$.

Figure 9.34 shows calculated noise ratios of a uniform cylindrical phantom for the helical extrapolative and helical interpolative algorithms. For the calculations, we assumed the projection noise was dominated by the x-ray photon statistics. Therefore, the standard deviation in the projection, after the logarithm operation, was roughly inversely proportional to the detected x-ray photon flux. In this figure, the x-ray tube is in the twelve o'clock position at the start of the helical dataset. The noise near the twelve o'clock position is less than that of the conventional reconstruction, while the noise near the six o'clock position is higher than that of the conventional reconstruction. These ratios indicate that the noise distribution of the helically reconstructed images is highly inhomogeneous. To further quantify our analysis, we measured the minimum and maximum values of $\xi(r, \phi)$ and obtained 0.60 and 1.69, respectively. This is a factor of 2.82

across the reconstruction FOV! To explain this phenomenon, consider the impact of the scaling factor L^{-2} in the backprojection process. When the x-ray source is near the twelve o'clock position, the magnitude of L^{-2} decreases quickly from the top of the figure to the bottom. This means that these projections contribute preferentially to the noise near the top of the figure. Similarly, the magnitude of L^{-2} increases quickly from top to bottom when the x-ray tube is in the six o'clock position. This indicates that these projections contribute preferentially to the noise near the bottom of the figure. When projections are not weighted, the noise more or less balances out with the 2π projection. For helical reconstruction, however, $w(\gamma, \beta)$ approaches zero for the projections near the twelve o'clock position (the starting angle of the dataset). Consequently, the noise contribution from these projections is largely suppressed (noise in the top portion of the image is not significantly increased). On the other hand, $w(\gamma, \beta)$ reaches its peak value of unity for projections near the six o'clock position. As a result, the noise in the bottom portion of the image is substantially increased. Since the top portion of the image receives less contribution when the noise is high and more contribution when noise is low, the noise ratio becomes less than unity.

As mentioned previously, the noise pattern follows the initial angle of the projection β_0. For example, if a helical image is reconstructed from a dataset in which the x-ray tube starts in the three o'clock position, the noise ratio $\xi(r, \phi)$ can be obtained by simply rotating clockwise by 90 deg the noise ratios depicted in Fig. 9.34(a) and (b).

To establish the link between the noise inhomogeneity and the image artifacts in MIP images, we will first describe the MIP generation process. The MIP process is similar to forward projection in that each pixel value in an MIP

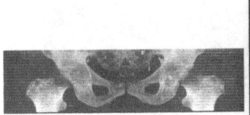
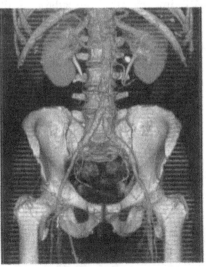

(a)   (b)

Figure 9.33 Example of a noise-induced artifact (a) in an MIP and (b) in a VR image.

image is calculated based on the intensities of all the reconstructed image voxels that intersect the ray. This is an imaginary ray that irradiates from a source and intersects the MIP image pixel. For convenience, a set of parallel rays is often selected. The incident angle of the ray can change to produce different views. Instead of performing integrations along the ray path (as is the case for forward-projection), the intensity of each pixel in the MIP is determined based on the maximum voxel intensity value along the ray, as shown in Fig. 9.35. If we consider a region where the average CT number is uniform, the MIP should produce a reasonably flat image since the maximum intensities along any ray path are essentially identical. However, when nonhomogeneous noise is added to the reconstructed volume, the rays that pass through regions with a higher standard deviation tend to produce higher values, since the data with a higher standard deviation have more fluctuations (although the mean value remains the same). On the other hand, the regions with a low variance tend to produce lower MIP readings. Since the noise pattern produced by the helical weights follows the starting angle of the projection dataset, the MIP values modulate between high and low intensities in a periodic pattern. This artifact is often called the zebra artifact or Venetian blind artifact. Figure 9.36(a) illustrates an example of the zebra artifact with a phantom experiment in which a 48-in uniform polyethylene phantom was scanned in a helical mode and reconstructed with the helical extrapolative algorithm. Horizontal bright and dark bands are clearly visible in the MIP image shown in Fig. 9.36(a). Similar behavior can also be observed in the patient body scan shown in Fig. 9.37(a).

Because this noise pattern is well behaved and can be predicted prior to the image reconstruction, zebra artifacts can be corrected.[19] For example, we can devise an adaptive filtering scheme based on the noise distribution predicted by Eq. (9.23) to reduce the noise in-homogeneity. Figures 9.36(b) and 9.37(b) depict MIP images after the adaptive filtering correction. The horizontal strips are nearly completely eliminated.

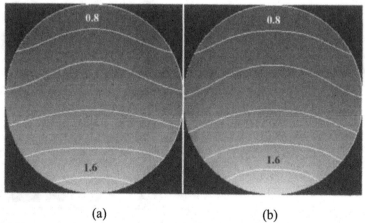

(a)   (b)

Figure 9.34 Noise ratios for fan-beam (a) helical interpolative and (b) helical extrapolative algorithms.

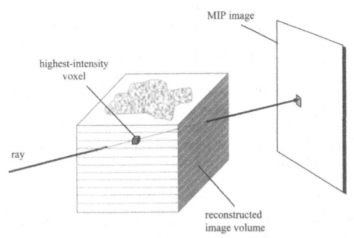

Figure 9.35 Illustration of the MIP generation process. An imaginary ray is cast through the reconstructed volume and projected onto the MIP image plane. The intensity of each pixel in the MIP is equal to the highest voxel intensity that intersects the ray.

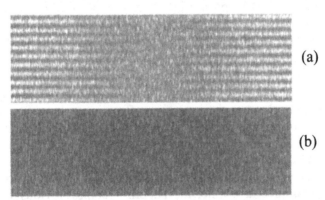

Figure 9.36 Maximum-intensity projection images of a 48-in uniform polyethylene phantom. (a) Original MIP without noise compensation and (b) MIP with noise correction.

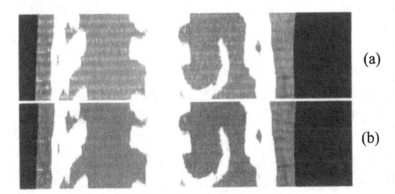

Figure 9.37 Maximum-intensity projection images of a patient scan. (a) Original MIP without noise compensation and (b) MIP with noise correction.

Alternatively, we can rebin the original fan-beam helical dataset to a parallel-beam helical dataset and perform a parallel-beam-based reconstruction. Because the L^{-2} factor is removed in the backprojection process, noise uniformity in the reconstructed images should be significantly improved. Since the root cause of the Venetian blind artifact is noise nonuniformity, a parallel-beam-based reconstruction should be effective at suppressing this artifact. Indeed, Figs. 9.38(a) and (b) show a pair of MIP images generated from the fan-beam- and parallel-beam-based images, and the Venetian artifact shown in Fig. 9.38(a) is nearly eliminated in Fig. 9.38(b). Although the artifact-reduction examples given here are MIP images, the effectiveness of these corrections is similar for VR images.

9.4.3 System-misalignment-induced artifacts

In many cases, image artifacts are caused by the interaction between the system alignment and the reconstruction algorithm. One such example is the image artifact produced in 3D-formatted images when the iso-center of the system is slightly misaligned.[20] For a conventional CT scan, the impact of iso-center misalignment (when the true location of the projection ray deviates from the ray position assumed by the backprojection) is loss of spatial resolution. Because the x-ray source trajectory is circular, the impact of the deviation is a smearing effect rather than a linear shift. When the amount of misalignment is small, its effect is difficult to detect in a single-slice image.

For 3D applications (e.g., multiplanar reformation), the artifact sensitivity is much greater, due mainly to two factors. First, because the weighting function is applied in the helical scan, the smearing effect is no longer symmetrical. It becomes biased toward the views with higher weights. Second, the amount of bias follows the starting angle of the x-ray source (the weighting function is specified relative to the starting angle). As a result, the bias becomes periodical. Although this periodical bias is hardly noticeable in the reconstructed images when viewed one slice at a time, the pattern becomes obvious when a set of

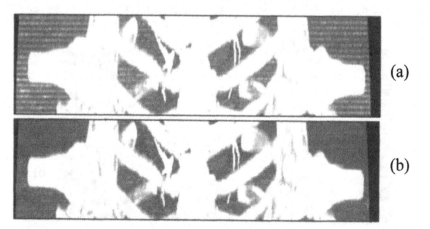

Figure 9.38 Maximum-intensity projection of a chest phantom scan. (a) Fan-beam helical reconstruction. (b) Parallel-beam helical reconstruction.

images is stacked to form a reformatted image. Figure 9.39 shows a reformatted image of a reconstructed cylindrical water phantom. The location of reformation relative to the phantom cross-section is illustrated in the figure. The reformatted edges of the phantom are supposed to be straight lines, but both edges show a zigzag pattern.

To fully understand the impact of the projection weighting function and iso-center misalignment on reconstructed images, we need to develop an analytical model to characterize or predict the amount of bias in the resulting image. Since most of the zigzag artifacts occur when a large contrast exists between the object and its background, we approximate the profile of the object across its boundary by a step function. For the boundary profiles that cannot be approximated by the step function, we can approximate the profile by the convolution of a step function with another function. Since the convolution operation is linear and commutable, the derivations and conclusions derived from the step function can be extended to these cases. If we use the radial symmetry of the reconstruction process, the analysis can be carried out only to the object boundaries that are perpendicular to the horizontal (x) axis. The boundary profile can be described by the following equation:

$$s(x) = \begin{cases} 0, & x < x_0, \\ h, & x \geq x_0, \end{cases} \tag{9.27}$$

where x_0 is the boundary location. When a projection is generated, the object boundary is sampled with an x-ray beam of finite width. The shape of the x-ray beam is determined by both the detector cell response and the x-ray focal spot size. If we assume a uniform response for both the detector and the x-ray focal spot, the x-ray beam profile can be approximated by the convolution of the projected detector response and the projected focal spot response at the location

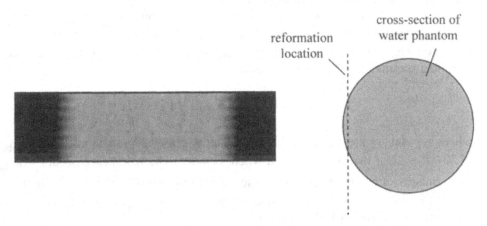

reformation
location

cross-section of
water phantom

Figure 9.39 Reformatted image of a reconstructed cylindrical water phantom.

(for a detailed discussion on the topic, see Chapter 8). The resulting object boundary $p(x)$ can be shown to be

$$p(x) = s(x) \otimes d(x)$$

$$
= \begin{cases}
\dfrac{\left[x - x_0 + 0.5\left(u_1 + u_2\right) \right]^2}{2u_1 u_2} & x_0 - \dfrac{u_1 + u_2}{2} < x \le x_0 - \dfrac{u_1 - u_2}{2} \\[2em]
0.5 + \dfrac{x - x_0}{u_1} & x_0 - \dfrac{u_1 - u_2}{2} < x \le x_0 + \dfrac{u_1 - u_2}{2} \quad , \\[2em]
1 - \dfrac{u_2}{2u_1} + \dfrac{\left[x - x_0 - 0.5\left(u_1 - u_2\right) \right]\left(x_0 + u_2 + 0.5u_1 - x\right)}{2u_1 u_2}, & x + \dfrac{u_1 - u_2}{2} < x \le x_0 + \dfrac{u_1 + u_2}{2}, \\[2em]
0 & \text{otherwise.}
\end{cases}
$$

$$\tag{9.28}$$

where u_1 and u_2 are the projected detector width and focal spot width. Since each projection needs to be filtered and backprojected to produce the final image, the reconstructed object boundary $\mu(x)$ is the convolution of the reconstruction PSF, $\Gamma(x)$, with $p(x)$,

$$\mu(x) = \Gamma(x) \otimes p(x). \tag{9.29}$$

In this equation, a stationary PSF is assumed. Strictly speaking, however, $\Gamma(x)$ is nonstationary and varies as a function of its distance from the iso-center. Because the variation is quite small, we will ignore its impact.

Note that the reconstructed object boundary is defined primarily by the projections that are perpendicular to the boundary. Therefore, we will focus our attention to the two views that satisfy this condition. When the iso-center misalignment equals Δx, the conjugate ray pair (180 deg apart) is shifted by Δx and $-\Delta x$, respectively. The resulting reconstructed boundary $f(x)$ is then the weighted summation of the two responses:

$$f(x) = w_1(x)\mu(x + \Delta x) + w_2(x)\mu(x - \Delta x), \tag{9.30}$$

where w_1 and w_2 are the helical weights applied to the projections. From Eq. (9.30), it is clear that $f(x)$ is shifted when $w_1 \ne w_2$. The worst case occurs when two images are generated with the weighting pairs $(w_1 = 2, w_2 = 0)$ and $(w_1 = 0, w_2 = 2)$. In a CT system, this occurs when the object is scanned twice with the starting angles 180 deg apart.

For illustration, we generated projections of a 20-cm-diameter uniform cylindrical phantom centered at the iso-center. The source-to-iso and source-to-detector distances were 630 mm and 1100 mm, respectively. The detector cell width was 1 mm, and the focal spot width was 0.9 mm. The shift in the iso-center was 0.036 mm. Two projection datasets were generated with the starting projection angles at the twelve o'clock and six o'clock positions, respectively. For each dataset, an image was reconstructed with an underscan weight. The difference image of the two reconstructed images is shown in Fig. 9.40. A shift along the horizontal (x) axis is clear. For quantitative analysis, we plotted the reconstructed edge profiles (along x) for the two images, as shown by the dotted lines in Fig. 9.41. The predicted object boundaries were calculated based on Eq. (9.30) and are plotted with the solid lines in Fig. 9.41. A nice match between the measurement and prediction was obtained.

Similar to the zebra artifact, the saw-toothed artifact can be corrected by software algorithms.[20] For example, since we know *a priori* the amount of shift in the image, we can adjust the object boundaries during either the image generation process or during the reformation production. For illustration, we repeated the reconstruction process for the phantom scan shown in Fig. 9.39 and produced the artifact MPR image shown in Fig. 9.42(a). Then we dynamically adjusted the object boundary and produced an MPR image of the same phantom scan, as shown in Fig. 9.42(c). The zigzag pattern was completely eliminated. Alternatively, we can force the starting angles for all reconstructed images to be identical so they are all biased in the same direction; the result in Fig. 9.42(b) shows that the sawtooth pattern was again completely eliminated. Of course, the best solution is to readjust the system so that its iso-center is in perfect alignment.

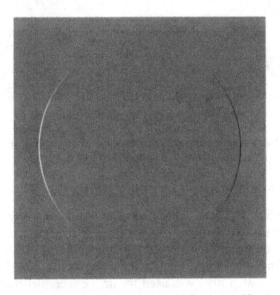

Figure 9.40 Difference image of two reconstructed images with starting angles at 12 o'clock and 6 o'clock positions. The iso-center shift is 0.036 mm. Underscan weight was used for reconstruction (WW = 200).

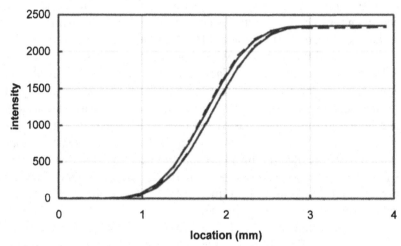

Figure 9.41 Profiles of reconstructed phantom edges. Dotted lines represent computer simulation results and solid lines represent theoretical prediction.

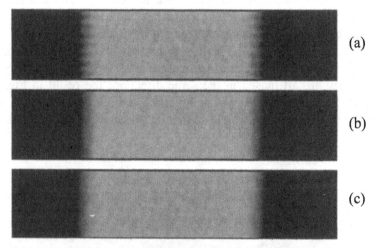

(a)

(b)

(c)

Figure 9.42 Multiplanar reformatted images of a water phantom. (a) Starting angles for consecutive images are 180 deg apart. MPR image was formed without any correction. (b) MPR image was formed with reconstructed images of the same start angle. (c) MPR image was formed with adaptive shift and filtering scheme.

9.4.4 Helical artifacts caused by object slope

In some cases, helically related 3D artifacts can occur even when the system is in perfect alignment. The artifact appears as periodical rings or grooves superimposed on the surface of the object, as shown in Fig. 9.43(a). This image is a surface rendering of a reconstructed soda pop bottle near its shoulder. The helical extrapolative algorithm was used for image reconstruction. The culprit of the artifact is the interaction between helical sampling of a sloped object and the reconstruction algorithm. For the convenience of our discussion, we divided a

2π–projection dataset into three types of zones, as shown in Fig. 9.44. The conjugate samples corresponding to zone A are located inside zone B, and the conjugate samples of zone C are located inside zone C'. In the 2π–based single-slice helical CT scan, the plane of reconstruction is generally selected inside zone B. The nature of the linear interpolation requires that the projections in zone B carry higher weights. On the other hand, the weights for samples in zone A are near zero, since they are located the farthest from the plane of reconstruction. Therefore, the portion of the object boundary tangential to these projections (zones A and B) is determined primarily by projections in zone B. For the projections that are located inside zone C, the weights are roughly the same, since they are approximately equidistant from the reconstruction plane. As a result, the boundary formed by these projections is determined by two samples that are 180 deg apart. When the shape of the scanned object does not vary significantly along the z axis (table translation direction), the composite boundary formed by two samples 180 deg apart (zone C) is nearly the same as the boundary formed by a single sample (zones A and B). When the object shape changes significantly along the z axis (as in the case of the soda pop bottle shoulder), the boundary formed by projections in zone C has a different edge response than the one formed by zones A and B. Figure 9.45 depicts the edge responses of four regions: A, B, C, and C'. We assumed that the edge of the object shifted toward the right as the scan progressed, to represent the scan of a sloped object in helical mode as the object was constantly translated during the scan (the distance between the responses in the figure is exaggerated for illustration). As discussed previously, the composite edge response for regions A and B is nearly the same as the response from region B alone, since the weight for region A is nearly zero. This is depicted by the thin gray solid line in the figure. On the other hand, the composite edge response from regions C and C' is the average of the two responses, as shown by the thick solid line in Fig. 9.45. It is clear that the edge formed by zones A and B is sharper than the edge formed by zone C. Although their average locations specified by the 50% peak intensity point are the same, the edge locations deviate gradually as we move away from the 50% point. For example, the edge difference at 90% is depicted in the figure.

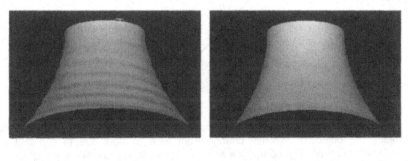

(a)   (b)

Figure 9.43 Shaded surface display of a scanned soda pop bottle. (a) Nonoptimal threshold of 10% and (b) optimal threshold of 50%.

In the 3D surface rendering process, the first step is to select a threshold to locate the object boundaries. If the selected threshold deviates significantly from the 50% peak intensity (e.g., 10%, or 90% of the peak intensity), a large discrepancy between the edges formed by different zones will produce groove artifacts, as shown in Fig. 9.43(a). If the selected threshold happens to be near the 50% peak intensity point, the groove artifacts can be significantly reduced, since edges defined by different zones become more consistent. This is shown in Fig. 9.43(b). Of course, modified helical weights that produce more-uniform boundary responses can also be employed to suppress this type of artifact.

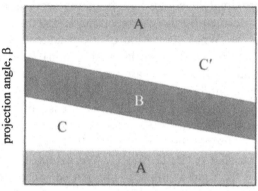

Figure 9.44 Illustration of three types of zones in projection sampling. The weights in zone A are typically small since they are farther away from the reconstruction plane. The weights in zone B are high due to their close proximity to the reconstruction plane. The weights in C and C' are balanced.

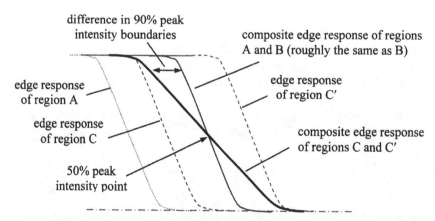

Figure 9.45 Illustration of weighting function impact on object boundaries. When scanning a sloped object, the object boundaries change from view to view due to constant object translation in helical mode. The composite edge response from projections in regions A and B is nearly the same as the response from region B projections alone, as shown by the thin solid line. The composite edge response of regions C and C' is the average of the two responses as shown by the thick solid line. Although the two composite edge responses overlap at 50% peak intensity point, they deviate significantly at other intensity points.

9.5 Problems

9-1   Section 9.1 discussed the need to perform a 50% overlapped reconstruction to satisfy the Nyquist sampling criteria. Given the SSP of the helical reconstruction shown by Eq. (9.17) and in Fig. 9.25, do you recommend changing the image overlap requirement? If so, what is your recommended overlap?

9-2   Figure 9.2 shows the advantage of centering the object to the reconstruction plane to maximize object contrast in a helical scan. If the helical pitch = 1, how much is the contrast improvement of a spherical object centered at the reconstruction plane versus the offset by the radius of the sphere? Assume the SSP for the step-and-shoot mode is a rectangular function with a width equal to the diameter of the sphere.

9-3   Repeat problem 9-2 for a helical pitch of 1.5. Is the amount of improvement larger or smaller than the pitch = 1 case? What conclusion can you draw?

9-4   In clinical practice, trying to center a small object of interest to the center of the reconstruction plane can be time consuming and impractical. What practical solutions can you provide to improve the object contrast?

9-5   Equation (9.1) gives the definition of a helical pitch. A CT system employs a postpatient collimator in z to improve the SSP of the system. For this system, should we use the prepatient collimator aperture or the postpatient collimator aperture at the iso-center for the value of s? Justify your answer.

9-6   Figure 9.6 illustrates the gaps in helical sampling. What approach can you take during the reconstruction process to consider the shape of the x-ray beam?

9-7   Equation (9.3) shows a general helical reconstruction algorithm for a fan-beam CT system. For the two helical weighting functions discussed [Eqs. (9.4) and (9.12)], what is the impact on image quality if γ_m is increased (that is, more detector channels are added to increase the scan FOV)?

9-8   Show that when helical data are collected with a parallel x-ray source positioned parallel to the x-y plane, the weighting function for the reconstruction can be obtained by setting $\gamma = 0$ in Eq. (9.4) or Eq. (9.12).

9-9   Fan-beam helical data of a 2π rotation contains 1000 views and 890 detector channels with the iso-channel located at 445.75. The source-to-detector distance is 950 mm and the detector spacing is 1 mm. Calculate the number of rebinned parallel views, assuming the view increments are kept the same ($\Delta\beta = \Delta\theta$).

9-10   Derive a helical weighting function for a rebinned parallel dataset based on the fan-beam helical extrapolative weighting function.

9-11   Estimate numerically the standard deviation ratio for the rebinned parallel-beam helical interpolative algorithm [Eq. (9.15)] over a 50-cm FOV.

9-12   Repeat problem 9-11 for the rebinned parallel-beam helical extrapolative case.

9-13   To overcome the Venetian blind artifact and to satisfy the Nyquist sampling criteria, a student proposes to scan patients with a 0.5 helical pitch and perform a 50% overlapped fan-beam reconstruction (without postprocessing compensation). Will this approach work? What are the potential issues?

9-14   Figure 9.23 depicts the relationship between the SSP in z and the reconstructed profile in x. An engineer is concerned about the influence of the reconstruction filter kernel on the accuracy of the SSP measurement. The engineer argues that a high cutoff frequency reconstruction kernel produces a sharper reconstructed profile and gives a better SSP value. Is the engineer's concern justified? If so, how do you design the measurement process to minimize the impact?

9-15   A CT system shown in Fig. 9.24 has the following parameters: $d = 250$ mm, $D = 630$ mm, and $\Delta = 0.7$ mm. The FWHM setting is based on the value at the iso-center in step-and-shoot mode. Calculate the SSP for a helical scan with FWHM = 2 mm at a helical pitch of 1.5. Repeat the calculation of SSP for a helical scan with FWHM = 3 mm at a pitch of 1. What conclusion can you draw comparing the results?

9-16   Can the Venetian blind artifact be reduced with the thin-slab MIP described in Chapter 4? Explain.

9-17   Can an MIP taken in a direction that forms a nonzero angle with respect to the x-y plane help to reduce the Venetian blind artifact? If so, how do you determine the best angle?

References

1. J. R. Mayo, N. L. Muller, and R. M. Henkelman, "The double-fissure sign: a motion artifact on thin section CT scans," *Radiology* **165**, 580–581 (1987).

2. R. D. Tarver, D. L. Conces, and J. D. Godwin, "Motion artifacts on CT simulate bronchiectasis," *Am. J. Roentgenol.* **151**, 1117–1119 (1988).

3. I. Mori, "Computerized tomographic apparatus utilizing a radiation source," U.S. Patent No. 4,630,202 (1986).

4. H. Nishimura and O. Miyazaki, "CT system for spirally scanning subject on a movable bed synchronized to x-ray tube revolution," U.S. Patent No. 4,789,929 (1988).

5. W. A. Kallender, W. Seissler, and P. Vock, "Single-breath-hold spiral volumetric CT by continuous patient translation and scanner rotation," *Radiology* **173P**, 414 (1989).

6. P. Vock, H. Jung, and W. Kallender, "Single-breath-hold spiral volumetric CT of the hepatobillary system," *Radiology* **173P**, 377 (1989).

7. C. R. Crawford and K. King, "Computed tomography scanning with simultaneous patient translation," *Med. Phys.* **17**, 967–982 (1990).

8. G. Wang and M. W. Vannier, "Longitudinal resolution in volumetric x-ray computerized tomography-analytical comparison between conventional and helical computerized tomography," *Med. Phys.* **21**, 429–433 (1994).

9. J. Hsieh, "A general approach to the reconstruction of x-ray helical computed tomography," *Med. Phys.* **23**, 221–229 (1996).

10. K. F. King and C. R. Crawford, "Extrapolative reconstruction method for helical scanning," U.S. Patent No. 5233518 (1993).

11. G. Besson, "New classes of helical weighting algorithms with applications to fast CT reconstruction," *Med. Phys.* **25**, 1521–1532 (1998).

12. W. A. Kalender, "Principles and performance of spiral CT," in *Medical CT and Ultrasound: Current Technology and Applications*, L. W. Goldman and J. B. Fowlkes, Eds., Advanced Medical Publishing, Madison, WI (1995).

13. D. J. Heuscher and M. Vembar, "Improved 3D and CT angiography with spiral CT," *Radiology* **197**, 222 (1995).

14. P. J. La Riviere and X. Pan, "Fourier-based approach to interpolation in single-slice helical CT," *Med. Phys.*, **28**, 381–392 (2001).

15. J. Hsieh, "Investigation of the slice sensitivity profile for step-and-shoot mode multi-slice computed tomography," *Med. Phys.* **28**, 491–500 (2001).

16. D. J. Goodenough, K. E. Weaver, and D. O. Davis, "Development of a phantom for evaluation and assurance of image quality in CT scanning," *Opt. Eng.* **16**, 52–65 (1977).

17. H. Hu and S. H. Fox, "The effect of helical pitch and beam collimation on the lesion contrast and slice profile in helical CT imaging," *Med. Phys.* **23**, 1943–1954 (1996).

18. H. Hu and Y. Shen, "Helical reconstruction algorithm with user-selectable section profiles," *Radiology* **189**, 335 (1996).

19. J. Hsieh, "Nonstationary noise characteristics of the helical scan and its impact on image quality and artifacts," *Med. Phys.* **24**(9), 1375–1384 (1997).

20. J. Hsieh, "Three-dimensional artifact induced by projection weighting and misalignment," *IEEE Trans. Med. Imag.* **18**(4), 364–368 (1999).

Chapter 10
Multislice CT

10.1 The Need for Multislice CT

So far, this book has focused on the detector configuration in which a single row of a detector is irradiated by the x-ray beam. Although helical (spiral) CT has significantly improved volume coverage, many clinical applications demand an even greater volume coverage and thinner slices. A good example is CT angiography.[1] For this application, a rapid volume acquisition is needed during the plateau phase of contrast enhancement. For thoraco-abdominal aorta studies, the volume includes the whole chest and abdomen, which can reach between 45 and 60 cm along the patient (z) axis. A more demanding case is the runoff study of the abdominal aorta and the legs, which covers the area from the celiac artery to the calves, with a typical z coverage between 90 and 120 cm.

To ensure optimal contrast enhancement and minimal patient respiratory motion, an entire study should be completed in less than 20 sec. For instance, to cover 40 cm in 20 sec, the patient table must travel at a speed of 2 cm/sec. With a single-slice scanner rotating at 0.5 sec per revolution, it is difficult to obtain a slice thickness of 5 mm or less at this coverage. For example, consider a likely scanning protocol of 5-mm collimation at a 2:1 helical pitch. As discussed in Chapter 9, the effective slice thickness of a reconstructed image increases quickly with an increase in helical pitch. Consequently, the slice thickness of this protocol is 6.4 mm at FWHM and 11.1 mm at FWTM. This is clearly suboptimal for the visualization of small vascular structures. If we use a thinner collimation or slower helical pitch, the time to cover the same volume increases, which leads to several complications. First, we may lose the optimal timing for contrast enhancement. When the scan time is significantly lengthened, the contrast washout effect becomes significant. Although we can partially compensate for the effect by prolonging the contrast injection time, we often run into limitations such as the total amount of allowable contrast administered to the patient. Prolonged scan time and contrast injection also preclude the examination of organs in their arterial (free of venous) phase.

The second problem with a thinner collimation or slower helical pitch is patient motion. This is particularly important for chest and abdominal studies. A typical patient can hold his or her breath for approximately 20 to 30 sec. For studies that extend beyond this range, multiple breath-holds are required.

Although images reconstructed within each breath-hold are free of respiratory motion artifacts, misregistration between breath-holds is often unavoidable (it is nearly impossible for a patient to repeat exactly the same breath-holding level). Since many studies are viewed in 3D (multiplanar reformat, MIP, or VR), misregistration between adjacent volumes is highly objectionable.

The third issue with thin collimation is tube utilization. When a thinner collimation is used, a larger number of the produced x-ray photons are blocked by the collimator and never reach the patient. For example, the total number of available x-ray photons produced with a 1-mm collimation is roughly one-tenth of the available x-ray photons with 10-mm collimation. To compensate for the reduced x-ray flux (to maintain the same noise level in the image), we must increase the x-ray tube current. This leads to a shorter scan time between forced x-ray tube cooling periods. For contrast-enhanced studies, interruptions in the scanning operation are unacceptable.

The root cause of this problem is the unavoidable tradeoff between slice thickness and volume coverage for single-slice scanners. Because the detector has no resolution capability in z (the detector has no way of finding the location of an x-ray photon hitting the detector), the spatial resolution of the scanner (in z) is determined primarily by the profile or the width of the x-ray beam. If the detector can provide z resolution, the width of the x-ray beam in z can be independent of the spatial resolution or the slice thickness. In other words, a wide x-ray beam (with better x-ray tube utilization and larger volume coverage) can be used to achieve a thin slice thickness. The desire to provide resolution along the detector z-axis is what led to the development of the multislice scanner.[2–7]

Two clinical examples illustrate the clinical benefits of multislice CT. Figures 10.1(a) and (b) show VR images of neural and abdominal CT angiogram studies performed on a Discovery™ CT750 HD scanner (GE Healthcare, Waukesha, WI). To visualize the fine vascular structures, the slice thickness must be small while the entire head or abdomen area must be covered in a short period of time to produce the best contrast. Such a level of image quality would be very difficult to obtain with a single-slice scanner. Although this particular study was performed on a 64-slice scanner, the clinical benefits of the multislice concept can be clearly demonstrated even on older vintage scanners. Figures 10.2(a) and (b) show a reformatted image and a VR image of runoff studies with a large coverage in z (from neck to foot). This study was performed on an 8-slice LightSpeed™ scanner (GE Healthcare, Waukesha, WI). Because of the high helical pitch and faster scan speed (0.5 sec), both studies were completed in less than 20 sec with a 2.5-mm aperture. At this scan speed and volume coverage, repeated scans of an organ at different contrast phases becomes an easy task. For example, many triple-phase liver studies (early arterial phase, late arterial phase, and venous phase) have been reported[8–10] since the introduction of multislice scanners.

There are many multislice scanners on the market, ranging from twin (2-slice) to 4-, 8-, 16-, 32-, 64-, 128-, and 320-slice scanners. During the

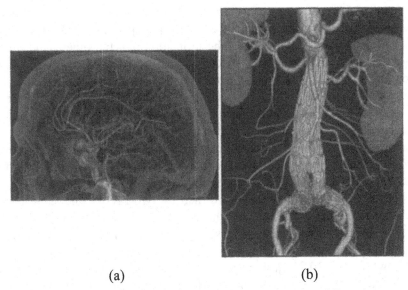

(a)   (b)

Figure 10.1 Volume-rendered images of CTA studies. (a) Neural CT angiogram and (b) abdominal CT angiogram.

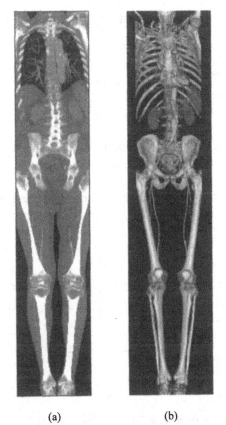

(a)   (b)

Figure 10.2 Images of a runoff study. (a) Multiplanar reformed and (b) volume rendered.

development of CT scanners from twin-slice to 64-slice, more slices generally lead to better volume coverage, increased iso-tropic spatial resolution, improved contrast efficiency, and better x-ray tube utilization. On the other hand, thinner detector cell sizes also potentially lead to the loss of the detector's geometric detection efficiency. As the detector coverage increases, many other factors, such as cone-beam and longitudinal truncation artifacts, can also complicate the performance of a multislice scanner. Consequently, one must examine these scanners on a case-by-case basis to judge their overall performance.

Some of the terminology that will be used throughout this chapter should be clarified. In the early days of multislice CT development, there was little doubt that a 4-slice scanner should be called multislice CT, not cone-beam CT. As the number of slices increased, the angle formed by two rays of the same channel location (the same distance to the iso-channel along the detector arc) of the first and last rows of the detector also increased, as illustrated in Fig. 10.3(a)–(c). This angle is typically called the cone angle. When the cone angle is large [e.g., 30 deg as shown in Fig. 10.3(c)], there is little dispute that it should be called cone-beam CT. But what about the scanners whose cone angles are in between? To avoid confusion, we use "multislice" and "cone-beam" interchangeably throughout this chapter.

10.2 Detector Configurations of Multislice CT

Figure 10.4 presents a schematic diagram of a multislice CT system. At a 3,000-ft level, the system configuration of a multislice CT system is nearly identical to that of a single-slice CT, with the exception of a pixilated detector in z (in a single-slice CT system, the detector is diced only in x along the arc). A photodiode cell is coupled to each pixilated detector cell so that signals can be read individually.

Two methodologies are used to obtain different slice thicknesses: prepatient collimation and detector collimation. As discussed in previous chapters, prepatient collimation is used primarily in single-slice CT scanners to narrow the x-ray beam exposure in z and achieve the desired slice thickness. For ease of comparison with detector collimation, we will briefly repeat the description here;

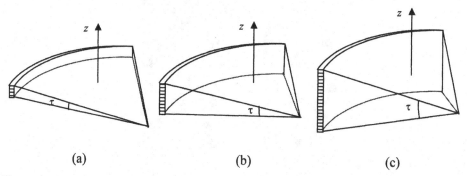

(a)   (b)   (c)

Figure 10.3 Transition from multislice CT to cone-beam CT. (a), (b), and (c) illustrate different z coverages and cone angles τ.

see also Fig. 10.5. The patient long axis (z axis) is in the horizontal direction. By mechanically adjusting the distance between the two collimator blades, the width of the x-ray beam on the detector (and therefore on the patient) can be changed. The figure depicts two collimator positions (one by dotted rectangles and the other by solid rectangles) and the corresponding x-ray beam widths (one by the dotted triangle and the other by the solid triangle). Because the source-to-detector distance is large (roughly 1 m) and the slice thickness is relatively small (roughly 10 mm or less), the shape of the x-ray beam is substantially more parallel than is illustrated by the figure.

In the detector-collimation approach, the slice thickness is defined by the detector cell size along the z axis, as shown in Fig. 10.6. For this example, the detector is subdivided into 16 detector cells in z (16 rows), and the width of each cell at the iso-center is 1.25 mm. Although the entire x-ray beam is significantly wider than 1.25 mm (16 × 1.25 mm = 20 mm, as defined by the lightly shaded triangle), the signal collected by a single detector cell (dark shaded triangle) corresponds to a slice thickness of only 1.25 mm. Therefore, the slice thickness becomes independent of the prepatient collimation.

There are three different types of detector designs for multislice scanners: matrix detector, adaptive array detector, and mixed matrix detector. Although these types of detector designs were introduced in the early stage of multislice CT development, newer detector designs of state-of-the-art scanners still fall into one of these design categories. All three configurations are presented in Fig. 10.7(a)–(c), using 4-slice scanners as examples. In a matrix detector

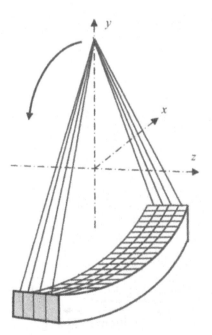

Figure 10.4 Geometrical relationship of a 4-slice CT scanner. At each projection angle, four sets of projections are collected simultaneously.

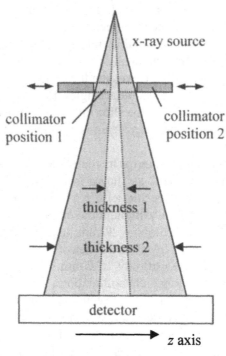

Figure 10.5 Illustration of prepatient collimation. Slice thickness is defined by the prepatient collimator.

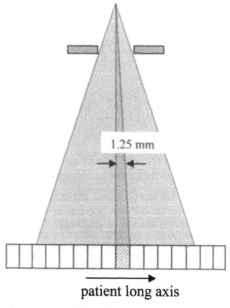

Figure 10.6 Illustration of detector collimation. Slice thickness is defined by the detector cell size.

configuration, the detector is diced uniformly along the z axis. For example, the detector of the LightSpeed™ 4-slice scanner (GE Medical Systems) is divided into 16 uniformly spaced cells, with a projected width of 1.25 mm at the iso-center. The projected width is the x-ray beam width at the iso-center if an ideal point source is used (see Chapter 8 for a detailed discussion). Because of the magnification factor, the physical size of the detector cell is significantly larger than 1.25 mm. Throughout this chapter, we will follow the industry-wide convention of using the projected detector width to describe the detector configuration.

Slice thickness other than the native detector cell size can be obtained by combining groups of cells to form a single cell. For example, in a 4 × 1.25-mm mode, only the center four cells are active and four individual signals are collected [top row of Fig. 10.7(a)]. Note that in this configuration, the prepatient collimator is configured so that the umbra of the beam (roughly 5 mm) is centered on the four cells to avoid an unnecessary dose to the patient. When operating in a 4 × 2.5-mm mode, two of the adjacent detector cells are combined to provide a single output, as shown by the second row in Fig. 10.7(a). Again, a prepatient collimator is used to limit the x-ray beam width to the center 10-mm region. Similarly, three neighboring cells are grouped to form a 4 × 3.75-mm mode [third row in Fig. 10.7(a)], and four neighboring cells are combined to form a 4 × 5-mm mode [fourth row in Fig. 10.7(a)]. Clearly, this type of scanner is "detector collimated." A detector aperture is defined by the projected detector cell width in z at the iso-center, not by the physical detector cell size. The actual detector cell size is larger than the detector aperture.

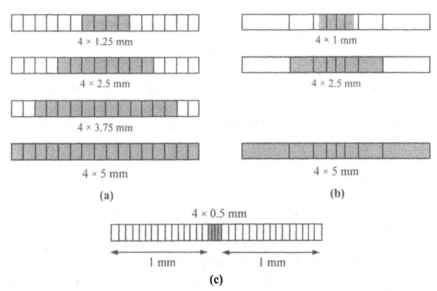

Figure 10.7 Different detector configurations. (a) Matrix detector with an identical row width of 1.25 mm. (b) Adaptive array detector with a variable row width (from left to right 5.0, 2.5, 1.0, 1.0, 0.5, 0.5, 1.0, 1.0, 2.5, and 5.0 mm). (c) Mixed matrix detector with two row widths. The center four rows are 0.5 mm and the outer thirty rows are 1 mm.

In the adaptive array detector, individual cells are divided into different sizes. For example, in the SOMATOM Volume Zoom scanner (Siemens Healthcare) and Mx8000 scanner (Phillips Healthcare), the detector cells are divided into four different sizes: 5.0, 2.5, 1.5, and 1.0 mm, as shown in Fig. 10.7(b). The slice thickness is determined by combining the prepatient collimation and detector collimation. For example, in a 4 × 1-mm mode [first row in Fig. 10.7(b)], the two center slices are defined by the detector cells (1-mm wide each). The definition of the outer two slices is slightly more complicated, because their inner edges are defined by the detector cell boundary, while the outer edges are defined by the prepatient collimator.

The mixed matrix detector design, used by Toshiba Medical Systems, is shown in Fig. 10.7(c). In this configuration, the center four rows are 0.5 mm wide, and fifteen 1-mm detector cells are placed on both sides. There are 34 detector elements in total. As in other detector configurations, only four channel signals are read out at a time. Like the matrix detector, the slice width is selected by grouping several cells together.

There are pros and cons for each of these detector configurations. The advantage of the adaptive array design is its superior dose utilization, since the intercell gaps are no longer present in the outer cells. However, the dose advantage is limited to thicker-slice modes, since the center cells (where thin-slice modes are operating) are still sliced and diced. On the other hand, this design lacks flexibility and future scalability. For example, it is difficult to obtain an intermediate thickness of 3.75 mm, and it is difficult to use the same detector for 8- or 16-slice configurations.

Although the number of detector rows has increased rapidly in recent years, the basic detector configuration is much the same as that of the original 4-slice scanners. For example, the 8-slice LightSpeed™ scanner uses the same detector modules with twice the number of DAS channels, as shown in Fig. 10.8(a). Two detector configurations are shown: 8 × 1.25 mm (left side) and 8 × 2.5 mm (right side). Figure 10.8(b) illustrates the similar concept of the 16-slice detector design. The center detector rows are further diced into a 0.625-mm detector width, while the outer rows are 1.25 mm. The tradition is carried forward with the 32-slice configuration in Fig. 10.8(c), in which the center 32 rows are 0.625 mm in width and the outer 16 rows are 1.25 mm. Two detector configurations each are shown for the 16- and 32-slice configurations. Figure 10.8(d) shows the LightSpeed VCT (64-slice) matrix detector, in which all detector cells are diced into a 0.625-mm width.

Because of the separation between slice thickness and x-ray beam width, the definition of helical pitch became somewhat confusing in the early days of multislice CT. If the conventional helical pitch definition is extended relative to the total x-ray beam width in z, the helical pitch should be calculated as the ratio of the distance the patient table travels in one gantry rotation over N (where N is the number of detector rows) times the single detector aperture. On the other hand, if the conventional helical pitch definition is extended relative to the slice thickness, the helical pitch is determined by the ratio of the distance the patient table travels in one gantry rotation over a single detector aperture. As an example, consider a 4-slice scanner with a 2.5-mm detector aperture. The ideal x-ray beam width for this mode is 4 × 2.5 mm = 10 mm. If the patient table travels at 7.5 mm per gantry rotation, then the helical pitch is 0.75 (7.5/10) based on the x-ray beam width definition. For the same configuration, the helical pitch is 3 (7.5/2.5) based on the detector aperture. Each definition has pros and cons. In recent years, the CT community has standardized the helical pitch definition based on the x-ray beam width instead of the detector aperture.

Prepatient collimation is used in two different ways. One way uses the collimator to achieve a slice thickness that is smaller than the elementary cell size.[11] For example, consider a matrix detector configuration in which the detector cell is divided into eight uniformly spaced cells of 1.25 mm each in z. In many clinical applications (e.g., cervical spine and IAC studies), a submillimeter slice thickness is desired to examine fine structures. To achieve submillimeter slice thickness with a 1.25-mm detector aperture, a prepatient collimator is positioned such that only a portion of the center two detector rows are illuminated by the x ray, as shown in Fig. 10.9(a). Figure 10.10 shows a clinical example of a "thin-twin" scan of an ankle. The image was reformatted from a helical scan to illustrate the spatial resolution in z (vertical direction).

The second way uses the prepatient collimator to block x-ray photons (in the x-y plane) from the region of the FOV that is not covered by the detector, as illustrated in Fig. 10.9(b). In this configuration, two sets of x-ray tubes and detectors are used to improve the temporal resolution. Because of the limited space on the gantry, the second detector has a smaller FOV than the first detector. To ensure that the patient is not exposed to any x-ray radiation that is not contributing to the formation of the final image, the prepatient collimator can be used (in the x-y plane) to match the exposed area to the size of the detector. A detailed discussion on this configuration is presented in Chapter 12.

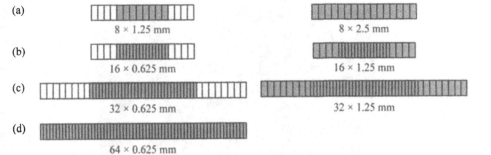

(a)

8 × 1.25 mm   8 × 2.5 mm

(b)

16 × 0.625 mm   16 × 1.25 mm

(c)

32 × 0.625 mm   32 × 1.25 mm

(d)

64 × 0.625 mm

Figure 10.8 Examples of detector configurations for 8, 16, 42, and 64 slices. (a) 8-slice configuration with 1.25-mm mode (left) and 2.5-mm mode (right). (b) 16-slice configuration with 0.625-mm mode (left) and 1.25-mm mode (right). (c) 32-slice configuration with 0.625-mm mode (left) and 1.25-mm mode (right). (d) 64-slice configuration with 0.625-mm mode.

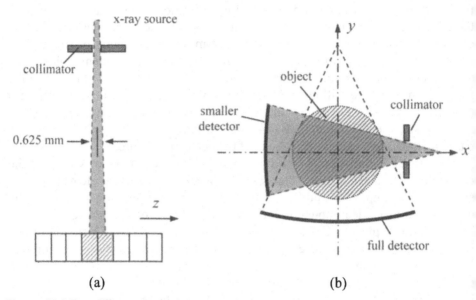

Figure 10.9 Two different uses of the prepatient collimator. (a) A collimator used to define slice thickness thinner than detector cell. (b) A collimator used to restrict x-ray exposure for smaller FOVs.

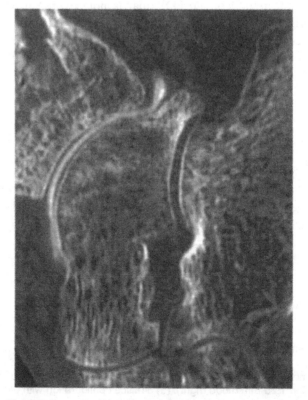

Figure 10.10 Reformatted image of an ankle scanned in thin-twin mode.

10.3 Nonhelical Mode of Reconstruction

Our discussion of projections in Chapter 3 was limited to the case of one dimension, where projections represent the line integrals of a single slice in an object; with the introduction of multislice CT, the detector z dimension also must be considered during the reconstruction. Like the equiangular and equal-spaced fan-beam configurations, the detector shape in a cone-beam scanner can be either curved or flat. As one may expect, compared to the fan-beam algorithms, cone-beam reconstruction is much more complex. Many research papers, conferences, and books have been dedicated to this subject.[12–35,44,45,48,57,81] Thus, given the rich nature of the research activities in this field, it is impossible to cover the topic thoroughly in a single section of a book.

One of the most popular reconstruction algorithms exhibits a striking resemblance to the fan-beam reconstruction formulas. This algorithm is the well-known Feldkamp-Davis-Kress (FDK) cone-beam reconstruction for the flat detector configuration.[12] Figure 10.11 shows a detailed description of a cone-beam sampling geometry. As in the equal-spaced fan-beam case, we can establish an imaginary detector at the iso-center to simplify the derivation. Note that the imaginary detector is simply the real detector reduced by the magnification factor (the source-to-detector distance over the source-to-iso-center distance). The iso-center is the rotation axis of the cone-beam system. The original derivation of the FDK algorithm explores the difference between the fan-beam and cone-beam reconstructions by using the relationship between the incremental view angle $\Delta\beta$ relative to the x-y plane and the incremental view angle $\Delta\beta'$ relative to the tilted fan-beam plane that contains the point of reconstruction. This section will present a different but more intuitive derivation of the same algorithm.

In Fig. 10.11, a rotated coordinate system is defined by (x', y', z') in which the x-ray detector is parallel to the x' axis. The point to be reconstructed (x', y', z') is mapped to a location (s, v) on the imaginary detector. The location is obtained

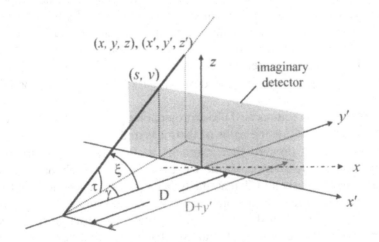

Figure 10.11 Geometric relationship of a cone-beam sampling geometry.

by calculating the intersection of the imaginary detector with a straight line that connects the x-ray source and the point (x', y', z'). In the flat detector configuration, the spacing between detector cells is uniform. The contribution of a detector pixel (s, v) to a reconstructed voxel in the image (x', y', z') is reduced with the increase in v, because the cross-sectional area of the detector cell to the point is proportional to $\cos(\tau)$, where τ is the tilting angle of the x-ray path with respect to the x'-y' plane, as shown in Fig. 10.11. Therefore, the scaling factor $\cos(\gamma) = D / \sqrt{D^2 + s^2}$ in Eq. (3.59) must be multiplied by $\cos(\tau)$ prior to the filtering operation. The combined scaling factor is simply the cosine of the angle ξ formed by the ray with the iso-ray (shown in the figure):

$$\cos(\gamma)\cos(\tau) = \frac{D}{\sqrt{D^2 + s^2}} \frac{\sqrt{D^2 + s^2}}{\sqrt{D^2 + s^2 + v^2}} = \frac{D}{\sqrt{D^2 + s^2 + v^2}} = \cos(\xi). \quad (10.1)$$

Now we will examine the weighting factor for backprojection. Section 3.7.3 stated that the backprojection weighting function is determined by the source-to-iso distance D and the projected distance between the source and the reconstructed point on the iso-ray. Because of the similar triangles formed in the tilted fan-beam and nontilted fan-beam geometries (Fig. 10.11), the ratio of the two distances remains the same for the cone-beam case. By substituting these relationships in Eq. (3.59), we obtain the FDK formula for a cone-beam reconstruction:

$$f(x,y,z) = \frac{1}{2} \int_0^{2\pi} \left(\frac{D}{D+y'}\right)^2 d\beta \int_{-\infty}^{\infty} \cos(\xi)q(s,v,\beta)h(s'-s)ds. \quad (10.2)$$

Like the fan-to-parallel-beam approaches discussed in Chapter 3, a similar row-wise fan-to-parallel rebinning approach can be used in the development of a cone-beam reconstruction. The rebinning process is carried out on a row-by-row basis in exactly the same manner as if it were a fan-beam dataset. After rebinning, each projection becomes a set of tilted parallel beams with different tilting angles for different detector rows, as illustrated in Figs. 10.12(a) and (b). Reconstruction algorithms can be derived based on the tilted parallel-beam geometry.

Equation (10.2) requires a 3D backprojection operation in which the paths of the backprojection follow the paths of the measured x-ray beams. Compared to a 2D backprojection (as in the case of a fan-beam reconstruction), 3D backprojection is computationally more expensive. In the early days of multislice CT development, nearly all CT manufacturers used 2D backprojection to perform reconstructions to meet the reconstruction speed demand. This approximation performs fairly well when the cone angle is small. Two images of an oval body phantom reconstructed with 2D and 3D backprojection are shown in Figs. 10.13(a) and (b), respectively. The phantom was formed with high-density rods placed at a steep angle with respect to the z axis, so image artifacts should appear when a large error is introduced in the backprojection process. Both images were reconstructed 2.2 mm off the x-ray plane (detector center plane), corresponding to a cone angle of 0.23 deg at the iso-center. At such a small cone angle, the images are visually indistinguishable from an image artifact point of view.

When the cone angle becomes large, the 2D approximation is no longer accurate enough to prevent artifact production. Figures 10.14(a) and (b) show reconstructed images located 9.1 mm off the x-ray plane (with a 0.96-deg cone angle at the iso-center) using 2D and 3D backprojection, respectively. Distortions of the high-density rods for the 2D backprojection are obvious. This clearly illustrates the need for a 3D backprojection when the cone angle becomes significant.

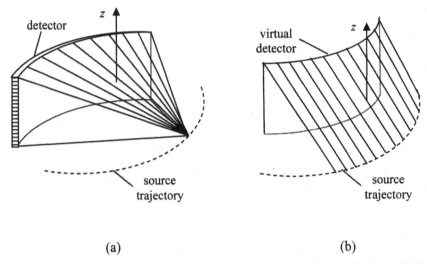

(a)   (b)

Figure 10.12 Illustration of (a) cone-beam geometry and (b) rebinned tilted parallel-beam geometry.

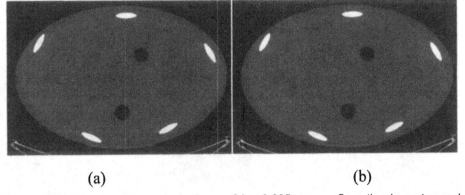

(a)   (b)

Figure 10.13 Phantom scan acquired on a 64 × 0.625-mm configuration in a step-and-shoot mode. Images were reconstructed 2.2 mm off the x-y plane (WW = 300). When the cone angle is small, the impact of 2D versus 3D backprojection is not obvious. (a) 2D backprojection and (b) 3D backprojection.

The FDK reconstruction algorithm is only an approximate formula for cone-beam reconstruction. It can be shown that a single circular trajectory does not provide sufficient sampling for an exact cone-beam reconstruction. Therefore, the algorithm performs reasonably well when the reconstruction plane is not far from the x-y plane. As the distance to the x-y plane increases, shading and streaking artifacts may be produced. The shape of the reconstructed object can also be distorted. Figure 10.15 shows an example of the distortion caused by an 8-deg cone beam. The phantom is made of a regular set of computer compact discs (CDs) stacked with robber washers to create air gaps between adjacent CDs [Fig. 10.15(a)]. Without artifacts, the coronal image of the reconstructed images should show parallel bars representing the cross-sections of the CDs. But because of the cone-beam artifacts, the CD cross-sections are distorted and cannot be resolved for images away from the center plane, as illustrated by Fig. 10.15(b).

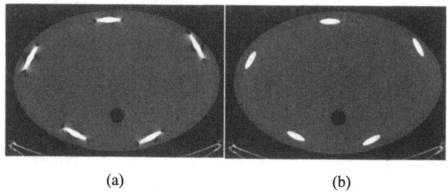

(a)   (b)

Figure 10.14 Impact of 3D backprojection (WW = 300). Images of a body phantom acquired on a 64 × 0.625-mm detector configuration in a step-and-shoot mode. Images are 9.1 mm off the x-y plane. (a) 2D backprojection and (b) 3D backprojection.

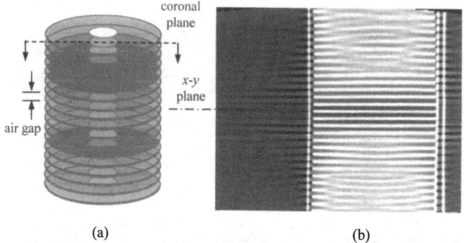

(a)   (b)

Figure 10.15 Illustration of artifacts for an 8-deg cone angle. (a) A set of CDs stacked with air gaps. (b) Coronal view of the reconstructed images.

Quite often, the cone-beam effect is identified as the culprit of all artifacts, since the cone angle increases as the reconstruction plane moves farther away from the x-y plane. However, the increase in cone angle is only part of the problem. For illustration, Figs. 10.16(a) and (b) show the reconstructed images of the same body phantom scan at the same z location relative to the x-y plane (9.1 mm off the x-y plane) using the FDK algorithm. Figure 10.16(a) was reconstructed with all 64 detector rows, while Fig. 10.16(b) was reconstructed with only the center 32 rows to simulate a 32-slice data acquisition. A 32 × 0.625-mm detector has a z coverage of 20 mm at the iso-center, and 9.1 mm is within the z coverage at iso. Fig. 10.16(b) exhibits a significantly higher level of artifacts compared to Fig. 10.16(a), while the cone angles for both images are identical. This is a clear indication that factors other than the cone angle need to be considered.

One such factor that contributes to the production of artifacts is the longitudinal truncation. To understand this effect, Fig. 10.17(a) provides a 3D view and a cross-section view of the exposed cone-beam volume. Because of the large detector size in z and the small dimension of the x-ray source (assumed to be a point), the z coverage of a single projection is location dependent. The coverage in z increases as we move away from the x-ray source and decreases as we approach the source. In a typical CT reconstruction, the ROI is a cylinder with its height equal to the z-coverage at the iso-center, as shown in Fig. 10.17. Because of the nonuniform z coverage, a portion of the cylinder that is closer to the detector is fully sampled, as shown by the shaded region in the figure. The portion that is closer to the x-ray source is not irradiated by the x-ray source and is not sampled. Thus, the reconstructed pixels in this region will not be accurate, due to the lack of projection samples. The cross-section of the subvolume that is fully sampled by all projection views is shown by the shaded region in Fig. 10.17(b). By spinning the shaded region about the z axis, we can obtain the

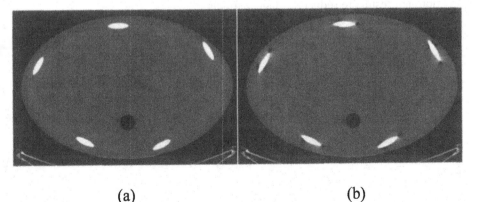

(a)   (b)

Figure 10.16 Impact of longitudinal truncation (WW = 300). Images of a body phantom acquired on a 64 × 0.625-mm detector configuration in a step-and-shoot mode. Images are 9.1 mm off the x-y plane. (a) All 64 rows of projection samples were used for reconstruction. (b) Only the center 32 rows of projection samples were used for reconstruction.

volume that is fully covered by all projections. In the example shown in Fig. 10.16(b), the reconstructed slice is near the top of the shade region for a 32 × 0.625-mm detector configuration, and a significant portion of the region is not fully sampled. For the 64 × 0.625-mm configuration in Fig. 10.16(a), the slice is well inside its fully sampled volume, so there is no impact from longitudinal truncation.

Several approaches have been investigated to overcome the FDK algorithm's shortcomings. In cone-beam sampling geometry, longitudinal truncation occurs at only a portion of the reconstructed image for a particular view;[19] that is, for the view shown in Fig. 10.17(a), the shaded portion of the cylinder is fully sampled. Therefore, as long as we avoid or minimize the contribution from the projections that do not properly sample the region of this shaded portion (e.g., the projections that are roughly 180 deg from this view), artifacts induced by the longitudinal truncation should be reduced. Since half-scans require only π plus the fan angle projections to reconstruct an image, we can use the half-scan concept to minimize the artifacts in a particular region, as illustrated in Fig. 10.18(a). In this figure, the x-ray tube position for the center view of a half-scan is at the 12 o'clock position. The portion of the image that is fully sampled in this view is at the bottom half of the ROI. Since more than half of a projection rotation is used in the reconstruction, the regions near the 3 o'clock and 9 o'clock positions contain contributions from projections that do not sample these regions well. Therefore, we expect a general degradation of image quality as we move away from the 6 o'clock position, as illustrated in the figure. The reconstructed image of the body phantom in Fig. 10.18(b) confirms this observation. The triangular region enclosed by the dotted lines shows a significantly reduced level of artifacts compared to the other regions. If we reconstruct multiple half-scan images with different center view angles, we can select from the good region of each image and form a composite image that contains a reduced level of artifacts. Figure 10.19(b) shows an example of a body phantom composed of four half-scan images. Comparing this image to the FDK-algorithm-reconstructed image in Fig. 10.19(a), it is clear that the multiple half-scan approach results in reduced image artifacts.

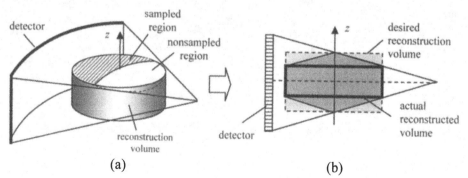

Figure 10.17 Illustration of the longitudinal truncation problem. (a) Schematic diagram of cone-beam sampling and reconstruction volume. (b) Cross-sectional view of the x-ray coverage (desired reconstruction volume shown by the dotted rectangle, and actual reconstructed volume shown by the solid rectangle).

This concept can be extended to optimize the contribution of each projection instead of the entire half-scan by enhancing the contribution of each view to image regions where it is fully sampled and suppressing its contribution to regions where it is not. This can be accomplished by weighting the projections prior to the backprojection step. This concept can be further extended to optimize the projection sample contribution based on its cone angle. If we denote two samples (from two projections) that intersect the same image pixel with a 180-deg difference in their fan angle γ (Fig. 10.20) as a conjugate sampling pair, we can examine each conjugate sampling pair in the cone-beam dataset to determine their relative contributions. For example, if one of the samples lands outside the active detector area—as in the case of longitudinal truncation—and the other is an actual sample, the one outside the detector area should receive a weight of 0 and the other a weight of 1. Similarly, the sample in the conjugate sample pair

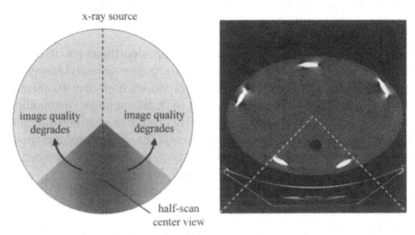

Figure 10.18 Illustration of image quality degradation as a function of the distance to the center-view angle for a half scan.

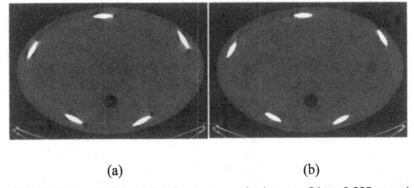

(a)   (b)

Figure 10.19 Images of a body phantom acquired on a 64 × 0.625-mm detector configuration in a step-and-shoot mode. Only the 32 center rows are used for reconstruction of images 9.1 mm off the x-y plane to simulate longitudinal truncation (WW = 300). (a) FDK algorithm. (b) Combination of four half-scan images.

with a smaller cone angle receives a higher weight than the sample with a larger cone angle, as shown in Fig. 10.20. Of course, the overall contribution from the conjugate pair needs to be normalized. In general, the reconstruction algorithm can be described by the following expression:[26]

$$f(x,y,z) = \frac{1}{2} \int_0^{2\pi} \left(\frac{D}{D+y'} \right)^2 d\beta \int_{-\infty}^{\infty} \cos(\xi) w(s,v,\beta) q(s,v,\beta) h(s'-s) ds. \quad (10.3)$$

In Ref. 26, a row-wise fan-to-parallel rebinning was first performed prior to the above reconstruction; however, the general idea is the same in both cases. Figures 10.21(a) and (b) show two images reconstructed with the original FDK algorithm and a modified FDK algorithm using the weighting function discussed above. When the reconstructed image is 17.2 mm off the x-y plane, longitudinal and cone-beam artifacts are obvious in the FDK reconstructed image [Fig. 10.21(a)]. With a properly designed weighting function, artifacts are significantly reduced [Fig. 10.21(b)].

Next, we will present a two-pass cone-beam algorithm that is based on a different observation: that the majority of the image artifacts caused by the incomplete cone-beam geometry occurs near or around high-density objects (e.g., bones).[15,18] These artifacts appear as shading or streaking in the surrounding low-density objects (e.g., soft tissues). High-density objects reconstructed by the FDK algorithm are accurate to a first-order approximation. Therefore, the cone-beam artifacts can be reproduced by synthesizing the cone-beam projection and reconstruction process of the reconstructed high-density objects.

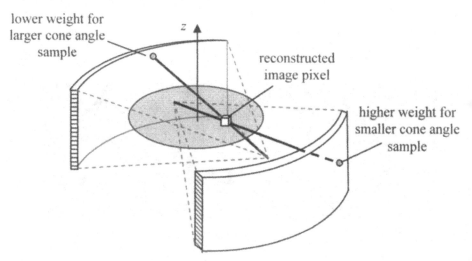

Figure 10.20 Illustration of two conjugate projection samples (180-deg fan-angle difference) with different cone angles. The sample with a smaller cone angle is assigned a larger weight than the sample with a larger cone angle.

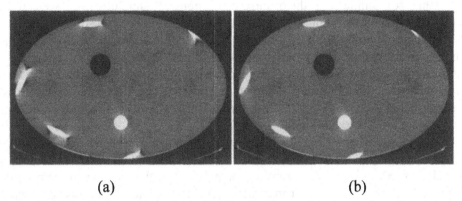

(a)   (b)

Figure 10.21 Body phantom acquired with 64 × 0.625-mm detector configuration. Images were reconstructed 17.2 mm off the x-y plane (WW = 250). (a) Original FDK algorithm and (b) modified FDK algorithm.

First, we segment the FDK (or modified FDK) reconstructed images into multiple classes. For most applications, we can simply segment the reconstructed volume into soft tissue and bones. Because there is a one-to-one correspondence between materials and their CT number ranges, object classification can be performed by simple thresholding. We denote the original reconstructed image by R_o, and the image that contains only the bones (or other high-density objects) by R_b. The next step is to produce a set of "bone" projections based on R_b (by performing forward projection). A new bone image R_n is then obtained by reconstructing the bone projections with the FDK (or modified FDK) algorithm. The difference between R_b and R_n is predominately the error E in FDK cone-beam reconstruction:

$$E = R_n - f(R_b), \qquad (10.4)$$

where f is a filtering function to model the PSF of the forward projection and reconstruction process. The final corrected image R_c is obtained by removing error image E from the original image R_o:

$$R_c = R_o - g(E), \qquad (10.5)$$

where g is an optional filtering operator for noise reduction. This approach requires one forward cone-beam projection and two cone-beam reconstructions. Computationally speaking, this approach is expensive. To reduce the computational complexity of the algorithm, the following set of simplifications can be used.

Because of the low-frequency nature of the error E, the number of channels and number of views can be reduced in the forward projection and reconstruction process. For example, if the original projection contains 400 detector channels and 400 detector rows, 200 × 200 detector elements can be used in the forward-projection process (each detector element is four times the original detector element size). Similarly, if the original projections contain 360 views, we can use only 200 views to generate the forward projection. This simplification reduces the amount of computation by a factor of 7.2. Additional simplifications, such as the use of a tilted parallel geometry instead of cone-beam geometry, can also be employed to further reduce the computational complexity. The result of the two-pass reconstruction algorithm is shown in Fig. 10.22(b). Compared to the FDK-reconstructed image in Fig. 10.22(a), the two-pass algorithm provides significant cone-beam artifact reduction.

In a clinical environment, the CT gantry is often tilted during the data acquisition so that the gantry plane is no longer perpendicular to the patient table. This is done to avoid direct x-ray exposure to sensitive organs such as the retina or to optimize the image quality by aligning the scan plane to the object of interest, such as vertebrae. If proper attention is not paid during the reconstruction process, a distorted object shape can result, as shown by the phantom scan in Fig. 10.23(b). In this study, the phantom is a uniform cylinder with square-shaped airholes. Because the CT gantry is tilted, the reconstructed images are parallel to the gantry plane, as shown by the x'-y'-z' coordinate system in Fig. 10.24, and the iso-centers of these images are along the z' axis. The phantom was placed along the z axis in the nontilted x-y-z coordinate system. The z' and z axes intersect at the origin and deviate as we move away from that point. If we perform an oblique reformation of the reconstructed volume so the reformatted images are parallel to the x-y plane, the images are shifted in y relative to each other since the iso-centers of the images are on the z' axis. This causes distortion in the reformatted object, as demonstrated by the elongated airhole shape in Fig. 10.23(b). Therefore, during the reconstruction process, the iso-center for each reconstructed image must be shifted by Δy:[7,15]

$$\Delta y' = z' \tan(\phi). \qquad (10.6)$$

Figure 10.23(a) shows the reformatted image with the compensation and shape distortion removed.

As mentioned at the beginning of this section, there is a rich body of research papers on the step-and-shoot mode reconstruction.[12–21, 26,30,31,33–35] To overcome

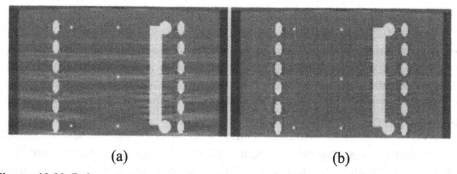

(a)   (b)

Figure 10.22 Reformatted images along the x-z plane of a simulated noiseless chest phantom reconstruction with a high-resolution kernel. (a) FDK reconstruction and (b) two-pass reconstruction (WW = 1000, WL = 0).

the incomplete sampling of the single circle trajectory, modifications to this trajectory that allow exact reconstruction have been proposed.[22–25,27–29,32,44,45,48,57,81] A few examples of these trajectories are shown in Figs. 10.25(a)–(c). It can be shown that these trajectories satisfy the famous Tuy condition so that an exact reconstruction of the object is feasible.[40] Given the limited scope of this chapter, we will not discuss in detail the reconstruction algorithms associated with these trajectories. Interested readers can refer to the references for further discussions on the subject.

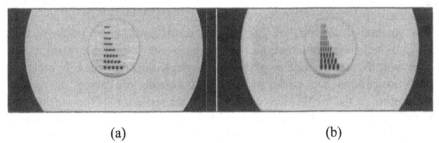

(a)   (b)

Figure 10.23 Impact of gantry tilt on multislice step-and-shoot acquisition. Phantom acquired with 16 × 0.626-mm configuration. (a) With row-dependent compensation and (b) without compensation.

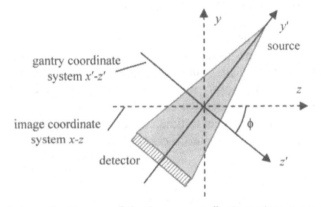

Figure 10.24 Schematic diagram of the image coordinate system x-y-z and the gantry coordinate system x'-y'-z'.

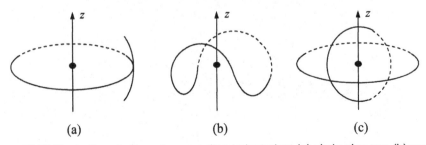

(a)   (b)   (c)

Figure 10.25 Illustration of alternative sampling trajectories: (a) circle plus arc, (b) saddle, and (c) double circle.

10.4 Multislice Helical Reconstruction

The multislice helical reconstruction approaches can be differentiated by the size of the cone angle. As discussed in Section 10.3, for a small cone angle, algorithms that use 2D backprojection are sufficient to suppress image artifacts and produce good image quality for clinical use. This type of algorithm (called a 2D algorithm) is attractive not only in terms of computational efficiency, but also from an image quality viewpoint due to the elimination of the projection interpolation step in z.[2,3,36,37] The first part of this section will focus on this type of algorithm; the latter part will address other reconstruction algorithms for large cone angles.

The general philosophy of 2D helical reconstruction is similar to that of single-slice helical reconstruction. In essence, we try to estimate a set of projections at the reconstruction plane. As in single-slice helical CT, most of the sampled projections are taken at locations outside the plane or region of reconstruction. In the single-slice case, we use interpolation of the measured data from different views to estimate a set of projections at the plane of reconstruction. All of the measured samples are taken from a single detector row. For the multislice case, however, additional samples are available from different detector rows, as illustrated by Fig. 10.26 for a four-slice helical case. As in the single-slice case, we are faced with the selection of the reconstruction regions, interpolating samples, and interpolation algorithms.

To describe the basic approach, we will derive 2D reconstruction algorithms for a 4-slice scanner. We assume that the cone-angle effect of the system can be ignored; that is, at any time instant, we collect four sets of fan-beam projections that are parallel to the x-y plane. Since the cone angles of the 4-slice CT scanner are less than 1 deg, this approximation is quite reasonable. Since data are collected in helical mode, the location of the detector rows (relative to the patient in z) changes linearly with respect to time, and therefore, projection angle β. This is depicted in Fig. 10.27. The reconstruction plane is shown by the vertical line

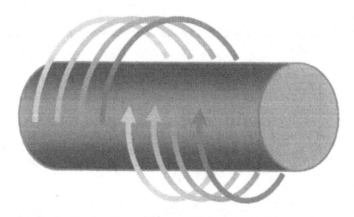

Figure 10.26 Multislice helical scans with a four-row detector.

located at z_0. For a four-row detector, we divide the measured projections into four categories or regions. In region 1, the reconstruction plane is located between detector rows A and B. In region 2, the plane is located between detector rows B and C. region 3 includes projections where the plane is between detector rows C and D. Region 4 covers all projections that are outside the reconstruction plane. Projections in region 4 are ignored for the reconstruction of this particular image. Linear interpolation takes place in regions 1 to 3. The formulation of linear interpolation is

$$p'(\gamma,\beta) = \left(\frac{d-x}{d}\right)p(\gamma,\beta,n) + \left(\frac{x}{d}\right)p(\gamma,\beta,n'), \qquad (10.7)$$

where $p(\gamma, \beta, n)$ represents projections collected on detector row n, x is the distance between the reconstruction plane and detector row n, d is the projected spacing at the iso-center between adjacent detector rows n and n', and $p'(\gamma, \beta)$ is the interpolated projection. For all practical purposes, d equals the detector aperture. In the above equation, the interpolation coefficients are γ independent, because the projections collected on two detector rows have identical projection angles. After linear interpolation, we obtain a set of fan-beam projections located at z_0. Depending on the selected helical pitch, the interpolated projections cover different angular ranges. For N-slice detectors, a pitch of $(N-1)/N$ produces a set

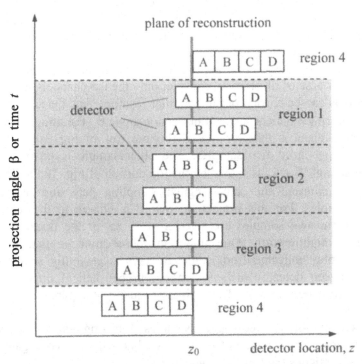

Figure 10.27 Illustration of different projection regions for a 4-slice scanner.

of projections over the range of 2π. A pitch larger than $(N-1)/N$ produces less than 2π interpolated projections, and a pitch less than $(N-1)/N$ produces more than 2π interpolated projections. Consequently, an additional weighting function is needed to handle the redundant samples and avoid shading artifacts. For example, for the interpolated dataset of more than 2π, overscan weights can be used. On the other hand, generalized half-scans can be applied to the interpolated dataset of less than 2π.

The advantages of this algorithm are its simplicity and flexibility. Since linear interpolation occurs on a row-by-row basis, the interpolation coefficients need to be calculated only once for each view. In addition, the above algorithm is applicable to projections collected at arbitrary helical pitches. The drawback is its suboptimal performance on image quality in terms of SSP and artifact suppression. More sophisticated reconstruction techniques can compensate for these drawbacks. The remainder of this chapter will discuss some representative approaches.

10.4.1 Selection of interpolation samples

In the previous example, linear interpolation between adjacent detector rows is used for projection estimation. This type of process is often called inter-row interpolation. Although in the example we used two neighboring detector rows, it should be noted that inter-row interpolation can be extended to samples beyond the nearest two neighboring rows. For example, we can use the nearest three rows for higher-order interpolations. The smallest distance between the interpolating samples, however, always equals the detector aperture. As discussed in Chapter 9, the distance between measured samples often limits the SSP of the reconstructed image.

An alternative interpolation approach uses conjugate samples, as shown in Fig. 10.28. For ease of illustration, consider only the iso-channels (the analysis of other channels can be carried out in a similar fashion). The detector moves from left to right during a helical scan. At projection angle β_1, the sample closest to the reconstruction plane is detector row C (bottom row of Fig. 10.28). For linear interpolation, we need two samples. The other sample is obtained from the projection at angle $\beta + \pi$, as shown by the top portion of Fig. 10.28. Samples that are π apart represent the same angular sampling path and therefore form conjugate samples. For this projection, row B is closest to the reconstruction plane. Once the two samples are selected, the rest of the linear interpolation process is straightforward. One important observation is that the distance between samples collected from conjugate rays is generally smaller than the distance between detector rows (when the proper helical pitch is selected). Therefore, we expect improved interpolation quality compared to the inter-row interpolation.

The search for the conjugate samples for single-slice helical CT is straightforward. As discussed in Chapter 9, we know that the conjugate sample to (γ, β), (γ', β') satisfies the following equation:

$$\begin{cases} \gamma' = -\gamma \\ \beta' = \beta \pm \pi - 2\gamma. \end{cases} \quad (10.8)$$

For multislice helical CT, the process is somewhat more complicated. Although the relationship between the detector angle γ and the projection angle β is unchanged, multiple samples can simultaneously satisfy this condition. For an N-slice scanner, each (γ, β) contains N samples. For interpolation, we must identify the samples that are closest to the plane or region of reconstruction. One way to simplify the search process is to first focus only on the iso-channels of the N-slice scanner. Once the conjugate relationship is identified for the iso-channels, other channels can be extended to connected regions, since the relationship of conjugate samples does not change abruptly from channel to channel.

For iso-channels, we use a 2D graph with the horizontal axis representing the detector cell location (relative to the patient in z) and the vertical axis representing the projection view angle β, as shown in Fig. 10.29. In this illustration, the configuration is a 4-slice scanner that operates at a 0.75 pitch (the table advances three detector widths in one gantry rotation). Because the patient table travels at a constant speed, each iso-channel maps to a straight line in the z-β graph. The horizontal spacing between the lines equals the detector aperture. The slope of these lines is

$$\frac{2\pi}{\text{pitch} \times N \times d}, \quad (10.9)$$

where d is the detector aperture. Different detector rows intersect the plane of reconstruction at different projection angles (these angles are labeled β_A, β_B, β_C, and β_D, respectively). The conjugate samples are the samples that differ by either π or 2π in β and are highlighted with similar line patterns. For example, samples

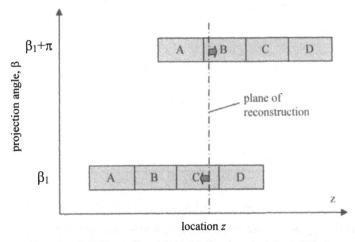

Figure 10.28 Illustration of conjugate-sample interpolation.

from rows A and D located to the left of the plane of reconstruction are painted with thick black lines, since these samples differ by a 2π projection angle. A set of samples from row C to the right of the plane of reconstruction is also painted with the same pattern, since these samples differ from the previous sets by π. To derive the reconstruction algorithm described in Ref. 3, we must identify the conjugate regions for all detector channels. Based on the iso-channel relationship described in Fig. 10.29, the other channels follow easily, as shown in Fig. 10.30. In this figure, the horizontal axis depicts the detector angle γ for detector rows A to D, and the vertical axis represents the projection angle β. We start by copying the four angles (from Fig. 10.29) at which the detector intersects the plane of reconstruction, β_A, β_B, β_C, and β_D. Since the plane of reconstruction is defined as a flat plane perpendicular to the z axis, all channels of the same row should have the same angle of intersection, so they form horizontal lines in the figure. Next, we locate the conjugate lines that correspond to these intersection lines. From Eq. (10.8), the conjugate line corresponding to the intersection line of detector row n (n = A, B, C, or D) is

$$\beta_{n\pm} = \beta_n \pm \pi - 2\gamma. \tag{10.10}$$

Each pair of intersection line and conjugate line defines a unique region in this graph. For example, the intersection line for detector row A (β_A) and the conjugate line for detector row C (β_{C+}) form a shaded trapezoidal region, as shown in Fig. 10.30. The conjugate regions are identified based on the conjugate samples of the iso-channel depicted in Fig. 10.29 (all regions that contain conjugate iso-channel samples form conjugate regions), and are shaded with the

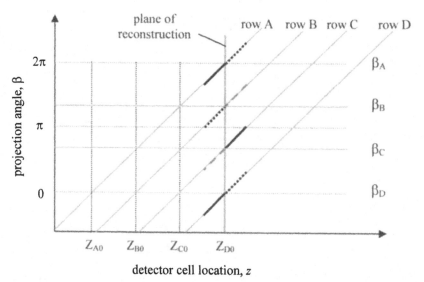

Figure 10.29 Conjugate samples of a 4-slice helical scan at a pitch of 0.75. Z_{A0}, Z_{B0}, Z_{C0}, and Z_{D0} depict the iso-channel locations of detector rows A, B, C, and D, respectively, at $\beta = 0$.

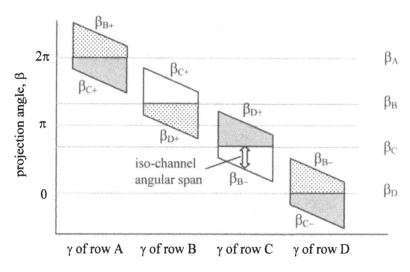

Figure 10.30 Illustration of conjugate regions for four-slice (pitch 0.75) helical scans.

same pattern in Fig. 10.30. As in the case of single-slice helical reconstruction, an interpolation between conjugate samples can be performed either by formulating a new set of projections or by weighting the existing projections. Because of the linearity of the FBP process, the two approaches are equivalent.

If we choose the projection weighting approach with linear interpolation, we assign unity weights to the projections at the lines of intersection (β_A, β_B, β_C, and β_D for detector rows A, B, C, and D, respectively) and zero at the conjugate lines defined by Eq. (10.10). Since rows A and D essentially sample identical ray paths (if we ignore the cone-angle effect), the weighting function can be shown to be

$$\beta_A(\gamma,\beta) = \begin{cases} \dfrac{\beta-\beta_{C+}}{2(\beta_A-\beta_{C+})}, & \beta_{C+} \leq \beta < \beta_A, \\[2ex] \dfrac{\beta_{B+}-\beta}{2(\beta_{B+}-\beta_A)}, & \beta_A \leq \beta < \beta_{B+}, \\[2ex] 0, & \text{otherwise.} \end{cases} \qquad (10.11)$$

$$\beta_B(\gamma,\beta) = \begin{cases} \dfrac{\beta-\beta_{D+}}{\beta_B-\beta_{D+}}, & \beta_{D+} \leq \beta < \beta_B, \\[2ex] \dfrac{\beta_{C+}-\beta}{\beta_{C+}-\beta_B}, & \beta_B \leq \beta < \beta_{C+}, \\[2ex] 0, & \text{otherwise.} \end{cases} \qquad (10.12)$$

$$\beta_C(\gamma,\beta) = \begin{cases} \dfrac{\beta-\beta_{B-}}{\beta_C-\beta_{B-}}, & \beta_{B-} \leq \beta < \beta_C, \\[2ex] \dfrac{\beta_{D+}-\beta}{\beta_{D+}-\beta_C}, & \beta_C \leq \beta < \beta_{D+}, \\[2ex] 0, & \text{otherwise.} \end{cases} \tag{10.13}$$

$$\beta_D(\gamma,\beta) = \begin{cases} \dfrac{\beta-\beta_{C-}}{2(\beta_D-\beta_{C-})}, & \beta_{C-} \leq \beta < \beta_D, \\[2ex] \dfrac{\beta_{B-}-\beta}{2(\beta_{B-}-\beta_D)}, & \beta_D \leq \beta < \beta_{B-}, \\[2ex] 0, & \text{otherwise.} \end{cases} \tag{10.14}$$

10.4.2 Selection of region of reconstruction

So far, our discussion has been limited to a reconstruction plane that is flat and perpendicular to the z axis [Fig. 10.31(a)]. As with the case of a single-slice helical scan, the selection of the plane or region of reconstruction for multislice CT is not unique. Chapter 9 discussed the pros and cons associated with the selection of a perpendicular flat plane. The major advantage of this selection is convenience; because images have been presented in this fashion since the early days of CT, the selection of this plane requires no special explanation or description. The drawback of this selection is the lack of flexibility, because optimizations on weighting functions or interpolations cannot be easily implemented.

An alternative approach is to define a generalized region of reconstruction.[36,37] This region can be a warped plane or a volume, and does not have to be perpendicular to the z axis. The helical weights derived from the more generalized region of reconstruction can possess the property of continuity and differentiability, and in many cases, make the weighting function derivation process much simpler. For example, Eqs. (10.11) through (10.14) require that the iso-channel angular span (shown in Fig. 10.30) is larger than $2\gamma_m$, where γ_m is the largest detector angle. This is mainly to avoid the intersection between the two boundary lines. The weighting needs to be unity at the plane of reconstruction line and needs to be zero at the conjugate line. When these lines intersect, an indeterministic solution results. This condition cannot be satisfied for 8- or 16-slice scanners with high helical pitches, a limitation that led to the development of a generalized region of reconstruction. One such example is shown by Fig. 10.31(b) as a double cone in image space. It is defined by a set of straight lines in the sinogram space (left side of Fig. 10.32). Since these lines are not parallel to the γ axis, each line forms a double-cone region in image space. At first glance, one may be concerned that the double cone diverges from the iso-center, because such an arrangement could lead to degraded slice profiles away from the iso-center. In reality, however, the distance between the double-cone surfaces is only a small fraction of the slice thickness (due to proper selection of the slope in the sinogram space), so its impact on the slice profile is minimal. As a matter of fact, its impact on the slice profile is less than the inaccuracy resulting from the approximation of the 2D backprojection. Although this approach was developed in order to solve the issues with 8- and 16-slice reconstructions, it offers an improved image quality even for the 4-slice case, as shown in Fig. 10.32. Note the reduction in shading artifacts near the bony structure identified with an arrow in Fig. 10.32(a).

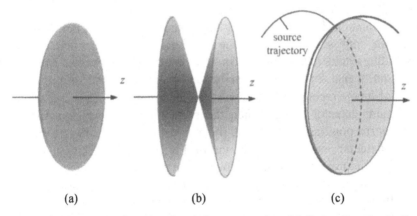

(a)   (b)   (c)

Figure 10.31 Selection of different regions of reconstruction. (a) Conventional selection of a flat plane perpendicular to the z-axis. (b) Double-cone-shaped region. (c) Tilted flat plane.

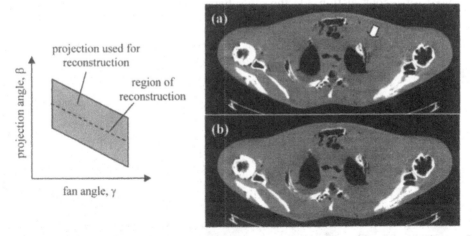

Figure 10.32 Left: illustration of the double-cone-shaped region of reconstruction and projection samples used for reconstruction. Right: reconstructed images of a shoulder phantom scanned on a 4-slice scanner at a pitch of 0.75. (a) Conventional plane-of-reconstruction shown in Fig. 10.31(a). (b) Double-cone-shaped region of reconstruction shown in Fig. 10.31(b).

Another selection of the reconstruction plane is based on minimizing the cone-beam error.[4,5,38,39] When cone-beam geometry is approximated by a stack of fan-beam projections (as in previous examples), an error is inherently present in the reconstruction process. For the configuration depicted in Fig. 10.33(a), the source trajectory intersects the plane of reconstruction at exactly one location in a dataset (the cone angle in the figure is exaggerated). The top (lightly shaded) portion of the image needs to be interpolated between projection samples of detector rows C and D. The bottom (darker shading) portion of the image must be obtained with projections from rows B and C. For a 2D backprojection approach, only one interpolated projection per view is used for image reconstruction. Therefore, an interpolation error is unavoidable. Even for an image that can be interpolated from the same set of detector rows, the distance between the sampling rays and reconstruction plane changes constantly. Therefore, location-dependent interpolation should be performed. For the 2D interpolation approach, we ignore these effects and assume that all rays are parallel to the reconstruction plane. Another source of error comes from the high helical pitch scans. For a fan-beam helical acquisition, the minimum dataset required for image reconstruction spans a projection angle of $\pi + 2\gamma_m$, where γ_m is the maximum detector angle. When the patient table moves too fast, sample locations corresponding to some of the reconstructed regions are outside the measured detector samples, as shown in Fig. 10.33(b). This is particularly problematic for projection samples located near both ends of the dataset, and often forces us to perform extrapolation (instead of interpolation) for these samples. In general, extrapolation is less stable than interpolation.

To overcome these shortcomings, a proposal was made to fit a reconstruction plane to the source helix,[38] as shown in Fig. 10.31(c). This approach minimizes the amount of interpolation location dependency and the amount of extrapolation required for reconstruction. On the other hand, since every reconstruction plane

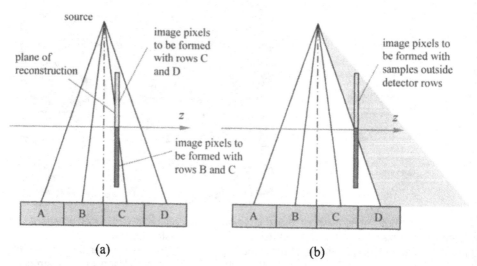

(a)   (b)

Figure 10.33 Illustration of errors introduced in the 2D filtered backprojection reconstruction.

is fitted to the source helix, the orientation of the reconstructed slices changes constantly from one image to another, and the reconstructed images are no longer parallel to each other. As a result, an additional interpolation step is needed to reinterpolate the slices to a set of parallel images.

10.4.3 Reconstruction algorithms with 3D backprojection

Section 10.4.2 discussed the class of algorithms that use 2D backprojection. As illustrated in Fig. 10.33, one major source of error is the approximate nature of the 2D backprojection. When the cone angle increases with the number of detector rows, the error also increases. Although an attempt has been made to minimize the impact of the cone-beam effect, such as selecting a tilted plane, as shown in Fig. 10.31(c), this approach is still an approximation. When the helical pitch increases, the accuracy of the fitting is reduced, since half of the helix forms a curved rather than flat plane. The best approach to reduce the error is to use 3D or cone-beam backprojection.

In recent years a large amount of research has been devoted to the subject of cone-beam helical reconstruction.[43,46,49–50,54–56,58–64,67–80,82–88] We can classify the helical cone-beam reconstruction algorithms into two major categories: exact reconstruction methods and approximate 3D FBP methods. The exact reconstruction methods include algorithms based on formulas that link cone-beam projections to the 3D Radon transform.[40–42,46–48] These algorithms try to derive an analytical expression for the exact solution to the reconstruction problem. One major limitation of these algorithms is their inability to handle long objects (objects that extend beyond the region of scanning). This is problematic for most medical applications, since the size of the patient is typically beyond the volume covered by the limited extent of the helical trajectory. This problem led to the investigation of various compensation schemes.[45,50,53,54,57,58,61,62] These schemes often require additional source trajectories, such as two circular scans at both ends of the helix, to enclose the dataset. To overcome the shortcomings of additional scans, quasi-exact and exact algorithms were later proposed for simple helical trajectories.[68] One of the most exciting developments in the past few years is the Katsevich algorithm.[69] This algorithm falls into the category of FBP with a shift-invariant 1D filter. Since this algorithm was introduced, many new algorithms have been proposed and developed to allow more general trajectories, more efficient use of the projection samples, and ROI reconstruction.[70–79,84,88]

The second category of algorithms is the approximate 3D FBP approach derived from the FDK algorithm.[33,44,51,80,82,83] Each projection is filtered on a row-by-row basis and is backprojected three-dimensionally into the object space. Although these algorithms are approximate in nature, they do offer some advantages that cannot be provided by the exact reconstruction algorithms. For example, the acquisition of a complete dataset is not always possible because of various clinical constraints. As will be discussed in Chapter 12, a cardiac scan often collects a half-scan helical dataset and requires the generation of an image volume with its z dimension substantially equal to the z coverage of the detector at the iso-center. Because of the flexibility and computational efficiencies of the approximate reconstruction algorithms, these algorithms are still the dominant force behind most commercial CT reconstruction engines.

Since it would be difficult to provide detailed descriptions of all cone beam-related algorithms, in this chapter we will look at one specific approximate reconstruction algorithm that leverages the discoveries made by the exact reconstruction algorithm (the Katsevich algorithm).

The Katsevich algorithm involves two major steps: filtering and backprojection. The filtering step includes taking a derivative with respect to the projection angle β, weighting it by the cosine of the cone angle, and filtering the data along the κ curves. If we normalize the detector width at the iso-center to 1, the κ curve expressed as the distance to the detector center h versus the detector fan angle γ can be expressed by the following formula:

$$h(\gamma, \psi) = \frac{p}{2\pi}\left(\psi \cos \gamma + \frac{\psi}{\tan \psi}\sin \gamma\right),\tag{10.15}$$

where p is the helical pitch, $-\gamma_m - \pi/2 \leq \psi \leq \gamma_m + \pi/2$ is a parameter that specifies a family of κ-curves, and γ_m is the maximum fan angle along the detector channel direction. Figures 10.34(a) and (b) show the κ curves for $p = 33/64$ and $p = 87/64$, respectively, with the vertical axes for both graphs covering the full z coverage of the detector [−0.5, 0.5]. The slope of the filtering lines for $p = 87/64$ is significantly higher than it is for $p = 33/64$. A close inspection of Eq. (10.15) shows that the slope at the detector iso-channel scales to the helical pitch p. Given the exact nature of the Katsevich reconstruction algorithm, it is logical to expect that part of the image quality improvement derived from the algorithm is the nontraditional filtering approach, in which the filtering operation is no longer carried out along the detector rows.

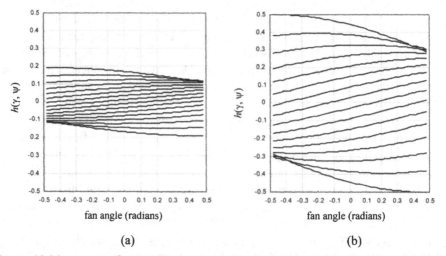

(a)   (b)

Figure 10.34 κ curves for the filtering operation in the Katsevich algorithm. (a) Helical pitch of 33/64 and (b) helical pitch of 87/64.

Based on the above discussion, we can resample the projection dataset along the κ curves prior to the filtering operation. On the other hand, since the orientation of these κ curves is roughly aligned with the tangential direction of the source helix, we can take advantage of the row-wise fan-to-parallel rebinning process. After the rebinning, the original cone-beam projection is transformed into a set of tilted parallel beams that follow the source helix, as shown by Figs. 10.35(a) and (b). Each "virtual row" in the rebinned projection falls along the tangential direction of the helix. When the filtering operation is carried out along the rows of the rebinned projection, a natural "tangential filtering" is accomplished. Figures 10.36(a) and (b) show the reconstructed Defrise phantom images without and with the tangential filtering. Both images were reconstructed with a modified FDK algorithm with projection-based weighting functions. Figure 10.36(a) was performed in the native cone-beam geometry and Fig. 10.36(b) was performed with a row-wise fan-to-parallel rebinning prior to the weighted filtered backprojection operation. The Defrise phantom is a set of thin discs stacked along the z axis with air gaps. The phantom was designed to enhance cone-beam artifacts by using objects of high-frequency variations in z. It is clear from the figure that by using tangential filtering, shading artifacts are eliminated. It should be stated that row-wise fan-to-parallel beam rebinning is only one of several approaches to implement the "tangential filtering" concept. Different realizations of such a concept were proposed.[36,85,86]

Similar to the discussion in Section 10.4.2 where a 3D weighting function was used on the step-and-shoot cone-beam dataset to suppress image artifacts, an optimized 3D weighting function can be used in the helical cone-beam case to improve image quality. The design of the weighting function is similar to that of the step-and-shoot case in which projection samples with smaller cone angles are

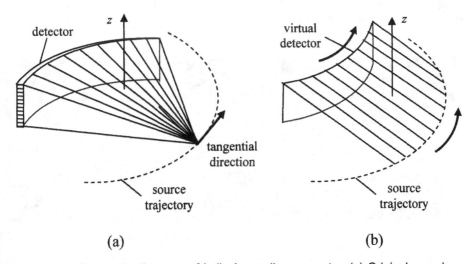

(a)   (b)

Figure 10.35 Schematic diagrams of helical sampling geometry. (a) Original cone-beam geometry with source trajectory. (b) Rebinned tilted parallel-beam geometry with a virtual detector row parallel to the source helix.

given higher weights than are given to the conjugate samples with larger cone angles. During the weighting function design process, a proper balance between image artifacts and noise can provide the best overall image quality. As an illustration, Figs. 10.37(a) and (b) show reconstructed images of a simulated body phantom scan with a 64 × 0.625-mm detector configuration and 63/64 helical pitch. Figure 10.37(a) was reconstructed with the approximate cone-beam algorithm discussed above, and Fig. 10.37(b) was reconstructed with the Katsevich algorithm. Comparable image quality was achieved for this study.

So far, we have focused our discussions on the helical cone-beam acquisition in which the helical pitch is constant throughout the scan. However, in many clinical settings, a nonconstant or variable-pitch helical scan is needed. For example, it is desirable to examine dynamically the contrast uptake in a patient's

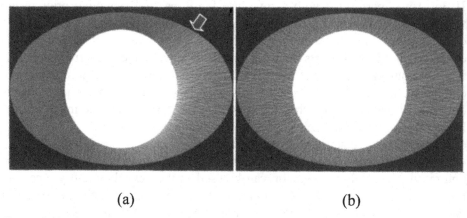

(a)   (b)

Figure 10.36 Illustration of the impact of tangential filtering with the Defrise phantom using simulated projection of a 64 × 0.625-mm detector configuration at 63/64 helical pitch. (a) Without tangential filtering and (b) with tangential filtering.

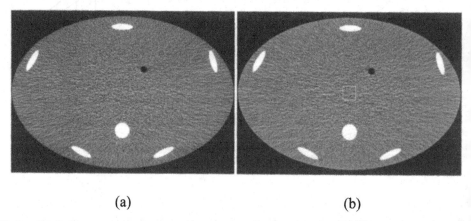

(a)   (b)

Figure 10.37 Reconstruction images of a simulated body phantom with 64 × 0.625-mm detector configuration and 63/64 helical pitch. (a) Modified FDK algorithm with tangential filtering and 3D weighting. (b) Katsevich reconstruction algorithm.

pulmonary vascular structure to detect abnormalities or pathologies. The typical dimension of a lung is 30 cm along z; no scanner is available today that is able to cover the entire organ in a single rotation. To enable such studies, the patient table must be shuttled back and forth during the data acquisition so the full coverage of the study is larger than the z coverage of the detector, as shown in Fig. 10.38. During the data acquisition, the helical pitch changes constantly. Near both ends of the scanning range, a helical pitch of zero is needed for a brief period of time so that the table can reverse its traveling direction. Consequently, the scan mode changes between a helical mode and a step-and-shoot mode (hence, the term "shuttle mode") during a single data acquisition.

The reconstruction algorithm for such a data acquisition is similar to that of a constant-pitch helical scan. The major difference is that the weighting function used in the reconstruction needs to change constantly from image to image to account for the different sampling patterns. Also, the optimization of the weighting function plays an even more important role in the case of a variable-pitch helical scan. For illustration, Figs. 10.39(a) and (b) depict two images of a liver phantom scan reconstructed with the same approximate cone-beam helical reconstruction algorithm. The only difference is the use of different weighting functions. Note the severe shading artifacts in the liver when the weighting function is not optimized [Fig. 10.39(a)].

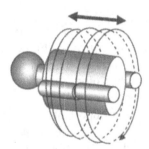

Figure 10.38 Illustration of a shuttle mode helical acquisition.

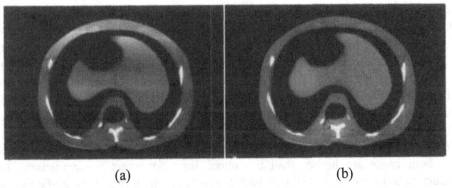

(a)   (b)

Figure 10.39 Impact of optimization of the weighting function for variable-pitch helical reconstructed images of a liver phantom. (a) Nonoptimized weighting function with shading artifacts. (b) Optimized weighting function.

10.5 Multislice Artifacts

10.5.1 General description

Many studies have been conducted in recent years to understand the artifacts produced by multislice CT.[6,89–100] For multislice helical CT with the FBP approach, there are two root causes of image artifacts. The first is related to the constant table translation of the helical acquisition. Because of the increased scanning volume, the table translation speed for multislice CT is typically higher than it is for single-slice CT. As a result, we expect more pronounced image artifacts related to the z interpolation. The second root cause of artifacts comes from the cone-beam effect. This effect generally increases with the increase in the detector coverage in z.

Let us focus our discussion on the second root cause of artifacts: the cone-beam effect. The sources of artifacts in helical cone-beam (3D) reconstruction can be categorized into three general areas: compromised selection of the data acquisition, suboptimal reconstruction algorithms, and limitations on design or quality control. Despite recent advances in the helical reconstruction algorithm, the data acquisition has certain limitations or constraints that are often ignored or forced to be ignored during the actual clinical CT design. For example, it is known that the helical pitch for the Katsevich algorithm must be less than 1.4 for most CT geometries to ensure accurate image reconstruction over the entire FOV. However, clinicians are often forced to increase the maximum helical pitch to cover the entire volume during the desired contrast uptake phase. For 4- or 8-slice scanners, helical pitches significantly higher than 1.6 are often selected because of the small detector coverage. Thus, tradeoffs must be made between image artifacts and suboptimal physiology.

Often the selection of the reconstruction algorithm and the selection of the data acquisition are closely linked. It is important to optimize the overall performance of the system rather than optimizing one particular parameter. Compared to the single-slice CT scanners, multislice CT offers much more flexibility in selecting acquisition and reconstruction protocols. For example, if a radiologist needed to acquire a set of CT images at a 2.5-mm slice thickness on a LightSpeed™ 16-slice scanner, the radiologist could acquire the projection data with either 2.5-, 1.25-, or 0.625-mm detector apertures. As a general rule of thumb, it is always more desirable from an image artifact point of view to reconstruct thick images from a thin-slice data acquisition than directly from a thick-slice data acquisition, as long as the coverage permits it (this will be explained in more detail later in this chapter). This is illustrated by a phantom example shown in Fig. 10.40. Although both images are 2.5 mm in slice thickness, Fig. 10.40(b), reconstructed from a 0.625-mm detector aperture, has a much reduced artifact level compared to Fig. 10.40(a), which was acquired with a 2.5-mm detector aperture. The helical pitches for both cases are higher than the upper limit for exact reconstruction. The reduction in image artifact of Fig. 10.40(b) is due to the fact that the artifact shown in Fig. 10.40(a) is caused mainly by the lack of adequate sampling in the z

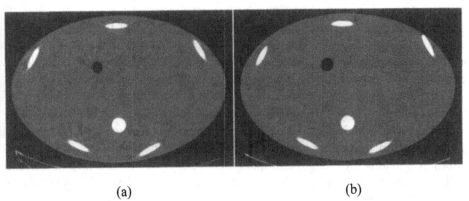

(a)   (b)

Figure 10.40 Illustration of thick-slice acquisition verses thick-slice reconstruction. (a) Reconstructed image of 2.5-mm slice thickness acquired with an 8 × 2.5-mm detector aperture at 1.675 helical pitch. (b) Reconstructed image of 2.5-mm slice thickness acquired with 16 × 0.625 mm at 28/16 helical pitch.

direction (z aliasing). Thin-slice acquisition (0.625 mm) with a thicker slice reconstruction (2.5 mm) sufficiently satisfies the sampling requirement and successfully eliminates the artifact. Therefore, prior to the data acquisition, the radiologist must look at the coverage requirement and select the detector configuration that provides the best image quality.

As detector coverage increases, the demand on the accuracy of other components also increases. The design specification that worked well for a system with a single or a small number of rows may not be adequate for a system with a higher number of rows. The increased accuracy requirement is not limited to the components that directly produce the projection data such as the detector, tube, or DAS, but also to the CT gantry and patient table. The remainder of this chapter provides examples of image artifacts that belong to different categories. Given the complexity of the topic, the examples given below are by no means an exhaustive list. They serve only as the starting point for further exploration and investigation.

10.5.2 Multislice CT cone-beam effects

Mathematically, the detector cone angle can be calculated by the following formula:

$$\theta = \tan^{-1}\left(\frac{s}{D}\right), \tag{10.16}$$

where s is the distance (at iso-center) of the outermost detector row to the center plane and D is the distance between the x-ray source and the iso-center. For example, for a scanner operating in a 0.625-mm detector aperture mode, the corresponding cone angles for the outermost row in 8-, 16-, and 32-slice configurations are 0.23, 0.50, and 1.03 deg, respectively (with a source-to-iso distance of 540 mm). Although these cone angles are small, their impact on image quality is not negligible. Consider a point located at the edge of a 50-cm FOV. A 0.23-deg cone angle leads to a 1-mm error, while a 1.03-deg cone angle leads to a 4.49-mm error between the backprojected position and its actual location (assuming a 2D backprojection is used for the approximation). Therefore, we expect backprojection-induced artifacts to increase with an increased cone angle of the acquisition. This is demonstrated in Fig. 10.41(a) to (c) (left column). These three images correspond to the outermost slices of 8 × 0.625-mm, 16 × 0.625-mm, and 32 × 0.625-mm configurations with a step-and-shoot data acquisition, respectively. The increase in image artifacts from top to bottom is obvious. Although the examples are shown for the step-and-shoot mode (to eliminate the impact of the helical pitch), the conclusion is applicable to helical cases as well.

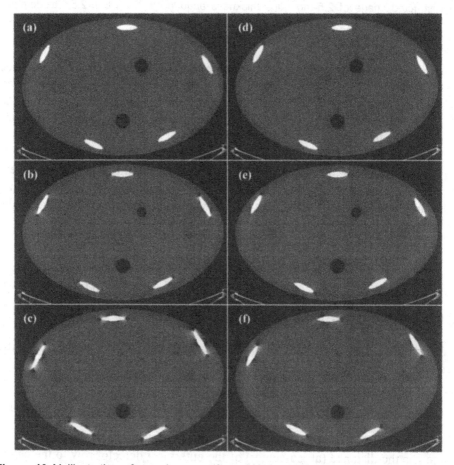

Figure 10.41 Illustration of cone-beam artifacts (WW = 300). An oval-body phantom was scanned in a step-and-shoot mode with 64 × 0.625-mm detector configuration. Different numbers of detector rows were selected to simulate 8-, 16-, and 32-slice detectors (top to bottom rows). 2D (left column) and 3D (right column) backprojections were used to demonstrate different sources of errors. The distances of the reconstructed images to the x-y plane are 2.2, 4.7, and 9.7 mm for the top, middle, and bottom rows, respectively. For this illustration, a weighting function was not used to enhance the artifact appearance.

When 3D backprojection is used, image artifacts come from two sources: longitudinal truncation and the cone-beam effect. At a moderate cone angle, the dominant source of error is longitudinal truncation. As the cone angle increases, the degree of longitudinal truncation also increases. [For a thorough explanation, refer to Fig. 10.17(b).] To reconstruct a voxel in the lightly shaded triangles at the upper- and lower-right of the "desired reconstruction volume," extrapolation has to be performed during the backprojection process, since a straight line connecting the x-ray source and the voxel intersects the detector plane either above or below the upper- and lower-bound of the detector. The extrapolation distance, and therefore the artifact, increases as the reconstructed slice is further away from the x-y plane (detector center plane.) This is demonstrated in Figs. 10.41(d) to (f) (right column). As the image location increases from 2.2 to 9.7 mm (from the top row to the bottom row), the image artifact level also increases.

If we compare the image artifacts of the 2D and 3D backprojections in the left and right columns in Fig. 10.41, we can conclude that when the cone angle is very small, both algorithms perform satisfactorily (top row). As the cone angle increases, the 3D backprojection generally produces less-severe artifacts than the 2D backprojection. For this study, weighting functions were not applied during the reconstruction process to enhance the appearance of the image artifacts. When an optimized weighting function is used, a significantly reduced level of image artifacts can be expected for all cases.

10.5.3 Interpolation-related image artifacts

Most helical reconstruction algorithms assume a linear model for the scanned object, so the projection value at a desired location is estimated by linearly interpolating adjacent readings. These readings are the conjugate projection rays collected over a 2π to 4π angular span. In most cases, linear interpolation performs quite well and produces a clinically acceptable image quality. When the distance between adjacent readings increases (e.g., when scans are taken with high helical pitches), the accuracy or effectiveness of the linear interpolation degrades. Interpolation degradation also occurs when high-density and high-frequency structures are present in the scanned object (e.g., the posterior fossa region in a human head).

Figure 10.42(a) shows a reconstructed image of an oval body phantom. The scan was collected on an 8-slice LightSpeed^{TM} scanner with a 2.5-mm detector aperture at a pitch of 1.675 (the table traveled 33.5 mm per gantry rotation). Image artifacts are clearly visible near the ellipsoids and the air pocket. To confirm that the artifacts are caused by inaccurate interpolation, we measured the angle α between the bright streaks around the air pocket, as shown in Fig. 10.43. The angle is roughly 27 deg, which equals the projection angle between consecutive detector rows that intersect the plane of reconstruction (360 deg/13.4). Note that the interpolated value is accurate only when the detector is located at the reconstruction plane (no interpolation takes place). The periodical artifact pattern is the result of the periodical projection estimation error of different detector rows.[96] This artifact can also be classified as the aliasing artifact in the z direction.[97]

As with single-slice helical CT, multislice image artifacts (cone-beam and interpolation artifacts) can be significantly reduced with z filtering, as explained in Chapter 9. By increasing the slice thickness roughly 40%, most of the artifacts shown in Fig. 10.42(a) can be eliminated, as shown in Fig. 10.42(b). To further preserve the SSP in the reconstructed images, adaptive z filtering can be used to create a good balance between artifact reduction and SSP preservation.[96]

Another approach to overcome the interpolation or aliasing artifacts is to increase the sampling frequency in the z direction, as is applied in the approaches for in-plane focal spot wobble discussed in Chapter 7.[100] In this approach, the x-ray focal spot is deflected back and forth between consecutive projection views

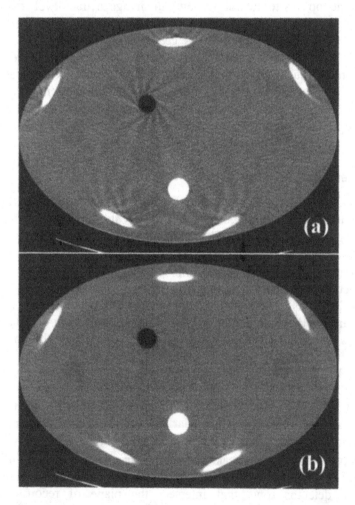

Figure 10.42 Reconstructed images of an oval phantom scanned in 8 × 2.5-mm mode with pitch 1.675 (table speed of 33.5 mm per gantry rotation) (WW = 300). (a) Reconstruction without z filtering. (b) Reconstruction with z filtering, resulting in a 40% broader slice thickness.

so that the sampling density at the iso-center is doubled, as illustrated in Fig. 10.44. One major difference between this approach and the in-plane focal spot wobble is the interlacing pattern. In the in-plane case, the pair of views with focal spots at two different locations interlace nearly perfectly throughout the entire FOV due to the rotational motion of the detector. In the z-wobble configuration, perfect interlacing is achieved only at the iso-center; the sampling pattern degrades as we move away from the iso-center. For example, the two sets of samples nearly completely overlap near the detector.

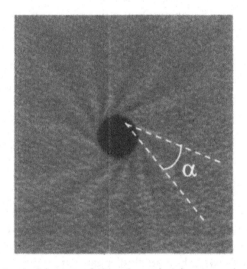

Figure 10.43 Reconstructed image of the air pocket in an oval phantom to calculate angles between streaks (α is approximately 27 deg).

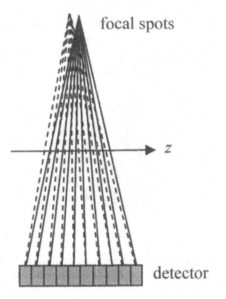

Figure 10.44 Illustration of the z-wobble concept.

10.5.4 Noise-induced multislice artifacts

Chapter 9 discussed the modulated noise pattern produced by single-slice helical CT. We stated that the noise pattern follows the starting angle of the dataset. More accurately, the noise pattern follows the minimum-maximum weighting function. For helical interpolative reconstruction, the weighting function at the start and end of a dataset is close to zero (minimum), and the weighting function roughly 180 deg away is close to unity (maximum). The highest noise gradient in the image follows the direction between the projection angle with a minimum weight to the projection angle with a maximum weight (min-max). For a single-slice CT using a 2π-based reconstruction, min-max occurs at one cycle per gantry revolution. However, for multislice CT, several min-max weighting cycles occur per gantry rotation. In fact, the weighting function cycle repeats whenever interpolation is switched from one row to another. Consequently, we would expect to observe similar image artifacts associated with single-slice noise modulation. Figures 10.45(a) and (b) show MIP images of a patient scan. A zebra artifact (horizontal stripes) can be clearly observed in Fig. 10.45(a). Of course, approaches similar to those used for single-slice CT can be used to combat zebra artifacts in multislice reconstructions, as shown in Fig. 10.45(b).

10.5.5 Tilt artifacts in multislice helical CT

For single-slice helical CT, gantry tilt does not impact the performance of helical reconstruction algorithms. For multislice helical CT, however, this nice property does not exist. If a proper correction is not rendered, severe image artifacts will result. Figure 10.46 shows a human skull phantom study performed on a 4-slice scanner with a 2.5-mm detector aperture. A helical pitch of 0.75 was used with the gantry tilted at 20 deg. Severe image artifacts can be observed in Fig. 10.46(a). For example, the left sphenoid bone appears in duplicates, as indicated by the arrow near the upper-left corner of the figure. In addition, many fine structures such as the mastoid air cells are nearly obliterated. For reference, Fig. 10.46(c) shows the same phantom scanned in a step-and-shoot mode at the same gantry tilt angle and detector configuration.

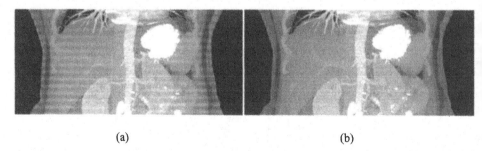

(a)   (b)

Figure 10.45 Image artifacts due to noise modulation (a) cone-beam reconstruction without noise compensation and (b) reconstruction with row-wise fan-to-parallel rebinning.

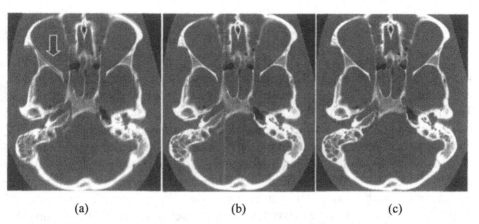

(a)   (b)   (c)

Figure 10.46 Reconstructed images of a human skull phantom scanned with a 4 × 2.5-mm detector aperture (WW = 1000). Images (a) and (b) were acquired with a helical mode at a pitch of 0.75 with a gantry tilt of 20 deg. Image (c) was acquired with step-and-shoot mode at the same detector configuration and tilt angle. (a) Helical reconstruction without compensation. (b) Helical reconstruction with compensation. (c) The step-and-shoot image as the gold standard.

To fully understand the root cause of tilted helical image artifacts, we can examine the scan geometry. In Fig. 10.47, we denote the patient axis along which the table is translated by z, the tilt angle formed by the gantry plane and the y axis by α, and a tilted coordinate system by y'-z'. In this system, the y' axis is parallel to the tilted gantry plane and coplanar with the y-z plane. The z' axis is selected to coincide with the gantry rotating axis. For a multislice CT system, each detector row forms a fan-beam plane with the x-ray source. The gantry iso-center for each detector row is the intersection of the corresponding fan-beam plane with the z' axis.

From an image reconstruction point of view, the iso-center should be located along the patient (z) axis, as shown by point B in Fig. 10.47, because each image in multislice CT reconstruction is formed with projections from multiple detector rows. As a result, the iso-center definition must be consistent and independent of the detector row location. Consider the reconstruction of the center plane defined as the x'-y' plane in Fig. 10.47. The reconstruction iso-center for this plane is naturally the origin of the y'-z' coordinate system. Since the patient is translated along the z axis during the scan, the reconstruction iso-center for the center plane also travels along the z axis. Therefore, the z axis is the reconstruction iso-center for the entire CT system.

For detector rows that are located off the center plane, the gantry iso-center and the reconstruction iso-center are not co-located. If the projection dataset is not properly adjusted, the reconstructed images will have multiple iso-centers, depending on the rows used to form the image. As a result, some reconstructed structures will appear shifted and overlapped and other structures smeared and blurred.

If we denote by ψ the angular difference between two rays that pass through two iso-centers, it can be shown that the following relationship exists:[7]

$$\psi = \tan^{-1}\left(\frac{rt \cdot \tan\alpha \cdot \sin\beta}{D + rt \cdot \tan\alpha - rt \cdot \tan\alpha \cdot \cos\beta}\right), \qquad (10.17)$$

where D is the distance from the source to the iso-center, β is the projection angle, α is the tilting angle of the gantry, and rt is the distance of a row from the center plane. The amount of angular difference (and therefore the projected difference between the two iso-centers) depends on the table translation direction, the amount of gantry tilt, the detector aperture size, the detector row location, and the projection view angle.

Based on the angular deviation formula [Eq. (10.17)], compensation schemes can be developed to reduce or eliminate the impact of the iso-center misalignment. In addition to the iso-center adjustment, there is a less-obvious effect of magnification change (where magnification is defined as the ratio of the source-to-iso distance over the point-to-source distance). A detailed analysis has shown that the amount of magnification change is very small and can be safely ignored.

One compensation approach for the iso-center deviation is to perform an iso-center adjustment during the backprojection process, so the iso-center used in the backprojection operation is dynamically adjusted based on Eq. (10.17). Alternatively, we can shift the projections dynamically by an amount described by Eq. (10.17) prior to the FBP. The shift can be performed either in the spatial or the frequency domain. Figure 10.46(b) depicts the same phantom experiment reconstructed with a projection shift approach. It is clear from the image that the gantry-tilt-induced image artifacts are significantly reduced or eliminated; note that the shape of the sphenoid bone is restored and the sharpness of the structure is preserved. A similar approach can be applied to tilted cone-beam helical data with row-wise fan-to-parallel rebinning.[20] An approach that enables the application of both approximate and exact reconstruction algorithms by first rebinning the tilted helical data into non-tilted data of a virtual object was also proposed.[87]

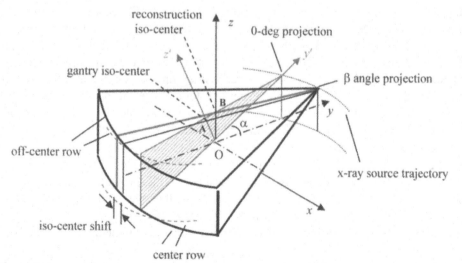

Figure 10.47 Geometrical relationship between the gantry iso-center and the reconstruction iso-center.

10.5.6 Distortion in step-and-shoot mode SSP

In a multislice CT detector configuration, small gaps are provided between scintillator cells so that highly reflective material can be placed between the cells. The reflective material serves to reflect or channel the light photons generated by the scintillator to the photodiodes at the bottom of the detector, as shown in Fig. 10.48. To prevent direct exposure of the reflective material and photodiode to x-ray photons, a small block of tungsten material is placed on the x-ray photon side of the detector to block incoming x-rays. Since no signal is produced at these gaps, dips are likely to be present in the detector SSP.[89]

For confirmation, we can use the thin-film technique discussed in Chapters 5 and 9 to measure the SSP. We scanned a 0.2-mm-thick x-ray film at a 5.36-deg angle with respect to the CT x-y plane. The scan was taken in a step-and-shoot mode with a 2.5-mm slice thickness and a small x-ray focal spot (0.3 mm in z projected at the iso-center). Images were reconstructed with the standard reconstruction algorithm at a 30-cm display FOV. The intensity of the reconstructed image is plotted in Fig. 10.49, which shows an intensity dip at the center of the 2.5-mm slice. Since a 2.5-mm slice is formed by summing the signals from two 1.25-mm cells, the dip in the profile corresponds to the gap between the two cells. Figure 10.50 shows four reconstructed images of scans taken with 1.25-, 2.5-, 3.75-, and 5.0-mm slice thicknesses in a step-and-shoot mode.

Further analyses indicate that the relative magnitude of the dips in the SSP is independent of the slice thickness. The magnitude, however, reduces with the increase in x-ray focal spot size and object thickness.[89] Results of these analyses indicate that at larger cone angles, the slice profile is further degraded and ill-defined. Therefore, when evaluating a multislice SSP, special attention must be paid to the data acquisition condition, since SSP can vary not only within the reconstruction plane (which is well known), but also varies significantly from detector row to detector row. The clinical impact of the SSP variation can be observed when imaging small vessels that form a small angle with respect to the scan plane. An intensity modulation of the vessel structure could potentially lead to a misinterpretation of the clinical results.

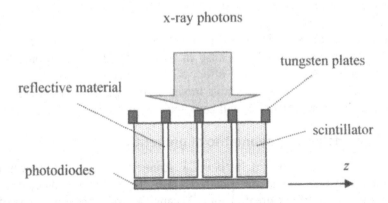

Figure 10.48 Detector cell configuration.

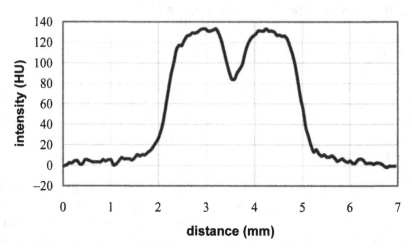

Figure 10.49 Measured intensity profile of a reconstructed film. Scans were taken in step-and-shoot mode with a 2.5-mm detector aperture.

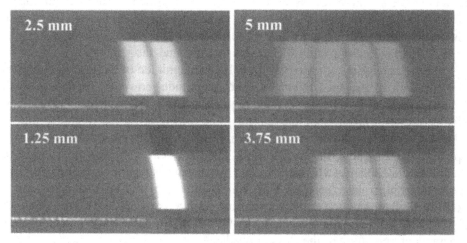

Figure 10.50 Images of an x-ray film in a step-and-shoot mode (WW = 300). Scans were collected with 5-, 3.75-, 2.5-, and 1.25-mm slice thickness.

The dips in the SSP disappear if the data acquisition is performed with a reasonably large helical pitch, because in a helical data acquisition, the object moves constantly. As a result, the effect of the intercell gaps is averaged over a large area. When the distance the table travels in one gantry rotation is larger than the cell size, the averaging effect will completely remove the dips in the SSP.

10.5.7 Artifacts due to geometric alignment

Although in the design of single-slice CT scanners some attention was paid to ensure the CT system was geometrically aligned, the precision and accuracy of the alignment is generally not great. This is because artifacts are not observed in the reconstructed images unless the system is grossly out of alignment. For a multislice

CT scanner, however, the importance of accurate system alignment increases significantly. For example, one of the requirements for a CT system is the perpendicular alignment of the patient table and the x-y plane of the gantry, as shown in Fig. 10.51. When a single-slice scanner is slightly out of alignment, the resulting image artifact is nearly invisible. However, the same misalignment error in a multislice system can produce significant image artifacts, as shown in Fig. 10.52. In this study, a set of thin wires were placed 45 deg with respect to the z axis in the x-z plane of a misaligned system. Multiple step-and-shoot data acquisitions were performed with a 64 × 0.625-mm detector configuration and a 40-mm table increment between scans. A discontinuity of the wires can be clearly observed in the generated coronal image, as shown by the arrow in Fig. 10.52(a). These artifacts disappeared when the system was properly aligned. The tolerance for alignment accuracy generally increases with the detector z coverage. Similar alignment requirements are needed for other geometric alignments, such as accuracy of the gantry tilt and horizontal leveling of the patient table.

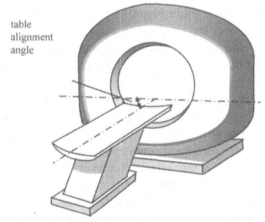

table alignment angle

Figure 10.51 Schematic diagram of the geometric relationship between the gantry and the patient table.

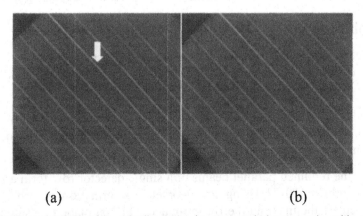

(a)   (b)

Figure 10.52 Impact of table misalignment in step-and-shoot mode with a detector configuration of 64 × 0.625 mm with (a) table skewed and (b) table properly aligned.

10.5.8 Comparison of multislice and single-slice helical CT

For multislice helical CT with small cone angles (< 2 deg of full cone angle), image artifacts are caused predominately by inaccurate projection interpolation. Therefore, the similarities between single-slice helical artifacts and multislice helical artifacts should not be surprising. For illustration, we scanned an oval phantom on both types of scanners at different helical pitches. The results are shown in Fig. 10.53. Figure 10.53(a) depicts a reconstructed image of a single-slice step-and-shoot mode scan, which serves as the gold standard. Figure 10.53(b) is a reconstructed image acquired with a single-slice helical mode at a pitch of 1; other than minor image artifacts, the overall image quality is comparable to that of the step-and-shoot mode acquisition. Figures 10.53(c) and (d) were acquired with a 4-slice scanner at a 0.75 pitch and a single-slice scanner at a 1.5 pitch, respectively. The shape and intensity of the shading artifacts in both images are quite similar.

Figures 10.53(e) and (f) illustrate the differences in artifact appearance between single-slice and multislice scanners. Figure 10.53(e) was acquired with a 4-slice scanner at a 1.5 pitch while Fig. 10.53(f) was acquired with a single-slice scanner at a pitch of 2. For the single-slice acquisition, the distance between the measured projection and the reconstruction plane varies one cycle per gantry rotation. This is well illustrated by the single-cycle distortion pattern of the air pocket in Fig. 10.53(f). The multislice case is not limited to the measurements from a single detector row, and the plane of reconstruction intersects different rows during a helical rotation. Therefore, the interpolation samples shift from one detector row to the next during the rotation to satisfy the proximity condition. The impact is that the distance between the measured projection and the plane of reconstruction is no longer a monotonic increasing or decreasing function as in the single-slice CT case (if the angular range of the projection dataset is divided into multiple subregions, the monotonicity still exists). As a result, the amount of projection error (and therefore the level of image artifacts) increases and decreases several cycles within a single dataset. This leads to the appearance of multiple-cycle artifacts in the reconstructed air pocket in Fig. 10.53(e).

10.6 Problems

10-1   Derive the FDK formula for the equiangular projection, assuming the source-to-iso distance is D, the source-to-detector distance is R, and a projection sample is $p(\gamma, \beta, z)$.

10-2   Calculate the shape and location of the virtual detector shown in Fig. 10.12, assuming the source-to-detector distance is 950 mm and the source-to-iso distance is 550 mm.

10-3   Do the rebinned parallel beams of a single detector row form a flat plane in problem 10-2? Using the geometric information in problem 10-2, calculate the maximum error introduced if a flat plane is assumed during

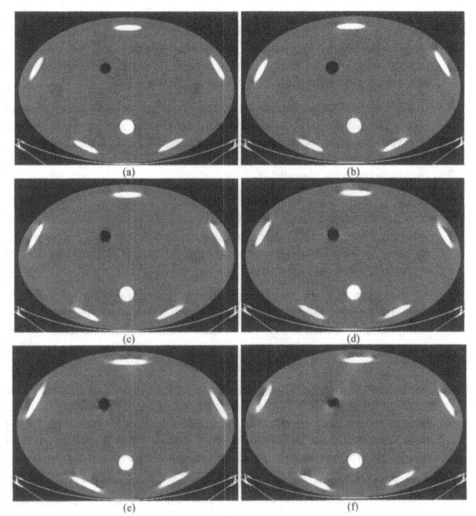

Figure 10.53 Comparison of single-slice versus multislice CT (WW=400). An oval phantom was scanned with 3.75-mm aperture. (a) Step-and-shoot mode. (b) Single-slice helical at pitch of 1. (c) Four-slice helical at pitch of 0.75. (d) Single-slice helical at pitch of 1.5. (e) Four-slice helical at pitch of 1.5. (f) Single-slice helical at pitch of 2.

the backprojection process over a 50-cm FOV, and the detector row is 20 mm off the detector center plane.

10-4   Derive the equivalent of the FDK algorithm for the rebinned tilted parallel geometry.

10-5   Calculate the maximum amount of error caused by the 2D fan-beam backprojection for a 50-cm FOV image located 2 mm off the x-y plane, assuming that the source-to-detector distance is 950 mm and the source-to-iso distance is 550 mm. Repeat the calculation when the image is 9 mm off the x-ray plane.

10-6   For a 50-cm FOV image located 19.75 mm off the x-y plane, calculate the percentage of the area that is not irradiated by the x-rays of a single projection due to longitudinal truncation. Assume that the source-to-detector distance is 1000 mm, the source-to-iso distance is 500 mm, and the detector configuration (in z) is 100×0.5 mm.

10-7   For the virtual detector in problem 10-2 and a 50-cm FOV image located 24.75 mm off the x-ray plane, calculate the percent area irradiated by the x rays from either of two projections that are 180 deg apart, assuming a 100×0.5-mm detector configuration (in z). What percent of the area is irradiated by both views?

10-8   What is the minimum number of detector rows needed to completely avoid longitudinal truncation (for each projection) over a 50-cm FOV, assuming the source-to-detector distance is 100 cm, the source-to-iso distance is 50 cm, the detector aperture is 0.5 mm at iso, and the desired volume coverage is 160 mm in z?

10-9   A student proposes a more aggressive weighting function for the step-and-shoot cone-beam reconstruction using Eq. (10.3). For each conjugate sampling pair shown in Fig. 10.20, the student assigns a value of 1 to the sample with the smaller cone angle and 0 to the other sample. What potential issues may the student encounter?

10-10   In a step-and-shoot mode data acquisition in which the detector configuration is 64×0.625 mm (in z), what is the table angular alignment accuracy required with respect to the gantry plane (Fig. 10.51) that ensures that the maximum shifting between two images in adjacent acquisitions is less than 0.2 mm over a 50-cm FOV?

10-11   A CT gantry has a specified tilt angular accuracy of ± 0.3 deg. A thin wire is placed parallel to the z axis, and two step-and-shoot data acquisitions are acquired with a 64×0.625-mm detector configuration and a 40-mm table increment. The gantry tilt angle indicator shows 0 deg. What type of image artifact do you expect to see if the gantry is at its maximum specification limit? What reconstruction parameters do you recommend to enhance the appearance of the artifact?

10-12   In the step-and-shoot data acquisition, the table increment is typically equal to the detector z coverage at the iso-center. For example, for a 64×0.625-mm detector configuration, a 40-mm table increment is used between adjacent data acquisitions. When a CT gantry is tilted at 30 deg, what is the appropriate table increment for a 64×0.625-mm acquisition?

10-13   How do you implement the double-circle source trajectory shown in Fig. 10.25(c) with a clinical multislice CT scanner? The source-to-iso distance of the scanner is 600 mm, and the whole-body scan needs to be performed on an average-sized patient.

10-14   All commercially available multislice CT scanners have an even number of detector rows (4-, 8-, 16-slice, etc.). Is there any advantage to a scanner with an odd number of detector rows?

10-15   The detector configuration of all commercially available multislice scanners has an identical detector thickness (in z) for all rows (e.g., 4 × 1.25 mm, 4 × 2.5 mm, etc.). Is there any advantage to having different detector thicknesses for different rows?

10-16   Write a computer program to interpolate images reconstructed with a tilted flat plane reconstruction [Fig. 10.31(c)], assuming a detector configuration of 8 × 1.2 mm, a helical pitch of 1, and a source-to-iso distance of 50 cm. The tilted-plane images are reconstructed at a 0.6-mm spacing at the iso-center.

10-17   A 16-slice scanner has two detector apertures, 0.625 mm and 1.25 mm, and four different helical pitches: 9/16, 15/16, 22/16, and 28/16. It has four different rotation speeds: 1, 0.8, 0.5, and 0.4 sec per gantry rotation. A CT operator needs to cover a 17-cm volume in z in less than 5 sec with a 1.5-mm slice thickness. What scan protocol should the operator use and why?

10-18   The source-to-iso distance of a CT scanner with a 16 × 1.25-mm detector configuration is 550 mm. What is the biggest angular difference in terms of cone angle between two conjugate rays inside a 50-cm FOV in the step-and-shoot mode acquisition shown in Fig. 10.20?

10-19   A CT scanner does not have tilt compensation during image reconstruction for the step-and-shoot mode acquisition. Can postprocessing compensate for this deficiency? If such an approach exists, what are the drawbacks?

10-20   Describe an approach to compensate for tilted cone-beam helical artifacts when row-wise fan-to-parallel rebinning is used.

10-21   Describe at least two postprocessing approaches to reduce the Venetian blind or zebra artifact shown in Fig. 10.45.

References

1. M. Prokop, "CT angiography," in *Categorical Course in Diagnostic Radiology Physics: CT and US Cross-Sectional Imaging*, L. W. Goldman and J. B. Fowlkes, Eds., Radiological Society of North America, Inc., Oakbrook, IL, 143–157 (2000).

2. K. Taguchi and H. Aradate, "Algorithm for image reconstruction in multi-slice helical CT," *Med. Phys.* **25**, 550–561 (1998).

3. H. Hu, "Multi-slice helical CT: Scan and reconstruction," *Med. Phys.* **26**, 1–18 (1999).

4. H. Bruder, M. Kachelrieß, S. Schaller, K. Stierstorfer, and T. Flohr, "Single-slice rebinning reconstruction in spiral cone-beam computed tomography," *IEEE Trans. Med. Imag.* **19**, 873–887 (2000).

5. R. Proksa, Th. Kohler, M. Grass, and J. Timmer, "The n-PI-method for helical cone-beam CT," *IEEE Trans. Med. Imag.* **19**, 848–863 (2000).

6. J. Hsieh, "CT image reconstruction," in *Categorical Course in Diagnostic Radiology Physics: CT and US Cross-Sectional Imaging*, L. W. Goldman and J. B. Fowlkes, Eds., Radiological Society of North America, Inc., Oakbrook, IL, 53–64 (2000).

7. J. Hsieh, "Tomographic reconstruction for tilted helical multislice CT," *IEEE Trans. Med. Imag.* **19**, 864–872 (2000).

8. W. D. Foley, T. A. Mallisee, M. D. Hohenwater, C. R. Wilson, F. A. Quiroz, and A. J. Taylor, "Multiphase hepatic CT with a multirow detector CT scanner," *Am. J. Roentgen.* **175**, 679–685 (2000).

9. A. Laghi, R. Lannaccone, P. Rossi, L. Carbone, R. Ferrari, F. Mangiapane, I. Nofroni, and R. Passariello, "Hepatocellular carcinoma detection with triple-phase multi-detector row helical CT in patients with chronic hepatitis," *Radiol.* **226**, 543–549 (2003).

10. A. F. Kopp, M. Heuschmid, and C. D. Claussen,"Multidetector helical CT of the liver for tumor detection and characterization," *Europ. Radiol.* **12**(b), 745–752 (2002).

11. J. Hsieh and J. Li, "A detector response adaptive helical reconstruction algorithm," in *IEEE Trans. Nucl.Sci.,* **49**(1), 215–219 (2002).

12. L. A. Feldkamp, L. C. Davis, and J. W. Kress, "Practical cone-beam algorithm," *J. Opt. Soc. Am.* **1**(6), 612–619 (1984).

13. H. K. Tuy, "An inversion formula for cone-beam reconstruction," *SIAM J. Appl. Math.* **43**(3), 546–552 (1983).

14. J. Hsieh, "A two-pass algorithm for cone beam reconstruction," *Proc. SPIE* **3979**, 533–540 (2000).

15. J. Hsieh, "A practical cone beam artifact correction algorithm," *Proc. IEEE Nuclear Science Symposium and Medical Imaging Conference*, **15**, 71–74 (2000).

16. S. Bartolac, R. Clackdoyle, F. Noo, J. Siewerdsen, D. Moseley, D. Jaffray, "A local shift-variant Fourier model and experimental validation of circular cone-beam computed tomography artifacts," *Med. Phys.* **36**(2), 500–512 (2009).

17. E. Y. Sidky and X. Pan, "Image reconstruction in circular cone-beam computed tomography by constrained total-variation minimization," *Phys. Med. Biol.* **53**(17), 4777–4807 (2008).

18. J. Hsieh, W. Lin, N. A. Ishaque, P. M. Edic, and M. Yavuz, "An iterative cone beam artifact correction scheme," *Radiology* **217**(p), 404 (2000).

19. J. Hsieh, "Reconstruction algorithm for single circular orbit cone beam scan," *Proc. IEEE Intern. Symp. on Biomedical Imaging,* 836–838 (2002).

20. J. Hsieh and X. Tang, "Tilted cone-beam reconstruction with row-wise fanto-parallel rebinning," *Phys. Med. Biol.* **51**, 5259–5276 (2006).

21. Th. Kohler, R. Proksa, and M. Grass, "A fast and efficient method for sequential cone-beam tomography," *Med. Phys.* **28**(11), 2318–2327 (2001).

22. X. Wang and R. Ning, "A cone-beam reconstruction algorithm for circleplus-arc data-acquisition geometry," *IEEE Trans. Med. Imag.* **18**(9), 815–824 (1999).

23. X. Tang and R. Ning, "A cone beam filtered back-projection (CB-FBP) reconstruction algorithm for a circle-plus-two-arc orbit," *Med. Phys.* **28**(6), 1042–1055 (2001).

24. A. A. Zamyatin, A. Katsevich, and B. S. Chiang, "Exact image reconstruction for a circle and line trajectory with a gantry tilt," *Phys. Med. Biol.* **53**(23), 423–435 (2008).

25. J. D. Pack and F. Noo, "Cone-beam reconstruction using 1D filtering along the projection of M-lines," *Inverse Problems* **21**, 1105–1120 (2005).

26. X. Tang, J. Hsieh, A. Hagiwara, R. A. Nilsen, J.-B. Thibault, and E. Drapkin, "A three-dimensional weighted cone beam filtered backprojection (CB-FBP) algorithm for image reconstruction in volumetric CT under a circular source trajectory," *Phys. Med. Biol.* **50**(16), 3889–3905 (2005).

27. H. Yang, M. Li, K. Koizumi, and H. Kudo, "View-independent reconstruction algorithms for cone beam CT with general saddle trajectory," *Phys. Med. Biol.* **51**, 3865–3884 (2006).

28. Y. Lu, J. Zhao, G. Wang, "Exact image reconstruction with triple-source saddle-curve cone-beam scanning," *Phys. Med. Biol.* **54**(10), 2971–2991 (2009).

29. T. Zhuang, B. E. Nett, S. Leng, and G. Chen, "A shift-invariant filtered backprojection (FBP) cone-beam reconstruction algorithm for the source trajectory of two concentric circles using an equal weighting scheme," *Phys. Med. Biol.* **51**, 3189–3210 (2006).

30. S. Cho, E. Pearson, C. A. Pelizzari, X. Pan, "Region-of-interest image reconstruction with intensity weighting in circular cone-beam CT for imageguided radiation therapy," *Med. Phys.* **36**(4), 1184–1192 (2009).

31. T. Rodet, F. Noo, and M. Defrise, "The cone-beam algorithm of Feldkamp, Davis, and Kress preserves oblique line integrals," *Med. Phys.* **31**(7), 1972–1975 (2004).

32. A. Katsevich, "Image reconstruction for a general circle-plus trajectory," *Inverse Problems* **23**(5), 2223–2230 (2007).

33. G. Wang, T. H. Lin, P. Cheng, and D. M. Shinozaki, "A general cone-beam reconstruction algorithm," *IEEE Trans. Med. Imag.* **12**, 486–496 (1993).

34. H. Hu, "An improved cone-beam reconstruction algorithm for the circular orbit," *Scanning* **18**, 572–581 (1996).

35. K. Zeng, Z. Chen, L. Zhang, and G. Wang, "An error-reduction-based algorithm for cone-beam computed tomography," *Med. Phys.* **31**(12), 3206–3212 (2004).

36. J. Hsieh, "A nonlinear helical reconstruction algorithm for multislice CT," *IEEE Trans. Nucl. Sci.* **49**(3), 740–744 (2002).

37. J. Hsieh, "A generalized helical reconstruction algorithm for multislice CT," *Radiology* **217**, 565 (2000).

38. G. Larson, C. Ruth, and C. Crawford, "Nutating slice CT image reconstruction apparatus and method," U.S. Patent No. 5802134 (1998).

39. D. J. Heuscher, "Helical cone beam scans using oblique 2D surface reconstructions," in *Proc. 1999 Int. Meeting Fully Three-dimensional Image Reconstruction in Radiology and Nuclear Medicine*, 204–207 (1999).

40. H. K. Tuy, "An inversion formula for cone-beam reconstruction," *SIAM J. Appl. Math.* **43**, 546–552 (1983).

41. B. D. Smith, "Image reconstruction from cone-beam projections: necessary and sufficient conditions and reconstruction methods," *IEEE Trans. Med. Imag.* **MI-4**, 14–25 (1985).

42. P. Grangeat, "Mathematical framework of conc-beam 3D reconstruction via the first derivative of the Radon transform," in *Mathematical Methods in Tomography*, G. T. Herman, A. K. Louis, and F. Natterer, Eds., (Lecture Notes in Mathematics **1497**), Springer, Berlin, 66–97 (1991).

43. H. Kudo and T. Saito, "Three-dimensional helical-scan computed tomography using cone-beam projections," *Sys. Comput. Japan* **23**(12), 75–82 (1992).

44. X. Yan and R. M. Leahy, "Cone-beam tomography with circular, elliptical, and spiral orbits," *Phys. Med. Biol.* **37**, 493–506 (1992).

45. G. L. Zeng and G. T. Gullberg, "A cone-beam tomography algorithm for orthogonal circle-and-line orbit," *Phys. Med. Biol.* **37**, 563–577 (1992).

46. Y. Wang, G. L. Zeng, and G. T. Gullberg, "A reconstruction algorithm for helical cone-beam SPECT," *IEEE Trans. Nucl. Sci.* **40**, 1092–1101 (1993).

47. M. Defrise and R. Clack, "A cone-beam reconstruction algorithm using shift-variant filtering and cone-beam backprojection," *IEEE Trans. Med. Imag.* **13**, 186–195 (1994).

48. H. Kudo and T. Saito, "Derivation and implementation of a cone-beam reconstruction algorithm for nonplanar orbits," *IEEE Trans. Med. Imag.* **13**, 196–211 (1994).

49. R. Clark and M. Defrise, "Overview of reconstruction algorithm for exact cone-beam tomography," *Proc. SPIE* **2299**, 230–241 (1994).

50. H. Kudo and T. Saito, "An extended completeness condition for exact cone-beam reconstruction and its application," in *Conf. Record 1994 IEEE Medical Imaging Conf.*, 1710–1714 (1994).

51. G. Wang, Y. Liu, T. H. Lin, and P. Cheng, "Half-scan cone-beam x-ray microtomography formula," *Scanning* **16**, 216–220 (1994).

52. F. Noo, R. Clack, and M. Defrise, "Cone-beam reconstruction from general discrete vertex sets using Radon rebinning algorithms," *IEEE Trans. Nucl. Sci.* **44**, 1517–1537 (1995).

53. K. C. Tam, "Method and apparatus for acquiring complete Radon data for exactly reconstructing a three-dimensional computerized tomography image of a portion of an object irradiated by a cone-beam source," U.S. Patent No. 5383119 (1995).

54. P. E. Danielsson, P. Edholm, J. Eriksson, and M. M. Seger, "Toward exact reconstruction for helical cone-beam scanning of long objects: A new detector arrangement and a new completeness condition," in *Proc. 1997 Meet. Fully 3D Image Reconstruction Radiol. Nuclear Med.*, 141–144 (1997).

55. S. Schaller, T. Flohr, and P. Steffen, "New efficient Fourier-reconstruction method for approximate image reconstruction in spiral cone-beam CT at small cone-angles," *Proc. SPIE* **3032**, 213–224 (1997).

56. M. Silver, "High-helical-pitch cone-beam computed tomography," *Phys. Med. Biol.* **43**, 847–855 (1998).

57. F. Noo, R. Clark, T. A. White, and T. J. Roney, "The dual-ellipse cross vertex path for exact reconstruction of long objects in cone-beam tomography," *Phys. Med. Biol.* **43**, 797–810 (1998).

58. K. C. Tam, S. Samarasekera, and F. Sauer, "Exact cone-beam CT with a spiral scan," *Phys. Med. Biol.* **43**, 1015–1024 (1998).

59. H. Kudo, S. Park, F. Noo, and M. Defrise, "Performance of quasiexact cone-beam filtered back-projection algorithm for axially truncated helical data," *IEEE Trans. Nucl. Sci.* **46**, 608–617 (1999).

60. H. Turbell and P. E. Danielsson, "Non-redundant data capture and highly efficient reconstruction for helical cone-beam CT," in *Proc. IEEE Conf. Rec. Nuclear Sci. Symp. Med. Imag. Conf.*, 3 1424–1425 (1999).

61. S. Schaller, F. Noo, F. Sauer, K. C. Tam, G. Lauritsch, and T. Flohr, "Exact radon rebinning algorithms using local region-of-interest for helical cone-bema CT," in *Proc. 1999 Meeting Fully 3-D Image Reconstruction in Radiology and Nuclear Medicine*, 11–14 (1999).

62. M. Defrise, F. Noo, and H. Kudo, "A solution to the long object problem in helical cone-beam tomography," *Phys. Med. Biol.* **45**, 1–21 (2000).

63. S. Schaller, F. Noo, F. Sauer, K. C. Tam, G. Lauritsch, and T. Flohr, "Exact Radon rebinning algorithm for the long object problem in helical cone-beam CT," *IEEE Trans. Med. Imag.* **19**, 361–375 (2000).

64. R. Proksa, Th. Kohler, M. Grass, and J. Timmer, "The n-PI-method for helical cone-beam CT," *IEEE Trans. Med. Imag.* **19**, 848–863 (2000).

65. R. Ning, B. Chen, R. Yu, D. Conover, X. Tang, and Y. Ning, "Flat panel detector-based cone-beam volume CT angiography imaging: system evaluation," *IEEE Trans. Med. Imag.* **19**, 949–963 (2000).

66. H. Bruder, M. Kachelrieb, S. Schaller, and T. Mertelmeier, "Performance of approximate cone-beam reconstruction in multi-slice computed tomography," *Proc. SPIE* **3979**, 541–553 (2000).

67. M. Defrise, F. Noo, and H. Kudo, "A solution to the long-object problem in helical cone-beam tomography," *Phys. Med. Biol.* **45**, 623–643 (2000).

68. H. Kudo, F. Noo, and M. Defrise, "Quasi-exact filtered backprojection algorithm for long-object problem in helical cone-beam tomography," *IEEE Trans. Med. Imag.* **19**, 902–921 (2000).

69. A. Katsevich, "Theoretically exact filtered backprojection-type inversion algorithm for spiral CT," *SIAM J. Appl. Math.* **62**(6), 2012–2026 (2002).

70. G. Chen. "An alternative derivation of Katsevich's cone-beam reconstruction formula," *Med. Phys.* **30**, 3217–3226 (2003).

71. A. Katsevich, "An improved exact filtered backprojection algorithm for spiral computed tomography," *Adv. Appl. Math.* **32**(4), 681–697 (2004).

72. X. Pan, D. Xia, Y. Zou, and L. Yu, "A unified analysis of FBP-based algorithms in helical cone-beam and circular cone- and fan-beam scans," *Phys. Med. Biol.* **49**, 4349–4369 (2004).

73. Y. Zou and X. Pan, "Exact image reconstruction on PI lines from minimum data in helical cone-beam CT," *Phys. Med. Biol.* **49**(6), 941–959 (2004).

74. Y. Ye, S. Zhao, H. Yu, and G. Wang, "A general exact reconstruction for cone-beam CT via backprojection-filtration," *IEEE Trans. Med. Imag.* **24**(9), 1190–1198 (2005).

75. J. Pack, F. Noo, R. Clackdoyle, "Cone-beam reconstruction using backprojection of locally filtered projections," *IEEE Trans. Med. Imag.* **24**, 1–16 (2005).

76. J. D. Pack and F. Noo, "Cone-beam reconstruction using 1D filtering along the projection of *M*-lines," *Inverse Problems* **21** (3), 1105–1120 (2005).

77. S. Zhao, H. Yu, and G. Wang, "A unified framework for exact cone-beam reconstruction formulas," *Med. Phys.* **32** (6), 1712–1721 (2005).

78. Y. Zou, X. Pan, and E. Y. Sidky, "Theory and algorithms for image reconstruction on chords and within regions of interest," *J. Opt. Soc. Am.* **22** (11), 2372–2384 (2005).

79. T. Zhuang and G. Chen, "New families of exact fan-beam and cone-beam image reconstruction formulae via filtering the backprojection image of

differentiated projection data along singly measured lines," *Inverse Problems* **22**, 991–1006 (2006).

80. X. Tang, J. Hsieh, R. Nilsen, S. Dutta, D. Samaonov, and A. Hagiwara, "A three-dimensional weighted cone beam filtered backprojection (CB-FBP) algorithm for image reconstruction in volumetric CT-helical scanning," *Phys. Med Biol.* **51**(4), 855–874 (2006).

81. A. Katsevich, "Image reconstruction for a general circle-plus trajectory," *Inverse Problems* **23**(5), 2223–2230 (2007).

82. J. Hsieh, X. Tang, J. Thibault, C. Shaughnessy, R. Nilsen, and E. Williams, "Conjugate cone beam reconstruction algorithm," *Opt. Eng.* **46**(6), 067001 (2007).

83. X. Tang and J. Hsieh, "Handling data redundancy in helical cone beam reconstruction with a cone-angle-based window function and its asymptotic approximation," *Med. Phys.* **34**(6), 1989–1998 (2007).

84. M. A. Anastasio, Y. Zou, E. Y. Sidky, and X. Pan, "Local cone-beam tomography image reconstruction on chords," *J. Opt. Soc. Am. A* **24**, 1569–1579 (2007).

85. M. Kachelriess, S. Schaller, and W. A. Kallender, "Advanced single-slice rebinning in cone-beam spiral CT," *Med. Phys.* **27**(4), 754–772 (2000).

86. X. Tang and J. Hsieh, "A filtered backprojection algorithm for cone beam reconstruction using rotational filtering under helical source trajectory," *Med. Phys.* **31**(11), 2949–2960 (2004).

87. F. Noo, M. Defrise, and H. Kudo, "General reconstruction theory for multislice x-ray computed tomography with a gantry tilt," *IEEE Trans. Med. Imaging,* **23**(9), 1109–1116, (2004).

88. G. Chen, T. Zhuang, S. Leng, and B. E. Nett, "Cone-beam filtered backprojection image reconstruction using a factorized weighting function," *Opt. Eng.* **46**(8), 87006 (2007).

89. J. Hsieh, "Investigation of the slice sensitivity profile for step-and-shoot mode multi-slice computed tomography," *Med. Phys.* **28**, 491–500 (2001).

90. J. Hsieh, "Image artifacts in CT," in *Categorical Course in Diagnostic Radiology Physics: CT and US Cross-Sectional Imaging,* L. W. Goldman and J. B. Fowlkes, Eds., Radiological Society of North America, Inc., Oakbrook, IL, 97–115 (2000).

91. C. H. McCollough and F. E. Zink, "Performance evaluation of a multi-slice CT system," *Med. Phys.* **26**(11), 2223–2230 (1999).

92. M. Kudo, N. Sato, and K. Fukuda, "Comparison of multi row detector and single row detector helical CT scanner: image quality (low contrast and high contrast detectability slice sensitivity profile)," *Radiology* **213**, 500 (1999).

93. K. Yamada, N. Hashimoto, M. Kawashima, M. Kimura, Y. Shioyama, and M. Sato, "The stair step artifact: Comparison between single and multi detector row CT," *Radiology* **213**, 504 (1999).

94. N. Sato, T. Pan, and K. Awai, "The study of cone-beam artifact at multi detector-row CT," *Radiology* **213**, 451 (1999).

95. K. Awai, N. Sato, K. Nakagawa, K. Osuga, S. Hori, and K. Ito, "Twister artifact: Cone beam artifact at the multi-detector-row CT (MDCT)," *Radiology* **213**, 503 (1999).

96. J. Hsieh, "Adaptive interpolation approach for multi-slice helical CT reconstruction," *Proc. SPIE* **5032**, 1876–1883 (2003)

97. K. Taguchi, H. Aradate, Y. Saito, I. Zmora, K. S. Han, and M. D. Silver, "The cause of the artifact in 4-slice helical computed tomography," *Med. Phys.* **31**(7), 2033–2037 (2004).

98. I. Hein, K. Taguchi, M. D. Silver, M. Kazama, and I. Mori, "Feldkamp-based cone-beam reconstruction for gantry-tilted helical multislice CT," *Med. Phys.* **30**(12), 3233–3242 (2003).

99. S. Leng, T. Zhuang, B. E. Nett, and G. Chen, "Helical cone-beam computed tomography image reconstruction algorithm for a tilted gantry with N-PI data acquisition," *Opt. Eng.* **46**(1), 1–14 (2007).

100. Y. Kyriakou, M. Kachelriess, M. Knaup, J. Krause, and W. A. Kalender, "Impact of the z-flying focal spot on resolution and artifact behavior for a 64-slice spiral CT scanner," *Euro. Radiol.* **16**(6), 1206–1215 (2006).

Chapter 11
X-ray Radiation and Dose-Reduction Techniques

When x-ray radiation penetrates an object, part of its energy is transferred to the object and causes changes in the object's material. During the energy transfer, an x-ray can indirectly produce ion pairs in the tissue. The ion pairs react with other chemical systems and cause radiation damage. Alternatively, the x rays may strike and break molecular bonds, such as those in DNA, and cause direct damage.[1] The level of acceptable dose for humans is greatly helped by the fact that humans have been exposed to natural radiation since the dawn of time.[2] There are three main sources of natural radiation: cosmic rays (charged particles from outer space), external gamma rays (radioactivity in the earth's crust), and internal radiation (radioactive material present in our body, such as K-40 and radon). On average, the human average exposure to manmade radiation is roughly equal to the total amount of natural radiation and therefore, doubles the radiation level that humans have been subjected to for centuries.[3]

A general sensitivity to and awareness of the x-ray dose delivered to patients has increased steadily over the years. Many studies have been conducted on the subject.[4-12] It is an understatement that the topic of CT x-ray dose has received much attention lately. Scientific journal publications, conference papers, and even articles in major newspapers present evidence and severe warnings about the adverse effects of CT radiation dose.[13-15] Others strongly defend the benefits of CT and raise serious concerns about research methodologies that lead to concerns about x-ray dosages.[16,17] Given the large number of CT procedures performed each year, it is not surprising that CT's impact on public health has received increasing scrutiny. It is estimated that about 62 million CT scans were performed in the U.S. alone in 2006, up from about 3 million in 1980. Regardless of the controversy, it is accurate to state that significant efforts have been made to reduce x-ray radiation from CT. These efforts come from three main sources: the research and clinical community, government agencies, and industry.

In recent years, the clinical community has gradually adopted the ALARA (as low as reasonably achievable) principle. ALARA is based on the conservative assumption that every radiation dose can produce some level of detrimental effect that may be manifested as an increased risk of genetic mutations or cancer.

433

Reducing the x-ray radiation to the lowest achievable level has become not only a sound safety principle, but also a regulatory requirement. The key is the word "reasonably." As explained in previous chapters, there is a relationship between the amount of x-ray radiation and the noise present in a reconstructed image. For a given algorithm, the image noise scales roughly inversely proportional to the square root of the x-ray tube current when other conditions are kept unchanged. When the x-ray tube current (and dose) is reduced, image noise increases. The ALARA principle strives to achieve the benefit of a lower radiation dose without impacting the diagnosis accuracy.

Later sections of this chapter will present various efforts made by the CT community to reduce x-ray radiation while maintaining image quality. Although these efforts have resulted in a major dose reduction to patients, the increased use of CT scanners in clinical practice and the publicity generated by the news media have increased public awareness of this issue, so dose reduction will continue to be a hot topic for years to come.

11.1 Biological Effects of X-ray Radiation

Radiation damage to cells may occur directly from a radiation hit on the critical target or indirectly from free radicals produced by the radiation. Since a large part of any living system is made of water, a significant portion of the energy transfer during the initial ionization and excitation events takes place within this molecule.[18] Several interactions are possible between the radicals produced in water and biologically important molecules. These include the extraction of hydrogen atoms, dissociative reactions, and addition reactions. Not all reactions are important in cellular alterations following irradiation, nor are all reactions important contributors to DNA damage. The relative importance of each is not yet fully understood.

Low-dose radiation can affect cells biologically, including cell killing, altered genes, or damaged chromosomes. It can permanently damage DNA in germ cells and cause gene mutations, which can be transmitted from one generation to another.

Three mechanisms allow molecules to be restored to a pre-irradiation condition: recombination, restitution, and enzymatic repair. Recombination takes place within the 10^{-11} sec after the irradiation event while the ion pairs or radical pairs are still very close together.[18] As the diffusion increases the separation distance of the ion pair, recombination becomes less likely. Restitution, in which a chemical restoration process brings the altered molecule to its original state, can take place on the time scale of milliseconds. For example, a DNA radical produced by irradiation can interact with another molecule in a radical exchange reaction that leaves the DNA molecule restored to its pre-irradiation condition. Enzymatic repair can occur on a longer time scale, in minutes to hours.

The chemical and biochemical repair or misrepair mechanism of single-strand and double-strand breaks in the DNA determine the viability or function of all living systems after the radiation exposure, as shown in Fig. 11.1(a)–(c). It is generally believed that single-strand breaks are common events in a cell, and that their repair is efficient and relatively free of error. It is also generally accepted that double-strand breaks are far more serious for a cell, and their repair is an error-prone process that frequently leads to mutation in the genome and loss of reproductive capacity.

For simple organisms such as bacteriophages and viruses, a quantitative relationship seems to exist between DNA damage and biological function. For higher organisms, the relationship between DNA damage and the loss of biological function is not quantitatively established. However, it is generally accepted that following exposure to ionizing radiation, damage to DNA may cause the acute effect of cell death and late effects such as malignant tumors. These findings are mainly based on absorbed doses above 0.3 Gy (grays).[19]

It is not at all straightforward to extrapolate from the radiation impact on simple organisms the health risks, such as cancer, to humans. Human bodies have biological defense mechanisms that prevent the ionization event from developing into a cancer. For example, low-level radiation can be shown to stimulate production of enzymes that repair DNA damage with high efficiency,[43,44] and stimulate apoptosis—a process by which damaged cells "commit suicide."[45] Because of the complexity, the principal data that have been cited by many researchers to estimate the risks of low-level radiation are from reports of solid tumors among Japanese atomic bomb survivors,[46] and findings of an International Association for Research on Cancer study of occupational radiation workers.[47] In order to extrapolate the low-dose risks from the high-dose data, many studies use a linear no-threshold model, which states that the cancer risk is proportional to the radiation dose. Due to its simplistic assumptions, the model itself has caused controversy regarding its accuracy.[48]

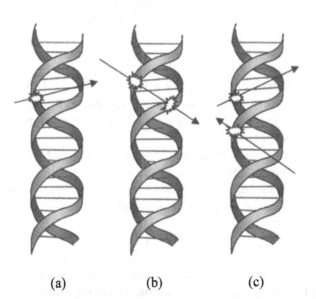

(a)   (b)   (c)

Figure 11.1 Illustration of radiation damage to DNA. (a) Single strand break. (b) Double-strand break by a single event. (c) Double-strand break by two independent events.

Despite the lack of accurate and quantitative risk analysis on low-dose radiation, it is generally accepted that the estimated x-ray radiation risk to children is significantly higher relative to adults, given the increased lifetime radiation risks for children. Based on a recent study, the estimated lifetime cancer mortality risk attributable to CT radiation for a one-year-old child is an order of magnitude higher than adults, although these figures still represent a small increase in cancer mortality over that from natural background radiation.[20] This is compounded by the fact that approximately 600,000 abdominal and head CT exams are performed annually on children under the age of 15.

11.2 Measurement of X-ray Dose

As CT technology evolved over the past thirty plus years, multiple dose measurement units and methodologies were introduced to address some of the shortcomings of the previous methodologies. When combined with proposals and recommendations from different agencies and organizations, the subject of dose measurement often causes confusion. For the ease of understanding, we approach this subject in the following manner. We will start with the dose measurement units and methods that have been generally accepted as the standard today. Once the reader gains a general understanding of the subject, we will briefly discuss other measurements units and methods.

11.2.1 Terminology and the measurement standard

Based on the previous discussion of the interaction between x-ray photons and matter, it is clear that during the interaction process, energies are transferred from the x-ray photons to the matter. Therefore, the x-ray radiation dose measurement is simply a measure of the amount of energy transferred per unit mass during the interaction. For clarity, let us first define a set of terminologies: x-ray exposure, absorbed dose, and equivalent dose.

The term "x-ray exposure" describes the x-ray's ability to ionize air, and is measured in roentgens (R).[21,22] It is defined as the amount of x rays required to produce an electrostatic charge of 2.58×10^{-4} coulomb (C) in 1 kilogram of air at standard temperature and pressure. Based on this definition, x-ray exposure describes only the amount of ionization, not the amount of energy absorbed by the tissues being irradiated.

Unlike the term "x-ray exposure," the term "absorbed radiation dose" (also known as the radiation dose) indicates the amount of energy absorbed per unit mass. The International Commission on Radiological Protection (ICRP) recommends that dose be measured in grays (Gy):

$$1 \text{ Gy} = 1 \text{ J/kg}. \tag{11.1}$$

To appreciate the magnitude of such a quantity, the equivalent energy of 1 Gy can sustain a 1-watt light bulb for 1 second. When the same equivalent energy is used to heat 1 gram of water, the water temperature will increase by 0.24 °C.

Our knowledge about the stochastic risk of radiation is based on whole-body exposure. For CT, however, the exposure is highly nonuniform and often involves a partial-body irradiation. Consequently, we need to map the stochastic risk of the partial-body CT scan to an equivalent whole-body exposure. In addition, we need to take into consideration the significantly different biological effect produced by different types of radiation (e.g., the biological effects of the absorbed dose deposited by alpha particles are greater than the same dose deposited by x rays). We also need to account for the different sensitivities of various organs to the radiation. To obtain a quantity that expresses the radiation damage on a common scale, the concept of effective dose E (formerly known as *dose equivalent*) was introduced:[21]

$$E = \sum_S w_S w_R D_{S,R}, \qquad (11.2)$$

where the subscript S represents each tissue type, the subscript R represents each radiation type, $D_{S,R}$ is the average absorbed dose to tissue S with radiation R, w_S is the tissue weighting factor, and w_R is the radiation-weighting factor. For diagnostic x rays, $w_R = 1$. The measurement for E is in sieverts (Sv), honoring the Swedish scientist who was active in the ICRP:

$$1 \text{ Sv} = 1 \text{ J/kg} \qquad (11.3)$$

For reference, the annual whole-body effective dose due to natural background radiation is about 3 mSv per year (1 mSv = 10^{-3} Sv), averaged over the population of the US, while because of the high altitude, people living in Denver receive a 0.5-mSv extra dose, as compared to people living in Los Angles.

Although these definitions are general and are not limited to x-ray CT, they provide little information on how dose should be measured. For a single CT scan taken with a step-and-shoot mode, nearly all of the primary radiation is confined to a thin cross-section of the nominal slice thickness T. Because of the beam divergence, the penumbra of the beam, and the scattered radiation, dose is also delivered to tissues outside the nominal imaging section. This results in a dose profile in z (perpendicular to the cross-section) with long tails, as illustrated in Fig. 11.2 for the dose profile of a 10-mm scan.

When multiple scans are performed in the adjacent region, x-ray dose from nearby scans also contributes to the dose to the current location, due to the long tails of the dose profile. If we combine the x-ray dose from all scans, we obtain a composite dose profile, as shown in Fig. 11.3. This figure illustrates the composite dose profile of seven scans acquired with 10-mm collimation at 10-mm increments (the table travels 10 mm between adjacent scans). Note that the dose at the center section is significantly higher than the single-slice dose profile. In this particular example, the average composite dose within the center region of width T is roughly 85% higher than the average dose of a single scan. Although this example is obtained with scans taken with step-and-shoot mode, similar conclusions can be obtained for the helical/spiral scan mode as well. In fact, the multiple-scan dose profiles for the helical mode are very similar to those of the step-and-shoot scans with the exception of inhomogeneities following the spiral pattern.[5]

To account for the fact that the majority of CT scans performed in a clinical environment consist of multiple scans, Computed Tomography Dose Index (CTDI) was proposed. The most commonly used index is $CTDI_{100}$, which refers to the dose absorbed in air, although it is measured in the standard polymethyl-methacrylate (PMMA) phantoms. For this index, the dose is integrated over a fixed length of 100 mm,

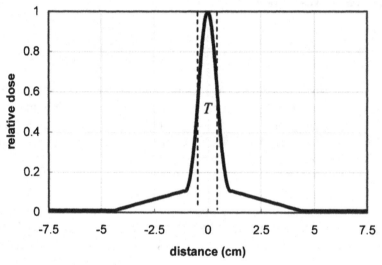

Figure 11.2 Example of single-scan dose profile for 10-mm slice thickness.

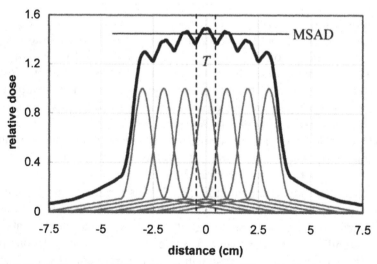

Figure 11.3 Illustration of multiple-scan dose contributions. Multiple-scan average dose (MSAD) is obtained by summing dose contributions from adjacent slices. The figure shows seven scans with 10-mm slice thickness at 10-mm increments.

$$\text{CTDI}_{100} = \frac{1}{nT} \int_{-50\,\text{mm}}^{50\,\text{mm}} D_a(z)dz \;, \tag{11.4}$$

where $D_a(z)$ is the dose absorption distribution in z for a single axial scan, n is the number of detector rows used during the scan, and T is the nominal thickness of each row (not necessarily equal to the reconstructed slice thickness). The quantity nT, therefore, is the nominal x-ray beam width during the data acquisition. For example, if a 64 × 0.625-mm detector configuration is used for acquisition, nT = 64 × 0.625 mm = 40 mm. The length 100 mm was chosen mainly for the practical reason that most CT dose chambers have an active length of 100 mm. Based on the definition of Eq. (11.4), CTDI_{100} measures the dose per unit length in z.

Because CT scanners expose patients to x-ray radiation over 360 deg, the x-ray dose is significantly more homogeneous than in conventional x-ray. In conventional x-ray radiography, the skin at the x-ray entrance plane receives 100% dose, and the percentage decreases quickly with the penetration depth. The exit exposure is roughly 1% or less of the entrance exposure. For illustration, Fig. 11.4(a) shows an estimated dose distribution inside a uniform water phantom in a radiographic configuration when the x-ray tube is at the 12 o'clock position. Note that the front surface of the water phantom receives the highest x-ray dose because x-ray photons are not attenuated. The back of the phantom (near the 6 o'clock position) receives the least amount of radiation because of the large attenuation of the phantom to the x-ray photons. In CT scans, the portion of the phantom that directly faces the x-ray source changes constantly as the x-ray tube rotates about the patient. As a result, doses are distributed more evenly across the entire phantom, as shown by Fig. 11.4(b). A closer inspection of Fig. 11.4(b) shows that variation in dose still exists between the periphery of the phantom and the center of the phantom. Note that during the scan, the interior region of the phantom is always partially "shielded" by the outer portion of the phantom and is never exposed directly to the x-ray source. It is not difficult to understand that the degree of dose nonuniformity depends highly on the size, shape, and composition of the object. For CT head scans, for example, the center of the patient receives nearly as much radiation dose as the periphery. For body scans, the dose uniformity decreases with the patient size increase. For a 35-cm diameter body, the central dose is roughly one fifth to one third of the peripheral dose.[1] To account for the spatial variation of the dose, a weighted dose index, CTDI_w, which combines dose information at different locations, was proposed, and is calculated based on the following formula:

$$\text{CTDI}_w = \left(\frac{1}{3}\right)\text{CTDI}_{100}(\text{central}) + \left(\frac{2}{3}\right)\text{CTDI}_{100}(\text{peripheral})\,. \tag{11.5}$$

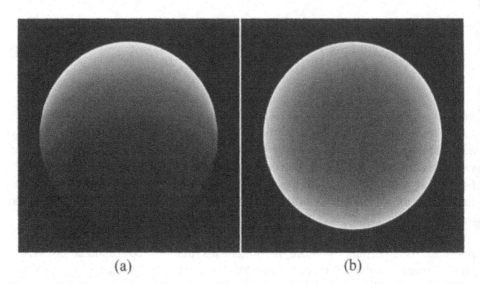

(a)   (b)

Figure 11.4 Comparison of x-ray radiography and CT dose distribution. (a) Dose distribution with the x-ray tube at the 12 o'clock position. (b) Dose distribution when the gantry rotates 360 deg.

U.S. federal regulations require CT manufactures to report CTDI values measured in phantoms that are at least 14 cm long with a diameter of 16 cm for head and 32 cm for body. The most commonly used phantoms for dosimetry are the PMMA phantoms. Figure 11.5 shows an example of the CT dose measurement setup with a 32-cm body phantom. The peripheral dose is based on the dose measurements of the ion chamber in the four predrilled peripheral holes near the rim of the phantom that is highlighted by the dotted circles in Fig. 11.5. The center dose is based on the measurement of the ion chamber in the center hole of the phantom that is highlighted by the solid circle in Fig. 11.5.

The definition of $CTDI_w$ considers only the x-ray exposure for a step-and-shoot scan, and does not take into account the x-ray dose received when a helical scan is performed. In helical or spiral scans, the patient table travels at a constant speed while data collection takes place. One parameter that describes how fast the patient table travels is the helical pitch, defined as the ratio of the table traveling distance in one gantry rotation over the nominal beam width. As the helical pitch increases (assuming all other parameters are kept constant), the same amount of x-ray radiation is distributed over a longer z. Since CTDI is defined as the dose per unit length in z, its value should reduce with the increase in helical pitch. This led to the introduction of $CTDI_{vol}$:[23]

$$CTDI_{vol} = \frac{CTDI_w}{pitch}.$$ (11.6)

This value is expressed in mGy and is displayed on most of the CT consoles during the scan prescription. In a clinical environment, however, the scan range in z can vary significantly depending on the clinical indications. For example, a

CTA runoff study of the entire vascular system covers over 1 m in z from the arch to the toes, while a chest scan may cover only 16 cm. If all other scanning parameters are kept the same, both studies will record the same $CTDI_{vol}$. However, the patient with the runoff study clearly receives significantly more radiation dose than the chest patient. To account for the integrated dose for the entire CT exam, a dose-length product (DLP) was established (with the unit in mGy × cm):[24]

$$DLP=CTDI_{vol} \times L \qquad (11.7)$$

where L is the scan length in z. Note that for a helical data acquisition, L is the total scan length, not the reconstructable length in z. Because the patient table travels constantly during a helical acquisition and there is a minimum angular range required to collect a complete dataset, the x-ray tube is turned on at some distance before the table reaches the starting location of the reconstructable volume, and is turned off at certain distance after the table passes the ending location of reconstruction. This is often called the "overbeam" or "overscan," which will be discussed later in the Section 11.3. The value of L should reflect both the overbeam range as well as the reconstructable length.

Different human organs have different sensitivities to the x-ray dose. That is, the same amount of x-ray radiation may present different risk levels to different

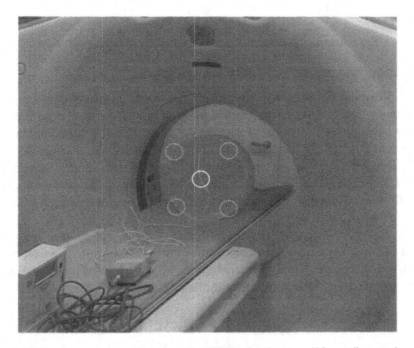

Figure 11.5 Dose measurement setup with CTDI body phantom (32-cm diameter) and ion chamber (100-mm length).

organs. To account for this effect, an *effective dose* concept was developed. Based on Eq. (11.2), the effective dose is calculated from information about dose to individual organs and the relative radiation risk assigned to each organ. Monte Carlo simulation is often used to determine specific organ doses by simulating the absorption and scattering of x-ray photons in various tissues using a mathematical model of the human body.[25,26] In a clinical environment, however, such calculation is too time consuming and not practical. Instead, a reasonable approximation of the effective dose is obtained using the equation:[24]

$$\text{Dose}_{\text{effective}} = k \times \text{DLP} \tag{11.8}$$

where k is a conversion factor in mSv $\times$ mGy^{-1} $\times$ cm^{-1} that varies depending on the imaged body region. Examples of some conversion factors are shown in Table 11.1.[3]

Table 11.1 Conversion factor for different anatomies.

Region of Body	K (mSv mGy^{-1} cm^{-1})
Head	0.0021
Neck	0.0059
Chest	0.014
Abdomen	0.015
Pelvis	0.015

11.2.2 Other measurement units and methods

In the previous section, we described the dose measurement units and measurement methodologies that are widely accepted today. These units and methods have evolved over the years as CT systems become more complex. For example, not long ago, the absorbed dose was measured in rads (radiation absorbed dose), which is the absorption of 1×10^{-5} J of energy per gram of matter. From Eq. (11.1), it is not difficult to come up with the conversion that 1 Gy = 100 rads. Similarly, the equivalent dose was measured in rems (Roentgen equivalent man), and 1 Sv = 100 rems.

Previously, we discussed the definition of CTDI$_{100}$, which was designed to account for the long tails in the dose profile (Fig. 11.2). Prior to the introduction of CTDI$_{100}$, a standard multiple-scan average dose (MSAD) was defined to account for the dose increase in the multiple-scan mode:[9]

$$\text{MSAD} = \frac{1}{I} \int_{-I/2}^{I/2} D_{N,I}(z)\,dz, \tag{11.9}$$

where the subscript N stands for N scans, I is the distance between adjacent scans, and $D_{N,I}(z)$ is the dose as a function of the distance z. This definition implies that scans beyond the two end scans are far from the central region and their dose contributions are not substantial. For illustration, MSAD is labeled in Fig. 11.3.

Although the most commonly used dose measurement today is the $CTDI_{100}$, the CTDI was originally defined by the U.S. Center for Devices and Radiological Health (CDRH) as the 14-slice average dose:

$$CTDI = \frac{1}{nT} \int_{-7T}^{7T} D(z)dz, \qquad (11.10)$$

where $D(z)$ is the dose function, T is the nominal tomographic section thickness, and n is the number of tomograms produced in a single rotation. For single-slice scanners, n is set to 1 for almost all commercially available scanners. For multislice scanners, n is commonly set to the number of slices. Although the CTDI definition of Eq. (11.10) is convenient for practical measurement, it is problematic when T is small. Note that the dose calculation based on this definition is limited to the width of 14 scans ($14T$), with the assumption that the region outside these limits does not contribute significantly to the dose calculation. For a small value of T or a large value of n, this assumption is not valid.

11.2.3 Issues with the current CTDI

$CTDI_{100}$ was defined for the convenience of practical measurements, since the active area of most dose chambers used to measure the x-ray dose is 100mm long. The original CTDI was defined over an infinite range,

$$CTDI_\infty = \frac{1}{T} \int_{-\infty}^{\infty} D(z)dz. \qquad (11.11)$$

Since the $CTDI_{100}$ integration distance is limited to 100 mm, there is a tendency to underestimate the x-ray dose. That is, if we measure the ratio of $CTDI_{100} / CTDI_\infty$, the value is significantly less than unity. A Monte Carlo simulation study has shown that the ratios for the 20- and 40-mm beam collimation are 0.63 and 0.62, respectively, for the center dose.[28] The ratios for the periphery dose are 0.88 and 0.87, respectively. As the detector z coverage increases, the dose underestimation becomes an even bigger issue.

Another issue with the CTDI is the phantom used in the dose measurement. Recall that the dose measurement of the two CTDI phantoms (16 and 32 cm) are used to provide the basis for the patient dose calculation. However, patients come in different shapes and sizes, and two round phantoms are an oversimplification of the patient population. In addition, patient organs consist of different tissue types and are not uniform across the entire FOV. This leads to a nonuniform dose distribution inside the patient, which is in sharp contrast to the uniform CTDI phantoms. These factors, no doubt, will introduce bias in the dose estimation for the patient.

CTDI$_{100}$ was defined many years ago and was based on the assumption that the x-ray beam width in z is substantially smaller than 100 mm so that the x-ray radiation outside this range can be ignored. In addition to the primary beam, leakage and scattered radiations also contribute to the overall dose to the patient. Leakage radiation refers to the x-ray photons that escape from the collimator assembly and interact with the patient [Fig. 11.6(a)]. Although there are specs governing the upper limit of leakage intensity outside the primary beam, the total amount of leakage radiation is not negligible. The scattered radiation accounts for both the primary scatter as well as the multiple scatter events. In recent years, the detector coverage in z (therefore the x-ray beam width) increases significantly. Some of the CT scanners on the market today are close to or exceed the 100-mm range for the primary beam [Fig. 11.6(b)]. Considering the additional x-ray radiation due to leakage and scatter, the large z-coverage systems have created two issues with dose measurements: dose measurement instrument (including the phantoms) and dose measurement methodology. In order to adequately simulate the x-ray scattering effects inside a human body, the phantom has to be significantly longer than the detector z coverage. Otherwise, scattered radiation can easily exit the phantom or can be outside the active range of the dose chamber, which leads to an underestimation of the true dose to patient [Fig. 11.6(b)]. A similar situation applies to the leakage radiation as well. On the other

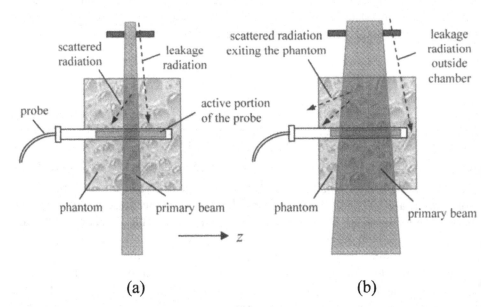

(a)   (b)

Figure 11.6 Impact of phantom size and probe size on dose measurement accuracy. (a) When the primary beam is significantly smaller than both the phantom and dose probe, nearly all leakage radiation and scattered radiation are captured by the probe. (b) When the primary beam is nearly the same size or bigger than the length of the probe, a significant portion of the leakage radiation, scattered radiation, or primary beam is not measured.

hand, a longer phantom makes the phantom handling more difficult than it is currently, and a longer dose-probe makes the measurement more expensive. Although many proposals have been made to overcome these difficulties, a new measurement standard has yet to be established to account for the much increased x-ray exposure in z.

11.3 Methodologies for Dose Reduction

Many factors impact the dose administered to a patient. For half-scan data acquisition (see Chapter 7 for a detailed description), the x-ray dose is not uniform since the x-ray is activated over only 180 deg plus a fan angle. The dose nonuniformity can often be used to reduce the dose to operators and to sensitive parts of the patient. For example, by turning on an x-ray when the x-ray tube is at the posterior side of the patient, the x-ray dose can be significantly reduced to the operator who performs the biopsy operations as well as to the patient's frontal surface, where the more sensitive organs are located.

Another important factor that impacts the dose is the x-ray beam quality. Most CT manufacturers provide flat filters to remove lower-energy x-ray photons that would otherwise be quickly absorbed by a patient, as shown in Fig. 11.7. The filter is commonly made of aluminum or copper. The goal of the filter design is to achieve a good compromise between radiation dose and low-contrast performance. To further reduce the skin dose to the patient (and to reduce the dynamic range of the data acquisition systems and improve the noise homogeneity), an additional bowtie filter can be used, as shown in Fig. 11.7. The bowtie filter is shaped to compensate for the variable path length of the patient across the scan FOV, and the filter's thickness increases quickly from the center to the outer edge, which significantly reduces the skin dose.

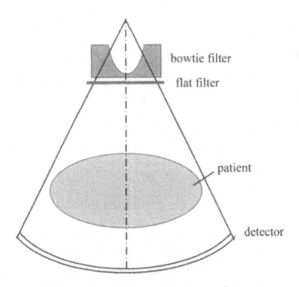

Figure 11.7 Illustration of a bowtie filter.

Most CT scanners used in a clinical environment are not dedicated to the scanning of a specific organ, so they must be able to scan a patient's head, chest, abdomen, or extremities. These different parts of the anatomy have significantly different sizes and shapes. For example, a head can easily fit inside a 25-cm FOV, while an abdomen often requires careful positioning to fit inside a 40-cm FOV. Because the bowtie filter was designed to compensate for the variable path length of a patient across the entire FOV, a single bowtie is unlikely to be optimal for all anatomical scans. Many CT scanners employ multiple bowties to account for different anatomies inside a patient and different patient sizes. These bowties deliver different dose amounts to patients even when all other conditions (e.g., tube voltage, current, scan time, helical pitch, etc.) are the same. For illustration, Fig. 11.8 plots a normalized dose as a function of the x-ray tube voltage for two different-sized-head bowtie filters. The average difference in dose between the two bowties is 13.7%.

11.3.1 Tube-current modulation

Another technique to reduce the x-ray dose to the patient while maintaining image quality is to use a modulated x-ray tube current based on the attenuation level of the patient.[12,29,30] Since the tube current is determined based on the attenuation characteristics of the patient at a particular location (rather than a predetermined value), an operator can select a desired noise level in the

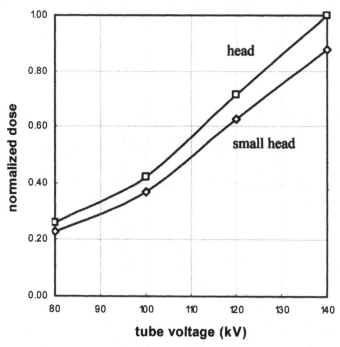

Figure 11.8 Illustration of the dose difference between two different bowties.

reconstructed image prior to the start of the scan and have the scanner determine the required tube current. The tube current generally varies with the projection angle β as well as the location z, as illustrated in Fig. 11.9. This approach not only helps to reduce the patient dose, but also provides better x-ray tube cooling performance.

There are two approaches to implementing this concept. One approach is to determine the patient attenuation variation prior to the CT scan based on the scout images. Since a scout image measures the patient attenuation from either the A-P direction (6 or 12 o'clock position) or lateral direction (with the tube in the 3 or 9 o'clock position), certain assumptions must be made about the patient to estimate the attenuation characteristics from other projection angles. On the other hand, because scout images are acquired ahead of CT data acquisition, the patient dose can be accurately calculated prior to the data acquisition, which allows the operator to make adjustments when necessary. The other approach is to modulate the tube current "on the fly" during data acquisition based on the actual projection measurement. Because of delays in transmitting and processing the measured signals, a tube current adjustment is typically based on the CT projections one revolution earlier. This approach does not require any modeling or assumptions about the patient. On the other hand, it makes it difficult to predict the total dose to the patient prior to the scan. In addition, its accuracy may be impacted by the detector-coverage increase, since the anatomy covered in the previous rotation can vary significantly. For illustration, Fig. 11.10 shows a scout image of a chest phantom and the corresponding average tube current per rotation as a function of z (horizontal direction). Note that the shoulder and upper arm regions produce significantly higher attenuation, and the corresponding average mA is significantly higher.

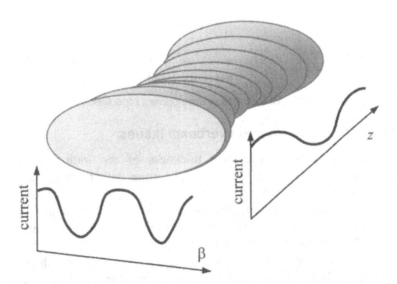

Figure 11.9 X-ray tube-current modulation with both the view angle β and location z based on the detected attenuation.

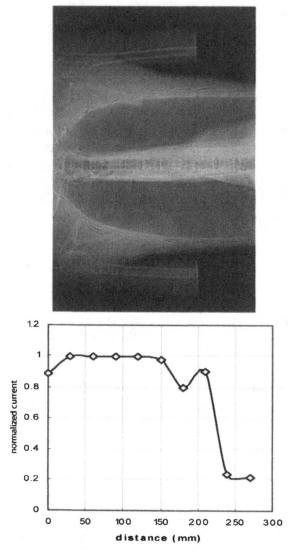

Figure 11.10 Example of tube-current modulation in *z*.

11.3.2 Umbra-penumbra and overbeam issues

As mentioned previously, the slice thickness of the multislice scanner is determined mainly by the detector cell aperture rather than the prepatient collimator aperture (as is the case for a single-slice scanner). This can lead to a compromise in dose utilization. For illustration, Fig. 11.11 depicts the cross-section of a 4-slice scanner. In early designs of the multislice scanner, the active cells of the detector were well within the umbra portion of the x-ray beam. (The umbra region can be defined as the zone in the detector in which the entire focal spot is visible.) This design helps to stabilize the detector readings, since the focal spot is likely to shift during data acquisition due to thermal expansion. By ensuring that the detector active region is inside the umbra portion of the beam with margins on both ends, small focal spot shifts in z will not influence the flux received by each detector cell. Unfortunately, x-ray radiation outside the umbra region irradiates the patient, as shown by the shaded triangles in Fig. 11.11(a). These areas, often called the penumbrae of the x-ray beam, can be defined as the detector zones in which only a portion of the x-ray focal spot is visible. Because these zones correspond to the inactive portion of the detector, the x-ray photons that pass through the patient are discarded. As a general rule, the umbra-to-penumbra ratio decreases quickly as the slice thickness decreases (the penumbra portion of the beam is held nearly constant for different slice thicknesses). If we define the dose efficiency of the system in z as the ratio of the detected x-ray photons to the total x-ray flux that irradiates the patient, the efficiency of the early multislice scanners is quite low.

To improve the dose efficiency, a dynamic x-ray beam-steering technique was developed.[11] As discussed previously, the active detector region is kept inside the umbra portion of the beam to ensure that small shifts in the focal spot do not impact the x-ray flux to each detector cell. If the x-ray flux can be kept stable even in the penumbra portion of the beam from a detector point of view, the penumbra beam can be included in the active detector cell region and improve dose efficiency. This is accomplished by dynamically adjusting the prepatient collimator so that it tracks the focal spot motion. If this is performed in real time during the scan, the x-ray beam "shadow" cast onto the detector can be kept nearly constant, as shown in Fig. 11.11(b). Consequently, a significant penumbra portion of the x-ray beam can be used during data acquisition.

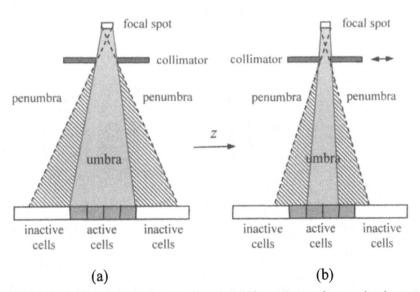

Figure 11.11 Illustration of dynamic collimator control for umbra and penumbra beams. (a) With no collimator control and the active detector region inside the umbra portion of the x-ray beam. (b) With dynamic collimator control and both umbra and penumbra portions of the beams inside the active detector region.

The significance of the penumbra utilization is reduced with an increase in the number of detector rows. As a hypothetical example, assume the width of each penumbra beam on a detector is 2 mm on each side, and its intensity profile is triangular in shape. The total contribution of the penumbra beams on both sides is roughly equal to a 2-mm umbra beam. For a 4 × 1.25-mm detector configuration, the umbra-to-penumbra ratio is 2.5. If only the umbra portion of the x-ray beam is used during data acquisition, the dose efficiency is roughly 71%. For the same system geometry, the umbra-to-penumbra ratio for a 64 × 0.625-mm detector configuration is 20. This corresponds to a 91% dose efficiency if only the umbra portion of the beam is used.

The concept of dynamic beam tracking can also be applied during a helical data acquisition. Because the patient table is moving constantly during data acquisition and the reconstruction process requires a minimum projection angular range for accurate reconstruction, the x-ray exposure range in z is typically longer than the range of reconstruction. At the start and end of a helical data acquisition, portions of the patient that are exposed to the x-ray radiation are not reconstructed, as shown in Fig. 11.12(a). This phenomenon is typically called "overbeam" or "overscan." The size of the overbeam region changes as a function of the detector configuration, helical pitch, and the reconstruction algorithm. Consequently, the amount of overbeam varies significantly for different scanners. The overbeam portion represents an unused x-ray dose and should be kept to a minimum. One method of reducing overbeam is to use dynamic collimator control. Since the geometric relationship between the reconstructed volume, patient table position, and helical pitch is well defined prior to scanning, the collimator aperture can be dynamically changed during the scan to reduce the overbeam. At the start of the scan, the left-hand side of the

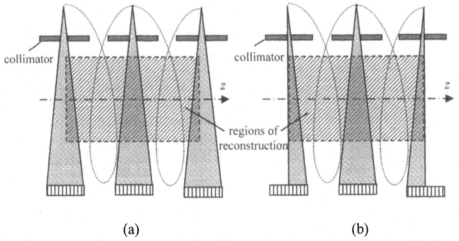

(a)   (b)

Figure 11.12 Illustration of dynamic collimator control for overbeam. (a) Collimator aperture is not adjusted during the helical scan and portions of the patient outside the reconstruction region are exposed to x rays. (b) Collimator aperture is dynamically adjusted so that only the reconstructed region is exposed to x rays.

beam (in z) is collimated out, and only the reconstructed region is exposed to the x ray, as shown in Fig. 11.12. As the patient table moves inside the reconstruction volume, the collimator aperture opens to allow the full x-ray beam to reach the patient. When the patient table approaches the end of the reconstruction volume, the right-hand side of the beam (in z) gradually collimates down. Since only the unused x-ray beams are collimated out, there is little impact on the reconstructed image quality.

11.3.3 Physiological gating

Most diagnostic CT scans are collected during patient breath-holding due to the short data acquisition time. However, with the introduction of PET-CT and SPECT-CT, came an increased need to acquire CT data in time-frames similar to PET or SPECT to properly model the patient's respiratory motion. When CT data are acquired and reconstructed in a different respiratory phase than the PET or SPECT data, artifacts can be introduced during the attenuation correction, as shown by the arrow in Fig. 11.13(a). This type of artifact can be effectively controlled by collecting the scans in multiple respiratory phases, then using either the average CT image as the basis of the attenuation correction or a gated PET data acquisition to match the respiratory phases of the CT and PET images [Fig. 11.13(b)]. One consequence of using a gated CT acquisition is the increased dose to patient, since CT projections need to be acquired over several respiratory motion cycles. To reduce the impact of the dose increase, extremely low-dose CT techniques can be used in combination with the projection filtration process to achieve a good balance between dose reduction and spatial resolution.[31] This approach takes advantage of the fact that the attenuation correction for PET is a low-spatial-resolution application.

Respiratory gated CT is not limited to PET-CT or SPECT-CT applications. It is also used in oncology applications to track a tumor location during treatment. Like the PET and SPECT acquisitions, the radiation treatment process is long, so

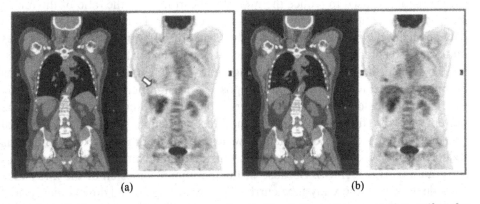

(a)   (b)

Figure 11.13 Illustration of respiratory-motion-induced attenuation correction artifact for (a) PET and (b) artifact reduction with gated CT (images courtesy of Dr. T. Pan of M.D. Anderson Cancer Center).

patient breath-holding is often difficult to implement. Under normal breathing conditions, many tumor motions follow the respiratory motion cycle. For oncology applications, the tumor must be placed inside the treatment zone to ensure adequate delivery of the radiation dose. Therefore, either the treatment margin needs to be increased to ensure the tumor always stays within the treatment volume, or the tumor motion needs to be tracked dynamically so the treatment zone can be adjusted during the process. The first approach is less desirable because it increases the collateral damage to the surrounding healthy tissues. To model and track the tumor motion during the treatment, four-dimensional (4D) CT is needed to characterize the tumor motion over several respiratory phases. Although for this application increased dose is not as critical as in the PET and SPECT cases, reducing the CT dose to a minimum is still desirable.

Another physiological gating often used in CT is cardiac gating. With the rapid increase of detector coverage in recent years, cardiac scanning has received increased attention. Although details of cardiac acquisition and reconstruction will be presented in Chapter 12, we want to mention here the important dose-reduction techniques for cardiac scanning.

When CT cardiac scanning was first introduced, the x-ray dose was high due to the nature of cardiac imaging: thin slice, fast scan speed, and low helical pitch. In typical cardiac CT imaging, the projection is acquired with the EKG signal and the reconstruction takes place during the heart's quiescent period. For a low heart rate, the quiescent period corresponds to the diastolic phase, and for a high heart rate, the systolic phase. Helical acquisition was typically used in the early days of cardiac imaging because of its small detector coverage. When the detector z coverage is less than 10% of the heart size in z, helical acquisition allows an efficient acquisition duty cycle (100%) and enables coverage of the entire heart within a single breath-hold. For cardiac acquisition, scanning speeds faster than 0.4 sec per gantry rotation are typically used. When the gantry rotation cycle is significantly smaller than the cardiac cycle, multiple gantry rotations take place within each heart cycle. Considering the fact that half-scan reconstruction is used to reduce the impact of heart motion, the ratio of the image reconstruction temporal window (less than 4 sec with a 1-cm detector) to the total data acquisition window [16 sec for a heart rate of 60 beats per min (bpm)] was quite small. Therefore, the dose efficiency was poor and patient dose was high.

To reduce the dose in cardiac imaging, x-ray tube-current modulation was proposed. The concept behind this proposal is to significantly reduce the x-ray tube current outside the image reconstruction window and keep the x-ray tube current at its full level inside the reconstruction window. A typical implementation uses 80% modulation, which means that the x-ray tube current is reduced to 20% of the peak current outside the reconstruction window, as illustrated in Fig. 11.14(a). Because of the thermionic emission of the x-ray tube cathode, however, the x-ray tube current does not ramp up and down quickly, and the dose reduction is compromised. To overcome the shortcomings of the helical acquisition, a prospectively gated step-and-shoot mode data acquisition was proposed.[32] This proposal is based on the observation that in recent years, the detector z coverage has increased significantly. For example, the detector z coverage of LightSpeed™ VCT scanners at the iso-center is 4 cm. Considering that the typical size of a human heart is less than 12 cm, this scanner needs only three data acquisitions to cover the entire heart. Therefore, the interscan delay due to the step-and-shoot acquisition is no longer a bottleneck in cardiac imaging. Since the x-ray tube can be turned on and off much faster than the time it takes to increase and decrease the current, gating is more efficient [Fig. 11.14(b)]. Recent studies have shown that up to an 83% dose reduction can be obtained with this approach.

Note that the concept of prospective gating can also be applied to the helical scan in order to reduce dose to the patient by turning the x-ray tube on and off during a low-pitch helical acquisition. Since the helical pitch is determined based on the measured heart rate prior to the data acquisition, the pitch is typically set conservatively to account for potential heart rate variation. In addition, the z coverage for a helical cardiac reconstruction during each cardiac cycle can be reduced by as much as a factor of $0.5(\pi+\gamma_m)p/\pi$ due to the continued table motion, where p is the helical pitch and γ_m is the full fan angle. As a result, the selected pitch is typically smaller than the "optimal" pitch value that avoids significant overlaps in the successive data acquisition, and results in a higher dose compared to the step-and-shoot mode of acquisition.

Recently, gated cardiac acquisition with high helical pitch has also been reported.[33,34] In such data acquisition, the entire heart is covered in a single

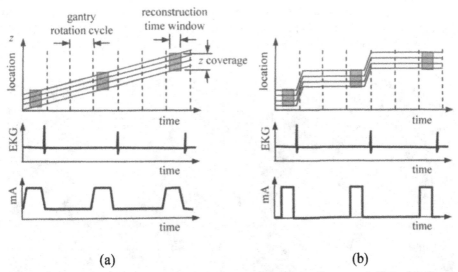

                (a)   (b)

Figure 11.14 Illustration of helical and step-and-shoot cardiac acquisition. (a) Cardiac helical acquisition in which the table translates at a constant rate. Because of the fast gantry speed relative to the heart rate, multiple gantry rotations take place within each cardiac cycle. X-ray tube current can be adjusted outside the reconstruction phase to reduce dose. (b) Step-and-shoot cardiac acquisition in which the table indexes to the next location only after completion of the acquisition. X-ray tube current is turned off outside the reconstruction window.

cardiac cycle with a helical pitch significantly higher than one. Given the definition of CTDI$_{vol}$ [Eq. (11.9)], the dose is inversely proportional to the helical pitch, and lower radiation dose is achieved at higher pitch values. Because the detector coverage is much smaller than the heart size in z, temporal skew is expected since the top portion of the heart is acquired at a different time (different cardiac phase) from the bottom portion of the heart.

11.3.4 Organ-specific dose reduction

In many clinical scans, organs that are not the target of the examination receive collateral x-ray radiation because of their proximity to the organ of interest. For example, the eyes often receive a significant amount of x-ray radiation during a head scan, even though the brain is the organ of interest [Fig. 11.15(a)]. Similarly, breasts are routinely exposed to x rays during a chest scan because it is impossible to avoid the breasts when the lung is the target organ [Fig. 11.15(b)].

It is known that the sensitivities of different organs to x-ray radiation vary significantly (Table 11.2). In recognition of this fact, different approaches have been made to minimize the dose to more sensitive organs. The best approach is sensitive-organ avoidance. For example, when performing brain scans, the CT gantry can be tilted with respect to the x-y plane to avoid direct x-ray exposure to the retina, as illustrated in Fig. 11.16 with a human skull phantom. Although this approach is effective, not all sensitive organs can be completely avoided due to the overlapping nature of organ locations. A good example is a chest scan. If an entire lung must be scanned, x-ray radiation exposure to the breasts seems unavoidable, but the dose to the breasts can be minimized. For example, x-ray tube-current modulation can be used to reduce the x-ray flux when the x-ray tube is in front of the patient, and the tube current can be increased when the tube is at

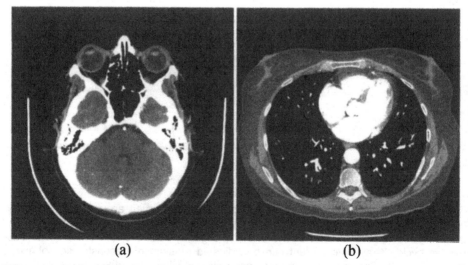

(a)   (b)

Figure 11.15 Examples of "collateral" CT dose to organs. (a) Lens of the eyes and (b) breasts.

Table 11.2 Weighting factor for organ dose (compiled based on ICRP 60 radiation weighting factors).

Organ	Weight
Gonads	0.2
Bone marrow (red)	0.12
Lung	0.12
Stomach	0.12
Colon	0.12
Bladder	0.05
Breast	0.05
Liver	0.05
Oesophagus	0.05
Thyroid	0.05
Skin	0.01
Bone surface	0.025

the patient's back. Figure 11.17(a) shows an estimated dose distribution of a uniform water phantom when a sinusoidal waveform is used as the tube-current modulation function. The peak-to-valley ratio of the current is 5:1 (80% modulation). It is clear from the figure that the x-ray dose is significantly reduced toward the front of the patient compared to the distribution without mA-modulation in Fig. 11.4(b). Because the breasts are located at the front of the patient and a minimum of 180 deg plus the fan angle is needed to reconstruct an image, the system can be designed such that the center view of the half-scan acquisition is in the 6 o'clock position and the x-ray tube is turned off beyond the half-scan region. A further reduction in x-ray dose toward the front of the patient can be realized, as shown in Fig. 11.17(b). The clinical protocol design, however, must achieve a proper balance between noise (noise uniformity) and the dose to sensitive organs.

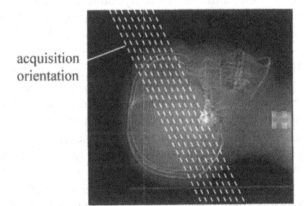

acquisition
orientation

Figure 11.16 Illustration of tilted gantry acquisition to avoid direct x-ray exposure to retina.

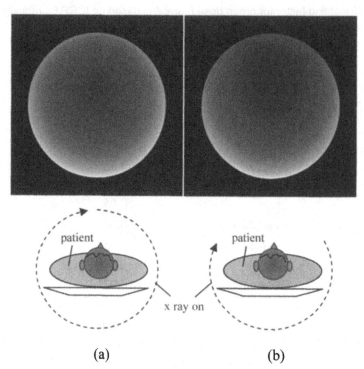

(a)   (b)

Figure 11.17 Estimated CT dose distribution of a uniform water phantom. (a) Full scan with mA modulation and 5-to-1 peak-to-valley sine curve. (b) Half-scan to avoid frontal disposure.

11.3.5 Protocol optimization and impact of the operator

The determination of a clinical protocol includes the selection of the x-ray tube voltage, tube current, tube-current modulation, detector configuration, gantry rotation speed, gantry tilt, contrast injection, helical pitch, reconstruction filter kernel, slice thickness, and the reconstruction FOV. Optimization of the protocol has a direct impact on the dose to the patient.

The operator has the ultimate authority on the protocol and technique selection. Although CT scanners can provide recommended protocols and scanning techniques, the operator must decide whether to follow the recommendation or overwrite the prescription. (The "operator" includes the technicians and the radiologists.) Since the operators have full knowledge of the specific pathology to be targeted and specific information about the patient, protocol optimization should be determined by the operators, not the scanner.

For illustration, we will examine the optimization of one protocol parameter: x-ray tube voltage. Chapter 7 discussed the energy-dependent characteristics of the attenuation coefficients for different materials (Fig. 7.66). Lower-energy x-ray photons provide better enhancement for low-contrast objects, and high-energy photons are more penetrating. On the other hand, a nonlinear relationship exists between the x-ray energy (and therefore the tube voltage) and the resulting dose to a patient. Figure 11.18 plots the normalized dose as a function of normalized kVp using 80 kVp as the reference point (tube current, scan time, collimator aperture, etc., are kept constant). The normalized dose is the ratio of the dose at the target kVp over the dose at 80 kVp. It can be shown that the normalized dose is related to the normalized kVp by the following equation:

$$\frac{\text{dose}(x\,\text{kV})}{\text{dose}(80\,\text{kVp})} = \left(\frac{x}{80}\right)^{2.47}. \tag{11.12}$$

It is clear from the figure that the dose does not increase linearly with the kVp. In the figure, the diamonds represent the measured dose ratios, and the solid line is the fitted curve.

Based on our previous analysis, we know that the variance in a reconstructed image decreases linearly with the x-ray tube current (ignoring the electronic noise), and the iodine-water contrast is reduced with an increased tube voltage. At the same time, the x-ray attenuation decreases exponentially with the increase in energy. Therefore, the optimal tube voltage selection is a multidimensional optimization problem. In general, for small-sized patients, a lower kVp setting should be selected since it provides a better contrast-to-noise ratio. On the other hand, when scanning a large patient, image noise is often the dominant factor so a higher kVp setting should be selected. For illustration, Figs. 11.19(a) and (b) show a shoulder phantom (45 cm wide) scanned in a step-and-shoot mode with two different voltages: 80 kVp and 140 kVp. For this study, the dose was kept constant (6.4 mGy). It is clear from the image that in order for an 80-kVp image to reach the same contrast-to-noise ratio, a significantly higher x-ray tube current is needed. Consequently, a higher dose is required to achieve similar image quality when scanning a large object.

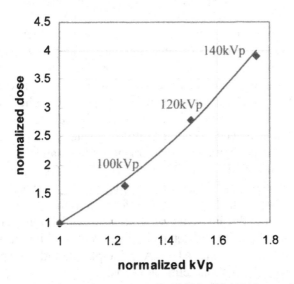

Figure 11.18 Normalized dose versus normalized kVp. 80 kVp is used as the reference point. Diamonds are the measured ratios and the solid line is the fitted curve.

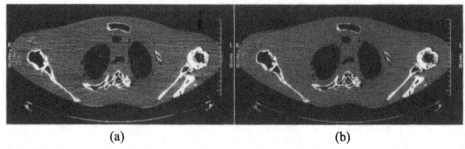

(a)   (b)

Figure 11.19 Impact of tube voltage selection on a shoulder phantom. (a) 80 kVp and (b) 140 kVp.

Another often overlooked factor in dose reduction is patient centering.[35] The design philosophy of the bowtie filter (Fig. 11.7) is to equalize the path length across the entire FOV. The shape of the bowtie filter assumes an oval-shaped and centered patient cross-section. When the patient is off-center, the long path length through the patient may line up with the long path length through the bowtie, and a significant increase in noise may result. For demonstration, the upper row of Fig. 11.20 shows a torso phantom scanned with the phantom centered [Fig. 11.20(a)] and 10 cm off-center [Fig. 11.20(b)]. For this study, all other scanning parameters were kept constant. The bottom row of the reconstructed images shows that the standard deviation of the off-center image is nearly twice that of the centered image ($\sigma = 53.8$ versus $\sigma = 28.8$). For this particular example, the x-ray tube current (and therefore the dose) must be increased $3.5 \times$ for the off-center case to achieve similar image quality.

Another factor that often impacts protocol selection is the noise level in the image. The noise level reduces proportionally to the square root of the x-ray tube current, while the x-ray dose increases directly proportionally to the current. Although a higher tube current generally results in better image quality, it may not always translate to a perceptible quality improvement because of the customary display window width selection. For example, the display window width of a midbrain study is typically between 80 and 100 HU, while the display window width for a lung examination is 1000 to 1500 HU. Therefore, the noise present in the lung image may be an order of magnitude higher than the noise in the brain image, while the perceived noise levels are the same for both. For demonstration, we scanned a water phantom repeatedly to produce the desired noise levels with image averaging. One image with a standard deviation of 26.5 HU is displayed with window width = 100 HU [Fig. 11.21(a)] and another image with a standard deviation of 2.65 HU is displayed with window width = 10 HU [Fig. 11.21(b)]. There is little perceived difference in the noise levels of the two images. Therefore, the display window width with which the images are viewed should be considered part of the determination process of the desired noise level in the image. In addition, the ALARA principle should be applied to avoid reducing the image noise beyond clinical needs.

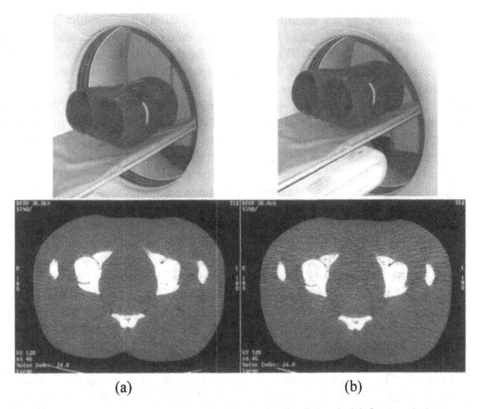

Figure 11.20 Impact of patient center on image noise (and dose). (a) Centered phantom and (b) 10-cm off-centered phantom.

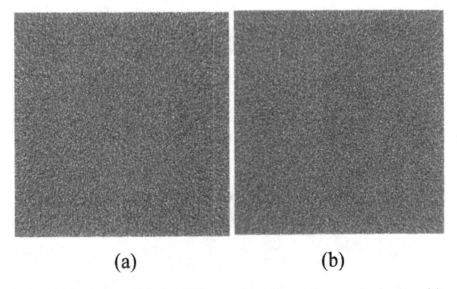

Figure 11.21 Illustration of display WW on noise with a uniform water phantom. (a) σ = 26.5 displayed with WW of 100 HU. (b) σ = 2.65 displayed with WW of 10 HU.

In summary, the CT scanning technique selection must consider both image quality and patient dose. The absolute best image quality does not meet the optimal selection criteria. The selection should be based on the clinically acceptable image quality that does not impact the accuracy of diagnosis and provides the least dose to the patient.[36]

The dose-reduction and protocol optimization effort is not limited to operators, researchers, and manufacturers. Professional societies and government agencies also play an important role. One example is the establishment of diagnostic reference levels for different CT procedures.[37] The reference levels are typically set at the 75th percentile of the dose distribution that results from dose surveys. Reference levels are neither the ideal or suggested doses, nor are they an absolute upper limit on the dose for a particular procedure. They intend to identify dose levels that are unnecessarily high so their reduction can still maintain the required level of image quality. The use of diagnostic reference levels is endorsed by many professional societies and regulatory organizations, such as the International Commission on Radiological Protection (ICRP), American College of Radiology (ACR), American Association of Physicists in Medicine (AAPM), Internal Atomic Energy Agency (IAEA), and European Commission (EC).[37] Beginning in 2002, the ACR Accreditation Program required all sites undergoing the accreditation process to measure and report the $CTDI_w$ and $CTDI_{vol}$ of their typical scanning protocols for both head and body phantoms. The use of reference levels in combination with the dose-efficiency improvements of CT scanners have resulted in an overall dose reduction. For example, the United Kingdom's national dose survey demonstrated a 30% reduction in the typical radiographic dose from 1984 to 1995, and a drop of 50% between 1985 and 2000.[38,39] A similar trend has been observed in the survey results of the ACR's CT Accreditation Program, as illustrated in Fig. 11.22.

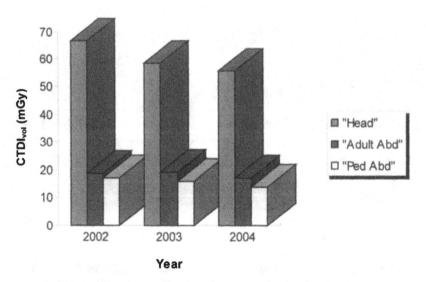

Figure 11.22 ACR CT Accreditation Program $CTDI_{vol}$ survey.

11.3.6 Postprocessing techniques

Postprocessing techniques, as the name implies, are applied after the tomographic reconstruction process. In recent years, many attempts have been made to use image processing or signal processing technologies to reduce image noise. This, in turn, has led to a lower-dose scanning technique because of the inverse relationship between the variance in the image and dose.

Postprocessing algorithms are typically adaptive in nature to minimize the impact on spatial resolution. For example, at each location the current pixel value is compared to its N neighbors. Figures 11.23(a) through (e) illustrate, respectively, 4-neighbor 2D processing, 8-neighbor 2D processing, 6-neighbor 3D processing, 18-neighbor 3D processing, and 26-neighbor 3D processing. The neighborhood selection can be separated into neighbors within the image slice (2D) or inside the image volume (3D), and by the maximum distance to its neighbors of 1 voxel, $\sqrt{2}$ voxel, or $\sqrt{3}$ voxels. In general, 3D algorithms are more robust, have less impact on spatial resolution, and have better noise-suppression capability. On the other hand, the 2D algorithms are computationally more efficient and flexible. These algorithms compare the intensity of the center voxel to its neighbors and determine whether the difference is due to noise or to the actual structure inside the object. In many cases, if the intensity of the center voxel is significantly different from the intensity of all of its neighbors, the chance is high that the difference is due to noise. Under this condition, the center voxel value is replaced by the weighted average of its neighbors. On the other hand, if the center voxel and a few of its neighbors exhibit different intensities compared to other neighbors, it is highly likely that a real structure will be encountered inside the object. The value of the center voxel should not be modified or modified only within a subgroup. For illustration, Figs. 11.24(a) through (c) show reconstructed images without postprocessing for a 70% dose and a 100% dose, and with postprocessing for a 70% dose, respectively. The image with postprocessing has a nearly identical noise level as the 100% dose without postprocessing image. For this example, a 30% dose reduction is achieved. Interested readers can refer to the references for additional information on postprocessing techniques.[40,41]

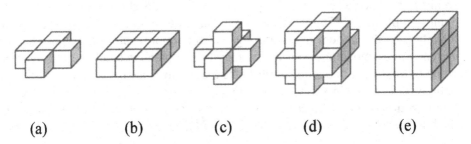

(a)   (b)   (c)   (d)   (e)

Figure 11.23 Illustration of *N*-neighborhood postprocessing. (a) 4-neighbor 2D processing, (b) 8-neighbor 2D processing, (c) 6-neighbor 3D processing, (d) 18-neighbor 3D processing, and (e) 26-neighbor 3D processing.

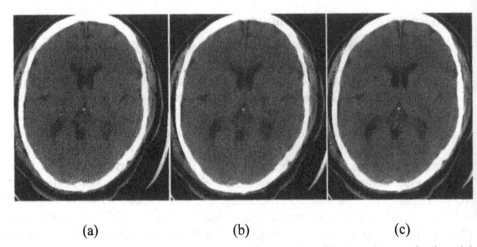

(a)   (b)   (c)

Figure 11.24 Illustration of the impact of postprocessing filter on dose reduction. (a) Original image with 70% dose. (b) Filtered image with 70% dose. (c) Original image with 100% dose.

11.3.7 Advanced reconstruction

Chapter 3 discussed a different class of reconstruction algorithm: iterative reconstruction (IR). The advantage of this algorithm is its ability to incorporate CT system optics models and statistical models into the reconstruction process. This approach overcomes the shortcomings of FBP algorithms in which many simplifications are used to make the mathematics more tractable. As mentioned in the previous discussion, the major drawback of the IR approach is its computational complexity, which leads to a long reconstruction time. With the introduction of multislice CT, the data acquisition time has been reduced significantly, and thin-slice data acquisition is routinely available. The number of images to be reconstructed for each patient examination has increased by more than an order of magnitude. Nowadays, many clinical applications demand several thousands of images for each patient study. Given this great demand and its impact on the workflow in a clinical environment, efforts must be made to ensure rapid image reconstruction.

Despite the rapid progress made in computer hardware and the advanced architectures such as cell or GPU processors, full IR is still out of reach for most clinical applications. The most time-consuming portion of the IR algorithm is the modeling of the system optics during the projection synthesizing process. The system optics modeling contributes mainly to spatial resolution improvement, while the statistical modeling contributes mainly to noise reduction. Note that the optimization problem shown in Eq. (3.71) can be rewritten in the form of the minimization of a cost function by utilizing the second-order Taylor expansion of the log-likelihood term [first term of Eq. (3.71)]:

$$\hat{\mu} = \arg\min_{\mu}\left\{\frac{1}{2}(p - A\mu)^{T} D(p - A\mu) + G(\mu)\right\}, \qquad (11.13)$$

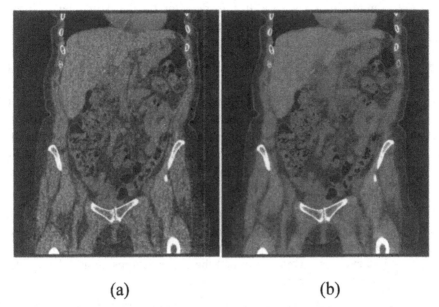

(a)   (b)

Figure 11.25 Illustration of a statistically based reconstruction algorithm. (a) FBP and (b) statistically based reconstruction algorithm.

where $G(\mu)$ is a scalar regularization term that is equal to log $Pr(\mu)$ within an additive constant, and D is a diagonal matrix. Our goal is to minimize the overall cost function shown in the curly bracket. During the iterative process, if we separate the minimization process for the two terms in the curly bracket, we can gain flexibility and computational efficiency.[42] Figure 11.25 shows an example of this approach. Compared to the FBP reconstruction in Fig. 11.25(a), the simplified IR algorithm in Fig. 11.25(b) exhibits significant noise reduction. This, in turn, can translate to a dose reduction during the acquisition.

11.4 Problems

11-1   Show that the penumbra portion of an x-ray beam is held nearly constant for different detector configurations.

11-2   For a CT system with a source-to-detector distance of 1000 mm, a source-to-iso distance of 550 mm, a source-to-collimator distance of 250 mm, and a focal spot size of 1 mm in z, calculate the dose efficiency in z for 4×1.25-mm, 4×2.5-mm, 4×3.75-mm, and 4×5-mm detector configurations if the active portion of the detector must be under the umbra portion of the x-ray beam.

11-3   Repeat the dose efficiency calculation in problem 11-2 if half of the penumbra portion of the beam is in the active portion of the detector cells.

11-4   Why is the amount of helical overbeam dependent on the helical pitch? Estimate the percentage of overbeam in a helical scan that covers a 20-

cm range along z with a 40-mm detector (at the iso-center in z) and a helical pitch of 0.5, assuming no dynamic collimator control is used. Repeat the analysis for a helical pitch of 1.

11-5   A student is asked to measure the amount of overbeam for 0.5-pitch helical scans on a CT scanner with a 40-mm detector in z. The CT console provides the following information during the prescription: tube voltage (kV), tube current (mA), gantry rotational speed (sec), table speed (mm/sec), detector z-coverage (mm), first and last image locations relative to the landmark (mm), bowtie selection (S/M/L), helical pitch, and total exposure time (sec). How does the student measure the amount of over-beam in millimeters?

11-6   Calculate the percentage dose savings over a nonmodulated acquisition for a helical cardiac scan with an 80% x-ray tube-current modulation. In this particular example, the reconstruction window width is 0.25 sec, the cardiac cycle is 60 bpm, and the tube current ramps up and down linearly over a 0.3-sec time interval.

11-7   Repeat the dose-saving calculation in problem 11-6 for a step-and-shoot mode data acquisition. All parameters remain the same except that the x-ray tube can be turned on and off in 10 ms.

11-8   Is the percentage of dose saving for the helical mode cardiac scan with tube-current modulation (problem 11-6) dependent on the heart rate? Does a similar relationship hold for the step-and-shoot mode cardiac acquisition (problem 11-7)?

11-9   The tube-current modulation (Section 11.3.1) for a general body scan is one method to reduce dose to a patient. Discuss the potential issues that may be encountered with the increase in detector coverage. Outline potential solutions to overcome these difficulties.

11-10   A low-dose scan is an important methodology in reducing x-ray dose to a patient (Section 11.3.5). How do you design an experiment that will demonstrate to radiologists and CT operators the diagnostic quality equivalency of two sets of exams collected at different dose levels, assuming several patients are scanned twice? Consider both the variations among different radiologists and variations of each radiologist over time.

11-11   To conduct the experiment in problem 11-10 without exposing the same patient to x-ray radiation multiple times, it is highly desirable to understand the relationship between diagnostic quality and dose. Describe in detail a computer simulation program to add noise to the projections to mimic a lower-dose examination. Does the noise addition need to take place before or after the minus logarithm operation?

11-12   An important method of dose reduction is optimization of the kVp setting based on the patient's size. Describe a set of experiments that will help to create a lookup table for CT operators when scanning different patient sizes and shapes.

11-13   One potential difficulty of patient centering may result from the significant variation in the patient's chest-to-abdomen profile. For example, if an operator adjusts the table height to center the chest, the abdomen may be significantly off center. What do you recommend to overcome this difficulty?

11-14   To compensate for the motion-induced attenuation correction artifact (Fig. 11.13), one student suggests replacing the normal CT (on the PET-CT scanner) with a CT scanner of slow rotation speed. The student believes that as long as the CT scanner's rotation speed is slower than the respiratory cycle, the mismatch between the PET and CT scanners should be minimized. Do you agree with the student's argument? What potential issues may arise?

References

1. L. E. Romans, *Introduction to Computed Tomography*, Williams & Wilkins, Baltimore (1995).

2. H. E. Johns and J. R. Cunningham, *The Physics of Radiology*, Charles C. Thomas Publisher, Ltd., Springfield, IL (1983).

3. "Ionizing radiation exposure of the population of the United States," NCRP Report No. 160 (prepublication copy), (2009).

4. L. N. Rothenberg and K. S. Pentlow, "CT dosimetry and radiation safety," in *Categorical Course in Diagnostic Radiology Physics: CT and US Cross-sectional Imaging*, L. W. Goldman and J. B. Fowlkes, Eds., Radiological Society of North America, Inc., Oak Brook, IL, 171–188 (2000).

5. L. N. Rothenberg and K. S. Pentlow, "CT dosimetry and radiation safety," in *Medical CT and Ultrasound: Current Technology and Applications*, L. W. Goldman and J. B. Fowlkes, Eds., Advanced Medical Publishing, Madison, WI, 519–553 (1995).

6. T. B. Shope, R. M. Gagne, and G. C. Johnson, "A method for describing the doses delivered by transmission x-ray computed tomography," *Med. Phys.* **8**, 488–495 (1981).

7. K. A. Jessen, P. C. Shrimpton, J. Geleijns, W. Panzer, and G. Tosi, "Dosimetry for optimisation of patient protection in computed tomography," *Appl. Radiation and Isotopes* **50**(1), 165–172 (1999).

8. K. A. Jessen, J. J. Christensen, J. Jørgensen, J. Petersen, and E.W. Sørensen, "Determination of collective effective dose equivalent due to computed tomography in Denmark in 1989," *Radiation Protection Dosimetry* **43**, 37–40 (1992).

9. American Association of Physics in Medicine (AAPM), "Standardized methods for measuring diagnostic x-ray exposures," AAPM Report No. 31, AAPM/AIP, New York (1990).

10. W. Huda, J. V. Atherton, D. E. Ware, and W. A. Cumming, "An approach for the estimation of effective radiation doses at CT in pediatric patients," *Radiology* **203**, 417–422 (1997).

11. T. L. Toth, N. B. Bromberg, T. Pan, J. Rabe, S. J. Woloschek, J. Li, and G. E. Seidenschnur, "A dose reduction x-ray beam positioning system for high-speed multislice CT," *Med. Phys.* **27**(12), 2659–2668 (2000).

12. J. Hsieh, "Method and apparatus for modulating x-ray tube current," U.S. Patent No. 5,696,807 (1997).

13. D. J. Brenner and E. J. Hall, "Computed tomography—An increasing source of radiation exposure," *New England J. Med.* **357**, 2277–2284 (2007).

14. S. Sternberg, "Study: Unnecessary CT scans exposing patient to excessive radiation," *USA Today*, Nov. 29, 2007.

15. A. Berenson and R. Abelson, "The evidence gap: Weighing the costs of a CT scan's look inside the heart," *New York Times*, June 29, 2008.

16. American Association of Physics in Medicine (AAPM), "The AAPM statement on radiation dose from computed tomography, in response to the Brenner and Hall NEJM article published Nov. 29, 2007," Public & General, AAPM, http://www.aapm.org/publicgeneral/CTScans.asp.

17. D. A. Bluemke, S. Achenbach, M. Budoff, T. C. Gerber, B. Gersh, L. D. Hillis, W. G. Hundley, W. J. Manning, B. F. Printz, M. Stuber, and P. K. Woodard, "Noninvasive coronary artery imaging. Magnetic resonance angiography and multidetector computed tomography angiography." A scientific statement from the American Heart Association Committee on Cardiovascular Imaging and Intervention of the Council on Cardiovascular Radiology and Intervention, and the Councils on Clinical Cardiology and Cardiovascular Disease in the Young, published online June 27, 2008: http://circ.ahajournals.org/cgi/content/abstract/CIRCULATIONAHA.108.18 9695v1.

18. E. L. Alpen, *Radiation Biophysics*, 2nd Ed., Academic Press, San Diego (1998).

19. L. E. Feinendegen, V. P. Bond, and C. A. Sondhaus, "The dual response to low-dose irradiation: Induction vs. prevention of DNA damage," in *Biological Effects of Low Dose Radiation*, T. Yamada, C. Mothersill, B. D. Michael, and C. S. Potten, Eds., Elsevier, St. Louis, MO, 3–18 (2000).

20. D. J. Brenner, C. D. Elliston, E. J. Hall, and W. E. Berdon, "Estimated risks of radiation-induced fatal cancer from pediatric CT," *Amer. J. Radiol.* **176**, 289–296 (2001).

21. J. T. Bushberg, J. A. Seibert, E. M. Leidholdt, J. M. Boone, *The Essential Physics of Medical Imaging*, 2nd ed., Lippincott Williams & Wilkins, Philadelphia (2001).

22. M. F. McNitt-Gray, "Radiation dose in CT," AAPM/RSNA Physics Tutorial for Residents: Topics in CT, *Radiographics* **22**, 1541–1553 (2002).

23. International Electrotechnical Commission (IEC), "Medical Electrical Equipment. Part 2–44: Particular Requirements for the Safety of X-ray Equipment for Computed Tomography," IEC Publication No. 60601-2-44, IEC Central Office, Geneva, Switzerland (2003).

24. "European guidelines on quality criteria for computed tomography," EUR 16262 EN 1999.

25. J. M. Boone and J. A. Seibert, "Monte Carlo simulation of the scattered radiation distribution in diagnostic radiology," *Med. Phys.* **15**, 713–720 (1988).

26. J. J. DeMarco, C. H. Cagnon, D. D. Cody, D. M. Stevens, C. H. McCollough, M. Zankl, E. Angel, and M. F. McNitt-Gray, "Estimating radiation doses from multidetector CT using Monte Carlo simulations: Effects of different size voxelized patient models on magnitudes of organ and effective dose," *Phys. Med. Biol.* **52**, 2583–2597 (2007).

27. C. H. McCollough, "CT accreditation program: image quality and dose measurements," RSNA refresher course 2008.

28. J. M. Boone, "The trouble with CTDI100", Med. Phys. 34, 1364-1372 (2007).

29. J. Hsieh, J. Mayo, K. Whittall, K. Kim, M. J. Brown, and O. Dake, "An algorithm for automatic prediction of optimal x-ray tube current," *Radiology* **205**, 352 (1997).

30. W. A. Kalender, H. Wolf, C. Suess, M. Gies, D. Hentschel, and W. A. Bautz, "Dose reduction in CT by anatomically adapted tube current modulation: Experimental results and first patient study," *Radiology* **205**, 471 (1997).

31. J. Colsher, J. Hsieh, J. Thibault, A. Lonn, T. Pan, S. Lokite, and T. Turkington, "Ultra low dose CT for attenuation correction in PET/CT," in *Proc. IEEE Med. Imag. Conf.*, Dresden, Germany (2008).

32. J. Hsieh, J. Londt, M. Vass, J. Li, X. Tang, and D. Okerlund, "Step-and-shoot data acquisition and reconstruction for cardiac x-ray computed tomography," *Med. Phys.* **33**(11), 4236–4248 (2006).

33. Y. Imai, R. Fujimoto, D. R. Okerlund, J. Hsieh, H. Yamaguchi, and T. Takahashi, "Potential of ECG-triggered high-pitch helical CT scanning within one heart beat," *Proc. RSNA 2008*, p. 692 (2008).

34. S. Achenbach, M. Marwan, T. Schepis, T. Pflederer, H. Bruder, T. Allmendinger, M. Petersilka, K. Anders, M. Lell, A. Kuettner, D. Ropers, W. G. Daniel, and T. Flohr, "High-pitch spiral acquisition: a new scan mode for

coronary CT angiography," *J. Cardiovasc. Comput. Tomogr.* **3**(2), 117–121 (2009).

35. T. Toth, et al., "The influence of patient centering on CT dose and image noise," *Med. Phys.* **34**, 3093–3101 (2007).

36. D. P. Frush, C. C. Slack, C. L. Hollingsworth, G. S. Bisset, L. F. Donnelly, J. Hsieh, T. Lavin-Wensell, and J. R. Mayo, "Computer-simulated radiation dose reduction for abdominal multidetector CT of pediatric patients," *Radiology* **202**(2), 453–457 (1997).

37. C. H. McCollough, "Dose assessments in clinical practice: What's typical? What's too much?" in *CT and Dose Considerations, Categorical Courses in Diagnostic Radiology Physics: CT and MR Imaging*, Radiological Society of North America, Annual Meeting 2008.

38. D. Hart and B. F. Wall, "UK population dose from medical x-ray examinations," *Euro. J. Radiology* **50**(3), 285–291 (2004).

39. P. C. Shrimpton, B. F. Wall, and D. Hart, "Diagnostic medical exposures in the U.K.," *Appl. Radiation and Isotopes* **50**(1), 261–269 (1999).

40. J. C. Russ, *The Image Processing Handbook*, 5th ed., CRC Press, Boca Raton, FL (2007).

41. R. C. Gonzalez and R. E.Woods, *Digital Image Processing*, 3rd ed., Prentice Hall, Upper Saddle River, NJ (2008).

42. J. Hsieh, F. Dong, J. Fan, M. Kulpins, J. Thibault, and X. Tang, "Advanced statistical reconstruction algorithm for CT dose reduction," in *Proc. European Congress of Radiology* (2009).

43. L. E. Feinendegen and M. Pollycove, "Biologic responses to low doses of ionizing radiation detriment versus hormesis. 1. Dose responses of cells and tissues," *J. Nucl. Med.* **42**(7), 17–27 (2001).

44. M. Pollycove and L. E. Feinendegen, "Biologic responses to low doses of ionizing radiation: detriment versus hormesis. 2. Dose response of organisms," *J. Nucl. Med.* **42**(9), 26–37 (2001).

45. S. Kondo, "Health effects of low level radiation," *Medical Physics*, 85–89, Madison, WI (1993).

46. D. A. Pierce, Y. Shimizu, D. L. Preston, M. Vaeth, and K. Mabuchi, "Studies of the mortality of atomic bomb survivors, report 12.1. Cancer: 1950–1990." *Radiat. Res.* **146**, 1–27 (1996).

47. B. L. Cohen, "The cancer risk from low level radiation," *Radiat. Res.* **149**, 525–526 (1998).

48. B. L. Cohen, "Cancer risk from low-level radiation," *Am. J. Roentgenol.* **170**, 1137–1143 (2002).

Chapter 12
Advanced CT Applications

12.1 Introduction

The clarity and accuracy of images produced by CT scanners have enabled CT to become one of the most widespread modalities for diagnostic imaging. Between November 1998 and October 1999, 26.3 million CT procedures were performed in the United States alone, based on a census survey of 90% of the nation's imaging centers.[1] This represents a 16% increase over the 1996 survey result. It was estimated that the number of CT procedures jumped to about 62 million in the United States in 2006. Many books are dedicated to the subject of the clinical utility of CT scanners, and these books cover different types of recommended clinical protocols for different human anatomies.[2–5] The goal of this chapter is not to replicate or summarize these protocols; instead, this chapter will address new clinical applications that have evolved in recent years and have had a significant impact on CT scanner design.

The definition of "new" clinical applications is vague. Many applications that were considered new and challenging to CT not long ago have become much less demanding on the scanners. One good example is CT angiography, the imaging of blood vessels opacified by a contrast medium. With helical and multislice CT, the scan of the entire vascular structure can be routinely completed in a single breath-hold. Many advanced visualization tools such as curved multiplanar reformation, MIP, surface rendering, and VR are readily available on many commercial scanners or workstations. The volume coverage speed is no longer the most important criterion for the performance evaluation of a CT scanner.

Three general trends have emerged in CT scanner usage. First, with increased scanner capability, radiologists are more inclined to use thinner slices for scanning. Thin slices have many technical advantages, such as reduced partial-volume effect and improved spatial resolution. In the past, thin slices were prohibitive due to limitations with the x-ray tube and volume coverage speed. For example, a conventional single-slice step-and-shoot scanner takes 60 sec to cover a 150-mm cervical spine with 1-mm collimation and 0.8-sec gantry speed.[6] The time span clearly exceeds a single breath-hold limit for most patients. The same volume can be covered by a 4-slice scanner with the same gantry speed in 17 sec, and by a 16-slice scanner in less than 2.3 sec. For a state-of-the-art 64-slice scanner, the same volume can be covered with submillimeter resolution in roughly 1 sec. For many clinical cases, the speed outpaces the flow of the contrast medium, which forces the operator to rethink the entire equation.

The second trend is the increased awareness of patient dose. In the past, more attention was paid to obtaining the best image quality at the lowest noise possible. This often led to the use of maximum scanning techniques. Many studies have been conducted in recent years to understand the relationship between the noise present in an image and the outcome of the diagnosis (not image quality). The awareness of the ALARA principle is becoming widespread, as discussed in Chapter 11. In Europe, more strict guidelines are placed on the use of CT scanners than are used in the United States. The current European guidelines are also more strict than they were in the past. Most major CT manufacturers also spend a significant amount of energy and resources to improve the dose efficiency of their scanners. At the same time, advanced features such as automatic adjustment of the x-ray tube current, color-coded protocols for pediatrics, and advanced reconstruction technologies are now available to help operators reduce the patient dose.

The third trend is the increased use of 3D visualization devices as the primary diagnostic tools. It is estimated that the number of images produced by a scanner increases at a rate of roughly 5 to 7 times faster than the rate of increase in the number of radiologists. Thus, the number of images that must be reviewed by each radiologist has increased significantly over the past few years. A typical study performed on an advanced multislice scanner contains several hundred to over 1000 images. Viewing these images one slice at a time is time consuming and easily leads to fatigue of radiologists. One way of reducing this information overload is to compress the 3D volume information to a small set of 2D images that represent the 3D structure of the data. Another factor that has increased the popularity of 3D tools is the cross-discipline use of CT data. For example, 3D VR images are often used by surgeons to plan surgeries.

CT images are also used with other modalities, such as SPECT-CT and PET-CT. CT images have been shown to provide valuable information for attenuation correction in both PET and SPECT. They also provide useful anatomical landmarks for these modalities. In recent years, many commercial devices have been designed to combine CT scanners with PET or SPECT. The advantages of the combined scanner are reduced examination time and better image registration. Since both exams can be conducted without moving the patient to a different location, the total study time is significantly reduced, and the registration process between the CT and the other modality images is somewhat easier since the patient stays on the same table for the entire study. Even with this added advantage, the image registration process between the two modalities is not trivial. CT images are typically acquired in a single breath-hold while PET and SPECT images are acquired in a much longer time period (5 to 40 min), which entails breathing motion (as discussed in Chapter 11). Figure 12.1 depicts an example of a registered PET-CT study.

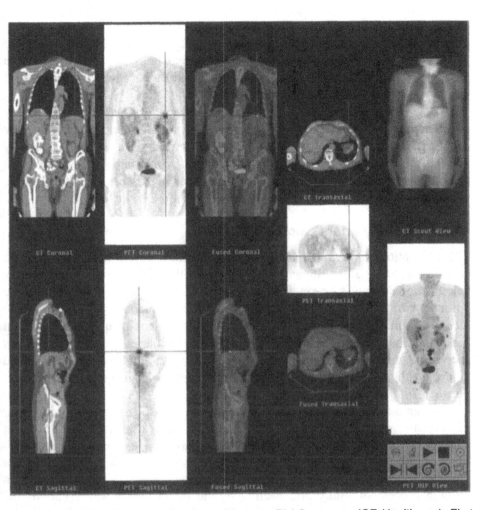

Figure 12.1 PET-CT image fusion on a Discovery™ LS scanner (GE Healthcare). First column: reformatted coronal (top) and sagittal (bottom) CT images; second column: coronal (top) and sagittal (bottom) PET images of the same patient; third column: fused coronal (top) and sagittal (bottom) images; fourth column: CT cross-sectional image (top), PET transaxial image (middle), and fused image (bottom); fifth column: CT scanogram (top) and PET MIP image (bottom) of the same patient.

12.2 Cardiac Imaging

Two major technological advances are key to the recent growth in cardiac applications with conventional CT scanners. The first major advance is faster scanning speed. Today's state-of-the-art scanners are capable of rotating at 0.35 sec per gantry rotation or faster. Compared with the typical scanning speed of 1 sec per gantry rotation nearly a decade ago, this represents an increase by a factor of 3 in the CT scanner's temporal resolution. Although the scanning speed is still slower than one could achieve with an electron-beam scanner, which is capable of collecting a complete dataset in 50 ms or less, more advanced reconstruction algorithms partially compensate for some deficiencies. (Since the focus of this chapter is on the use of conventional scanners for cardiac applications, we will not discuss electron-beam scanner technology.) The second key advance is the introduction of multislice CT. State-of-the-art scanners today collect 64 or more projections simultaneously. This technology provides improved cross-plane (z) spatial resolution and improved volume coverage.

There are two general types of cardiac applications: coronary artery calcification and coronary artery imaging. Strictly speaking, coronary artery calcification exams belong to the category of screening, since they generally involve the scanning of asymptomatic patients. We grouped this topic with coronary artery imaging since many important technical issues are shared by the two applications.

12.2.1 Coronary artery calcification (CAC)

The association between coronary artery calcification (CAC) and coronary death was described by John Hunter in 1775.[7] More than 200 years later, coronary calcification detected by x-ray fluoroscopy was shown to be predictive of a future coronary event.[8] However, because of the nature of fluoroscopy, this technique receives limited usage due to its limited sensitivity, lack of quantification, and lack of reproducibility.[9] The use of CT to detect calcium in coronary arteries has received a great deal of attention in recent years. A PubMed search revealed approximately 3100 articles written in the English language on this subject between 2000 and 2008.

The amount of calcium present in the arteries may be an important indicator of coronary artery disease and therefore one's risk of a heart attack. In recent studies, the negative predictive ability of CAC has been shown to be valuable— that is, the lack of artery calcification is a good predictor of a low probability of a coronary event. The presence of calcium is the nature of atherosclerosis: a chronic process of injury and healing of the blood vessel walls. Part of the healing process involves calcium deposition to the injured area. Figure 12.2 shows a VR CT image with calcium depicted as light-colored objects attached to the vessels. Histological studies have shown that approximately 20% of the plaque volume in coronary atherosclerosis is marked by a detectable level of calcium.[11] Studies have also shown that although soft plaques with no detectable calcium exist, 96% of asymptomatic patients with clinical coronary artery disease have detectable coronary calcium as demonstrated by angiography.[12] There is also increasing evidence that the deposition of calcium in atherosclerosis is an active metabolic process, which supports the hypothesis that the degree of CAC reflects the severity of atherosclerotic disease.[13]

During the CAC screening process, a series of low-dose CT images are reconstructed by centering on the diastole phase of the heart to minimize the impact of cardiac motion. This is generally accomplished with the aid of an EKG signal. A detailed discussion on the data acquisition and reconstruction can be found in Section 12.2.2 on coronary artery imaging. Figure 12.3 shows an example of a reconstructed CAC image. Using the reconstructed CT images,

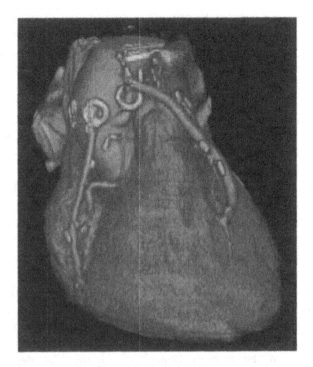

Figure 12.2 Volume-rendered image of a cardiac CT scan.

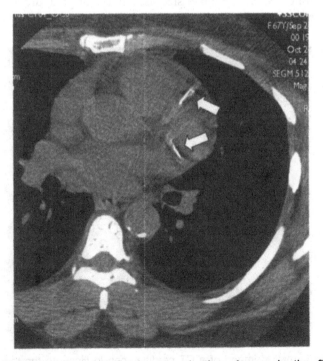

Figure 12.3 Coronary artery calcification examination. Arrows in the figure indicate locations of calcium.

calcium "scores" are derived to provide a quantitative estimate of the plaque burden. A score may contain individual scores for major arteries (e.g., left main artery, left anterior descending, left circumflex, right coronary artery, posterior descending artery), but a total score is generally given for the entire examination. A higher score usually implies a higher risk of subsequent cardiac events.[14] However, the interpretation of the score is complicated by its age dependency.

The current scoring of CAC is based on some variation of the Agatston method.[15] The original method requires the placement of an ROI over each calcified coronary artery. A simple threshold (e.g., 130 HU) is used to separate calcified and noncalcified tissues. The area of the plaque and the peak attenuation (the maximum CT number) within the ROI are then calculated. The peak attenuation is used primarily to determine a scaling factor. In the method described by Agatston et al., four peak attenuation ranges (131 to 200 HU, 201 to 300 HU, 301 to 400 HU, and 400+ HU) are used to assign four different scaling factors. A score is calculated by multiplying the measured area by the scaling factor. Individual scores measured within the region of the coronary arteries are then summed to arrive at a final calcification score. This method is called area scoring because the calcium score is based on the measured area of the calcium deposit.

In recognition of the fact that coronary calcification is a volume phenomenon, other scoring methods were investigated to account for the cross-plane dimension of the calcium deposits.[16] Figure 12.4 depicts an oblique reformatted image to illustrate the third dimension of the calcium (calcium deposits are white). Since the voxel size used in the original area-scoring method is larger in z than in x-y, the new scoring methods interpolate intermediate images

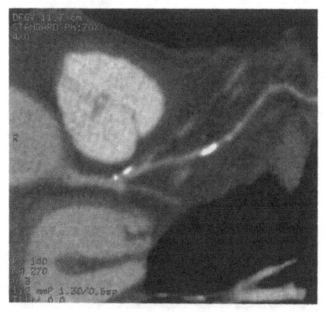

Figure 12.4 Reformatted image of a CT cardiac scan to illustrate the third dimension of the calcification.

in z so the voxel size is more or less iso-tropic. Threshold and 3D region-growing algorithms are then used to identify interconnected voxels with CT numbers greater than the threshold. The score represents the estimated volume as a fraction of a cubic centimeter, multiplied by 1000. This method is called volume scoring.

A third scoring method is based on the total mass of the calcium deposit. This mass is obtained by first multiplying the pixel volume by its CT number. The values are then summed over all pixels whose CT number exceeds a predetermined threshold in the distribution of the coronary arteries.[13] The value is divided by 100 to get in the range similar to that of area scoring or volume scoring. This method is called mass scoring.

Successful CAC scoring depends on its reproducibility, sensitivity, and consistency. Good reproducibility means that the CAC score of a patient who is scanned twice on the same scanner within a short period of time (e.g., a few hours) remains essentially the same. If the CAC scores vary significantly, it not only complicates the interpretation of the score, but also makes it difficult for follow-up studies to monitor the evolution of coronary atherosclerotic disease. Sensitivity means that a clinically significant change in a patient's condition should be reflected in the measured score. Consistency refers to interscanner variability and long-term variability; CAC scores should not change significantly when a patient is scanned on different scanners, and if the patient's condition remains unchanged over a long period of time, his or her CAC score should reflect the same. These seemingly trivial requirements are often challenging in a real clinical environment. The error comes mainly from four sources: protocols, scanner, scoring algorithm, and the patient.

Good clinical protocols play a significant role in the accuracy of the CAC score. For example, the area-scoring method relies on the maximum pixel intensity inside the ROI to determine the scaling factor and therefore the final score. Since the size of the calcified area is often small, the partial-volume effect is likely to affect the reconstructed object density. Although volume scoring and mass scoring have been shown to be superior to area scoring, they are not totally immune to this effect. Similarly, if the x-ray tube current is set too low, the accuracy of the reconstructed image will be degraded by statistical noise. This, in turn, affects the CAC scores, since all methods rely on accurate CT numbers.

The quality of the CT scanner can significantly impact reproducibility and consistency. For most commercial scanners, the reconstructed CT number can be influenced by patient positioning, the presence of foreign objects, the size of the patient, and calibration (a more detailed discussion is in Section 12.5.2). The operator must ensure that the variation is small compared to the accuracy requirement of the CAC score. The impact of scanner accuracy on the CAC score can be minimized by using a calibration phantom behind the patient.[17]

The goal of the three scoring algorithms is to achieve the desired reproducibility and sensitivity. For example, the volume-scoring and mass-scoring methods have been shown to have better reproducibility than the area-scoring techniques. Other modifications to the area-scoring algorithm, such as the use of a continuous scaling function instead of a step function, also improves the performance of this method. The separation of clustered calcium deposits and connectivity constraints has also been shown to produce better results. Another important factor that affects the performance of the scoring algorithms is the selection of the threshold to separate calcified and noncalcified tissues. If the selected threshold is too high, newly forming plaques may be excluded in the score, since these plaques often have an attenuation near or lower than 130 HU.[16] The optimal threshold value allows an improved sensitivity of the scoring method while reducing the number of false positives.

The fourth major error contributor is the patient. For example, variation in the cardiac cycle often results in CT images corrupted by cardiac motion artifacts.[18] These artifacts, as discussed in Chapter 7, lead to inaccurate CT numbers. Respiratory motion can also produce blurring and shading artifacts that impact the reconstructed calcium intensity.

CT calcification screening is not without controversy. Although this test has the potential for early detection of heart disease, it provides little value for men over 65 or women over 75 in the United States. Virtually everyone in these age groups has some coronary calcification, which makes it very difficult to distinguish people with heart disease from people without it. Another complicating factor is that although calcium deposition occurs relatively early in the atherosclerotic process, the plaque material is not initially calcified. Consequently, atherosclerosis in its very early stages may be undetected by this technique. This is important because noncalcified plaque can rupture and cause an unstable angina or acute myocardial infarction. In addition, physicians must recognize that the prevalence of calcification is much higher than the risk of suffering a clinical event, which means that only a small proportion of patients with any identifiable calcium will suffer a clinical event.[9] Debates and controversy over this subject are likely to continue. Academic and industry researchers continuously investigate new techniques to make this test more meaningful and reliable.

12.2.2 Coronary artery imaging (CAI)

The second type of CT cardiac application is coronary artery imaging (CAI).[19–27] The objective of CAI is to visualize the vascular structure of the heart, which allows physicians to detect stenosis (narrowing of a vessel) and plaque. It may also enable physicians to examine the dynamic motion of the muscles and detect abnormalities. This type of examination is typically conducted with contrast injection. From a purely technological point of view, the scanner's performance requirements for CAI are higher than those for CAC for two reasons. To visualize the narrowing of a small vessel, CT scans must not only freeze the cardiac motion (which requires a higher temporal resolution) but also accurately depict the size of the vessel (which requires a higher spatial resolution). For calcification screening, on the other hand, the scanner's ability to freeze cardiac motion is less important because the scores are averaged over a small region.

Figures 12.5(a) and (b) show examples of stenosis detected on an x-ray angiographic system and on a CT system, respectively. There are technical challenges associated with the accurate assessment of the degree of stenosis (the percentage of the vessel that has narrowed). The imaging device must be able to freeze the cardiac motion and have sufficient spatial resolution to perform a quantitative evaluation. Since a typical x-ray angiographic system samples at 30 frames/sec with 100- to 200-μm spatial resolution, this task can be easily accomplished from a technology point of view. However, the x-ray angiogram approach is an invasive procedure that requires an interventional radiologist to thread a catheter to the location of interest. The x-ray dose to both the patient and operator is high. In addition, the size and characteristics of the plaque are not visible, since the angiographic image is obtained by subtracting the precontrast angiogram from the postcontrast image.

Another example of the CAI application is the evaluation of cardiac function. Figures 12.6(a) and (b) show two CAI images reconstructed at two different

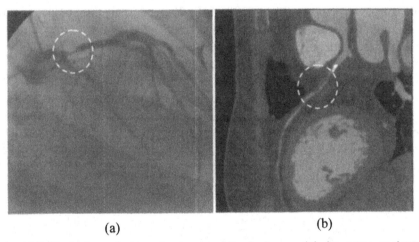

(a)   (b)

Figure 12.5 Illustration of stenosis as a result of plaque. (a) An x-ray angiographic example and (b) an example from CT CAI.

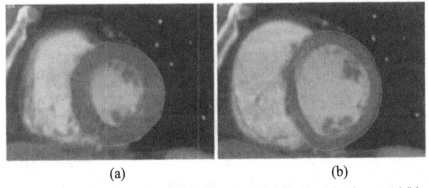

(a)   (b)

Figure 12.6 Calculation of ejection fraction from CAI. (a) End-systole phase and (b) end-diastole phase.

cardiac phases: end systole and end diastole. The end-systole phase occurs when the left ventricle (LV) is in full contraction and the end-diastole phase occurs when the LV is most expanded. The ejection fraction is often used to indicate the heart's "pump efficiency," which is defined as the ratio of the difference between the maximum and minimum LV volumes over the maximum LV volume. Given the accurate anatomical details provided by CT CAI, its potential to provide a reliable ejection fraction calculation is high.

12.2.2.1 Data acquisition and reconstruction

To reduce the impact of cardiac motion in CAI, the data acquisition typically relies on EKG signals to indicate the phase of the heart, as illustrated in Fig. 12.7. A cardiac cycle has two phases in which the heart motion is relatively small: the end-systolic and end-diastolic phases. During these phases, the heart undergoes quiescent periods of cardiac motion when the related artifacts and degradation can be minimized. On an EKG trace, the mid-diastolic phase generally corresponds to a region between 70 and 75% of the R-R interval, and the end-systolic phase is between 30 to 35% of the R-R interval. Studies have shown[28] that for patients with a relatively slow heart rate, data acquired during the diastolic phase generally provide better image quality, and for patients with a higher heart rate, the end-diastolic phase offers a better image quality.

For volumetric imaging (either reformatted or VR images), the successive 2D axial images must be acquired at the same phase of cardiac motion so that the shape of the heart does not change from one slice to another, which would result in image artifacts. Human heart sizes range from 10 to 16 cm in z. For a detector with a z coverage of 1 cm, at least 10 to 16 cardiac cycles are needed to cover the entire heart. This was the main obstacle encountered by 4- and 8-slice scanners.

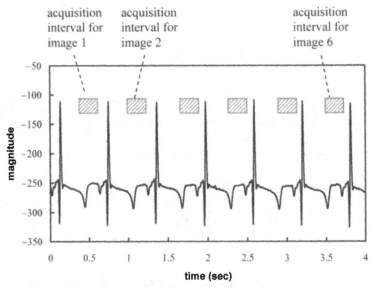

Figure 12.7 Illustration of single-cycle acquisition.

Even with 16-slice scanners, the detector z coverage is still no bigger than 1 cm, since submillimeter slice thickness is required to provide sufficient anatomical details for diagnosis. A typical cardiac scan still requires completion of 15 to 20 cardiac cycles on a 16-slice scanner (larger-coverage scanners will be discussed later). This is the main reason the cardiac scan mode was predominately helical in the early days of the multislice CT design—to avoid an excessive number of interscan delays if a step-and-shoot data acquisition was used. To ensure adequate and continuous coverage of the heart in z, helical pitches between 0.1 and 0.3 are typically used in clinical settings. The low helical pitches are driven by the reduced detector z coverage and the margin used to account for the variations in heart rates.

A common misperception is that a continuous data acquisition and reconstruction is used in helical mode; however, cardiac scans are reconstructed at discrete time and z locations, as illustrated in Fig. 12.8. Consider a case in which a patient with a heart rate of 57 bpm is scanned using a gantry rotation speed of 0.35 sec/revolution. After the first data acquisition, the gantry needs to rotate 3 rotations before the heart is in the same phase again for data reconstruction, as illustrated by the thick solid line in the figure. Although helical mode is used for the data acquisition, little advantage is gained by the helical reconstruction because the view depicted by the dotted line is not suitable for reconstruction. As a matter of fact, a small penalty is paid in terms of coverage due to the helical acquisition. Figure 12.9(b) plots the detector row locations as a function of time for helical mode. For clarity, only a four-row detector is shown in the figure, but the analysis is applicable to other detector configurations. The patient serves as the reference point in this analysis. In helical mode the patient table is translated at a constant speed during the data acquisition and the detector z location is translated at the same speed in the opposite direction relative to the patient. During the data acquisition process, a minimum number of views is required for the reconstruction, as depicted by the shaded box in the figure. Therefore, a portion of the region covered by the left-most detector during the data acquisition cannot be reconstructed if extrapolation during the reconstruction must be avoided. Similarly, a portion of the region covered by

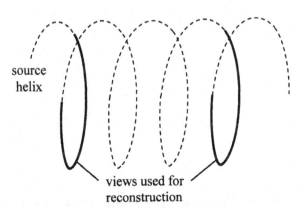

Figure 12.8 Illustration of segmented source helix for cardiac reconstruction.

the right-most detector row will not contribute to the final reconstructed volume. The actual reconstruction size in z is smaller than the detector coverage in z, as illustrated by the dotted vertical lines in the figure. The amount of volume reduction is linked directly to the helical pitch used for acquisition. In general, the higher the helical pitch, the smaller the reconstructed region. However, in the actual implementation of the cardiac scanning, some extrapolation is used during the reconstruction to increase the volume coverage in z. The allowable extrapolation is often determined by phantom and clinical experiments to ensure adequate image quality.

Another factor that limits the reconstructable volume is the cone-beam geometry, as depicted in Fig. 12.9(a). Because all x-rays are emitted from the x-ray focal spot, the z dimension covered by the cone-beam geometry is shaped as a trapezoid (shaded region). The side closest to the x-ray focal spot is smaller than the side closest to the detector. The detector z coverage commonly mentioned in the literature is in between these sides. The reduction of the actual z coverage is related to the distance between the reconstruction location and the iso-center. The larger the distance to the iso-center, the smaller is the actual coverage by the cone beam. To determine the minimum coverage, both the reconstruction FOV and the center of the reconstruction must be considered. In the actual implementation, the reconstructed volume is typically slightly larger than the minimum size irradiated by the x-ray beam (in z) to improve scanning efficiency.

The challenges associated with CAI reconstruction can be summarized as follows: How do we reconstruct an image at the appropriate cardiac phase and at the appropriate z location? For simplicity, consider first the case of a single-cycle data acquisition. In this acquisition mode, the entire dataset used to reconstruct an image is acquired within a single cardiac cycle (Fig. 12.7). To ensure that the reconstruction is performed at the same phase of the cardiac cycle, the data

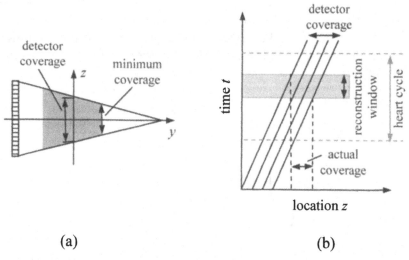

(a)   (b)

Figure 12.9 Reduction of volume coverage in cardiac helical mode. (a) Reduction due to cone-beam geometry and (b) reduction due to helical motion.

acquisition and reconstruction window (shown by the shaded rectangles in the figure) is selected at the same location relative to the R-to-R interval of the EKG. Considering only the phase requirement, one draws the conclusion that a faster gantry speed and helical pitch are always better, since the duration of data acquisition will be a smaller fraction of the cardiac cycle. In practice, however, it is important to consider other factors and to ensure that the CAI data are collected at the right location. Figure 12.10 plots the detector row position as a function of time. The cardiac cycles are separated by horizontal dotted lines, and the detector row locations are depicted by the solid diagonal lines. Every point on these lines represents a single-row projection collected at a certain z location and a particular time (and therefore a particular projection angle). Since a four-row detector system is assumed, four lines are depicted in the figure. The gray boxes in the figure show the reconstruction windows for the cardiac images. Consequently, these boxes depict a unique set of time intervals and z locations. The width of the box represents the volume in z that can be covered with reconstructions corresponding to a particular cardiac cycle, which is typically smaller than the detector z coverage (as discussed previously). The adjacent set of reconstructions takes place only after the heart reaches the same cardiac phase in the next cardiac cycle. If the combination of gantry speed and helical pitch is not properly selected (for a fixed slice thickness, the product of the gantry speed and the helical pitch determines the table traveling speed), the entire heart volume will not be uniformly covered in the reconstructed images. For example, Fig. 12.10(a) shows a case in which an overlapped cardiac volume is obtained due to a smaller-than-optimal helical pitch. Although at first glance it may not appear to be a problem to have multiple images covering the same volume, penalties are paid for in terms of total scan time and dose. If the optimal table

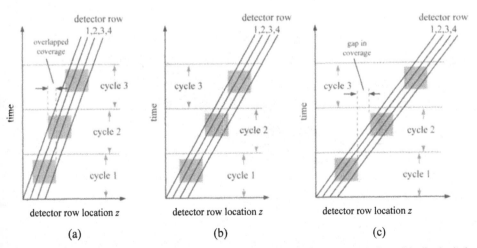

Figure 12.10 Phase and coverage interdependence. (a) Gantry speed and helical pitch selection is too slow, resulting in overlapped coverage occurring in the reconstructed images. (b) Proper gantry speed and helical pitch selection. (c) Gantry speed and helical pitch selection is too fast, leading to gaps in the coverage of reconstructed images.

speed had been prescribed, the entire study could have been significantly shortened. A longer scan time translates to a longer breath-hold period for the patient. Clinical experience shows that the heart rate often increases significantly at the end of a long breath-holding period, which presents a potential challenge to proper gating. On the other hand, if the table travels too fast, gaps will be present between adjacent volumes, as shown in Fig. 12.10(c). Although small gaps could be filled by image space interpolation, larger gaps will lead to discontinuities and artifacts in the VR images.

Figure 12.10(b) shows a theoretical "optimal" helical pitch selection, but in practice the helical pitch is selected to be somewhat smaller than the optimal value for two reasons. First, the cone-beam geometry and helical motion reduce the volume coverage. Generally speaking, the helical pitch is bounded by the following equation:

$$\frac{2\pi(D-r)s}{\left[(\pi+2\gamma_m)s+2\pi T\right]D} < p < \frac{s}{T} , \tag{12.1}$$

where D is the source-to-iso distance, r is the largest distance of the reconstructed pixel to the iso-center, s is the gantry rotation period in sec/revolution, γ_m is the maximum fan angle of the detector, and T is the duration of the cardiac cycle in seconds. Figure 12.11 depicts the pitch selection range as a function of heart rate in bpm based on Eq. (12.1). Note that the above discussion is applicable only to the case in which the entire cardiac data acquisition is completed in two or more cardiac cycles.

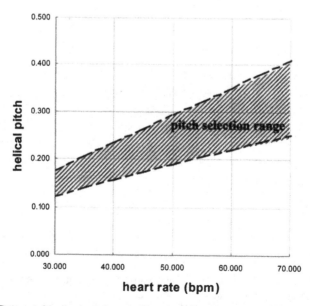

Figure 12.11 Range of helical pitch as a function of heart rate for $D = 550$ mm, centered reconstruction FOV = 250-mm diameter, gantry speed of 0.35 sec/revolution, and 55-deg fan angle.

The second reason for a reduced helical pitch is heart rate variation. During the data acquisition time period, the heart rate is likely to change. The helical pitch is calculated prior to the data acquisition and therefore is based on the heart rate preceding the start of the scan. If the heart rate increases slightly during the scan, the problem is not critical, since slightly more overlap takes place between adjacent acquisitions. However, if the heart rate slightly decreases during the scan, small gaps may be produced in the coverage. As a result, a small margin should be included in the selection of helical pitch. Given the range of helical pitch selection due to the coverage reduction shown in Fig. 12.11, both effects need to be combined in the pitch selection. The final pitch selection, of course, needs to be tested and optimized based on real clinical trials.

For single-cycle cardiac acquisitions, half-scan is often used for image reconstruction because of its better temporal resolution (see Chapter 7 for details). Since the data acquisition is in helical mode and the patient table travels constantly, some type of z interpolation is required to estimate a set of projections at the plane of reconstruction when 2D backprojection algorithms (for very small cone angles) is used. For simplicity and convenience, interpolation is typically carried out between samples of adjacent rows. Other more advanced interpolation schemes can also be employed. For illustration, Fig. 12.12 shows an exemplary sampling pattern for a four-row detector performing single-cycle data acquisition and reconstruction. This figure focuses on one of the cardiac cycles. Again, each point on the diagonal line represents a projection of a detector row collected at a specific z location at a particular time. To reconstruct an image at the plane of reconstruction shown by the vertical dashed line in the figure, we first perform row-to-row interpolation to estimate a set of projections at the reconstruction plane. For the projections represented by the top portion of the dashed line, measurements from detector rows 1 and 2 are used for linear interpolation (projections from more detector rows are needed for higher-order interpolations). Similarly, projections for the bottom portion of the dashed line are formulated with data from detector rows 2 and 3. Once a complete projection set at the plane of reconstruction is obtained, it is weighted by the half-scan weighting function and reconstructed with the conventional reconstruction algorithms. Of course, the z interpolation need not be a separate step; it can be combined with the half-scan weights to formulate a combined weighting function, as in the helical weighting approach discussed in Chapter 9.

If a 3D backprojection is used, the interpolation in the projection takes place automatically based on the intersection of the cone-beam ray that passes through a particular image voxel with the detector array. A row-wise half-scan weighting function can be applied prior to the filtering and cone-beam backprojection steps. If one chooses to first perform row-wise fan-to-parallel rebinning to improve noise uniformity, the half-scan weighting function can be applied either before or after the filtering step, as discussed in Chapter 10.

Figures 12.13(a) and (b) show reconstructed cross-sectional cardiac images obtained on an 8-slice LightSpeed™ scanner using the single-cycle half-scan technique. Because proper gating is provided for these scans, the arteries (left anterior descending coronary artery, left circumflex coronary artery, and right coronary artery) of the heart can be clearly visualized with little motion blurring. Figure 12.14 shows an oblique reformed image of a heart formulated from a set of 2D cardiac images. The clear definition of the vessels indicates that the phase gating is consistent from slice to slice.

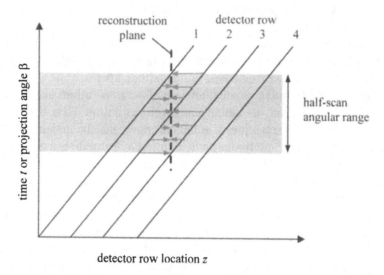

Figure 12.12 Illustration of row-to-row interpolation for cardiac single-cycle reconstruction. For the upper portion of the half-scan, projections collected from detector rows 1 and 2 are used to interpolate projections at the reconstruction plane. The bottom portion of the half-scan is obtained from projections from detector rows 2 and 3.

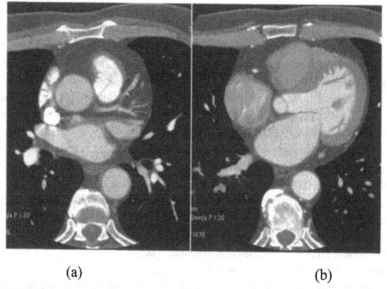

(a)   (b)

Figure 12.13 Reconstructed images of a cardiac scan with half-scan algorithm. (a) Left anterior descending coronary artery and left circumflex coronary artery. (b) Right coronary artery.

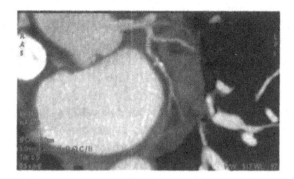

Figure 12.14 Reformatted image of a heart.

12.2.2.2 Temporal resolution improvement

The success rate of cardiac acquisition and reconstruction is less than 100%, especially on older vintage scanners. Operators often run into difficulties when the heart rate is either outside the scanner's optimal range or is irregular. For each scanner type, there is a limited heart rate range within which the reconstruction performs well. For example, on the LightSpeed™ Plus scanner (8 slice), a patient heart rate at or below 75 bpm is recommended. When the patient heart rate is too high, the coronary arteries can appear blurred, as shown in Figs. 12.15(a) and (b). Because of the nature of heart motion, the right coronary artery is especially prone to this type of artifact. One of the most commonly used approaches to combat blurring artifacts is to reduce the patient heart rate with medication prior to the data acquisition. For data collected with a high heart rate, one can also select a slightly different cardiac phase for reconstruction to reduce the blurring artifacts on a trial-and-error basis.

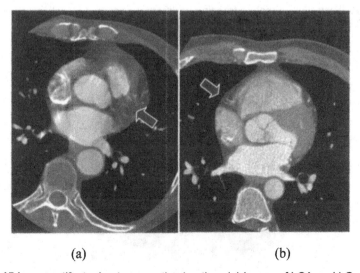

(a)   (b)

Figure 12.15 Image artifacts due to nonoptimal gating. (a) Image of LCA and LCx. (b) Image of RCA.

The optimal phase for CAI may be anatomy dependent. For example, the best phase to freeze motion in the right coronary artery (RCA) is often at the end-systole phase (35% to 45% R-R interval) instead of the end-diastole phase (70% to 75% R-R interval). Figures 12.16(a) and (b) show images reconstructed with the same patient scan at two different R-R intervals. Motion artifacts are clearly visible in the end-diastole phase shown by the dotted white circle in Fig. 12.16(a), and the artifact is eliminated in the end-systole phase in Fig. 12.16(b).

Often, metal objects (such as the lead in a pacemaker) or high-density objects (such as contrast-filled chambers) may be present that cause streaking artifacts. These artifacts often appear to be more severe than similar artifacts in other organs, due to rapid cardiac motion, as shown by the example in Fig. 12.17.

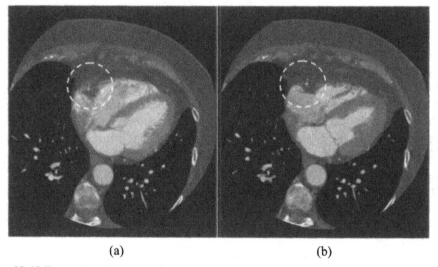

(a)   (b)

Figure 12.16 Illustration of motion freezing at different phases. (a) Motion appears in RCA at 75% R-R interval. (b) Motion is frozen at 45% R-R interval.

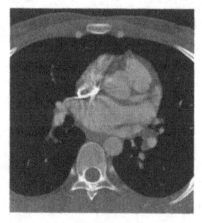

Figure 12.17 Image artifact due to the presence of either metal objects, dense contrast, or calcified deposits.

Since most CAI requires more than one cardiac cycle to complete, and the contrast agent is continuously pumped out and refilled, an inconsistent contrast level often leads to discontinuities in the reformatted images, as shown in Fig. 12.18(a). Suboptimal contrast injection can also cause streaking in the superior vena cava or right atrium, or an enhancement that is too high or too low in different portions of the heart, as illustrated in Fig. 12.18(b). Research has shown that image quality can be significantly improved by optimizing the contrast-injection protocols (e.g., following iodine contrast injection with a saline flush) to allow a steadier and more uniform flow of contrast.

Occasionally, image artifacts in the cardiac region are caused by respiratory motion instead of cardiac motion. In most cardiac protocols used today, the patient is asked to hold his or her breath during scanning. However, in many clinical practices it is left to the patient to conform to the operator's instructions. If a patient cannot hold his or her breath or if the operator starts the scan prematurely, motion artifacts will result, as shown in Fig. 12.19(b). Note that the chest wall appears as a double structure near the top of the image. When a later

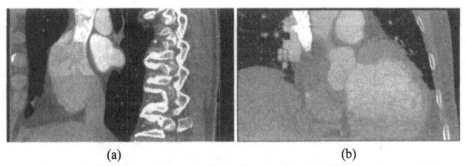

(a)   (b)

Figure 12.18 Discontinuities in the reformatted image due to (a) contrast inconsistency in different cardiac cycles, and (b) the lack of contrast enhancement due to suboptimal contrast injection protocol.

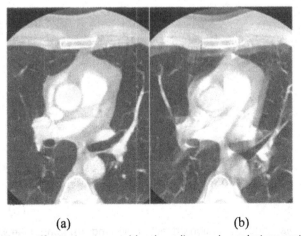

(a)   (b)

Figure 12.19 Motion artifacts due to combined cardiac and respiratory motion. (a) Properly gated and (b) improperly gated.

scan was collected of the same patient but with a proper breath-hold, a significantly improved image was produced, as shown in Fig. 12.19(a). The best approach for overcoming this deficiency is to employ a respiratory monitoring device in addition to the EKG monitor. A respiratory monitoring device provides the operator and the patient with timely feedback on the patient's breath-hold condition.

The impact of cardiac motion is not limited to heart imaging. Cardiac-induced motion artifacts frequently can be observed in other anatomies. For example, motion artifacts induced by the heart often appear in scans of the liver (Fig. 12.20) and lung [Fig. 12.21(b)]. An EKG-gated data acquisition can help to combat this artifact, as illustrated in Fig. 12.21(a).

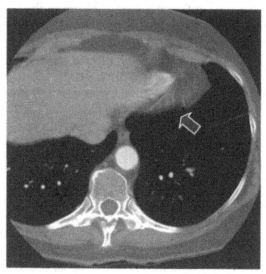

Figure 12.20 Motion artifact in the liver, induced by cardiac motion.

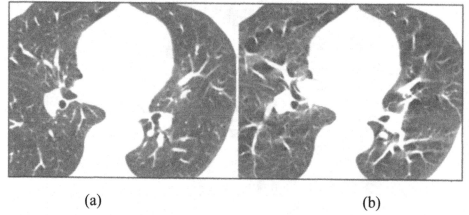

(a)   (b)

Figure 12.21 Cardiac-induced motion artifact in the lung. (a) Lung scan with proper EKG gating. (b) Lung scan without proper EKG gating.

Improving the temporal resolution of a CT system is very important for cardiac imaging, and the most straightforward way of doing so is to increase the gantry rotation speed. When the gantry rotation speed increases from 2 to 3 Hz (a 50% increase), the temporal resolution of the system increases by the same factor. Today's state-of-the-art CT systems typically operate around 3 Hz for cardiac imaging (0.35 sec or faster). However, the centrifugal force increases proportionally to the square of the speed ratio. For the 2- to 3-Hz example, the centrifugal force on the components mounted on the gantry increase more than twice (2.25 ×). These components are typically mounted more than a half-meter from the iso-center, and the centrifugal force increases proportionally to the square of the distance to the iso-center, as illustrated in Fig. 12.22. Since each CT gantry weighs well over 450 kg, the mechanical design for a faster CT system becomes increasingly difficult.

To further improve the temporal resolution of a CT system without redesigning the gantry, a multicycle or multisector data acquisition was proposed.[19–22,24–26] In this mode, a complete dataset for a single image reconstruction is collected over several cardiac cycles, as depicted in Fig. 12.23. For this example, a dataset was collected over three consecutive cardiac cycles. Because each data acquisition window was significantly less than that of a single-cycle data acquisition, improved temporal resolution can be expected. A detailed discussion of the temporal resolution of multisector reconstructions can be found in Chapter 5. To illustrate the image quality improvement obtainable with the multisector approach, Figs. 12.24(a) and (b) show the reconstructed images of a patient cardiac scan using single-cycle and four-cycle reconstructions. For this particular study, the patient heart rate was 100 bpm, well over the range of a single-sector reconstruction. Blurring, shading, and phase misregistration artifacts are clearly visible in the reformatted images [Fig. 12.24(a)]. These

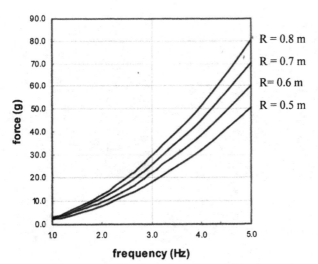

Figure 12.22 Centrifugal force as a function of gantry rotation frequency and distance to the iso-center.

artifacts are nearly completely eliminated with the multisector approach, as shown in Fig. 12.24(b). However, as the number of sectors increases, the reliability of the algorithm tends to decrease, because image reconstruction relies more and more on the consistency of the heart motion from one cycle to the next. Clinical experience has indicated that the heart motion is not highly uniform. In addition, when a contrast agent is used in a study (as in the majority of these cases), the contrast uniformity from cycle to cycle can be difficult to maintain. Consequently, most algorithms limit the number of sectors to less than four.

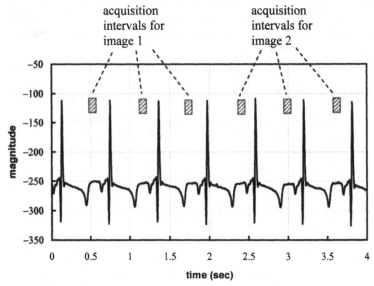

Figure 12.23 Illustration of multicycle acquisition.

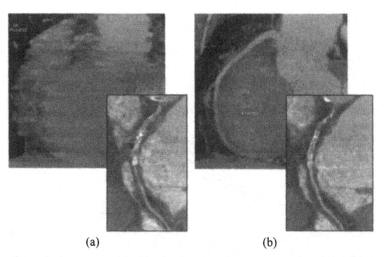

(a)   (b)

Figure 12.24 Cardiac scan performed on a patient with heart rate at 100 bpm. (a) Single-cycle half-scan reconstruction. (b) Four-cycle half-scan reconstruction.

Another method of improving the temporal resolution of the CT system is to add another tube and detector set to the original third-generation system design, as illustrated in Fig. 12.25(a).[27–29] The two sets of tube–detector pairs are offset by an angle so that each pair must only rotate significantly less than the $\pi + 2\gamma_m$ required by a half-scan reconstruction. One approach is to offset the two pairs by $\pi/2 + \gamma_m$ so that each pair must rotate only $\pi/2 + \gamma_m$ instead of $\pi + 2\gamma_m$. The combined projection datasets from both pairs form a complete half-scan acquisition. This is a nearly 50% reduction in terms of data acquisition time. Another approach is to offset the two pairs by $\pi/2$. In this case, to complete the total half-scan acquisition, each pair must travel $\pi/2 + 2\gamma_m$ to compensate for the total required angular coverage [Fig. 12.25(a)]. For typical commercial CT geometries, this represents a 38% to 44% improvement in temporal resolution.

One potential drawback of this approach is that the size of the second tube–detector pair must be reduced, due to the limited space on the gantry, as depicted in Fig. 12.25(a). For the smaller detector, the measured projection is essentially truncated. Some type of algorithmic compensation that leverages the information obtained from the full-size detector measurements must be used. Alternatively, one can increase both the offset angle between the two tube–detector pairs and the distance from the detector to the iso-center to enable two full detectors to coexist on the same gantry, as illustrated in Fig. 12.25(b). In a system with two tube–detector pairs, the level of scatter is likely to increase because both tubes are operating simultaneously during the cardiac data acquisition. In addition, the 90-degree offset configuration of the two tube-detector sets enhances the potential of the cross-scatter in which the scattered radiations produced by the x-ray photons from one tube-detector set are received by the other detector. [30]

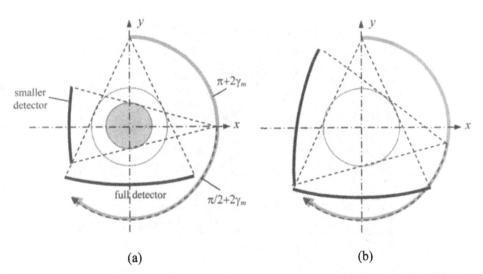

(a)   (b)

Figure 12.25 Illustration of a two-tube two-detector system in which the two tube–detector pairs are offset by 90 deg. (a) Configuration with a smaller detector. (b) Configuration with equal detector size.

12.2.2.3 Spatial resolution improvement

Although much attention has been paid to the temporal resolution aspect of cardiac imaging, spatial resolution plays an equally important role in CAI. Many applications require estimations of the percentage stenosis on relatively small vessels of 2 mm or less diameter. To differentiate between a 25%, 50%, and 75% stenosis, the system must be capable of resolving structures smaller than 0.5 mm. Similar requirements exist if one needs to assess the integrity of a stent that is only a few millimeters in diameter.

Spatial resolution plays an important role in reducing blooming artifacts when high-contrast objects are imaged. Blooming artifacts are high-density objects that appear larger than life-size in the reconstructed image. The intensity of these high-density objects "bleed" into their surroundings and make it difficult to determine the object boundary. A clearly visible object boundary is critical when trying to assess the narrowing of the lumen. When blooming from a stent invades into the luminal space, it is nearly impossible to accurately assess vessel integrity, as illustrated in Fig. 12.26(a), which was collected on a vintage CT scanner.

As discussed in previous chapters, the spatial resolution of a CT system is influenced by the detector size, focal spot size, system geometry, data sampling rate, and reconstruction algorithms. Cardiac imaging is particularly challenging because of the short scan time, high spatial resolution, and relatively narrow display window width. Previous discussions have shown the relationship between the spatial resolution and noise. Noise typically increases at a much faster rate than spatial resolution if the total x-ray flux is kept constant. Therefore, in pursuing a higher spatial resolution for CAI applications, special attention must be paid to achieving an optimal balance between the two. In addition to the

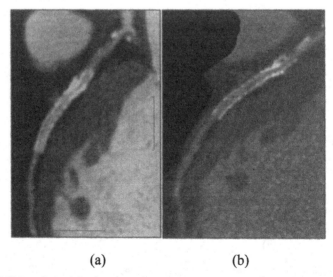

(a)   (b)

Figure 12.26 Illustration of the impact of spatial resolution on CAI. (a) Curved reformatted image of a patient scanned on an older vintage scanner. (b) The same patient scanned later on a Discovery™ CT750 HD scanner.

advanced hardware designs that enable high spatial resolution, advanced reconstruction algorithms are often required to keep image noise to a lower level. Figure 12.26(b) depicts the patient in Fig. 12.26(a) scanned on a Discovery™ CT750 HD. The improved visibility inside the lumen is obvious.

Figure 12.27 depicts another example of the benefit of an improved spatial resolution. Figure 12.27(a) was acquired on a LightSpeed™ 8-slice scanner. Since the slice thickness of the data acquisition was 1.25 mm, only large vessels are visible in the VR images. For comparison, Fig. 12.27(b) depicts a VR image acquired on a Discovery™ CT750 HD scanner with a slice thickness of 0.625 mm. The improved visibility of smaller vascular structures is evident.

12.2.2.4 Dose and coverage

Dose is yet another important aspect of CAI. Not long ago, CAI was one of the highest-dose CT applications, with an average dose well over 20 mSv. Despite CAI's benefits, the dose level became a significant concern for radiologists. The high dose level is linked to the high x-ray tube current and low-pitch helical scanning. As discussed in Section 12.2.2.1, a typical helical pitch used for CAI is between 0.1 and 0.3. At such a low pitch, 70% to 90% of the volume irradiated by the first gantry rotation is again irradiated by the second rotation [top portion of Fig. 12.28(a)]. In the central heart regions, the same location is often exposed to 3–10 rotations of the x-ray flux. Because a thin detector aperture is used for this application, the tube current is often set to its maximum to ensure an adequate signal-to-noise ratio.

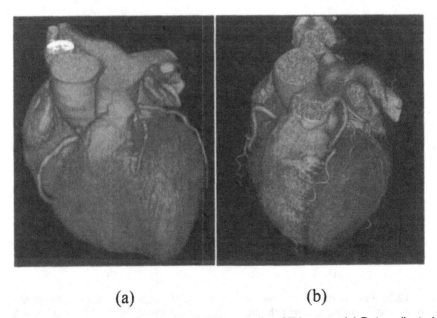

(a)   (b)

Figure 12.27 Examples of 3D volume-rendering cardiac CT images. (a) Data collected on a LightSpeed™ 8-slice scanner. (b) Data collected on a Discovery™ CT750 HD scanner.

X-ray tube-current modulation was introduced to reduce the x-ray dose to patients. In this mode, the x-ray tube current is reduced significantly outside the reconstruction window shown by the shaded boxes in Fig. 12.28(a). An 80% mA modulation means that the tube current is reduced to 20% of its peak value, and the overall dose to the patient is reduced. However, because of the x-ray cathode's thermionic emission process, the x-ray tube current cannot change instantly. The tube-current modulation curve is typically a smoothly varying function. This significantly limits the level of dose reduction achievable with such an approach.

To further reduce the dose, the use of step-and-shoot mode with prospective gating was introduced.[31] In this mode, the x-ray current is turned on only inside the reconstruction window and is otherwise turned off. To prevent a possible position deviation of the reconstruction window (since predicting the next cardiac cycle from previous cycles is rarely perfect), a small phase padding can be prescribed, as shown in Fig. 12.28(b). Because the x-ray tube can be turned on and off within a few milliseconds, the dose reduction is more efficient. The amount of table motion between adjacent scans is completely controlled by the scanner (instead of being dependent on the patient's heart rate, as in the case of helical mode), and little overlapping of the x-ray exposure is present between consecutive exposures. Phantom and clinical studies have shown that up to an 83% dose reduction can be achieved with this approach.[32]

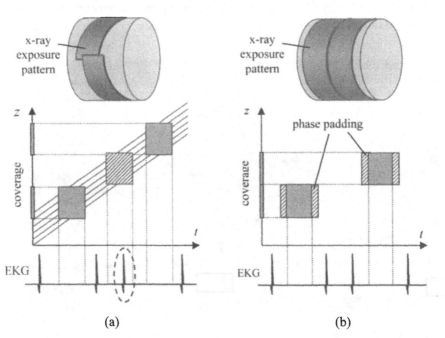

Figure 12.28 Two modes of cardiac acquisition. (a) Helical mode with highly overlapped exposure and handling of arrhythmia. (b) Step-and-shoot mode with little overlapped exposure and handling of arrhythmia.

An added advantage of the step-and-shoot mode cardiac acquisition is its flexibility in handling arrhythmia (irregular heart rate). When arrhythmia occurs, the heart is not in its normal state, and data collected during this cardiac cycle are likely to contain motion artifacts, as shown in Fig. 12.29. In the helical mode of data acquisition, the operator must use either this projection to reconstruct a set of images [diagonally shaded box in Fig. 12.28(a)] or adjacent acquisitions with suboptimal phases to fill in the gap (often called EKG editing), since the patient table moves at a constant speed during the data acquisition. Neither case is a preferred choice. In a step-and-shoot mode acquisition, when arrhythmia is detected, the patient table simply remains in its current position to wait for the next cardiac cycle and acquire projections.

Another commonly observed image artifact is phase misregistration. This is most pronounced when a reformatted coronal or sagittal image of the heart is displayed. The artifact appears as horizontal shifts or stair-steps in sections of the anatomy, as shown in Fig. 12.30(a). This is a curve-reformatted image generated by using a vessel analysis package to perform image reformation along the vessel. This artifact is most prominent in the midsection of the heart or the edge of the left ventricle. It can even be visible in axial images when the images are played back in a fast sequence. The cardiac structures appear to move from side to side or up and down, because the images are not reconstructed at exactly the same phase of the heart cycle. The combined effect of nonoptimal gating in the reconstruction algorithm and the inherent inadequate temporal resolution under nonuniform heart rate conditions creates this effect. Phase misregistration artifacts can sometimes be minimized by changing the phase at which the reconstruction takes place. Figure 12.30(b) shows the same scan reconstructed at a slightly different phase. Note that the misregistration artifacts are significantly reduced.

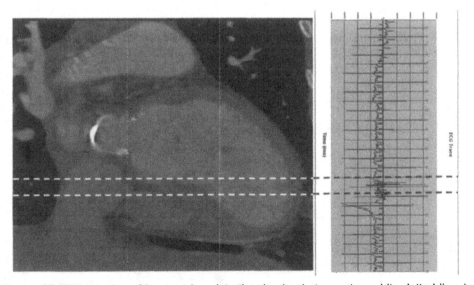

Figure 12.29 Illustration of image misregistration (region between two white dotted lines) due to arrhythmia (EKG signal on the right) on an older vintage scanner.

An alternate approach to overcome phase misregistration is to employ an array detector that is sufficiently large in z so that the entire heart is within its FOV (Fig. 12.31), allowing the entire organ to be scanned in a single acquisition. Since the detector size is comparable to the object of interest, there is little advantage to using the helical mode of data acquisition. The most efficient data acquisition is the step-and-shoot mode. One potential issue is the image artifacts resulting from the incomplete data sampling due to cone-beam geometry and longitudinal truncation. This has been discussed in detail in Chapter 10.

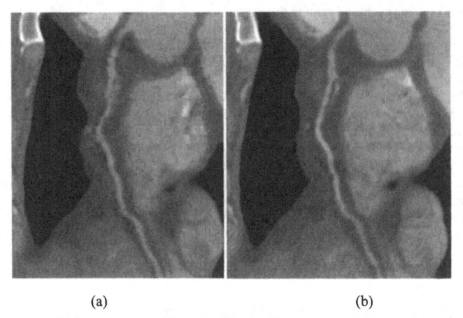

(a)   (b)

Figure 12.30 Misregistration artifacts in vessel-based reformatted images. (a) Image reconstructed with 70% R-R interval. (b) Image reconstructed with 80% R-R interval.

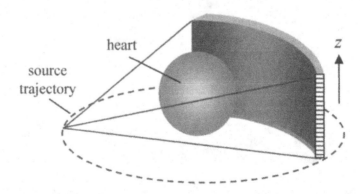

Figure 12.31 Illustration of a detector configuration to cover the entire heart.

12.3 CT Fluoroscopy

In recent years, interventionalists have increasingly recognized the value and advantages of CT for a variety of procedures. For example, CT has been used to guide biopsy procedures and cervical nerve root blocks, and in radio-frequency ablation of the osteoid osteoma. Similar to x-ray fluoroscopy, CT fluoroscopy provides near-real-time feedback to the operator. A survey result has shown that of the 26.3 million procedures performed between November 1998 and October 1999, 1.2 million (roughly 4%) were classified as guided procedures. These procedures include biopsy, abscess drainage, and interventional procedures.

The advantage of CT over other modalities, such as ultrasound or x-ray radiography, is the invaluable anatomical information that CT provides on the structures between the skin and the area of interest in complex procedures.[1] This information helps clinicians understand the best approach to introduce a needle or catheter. There are two basic modes of CT interventional operation. The first one uses the "real time" acquisition and reconstruction to provide guidance to the operator while the procedure is being performed, as illustrated by an artist's rendering of a CT operating suite in Fig. 12.32. In this setup, the monitor and scanner control is located inside the CT suite next to the patient table, unlike the conventional CT suite where the operator controls the entire process outside the room. During the procedure, CT images are provided to the operator on a "real time" basis to depict the location of the interventional instrument inside the patient. The instrument is dynamically adjusted to reach the target location and to avoid vital organs. This operating mode is often called CT fluoroscopy. To free

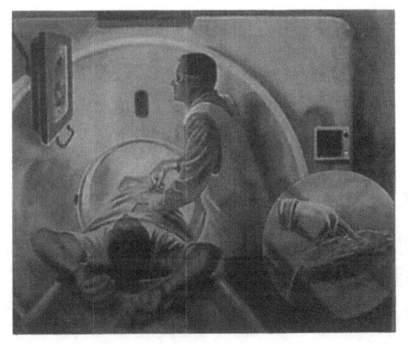

Figure 12.32 Artist's rendered version of CT fluoroscopy operating suite.

both of the operator's hands for the interventional operation, most fluoroscopy devices offer footpaddles to initiate and end the scan. For example, the SmartView™ scanner (GE Healthcare) has both scan and tap modes on a footpaddle so the operator can acquire either a single or multiple images. An imaging review of the operating sequence (in which the operator can review the acquired image during the procedure) is also available using a handheld control, as shown in Fig. 12.33. The monitor alerts the operator to the amount of x-ray exposure time that has elapsed since the start of the procedure. Figures 12.34(a)–(d) depict a biopsy needle placement sequence. The imaging location during the operation changes slightly from image to image (slightly different anatomical features) to account for the oblique insertion angle of the biopsy needle.

The CT fluoroscopy display should not be limited to conventional 2D CT images. A 3D image provides additional volumetric information about the location of the needle relative to other anatomical structures. This feature becomes feasible with multislice CT scanners, since a volume is covered at all times during the scan. One example of a 3D display of a chest biopsy is shown in Fig. 12.35.

Because images need to be reconstructed and displayed in real time to provide timely feedback to the operator, different reconstruction approaches must be employed to ensure fast image production. Only six years ago when the first edition of this book was published, the state-of-the-art scanners took roughly 0.2 to 0.5 sec to reconstruct a single image, which is equivalent to an image frame rate of 2 to 5 frames/sec. This is clearly inadequate for real-time applications, which require a frame rate between 6 and 16 frames/sec or higher. To enable

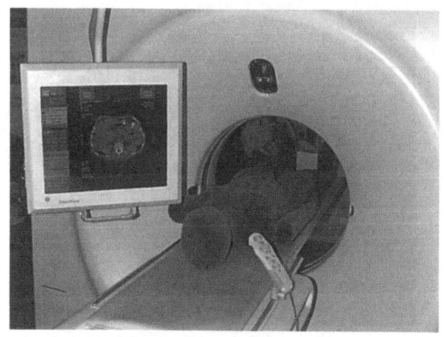

Figure 12.33 Photograph of a CT fluoroscopy device.

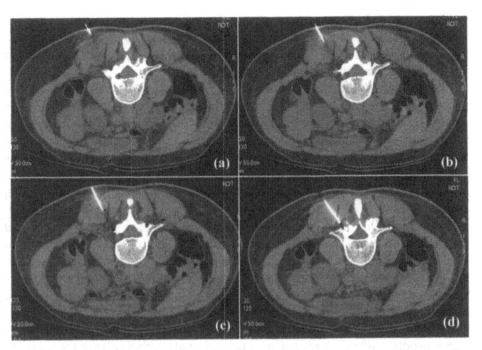

Figure 12.34 Examples of a biopsy operation. (a) Biopsy needle just entering the patient. (b) Biopsy needle en route to target. (c) Biopsy needle farther en route to target. (d) Biopsy needle reaching the target.

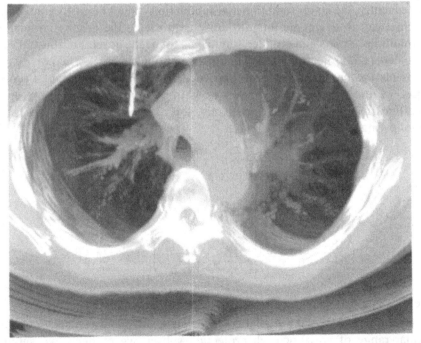

Figure 12.35 Depth-weighted volume-rendered view of a biopsy procedure.

rapid image production without a significant increase in the reconstruction hardware, more efficient reconstruction algorithms must be explored.[33-39] These algorithms make optimal use of previously reconstructed images to reduce the amount of redundancy in image reconstruction.

The following simple example will illustrate this concept. We want to reconstruct a series of images at 16 frames/sec on scans collected with 0.5 sec per gantry rotation. If we assume that we need a 360-deg rotation of data to produce a single image, the angular difference between the projections used in two consecutive fluoroscopy images is 45 deg (360 deg/8). Consequently, projections within the center 270 deg (360 deg − 2 × 45 deg) are common to both images. Alternatively, if we examine the projections used to produce the second image, only 45 deg of the projections are new. If we can avoid preprocessing, filtering, and backprojecting the older projections, we can save over 87% of the computation. This is not difficult if images were generated with a uniform weighting function (no weighting). However, for CT fluoroscopy applications, this approach has severe drawbacks. In most CT-guided procedures, the high-density interventional instrument is moving constantly for position adjustment. In addition, the patient table needs to be moved from time to time to ensure the tip of the instrument is within the scanning plane or scanning volume (for example, a biopsy needle is often introduced at an oblique angle with respect to the scanning plane to avoid vital organs). For many CT fluoroscopy devices, the patient table can be set to a "float" mode in which the table position can be easily adjusted by hand. To suppress motion artifacts caused by the needle and patient movement, projections must be properly weighted (see Chapter 7 for detailed discussions). Consequently, the simple addition and subtraction cannot be easily implemented.

One way to overcome this difficulty is to divide the original projection set into multiple subsets and produce subimages for each subset. For illustration, Fig. 12.36 depicts a projection dataset and its weighting function. The horizontal axis represents the projection number, and each small block represents a projection subset. In this example, an overscan weighting function is used, represented by thick lines in the figure. Mathematically, the overscan function is described by the following equation:

$$
w(\beta) = \begin{cases} 3\left(\dfrac{\beta}{\beta_0}\right)^2 - 2\left(\dfrac{\beta}{\beta_0}\right)^3, & 0 \le \beta < \beta_0, \\ 1, & \beta_0 \le \beta < 2\pi, \\ 3\left(\dfrac{2\pi + \beta_0 - \beta}{\beta_0}\right)^2 - 2\left(\dfrac{2\pi + \beta_0 - \beta}{\beta_0}\right)^3, & 2\pi \le \beta \le 2\pi + \beta_0, \end{cases} \tag{12.2}
$$

where β_0 is a parameter that describes the overscan angle. Typically, $\beta_0 = \pi/4$. In this example, we divided the entire $2\pi + \beta_0$ into nine subsets, each covering an angular range of $\pi/4$. For each projection subset, we produce two subimages:

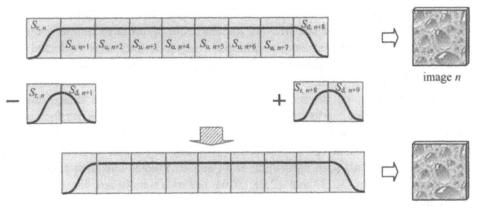

image n

image n+1

Figure 12.36 Recursive image generation for CT fluoroscopy. For each subframe, two subimages are reconstructed: $S_{r,n}$ and $S_{u,n}$. $S_{r,n}$ is a subimage reconstructed with an ascending weighting function and $S_{u,n}$ is a subimage with unity weighting function. $S_{d,n}$ can be obtained by $S_{u,n} - S_{r,n}$. To generate the next image, several subimages are added and subtracted from the previous image.

one reconstructed with a weighting function specified by the first row of Eq. (12.2) for $0 \le \beta < \beta_0$, and the other reconstructed with a uniform weight. These subimages are denoted by $S_{r,n}$ and $S_{u,n}$, respectively, where subscript "r" stands for a rising or ascending function over β, "u" stands for the uniform weight, and n stands for the nth projection subset. From $S_{r,n}$ and $S_{u,n}$, we can produce a subimage reconstructed with the descending weight, $S_{d,n}$ [last row in (Eq. 12.2)]:

$$S_{d,n} = S_{u,n} - S_{r,n}.$$ (12.3)

If we denote by I_n the nth full image produced for CT fluoroscopy, the $(n + 1)^{th}$ image, I_{n+1}, can be generated by the simple additions and subtractions of several subimages:

$$I_{n+1} = I_n - S_{r,n} - S_{d,n+1} + S_{r,n+8} + S_{d,n+9}.$$ (12.4)

To produce a new image, we need to filter only one and backproject two projection subsets. Since backprojection is often performed with fast computer chips, a factor of nearly 9 improvement in computational speed can be achieved compared to the brute-force reconstruction. This approach is called recursive reconstruction.

In the above example, the overscan weighting function is used mainly for its simplicity to describe the recursive reconstruction concept. In practice, however, there is an inherent disadvantage in using the overscan type of weighting, primarily due to temporal resolution considerations (see Chapter 5 for a discussion on temporal resolution for CT fluoroscopy). For illustration, we performed the following experiment. A biopsy needle was placed inside a glass tube that was positioned vertically in the scanner FOV. A remote triggering device was built so the needle free-fall motion could be initiated remotely during the scan. During the experiment, continuous cine-mode data were collected (the patient table remained stationary) at a 1-sec gantry speed to cover the entire sequence of needle motion. A series of images were reconstructed to examine the inherent time delay of the reconstruction process (we excluded delays due to computer and display hardware and examined only the time it took for the needle to appear resting at the bottom of the tube after the trigger release). Two sets of images were reconstructed: one with the overscan algorithm described above, and one with the half-scan algorithm. Both sets of images and their time scales are depicted in Figs. 12.37 and 12.38. If we use as a benchmark the first image frame in which the needle appears at the tube bottom without a "ghost" needle or artifacts inside the tube, the overscan reconstruction took 1.1 sec to observe the resting needle after the initial drop. On the other hand, the half-scan reconstruction took only 0.7 sec to reach the same status. Unfortunately, the half-scan weights vary with both the detector angle γ and the projection view angle β. An additional adjustment is needed to implement the recursive reconstruction. One such scheme is described in Ref. 37.

The approaches described above were developed when the reconstruction speed was limited by the computer hardware. Over the last few years, computing hardware has advanced considerably, largely driven by the gaming industry. New computer architectures, such as the GPU and cell processor, offer significant benefits in terms of massive parallel processing. The FBP reconstruction algorithm lends itself well to such architectures, and a reconstruction speed near

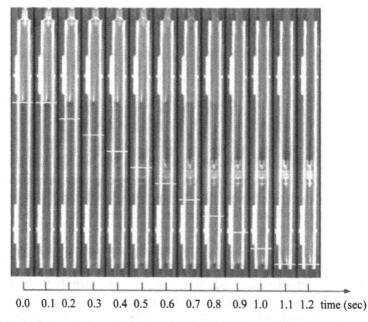

0.0   0.1 0.2   0.3   0.4 0.5   0.6   0.7 0.8   0.9 1.0   1.1 1.2 time (sec)

Figure 12.37 "Free-fall" needle experiment reconstructed with overscan weight.

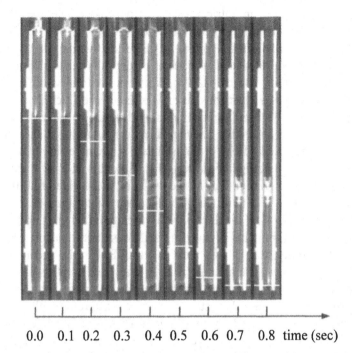

0.0   0.1 0.2   0.3   0.4 0.5   0.6 0.7   0.8   time (sec)

Figure 12.38 "Free-fall" needle experiment reconstructed with half-scan weight.

or exceeding real time (30 frames/sec) is commercially available. A reconstruction speed well over 200 frames/sec has been prototyped in a recent feasibility study. As a result, the CT fluoroscopy mode no longer needs special reconstruction algorithms to provide real-time images.

The second mode of interventional procedure uses software analysis tools to establish a virtual needle placement path using CT images acquired prior to the procedure. This is quite similar to treatment planning for oncology. Since this mode does not require special operations from a CT scanner point of view, we will omit further discussion of this subject.

12.4 CT Perfusion

Blood flow provides oxygen and nutrients to the brain to maintain its normal functions. The basic functions of the brain, such as synaptic transmission, membrane ion pumping, and energy metabolism, are interrupted at different blood flow levels. For example, if the blood flow falls below 16 to 18 ml/min per 100 g of brain tissue, synaptic transmission ceases to function.[40,41] When the cerebral blood flow (CBF) drops below 8 to 10 ml/min per 100 g of brain tissue, the cell membrane pumps fail, causing membrane depolarization.[41,42] This is followed by infarction, the irreversible destruction of cells. It was found that with the reduction of blood flow, the duration of ischemia required to induce infarction shortens.[43]

Because the brain has an intricate system to self-regulate the CBF within normal limits, the measurement of CBF alone is inadequate to assess brain tissue viability. When cerebral perfusion pressure decreases within the autoregulation limit, the blood vessels are dilated to reduce cerebral vascular resistance and to increase the cerebral blood volume (CBV), which is the total volume of blood in the large conductance vessels, arteries, arterioles, capillaries, venules, and sinuses. This phenomenon can be used to distinguish salvageable tissue from infarcted tissues. For the ischemic but viable (salvageable) tissue, autoregulation leads to an increased CBV and reduced CBF. For the ischemic but nonviable (infarcted) tissue, on the other hand, the autoregulation ceases to function and both the CBV and CBF are reduced. The distinction between viable and infarcted tissue is of important clinical value. If a particular brain region is salvageable, thrombolysis can be used to restore the CBF. If the region is already infarcted, thrombolysis can only increase the risk of intracranial bleeding and does little to restore the CBF.

One useful quantity that relates the CBF to CBV is the mean transit time (MTT). MTT is defined as the average time it takes for blood to travel from the arterial inlet to the venous outlet. In reality, since blood traverses through different paths of the vascular structure with different path lengths and resistances, as shown in Fig. 12.39, the transmit time is distributed over a range. Thus, MTT is simply the mean or average of this distribution. The relationship between CBV, CBF, and MTT is governed by the central volume principle:[44]

$$CBV = CBF \times MTT. \qquad (12.5)$$

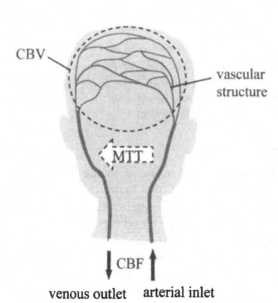

Figure 12.39 Illustration of variation of transit time due to different path lengths and the relationships among CBF, CBV, and MTT.

The use of CT to measure CBF, CBV, and MTT was first introduced in 1980.[45] Over the past two decades, various algorithms have been developed. These algorithms can be classified into two classes: direct measurement and deconvolution. The direct-measurement method is based on the Fick principle. If we denote the arterial concentration of the contrast medium by $C_a(t)$, the rate of influx of the contrast medium into the brain is simply $C_a(t)F$, where F is the CBF shown in Fig. 12.39. Similarly, the rate of efflux of the contrast medium is $C_v(t)F$, where $C_v(t)$ is the venous concentration of the contrast medium. Based on the conservation of contrast medium in the brain, the rate of contrast accumulation in the brain, $q(t)$, is the difference between the influx rate and the efflux rate of contrast. At time t, the total amount of contrast in the brain, $Q(t)$, is the difference between the total influx of contrast medium and the total efflux of contrast medium up to t. This relationship can be described by the following set of equations:

$$q(t) = \frac{dQ(t)}{dt} = F[C_a(t) - C_v(t)] \tag{12.6}$$

and

$$Q(t) = F\left[\int_0^t C_a(t)dt - \int_0^t C_v(t)dt \right]. \tag{12.7}$$

If we acquire a set of CT images at a fixed location (cine mode) over a period of time after the contrast injection, we obtain a measurement of contrast uptake in the brain. A time-density curve (TDC) can be calculated for every point inside the scanned volume. $C_a(t)$ is the TDC measured at an artery, and $Q(t)$ is the TDC measured for the entire brain. If we can identify "local" vessels that supply and drain blood for a subsection of the brain, Eqs. (12.6) and (12.7) are valid for the subsection.

Based on these equations, a method was proposed to calculate the CBF, F, assuming that all injected contrast medium remains inside the vascular structure and no venous contrast outflow occurs $[C_v(t) = 0]$.[46] Based on this assumption, equation (12.6) is reduced to the following expression:

$$q(t) = F \cdot C_a(t), \tag{12.8}$$

where $q(t)$ is the rate of accumulation of the contrast medium in the organ and $C_a(t)$ is the arterial TDC. It happens that when the arterial TDC reaches its maximum $C_{a\text{-max}}(t)$ the rate of contrast accumulation also peaks to $q_{max}(t)$. The cerebral blood flow F can then be readily calculated using the ratio of the maximum values:

$$F = \frac{q_{max}(t)}{C_{a-max}(t)}. \tag{12.8a}$$

Hence, the cerebral blood flow F is the ratio of the maximum slope of total contrast accumulation $Q(t)$, to the maximum arterial concentration $C_{a\text{-}max}(t)$. This approach is often called the "maximum slope method."

This method was extended to tissue perfusion measurement in abdominal organs and dynamic CT.[47,48] Although this method is conceptually simple, it has significant limitations due to the assumption of no venous outflow. In clinical settings, the assumption that no venous outflow occurs at the maximum tissue TDC slope is often violated. Consequently, the method can lead to underestimation of the blood flow.

To overcome this difficulty, we can either reduce the total contrast volume to the patient or increase the injection rate. A reduction of the total contrast volume can result in a poor signal-to-noise ratio for the TDCs, since it reduces the overall CT number enhancement of the organ. Although an increased injection rate approach has been tried,[49,50] this technique generally requires a large-gauge catheter and raises safety concerns for the patient. An alternative approach is to find a "local" vein that drains the tissues of interest and measures the actual venous outflow for the organ subregion.[51] This approach essentially divides the organ of interest into multiple subregions of interest and solves the blood flow problem on a local basis. The CBF, F, can then be calculated by rearranging Eq. (12.7):

$$F = \frac{Q(t)}{\int_0^t C_a(t)dt - \int_0^t C_v(t)dt}. \tag{12.9}$$

In this equation, $Q(t)$ is the height of the TDC for the tissues of interest at time t, the first integral in the denominator represents the area under the arterial TDC, and the second integral represents the area under the venous TDC. The advantage of this approach is again its conceptual simplicity. The difficulty that may arise with this approach is in identifying the local vein for the tissues of interest. Even with proper identification, the size of the local vein is usually too small compared to the spatial resolution of the scanner. This may lead to underestimation of the area under the venous TDC, and eventually to underestimation of the blood flow.

The class of deconvolution methods takes a slightly different approach.[52–58] These methods essentially treat the organ perfusion as a linear system problem. From linear system theory, we know that a linear system can be uniquely characterized by its impulse response, $h(t)$. An impulse function (also known as the Dirac delta function), $\delta(t)$, is defined as a mathematical function that has infinitely small width and unity area:

$$\delta(t) = 0, \text{ if } t \neq 0, \tag{12.10}$$

and

$$\int_{-\infty}^{\infty} \delta(t) = 1. \tag{12.11}$$

The system response to any input signal is simply the convolution of the input function with the system's impulse response. For organ perfusion, an arterial bolus of contrast medium with unity volume is injected over an extremely short period of time to approximate the Dirac delta function. The tissue TDC corresponding to the injection is called the tissue's impulse-residual function,[59] the analog to the impulse-response function of the linear system. For simplicity, we use $h(t)$ to represent the impulse-residual function. The term "residual" indicates that we are measuring the amount of contrast that remains in the tissue after the contrast injection. An exemplary impulse-residual function is shown in Fig. 12.40. If we take an infinitely small increment in $h(t)$, Δh, the size of Δh represents the fraction of the contrast medium that is washed out during a small time increment at t. Thus, Δh represents the fraction of the contrast that has transit time t. The average transit time (MTT) is therefore the summation of all transit time t weighted by $\Delta h(t)$. When Δh approaches zero, we have

$$\mathrm{MTT} = \int_0^1 t\,dh . \tag{12.12}$$

The integration represents the area under the impulse-residual function $h(t)$. Therefore, the area under the impulse-residual function represents the MTT. The plateau of $h(t)$ near $t = 0$ represents the fact that there is a minimum time interval for the contrast to travel from the arterial inlet to the venous outlet. Since the amount of contrast is of unit volume, the plateau of $h(t)$ is 1.

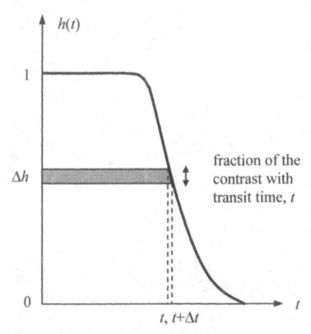

Figure 12.40 Representative impulse-residual function and its physical meaning.

Of course, the impulse-residual function cannot be measured directly in clinical practice. However, we can measure the tissue response $Q(t)$ to the influx contrast medium of $C_a(t)F$. Based on linear system theory, these quantities are related to $h(t)$ by the following equation:

$$Q(t) = C_a(t)F \otimes h(t) = C_a(t) \otimes g(t), \tag{12.13}$$

where

$$g(t) = h(t)F . \tag{12.14}$$

Here, F is roughly constant during the short scan time interval, $Q(t)$ is the tissue TDC, and $C_a(t)$ is the arterial TDC. Both quantities can be readily measured on the contrast-enhanced CT images. Since $Q(t)$ is the convolution of $C_a(t)$ and $g(t)$, the problem then becomes using deconvolution to obtain $g(t)$.

Now we will examine the quantity $g(t)$. Based on Eq. (12.14), $g(t)$ is the impulse-residual function scaled by the blood flow F. Since the plateau of $h(t)$ is unity, the height of the plateau of $g(t)$ is F. Previously, we stated that the area under $h(t)$ represents the system MTT. Based on the central volume principle stated in Eq. (12.5), the area under $g(t)$ then represents CBV, since it is the product of MTT and CBF. For illustration, Fig. 12.41 shows a brain perfusion study for a stroke patient, and Fig. 12.42 shows a body perfusion study for tumor characterization.

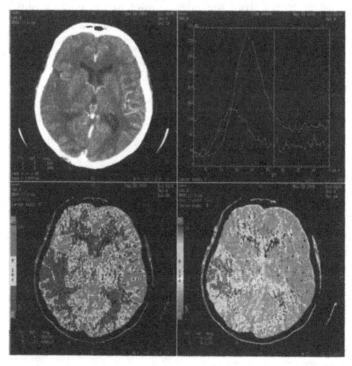

Figure 12.41 Perfusion images of a head scan. Upper left: CT image of a head scan with contrast injection. Upper right: contrast uptake curves in the vessels. Lower left: blood flow image. Lower right: mean-transit-time image.

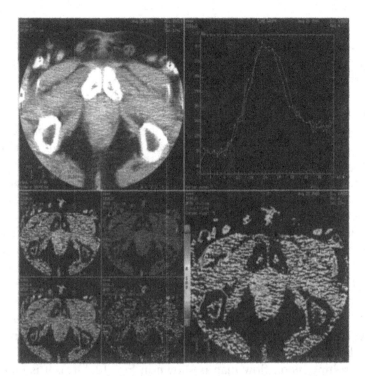

Figure 12.42 Perfusion images of a body scan. Upper left: original CT scan. Upper right: contrast uptake curves in the vessels. Lower left: blood volume, blood flow, mean transit time, and permeability surface images. Lower right: blood volume image.

The advantage of the deconvolution approach is that it does not make extensive assumptions about the underlying vascular structures. Consequently, its applicability to wider clinical conditions can be expected. The disadvantage of the approach is its relatively more complex calculation. Since deconvolution can be sensitive to noise, a highly stable deconvolution method must be used. In addition, image registration is needed to correct for patient motion during CT scans. Clinical tests have shown that these obstacles can be overcome to ensure accurate perfusion measurements.[54]

One important issue in CT perfusion is the x-ray dose. Because the x-ray beam is on continuously for an extended period of time (30 to 50 sec), the total amount of radiation dose is fairly high despite the low x-ray tube current used during the scan. One way of reducing the dose is to perform sparse samples in the temporal domain.[60,61] Instead of scanning the patient continuously, the x-ray tube is turned on over one gantry rotation and turned off for a short period of time. The x-ray tube is then turned on again and the entire process repeats. If the x-ray on-and-off periods are 1 sec each, the amount of dose savings is 50%. A longer off period provides a higher dose reduction. One key limiting factor is the temporal frequency contents of the contrast uptake curves. Based on the Nyquist sampling theory, the sampling frequency needs to be at least twice the highest frequency present in the signal. Phantom and clinical experiments have shown that the sampling intervals cannot be significantly larger than 3 sec to maintain the accuracy of the perfusion map in the brain. If we assume a rotation gantry speed of 1 sec per revolution, a 3-sec sampling cycle means that the x-ray needs to be on only one-third of the time. This represents a factor of 3 dose reduction. Figure 12.43 depicts two blood flow maps of the same patient. Image (a) was acquired and reconstructed with 1-sec time intervals as inputs to the perfusion algorithm (x-ray continuously on), and image (b) was reconstructed with 3-sec intervals of the same scan data (two-thirds of the scan data were not used). Despite a slight noise increase in the perfusion map, nearly identical results were obtained.

As discussed in Chapter 11, advanced reconstruction algorithms such as IR can be used to reduce the noise in reconstructed images. These algorithms also can be used to reduce the patient dose during the data acquisition. For demonstration, a clinical experiment was conducted on a LightSpeed™ VCT scanner in which a patient undergoing a perfusion study was scanned twice: once with a normal perfusion protocol (continuous scan), and once with the x-ray tube current reduced to 1/8 of the original dose (87.5% reduction). Using product software (FBP reconstruction) and a standard perfusion package, two sets of perfusion maps were obtained. Figures 12.44(a) and (b) show the corresponding blood flow maps. Not surprisingly, the blood flow map of image (b) is noisier than image (a) due to the reduced x-ray flux. When we use the simplified IR algorithm discussed in Chapter 11 to perform the reconstruction using a 1/8 dose scan, the resulting blood flow map is shown in Fig. 12.44(c). It is clear that the noise in the map is significantly improved over image (b). It can be argued that the noise in image (c) is even less than that of image (a). Next, we subsampled the data to simulate two cases in which perfusion scans were performed every 2 sec and every 3 sec. The resulting blood flow maps, with the advanced estimation algorithm, are shown in Figs. 12.44(d) and (e). Both quantitative and visual evaluations by experts show that the perfusion maps of images (d) and (e) are comparable to that of image (a). These maps represent 1/16 and 1/24 of the original dose, or dose reduction factors (respectively) of 93.75% and 95.8%!

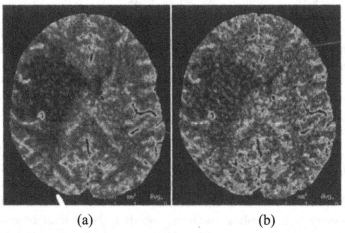

(a)   (b)

Figure 12.43 Blood flow map of a brain perfusion study with (a) x ray on continuously and (b) x ray on every three seconds.

Because of the relatively long acquisition time needed for perfusion CT, patient motion is naturally an issue. Perfusion maps are generated by calculating contrast uptake curves for each pixel location over the entire acquisition time interval. If the patient moves during the scan, the contrast uptake curve no longer represents the true contrast uptake at the same location and an inaccurate result will be produced. For brain perfusion, the correction is relatively simple since a rigid registration can be performed over all reconstructed images to ensure proper location registration. For body organ perfusion, on the other hand, the issue is more complex since the motion is typically nonrigid and 3D.

CT perfusion is still an active area of research. Although we discussed in general terms the pros and cons of different approaches, more refined and advanced algorithms are constantly being developed. For example, a recent study suggests the equivalency of the deconvolution approach and the maximum slope approach for brain perfusion studies.[62]

So far, we have discussed methodologies to perform quantitative perfusion. In some clinical applications, however, only qualitative perfusion is needed to assess the health status of the tissue. For example, in cardiac imaging it is often desirable to assess myocardial perfusion and detect microvascular obstructions. In such applications, a first pass followed by a delayed CT scan is used in clinical research. The delayed scan is performed several minutes after the initial injection of contrast bolus.[63] This technique has been used to identify myocardial infarction.

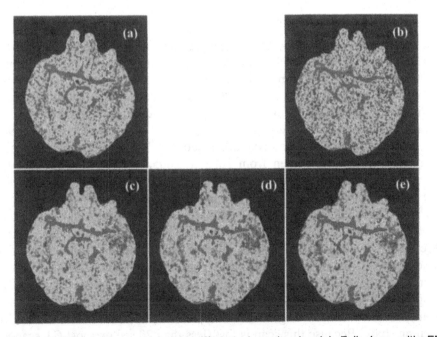

Figure 12.44 Blood flow maps with different dose levels. (a) Full dose with FBP reconstruction. (b) 1/8 dose with FBP reconstruction. (c) 1/8 dose with simplified iterative reconstruction. (d) 1/16 dose with simplified iterative reconstruction and sampling every 2 sec. (e) 1/24 dose with simplified iterative reconstruction and sampling at 3-sec intervals. (Patient scan data courtesy of Dr. T. Y. Lee, of St. Joseph's Healthcare, Canada.)

12.5 Screening and Quantitative CT

Slowly but surely, CT is moving into the area of screening. Screening implies that the patient under examination is asymptomatic. For CT to be successful in screening, two key enabling technologies must be developed and perfected. The first is low-dose scans. Since human exposure to x-rays is generally considered harmful, the risks of patient radiation exposure have to be significantly lower than the risks of undetected pathologies that can be treated at early stages to impact the outcome. In other words, the benefit of CT screening examinations has to outweigh the risks of the additional radiation exposure. Although many research studies have been conducted to minimize the x-ray dose to patients, more progress still needs to be, and can be, made in this area.

The second key technology is the handling of a large amount of data. Because screening implies that a significant portion of the population or a subpopulation is going to be examined, the number of images generated in the screening study is overwhelming. The ability of a limited number of radiologists to handle such a large data volume is certainly going to be a limiting factor. One good example is the widespread use of mammograms. Today, several systems and software packages are available to help radiologists by automatically identifying suspicious regions. These systems rely on segmentation tools and large databases to train various algorithms to differentiate abnormalities. Even with the aid of these tools, radiologists must contend with heavy workloads.

12.5.1 Lung cancer screening

Lung cancer screening has received significant attention in recent years. A study has shown that in 1998 in the United States alone, there were an estimated 160,000 deaths from lung cancer and an estimated 172,000 new cases detected. In 1999, lung cancer resulted in 158,900 deaths. Based on the data from Centers for Disease Control and Prevention, lung cancer incidence has decreased by 1.8% per year among men and increased by 0.5% per year among women from 1991 to 2005. Deaths from lung cancer have decreased by 1.9% per year among men and remained level among women from 2003 to 2005.[64-68] The cure rate for lung cancer is 12%, and the five-year survival rate is slightly higher. However, when a stage I cancerous tumor is resected, the five-year survival rate can be as high as 70%.[69] A study indicated that low-dose CT greatly increases the likelihood of detecting small noncalcified nodules compared with chest radiography.[70] In the entire study population, malignant disease was detected four times more frequently with low-dose CT than with chest radiography. Similarly, stage I tumors were detected six times more frequently with low-dose CT than with radiography. This means that CT is more likely to detect lung cancer at an earlier and more curable stage.[70] Figure 12.45 depicts an example of a low-dose lung screening study. Because this scan is fast (less than 20 sec on most CT scanners) and does not require contrast injection, it has generated significant interest in the radiology community.

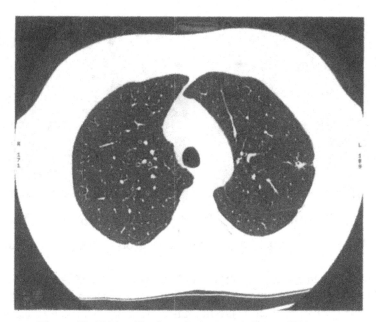

Figure 12.45 Identification of a lung nodule in a CT image.

Many studies have been conducted to determine optimal clinical protocols and parameters. Research activities also focus on the automatic detection, sizing, and characterization of lung nodules.[71-74] Several software packages are available to track the growth of lung nodules over time to aid in differentiating benign versus malignant nodules, since the primary indication of a nodule malignancy that is 5 mm or less in diameter is the growth measured with follow-up CT studies. The algorithms used in these methodologies all require the reliable and accurate detection of lung nodule boundaries and measurement of the lung nodule volumes. Figures 12.46(a) and (b) depict, respectively, an automatic nodule boundary detection, and a VR view of the nodule. This book will not discuss the pros and cons associated with these algorithms. Instead, we want to raise the awareness of researchers of the impact of different CT data acquisition and reconstruction techniques on the outcome of the boundary detection and volume measurement.

The shape and volume of a lung nodule change with the selection of different reconstruction algorithms. This clearly follows from our discussion in Chapter 3. The PSF of the system changes for different reconstruction kernels, since both the cutoff frequency and the frequency response of the filtered projections are modified by the filter kernel. To illustrate this effect, a simple experiment was performed. An irregularly shaped piece of chewing gum of roughly 6-mm diameter was scanned with a 0.63-mm collimator aperture in step-and-shoot mode. The chewing gum was selected for its close resemblance to lung nodule density and its flexibility to form into any shape. To simulate the lung background, the gum was placed inside a large piece of foam. Spacing between adjacent axial images was 0.63 mm. The images were reconstructed from the

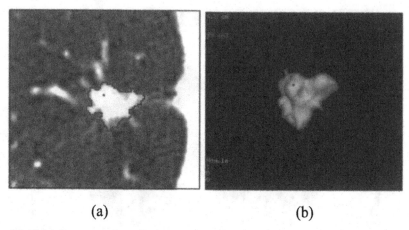

(a)   (b)

Figure 12.46 (a) Segmentation of a lung nodule and (b) 3D volume-rendered view.

same scan data with three different reconstruction kernels: standard, lung, and bone. Three separate volume measurements were performed on three sets of images. Since the CT number of the foam is close to that of air, a single threshold was used to segment the gum from its background. The results are shown in Table 12.1. It is clear from the table that significant variation exists between the measured volumes with different reconstruction algorithms (the standard algorithm result is used merely as a reference point, not as a standard). Figures 12.47(a)–(c) show surface-rendered images obtained from the reconstructed gum phantom. The shape differences in the reconstructed nodule are apparent. The above experiment was conducted with a nodule size of roughly 6-mm diameter; the percentage error in the measured volume would likely increase with a decrease in the nodule size.

Different acquisition protocols also play a significant role in the accuracy of the measured volume. For illustration, the same gum phantom was scanned with six different acquisition protocols ranging from thin-slice step-and-shoot mode to thin-slice helical mode and thick-slice step-and-shoot mode. The measured volumes and corresponding surface-rendered images are shown in Table 12.2 and Fig. 12.48, respectively. Large variations can be observed in the results. Small structures that can be observed in the surface-rendered images of the thin-slice acquisition modes disappear in the thick-slice data acquisition due to the partial-volume effect. For the classification algorithms that rely on the nodule "texture" to differentiate different classes of nodules, careful selection of appropriate acquisition protocols is crucial to their success.

For any screening applications, the x-ray dose is important because the patients to be scanned are asymptomatic. The dose-reduction techniques discussed in Chapter 11 are directly applicable to screening applications. More importantly, the implementation of the ALARA principle is critical to the success of CT screening.

Table 12.1 Volume measurement varies with the reconstruction kernel. All images were acquired with 0.63-mm axial mode at 0.63-mm spacing.

Reconstruction kernel	Standard	Lung	Bone
Measured volume (mm^3)	160.54	137.95	141.79
% variation to standard		−14.1%	−11.7%

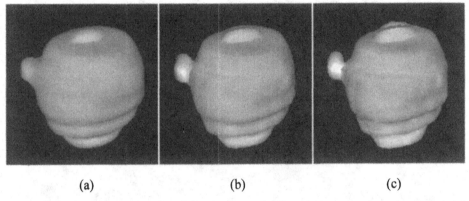

(a)   (b)   (c)

Figure 12.47 Surface-rendered nodule phantom of different reconstruction algorithms: (a) standard, (b) bone, and (c) lung.

Table 12.2 Measured volume varies with data acquisition mode. All images were reconstructed with the standard algorithm.

Acquisition mode	Image spacing	Measured volume (mm^3)	% variation to 0.63-mm axial
0.63-mm axial	0.63 mm	160.54	
1.25-mm axial	1.25 mm	166.31	3.7%
2.50-mm axial	2.5 mm	178.80	11.4%
5.00-mm axial	5.00 mm	173.04	7.8%
4 × 1.25 mm 1.5:1/helical	0.63 mm	173.71	8.2%
4 × 2.50 mm 1.5:1/helical	0.63 mm	193.99	20.8%

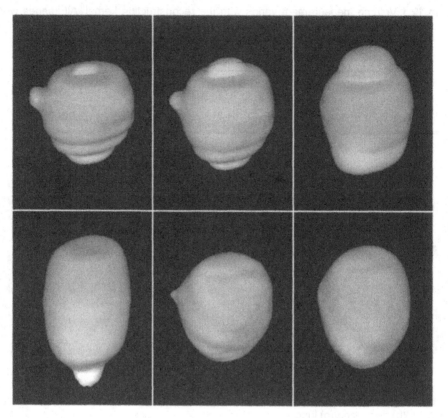

Figure 12.48 Top row, left to right: axial scan 0.63-mm at 0.63-mm spacing, 1.25 mm at 1.25-mm spacing, 2.50 mm at 2.50-mm spacing. Bottom row, left to right: axial scan 5.0 mm at 5.0-mm spacing, helical scan 4 × 1.25-mm HS mode at 0.63-mm spacing, helical 4 × 2.5-mm HS mode at 0.63-mm spacing.

12.5.2 Quantitative CT

The previous section discussed the importance of providing quantitative information for lung cancer screening. This quantification is in the form of a volume measurement, which requires the CT images to produce reliable and repeatable results. In other words, the CT values need to be both consistent and accurate. Consistency implies that if the same object is scanned over time, the resulting CT images should not change. Accuracy indicates that the reconstructed CT values need to be a true representation of the material and not impacted by the presence of other objects.

Consistency is determined primarily by the scanner's design, calibration, and maintenance. To ensure a consistent performance, the characteristics of all major components in a CT system should not drift or change over time. For example, the output spectrum of the x-ray tube should not change significantly as it ages, because a spectral change in the tube output makes the calibration process suboptimal. Sometimes, it is impossible to keep the components absolutely stable over time. One good example is electronic dark current. Regardless of the design, one characteristic of electronic dark current is that it changes with the ambient temperature. As long as the temperature fluctuates in the CT gantry, the detector offset will change. A good design will properly compensate for these fluctuations during the scan and make the compensation process transparent to the operator. For the detector offset, the electronic dark current is constantly measured to provide a correction vector during the reconstruction process. The scanner also needs to be properly maintained to ensure its performance. For example, most CT scanners require daily air calibration, in which air scans are collected and used to adjust the detector gain change.

The accuracy requirement is more challenging and is often limited by the fundamental physics, such as beam hardening. Thus, the CT values of an object may be influenced by the presence of other objects inside the scan FOV. For illustration, we scanned a round phantom with different density inserts on a vintage 4-slice scanner. The schematic diagram of the phantom and the reconstructed images are shown in Figs. 12.49(a) and (b). The CT numbers of these inserts range from 5.6 HU to 476.7 HU to represent a wide variety of materials; the measured CT numbers are shown in the row of Table 12.3 labeled "CT number (a)." The round phantom was then placed inside an oval phantom and the experiment was repeated [Fig. 12.49(c)]; the exact same scanning protocol was used to avoid factors that may influence the results. The measured CT numbers are shown in the bottom row of Table 12.3. The measured CT numbers increase for the lower-density inserts and decrease for higher-density objects. This is not surprising since the average photon energy passing through the round phantom increases with the placement of the oval phantom. This experiment serves as a caution to researchers who want to use the measured CT value as the absolute measurement to characterize a material.

Another factor that may impact the accuracy of the measured CT values is the location of the object. This is due mainly to the influence of the bowtie filter. For illustration, we scanned a 20-cm water phantom with a 4-slice scanner. The water phantom was initially positioned at the iso-center and gradually raised vertically. The average CT number inside an ROI at the center of the water phantom was measured, and the average CT number was plotted against the phantom position, as shown in Fig. 12.50. It is clear from the graph that the CT number does change with the phantom position. Therefore, when performing longitudinal studies, it is important to ensure that the patient is positioned similarly in all scans.

To overcome the variation in CT numbers, the use of a reference phantom was proposed.[75] When measuring bone-mineral density (BMD) with CT, a phantom with known density inserts is placed below the patient for calibration. The measured BMD is adjusted based on the phantom's values to maintain a consistent measurement over time. For cardiopulmonary disease (COPD) applications, researchers have used air pockets inside the airways as references to calibrate the lung-density measurement. Although these approaches help to reduce CT number variations, it has been shown that the calibrated result is influenced by the relative position of the phantom inserts to the patient's bone[76] or the relative position of the air pockets to the lung ROI. Section 12.6 will discuss the use of dual energy to reduce the beam-hardening impact. Although this technology is still under development, it has already shown great promise.

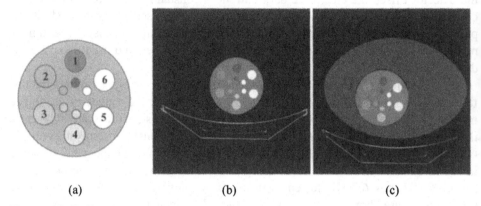

(a)   (b)   (c)

Figure 12.49 Density phantom experiment setup. (a) Schematic diagram of different density inserts. (b) Density phantom scanned by itself. (c) Density phantom placed inside an oval phantom.

Table 12.3 Average CT numbers for different density inserts (for phantom experiment shown in Fig. 12.49) measured on an older vintage four-slice scanner. CT number measurements for the phantom configuration of (a) Fig. 12.49(b) and (b) Fig. 12.49(c).

Insert	1	2	3	4	5	6
CT number (a)	5.62	108.89	147.17	243.08	361.55	476.75
CT number (b)	10.09	117.50	153.97	243.24	345.87	447.46

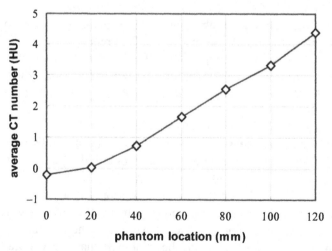

Figure 12.50 Average CT number for a water phantom located at different height distances from the iso-center.

12.5.3 CT colonography

Colorectal carcinoma was the third most commonly diagnosed cancer and the third leading cause of death from cancer in the U.S. in 2008. It was estimated that 148,810 people would be diagnosed with colorectal cancer and 49,960 would die of the disease in 2008 alone.[68] Most nonhereditary colorectal carcinomas arise from pre-existing adenomatous polyps. Early detection and polypectomy of these polyps significantly reduces morbidity and mortality.[77]

CT colonography is an exciting development that has the potential to revolutionize colorectal cancer screening.[78] The typical protocol uses helical CT with thin slices and requires a prescreening bowel cleaning, which is usually accomplished by administering a bowel preparation regimen the day before the screening. Before the data acquisition takes place, the patient's colon is insufflated with room air or CO_2 to provide optimal colonic distention.[79] Patients are scanned in both supine and prone positions. Figure 12.51 depicts a typical colonography scan. For the convenience of diagnosis, the prone image (right) is displayed upside down so the image's anatomical orientation is identical to the supine image (left). A clinical study has shown that the sensitivity and specificity of CT colonography averaged 75% and 90% in patients with adenomas 10 mm in diameter or larger.[80] The risk of malignancy is 10% for adenomas 10 to 20 mm in diameter and increases to at least 30% in adenomas larger than 20 mm in diameter. Because CT colonography offers a complete noninvasive examination of the colon, a short examination time, and does not require sedation, it could become the diagnostic examination of choice if it is proven to be more sensitive than other examination methods.

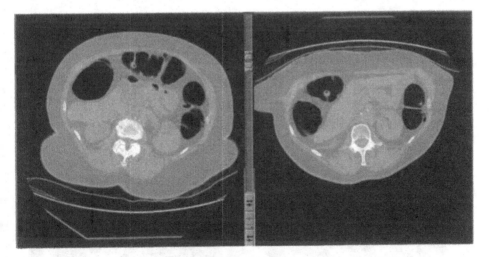

Figure 12.51 Examples of a CT colonography study. A patient is scanned in both supine (left) and prone (right) positions. (The prone image is shown upside down so that patient orientation is identical to the supine image.)

Typical CT colonography scans are acquired from the top of the colon through the rectum, and each examination produces up to 700 images per patient.[81] For this application to be a practical means of screening, various tools must be available to reduce the workload of radiologists. The four images in Fig. 12.52 depict several examples of different image presentation methods for CT colonography. In the upper-left image, a polyp in the reconstructed cross-sectional image is identified and the size of the polyp is measured, as shown by the dotted lines. In addition to the conventional cross-sectional image, a reformatted image, a surface-rendered image, and a VR image are also used to help radiologists identify polyps and reduce the amount of time for image viewing, as shown by the three other images in Fig. 12.52. The reformatted image in the upper-right provides size information in the orthogonal direction.

One of the most commonly used tools in CT colonography is the "fly-through" technique. This is often called "virtual endoscopy" since the generated 3D CT image mimics the appearance observed by an endoscope. For this type of

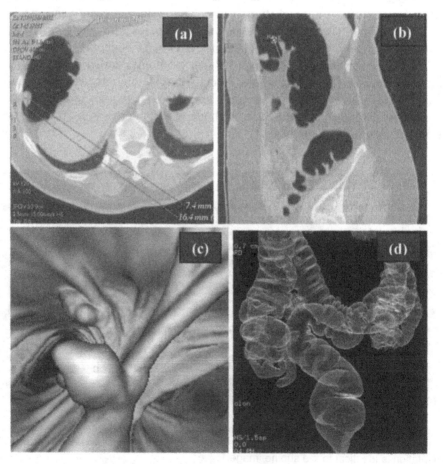

Figure 12.52 CT colonography images: (a) cross-sectional image, (b) reformatted image, (c), "fly through" image, and (d) volume-rendered image.

display tool, either surface rendering or VR is used. The methodology is similar to those described in Chapter 4, except that the imaginary light source and the viewing point are located inside the colon. In a study, VR techniques were found to be superior to surface rendering because the latter technique is sensitive to artifacts, noise, volume averaging, and reduced surface detail.[82] For example, due to the limiting frequency of the reconstruction kernel, a layer of 3 to 4 voxels is formed between the air and the colon wall. Surface rendering forces this layer to be included with either the wall, the lumen, or divided between the two, since a threshold is used to define the object boundary. Volume rendering, on the other hand, allows the transition zone to be reconstructed as a separate structure and enhances the mucosal and lesion details.

Figure 12.53 depicts a slightly different method of viewing the colon. The software tool automatically unwraps and straightens the colon to a flat structure. This provides added dimension to the fly-through technique and helps to identify other hidden polyps. Several research studies have been conducted to determine the optimal method of mapping the colon surface to a flat plane.[83,84] Similar to the case of lung nodule detection, these tools are designed to help radiologists view a large volume of data, identify highly suspicious regions, and perform size and characterization measurements.

As in the case of lung cancer screening, the accuracy and efficacy of these methods are largely influenced by the data acquisition and image reconstruction techniques. Therefore, the algorithms used for the polyp analyses must be optimized for the scanning protocols and reconstruction techniques. Analyses on the sensitivity of colonography tools to acquisition and reconstruction parameters need to be conducted in a manner similar to lung nodule detection.

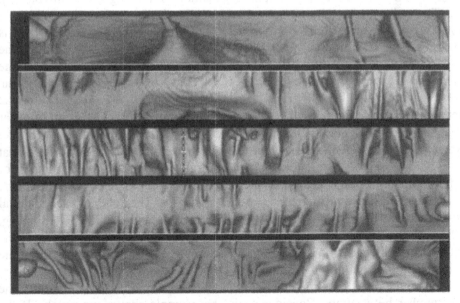

Figure 12.53 Flattened and straightened CT colonography image.

12.6 Dual-Energy CT

One of the hottest research areas in CT is energy resolution CT or dual-energy CT (DECT).[85–117] In previous discussions, we have treated all x-ray photon-matter interactions in an identical fashion, regardless of the photon energies. Other than causing image artifacts and calibration complexities, a wide energy spectrum generated by the x-ray tube does not provide additional useful information. DECT tries to exploit the different ways in which x-ray photons interact with matter and extract information that can help us to further pinpoint the characteristics of the scanned object.

DECT is not a new concept. In fact, not long after the introduction of the first commercial CT scanner, the DECT concept was discussed and investigated.[85,86] Nearly all approaches are based on the observation that in the diagnostic energy range, there are two dominant ways in which x-ray photons interact with matter: the photoelectric effect and Compton scatter (coherent scatter will be ignored because of its small contribution). Before discussing a more theoretical approach, we will present a more intuitive explanation on how dual-energy approaches provide additional information and help to identify different materials.

Let us assume that the mean energy of the x-ray photons produced on a CT scanner with a 120-kVp setting is 75 keV. For the ease of explanation, we will also assume that the linear attenuation coefficients for the material of interest reconstructed on this CT system are equivalent to the linear attenuation coefficients produced by a monoenergetic CT system with 75 keV, provided the beam-hardening correction is perfect. Clinical applications are concerned primarily with three materials: soft tissue (water), bone, and iodine. Iodine is typically diluted prior to being injected and undergoes an additional dilution when mixed with blood. At a certain concentration level (e.g., 1.3% in volume), the CT number of the mixture becomes 1550 HU, similar to the value of bone at 75 keV, so it is nearly impossible to determine whether the object of interest is made of bone or an iodine-water mixture. If we take another scan of the same object with an 80-kVp setting (with a mean x-ray photon energy of 55 keV), the measured CT number will be different. If the object of interest is made of bone, the CT number will be 2200 HU. If the object of interest is a mixture of iodine and water, the CT number will be 2800 HU due to the different behavior of linear attenuation coefficients as a function of energy for different materials. Therefore, we are able to differentiate different materials based on their behavior, and that behavior is based on high and low photon energies. If we plot the CT numbers of different materials on a 2D graph with the CT numbers for the two kVps as the axes, they will occupy different "zones" in this space. The zone for each material is characterized by the angle it formed with the two axes, as illustrated in Fig. 12.54. In this graph, the bone zone is designated with diagonal lines, and the iodine zone is shaded in solid gray. This simple example illustrates the basic principle of material decomposition with DECT.

Given the close proximity of different material zones, it is important that the CT number be accurate. Consider again the example shown previously. If the measured CT numbers are biased due to the beam-hardening effect and other artifacts (and considering the broad energy spectrum of the x-ray photons), we could erroneously classify the iodine-water mixture as bone.

With this intuitive explanation, let us now discuss the theoretical basis for the material decomposition with DECT. As discussed in Chapter 2, the mass attenuation coefficient of a particular material, $\mu/\rho(E)$, can be considered the composition of two components: attenuation due to the photoelectric effect, and attenuation due to Compton scatter. Both effects are functions of the x-ray photon energy E as discussed in Chapter 2. If we denote the attenuation function due to the photoelectric effect by $f_p(E)$ and the attenuation function due to Compton scatter by $f_c(E)$, the mass attenuation coefficient can be described as the combination of the two basis functions:

$$\left(\frac{\mu}{\rho}\right)(E) = \alpha_p f_p(E) + \alpha_c f_c(E), \qquad (12.15)$$

where α_p and α_c represent the contributions of the photoelectric effect and Compton scatter, respectively. The two functions $f_p(E)$ and $f_c(E)$ are known based on x-ray physics. With Eq. (12.15), we can now represent any material by its two basis functions, $f_p(E)$ and $f_c(E)$, instead of by the mass attenuation coefficient $\mu/\rho(E)$. To determine the unique contributions of the two effects, we need two measurements at two different photon energies: E_L and E_H, which represent the low and high energies, respectively:

$$\left(\frac{\mu}{\rho}\right)(E_L) = \alpha_p f_p(E_L) + \alpha_c f_c(E_L)$$

and

$$\left(\frac{\mu}{\rho}\right)(E_H) = \alpha_p f_p(E_H) + \alpha_c f_c(E_H). \qquad (12.16)$$

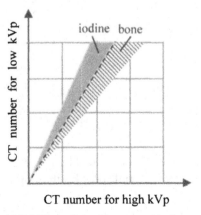

Figure 12.54 Illustration of bone and iodine zones.

Given two measurements and two unknowns, we can solve for α_p and α_c. Therefore, each object of interest is now uniquely represented by (α_p, α_c). For each cross-section of the object, we provide two reconstructed images, α_p and α_c, instead of a single image representing μ.

From a clinical point of view, the pure physics representations—the photoelectric and Compton effects—do not provide a direct link to the human anatomy, a pathology, or the contrast agent. Therefore, it is difficult for a radiologist to interpret the resulting images. Since one objective of DECT is to provide an easy way for radiologists to differentiate between different materials, it is more convenient to use two known materials (instead of two physics effects) as the bases of presentation. For example, if we use water and bone as the bases to produce a "water-equivalent-density" image and a "bone-equivalent-density" image, the bony structures inside the human body will be automatically removed in the "water-equivalent-density" image. Similarly, the bone-equivalent-density image should not contain any soft tissue. To accomplish this, we need to represent the functions $f_p(E)$ and $f_c(E)$ by the mass attenuation functions of the basis materials A and B, $(\mu/\rho)_A(E)$ and $(\mu/\rho)_B(E)$. From Eq. (12.15), the two sets of functions are related by

$$\left(\frac{\mu}{\rho}\right)_A (E) = \alpha_{A,p} f_p(E) + \alpha_{A,c} f_c(E)$$

and

$$\left(\frac{\mu}{\rho}\right)_B (E) = \alpha_{B,p} f_p(E) + \alpha_{B,c} f_c(E), \qquad (12.17)$$

where $\alpha_{A,p}$, $\alpha_{A,c}$, $\alpha_{B,p}$, and $\alpha_{B,c}$ represent the contributions of the photoelectric and Compton effects for materials A and B, respectively. From Eq. (12.17), it can be shown that

$$f_p(E) = \frac{\alpha_{B,c} \left(\dfrac{\mu}{\rho}\right)_A (E) - \alpha_{A,c} \left(\dfrac{\mu}{\rho}\right)_B (E)}{\alpha_{A,p}\alpha_{B,c} - \alpha_{B,p}\alpha_{A,c}}$$

and

$$f_c(E) = \frac{\alpha_{B,p} \left(\dfrac{\mu}{\rho}\right)_A (E) - \alpha_{A,p} \left(\dfrac{\mu}{\rho}\right)_B (E)}{\alpha_{A,c}\alpha_{B,p} - \alpha_{B,c}\alpha_{A,p}}. \qquad (12.18)$$

By substituting Eq. (12.18) back into Eq. (12.15), we can represent the mass attenuation coefficient of any material, μ/ρ, by the mass attenuation functions of materials A and B:

$$\left(\frac{\mu}{\rho}\right)(E) = \frac{\alpha_p \alpha_{B,c} - \alpha_c \alpha_{B,p}}{\alpha_{A,p} \alpha_{B,c} - \alpha_{B,p} \alpha_{A,c}} \left(\frac{\mu}{\rho}\right)_A (E) + \frac{\alpha_c \alpha_{A,p} - \alpha_p \alpha_{A,c}}{\alpha_{A,p} \alpha_{B,c} - \alpha_{B,p} \alpha_{A,c}} \left(\frac{\mu}{\rho}\right)_B (E)$$

$$= \beta_A \left(\frac{\mu}{\rho}\right)_A (E) + \beta_B \left(\frac{\mu}{\rho}\right)_B (E). \qquad (12.19)$$

Materials A and B are often called the basis materials. The two basis materials should be sufficiently different in their atomic number Z to ensure their different photoelectric and Compton attenuation characteristics. In general, the mass attenuation coefficient μ/ρ is location $\mathbf{r}$ dependent, due to the inhomogeneity of the scanned object, and Eq. (12.19) should be expressed as a function of both energy E and location $\mathbf{r}$. In CT, we measure the line integral of the linear attenuation coefficient, and this integral can then be expressed by the following equation:[94]

$$\int \mu(\mathbf{r}, E) ds = \eta_A \cdot \left(\frac{\mu}{\rho}\right)_A (E) + \eta_B \cdot \left(\frac{\mu}{\rho}\right)_B (E) \qquad (12.20)$$

where

$$\eta_A = \int \rho(\mathbf{r}) \beta_A(\mathbf{r}) ds \text{, and } \eta_B = \int \rho(\mathbf{r}) \beta_B(\mathbf{r}) ds. \qquad (12.21)$$

The variables $\rho(\mathbf{r})\beta_A(\mathbf{r})$ and $\rho(\mathbf{r})\beta_B(\mathbf{r})$ represent the local densities in g/cm^3 of basis material A and B, respectively. The quantities η_A and η_B represent the density projection in g/cm^2 of the basis materials. For locations where the material is neither A nor B, the densities are represented in "equivalent basis material densities," and the true physical density for the location is a linear combination of the two equivalent densities. Equation (12.20) essentially maps the original measured line integrals into the sum of two sets of line integrals. When CT data are collected with two different energy spectra, $\Psi_H(E)$ and $\Psi_L(E)$, the resulting measurements are:

$$I_L = \int \Psi_L(E) \cdot \exp\left[-\eta_A \cdot \left(\frac{\mu}{\rho}\right)_A (E) - \eta_B \cdot \left(\frac{\mu}{\rho}\right)_B (E)\right] dE \qquad (12.22)$$

and

$$I_H = \int \Psi_H(E) \cdot \exp\left[-\eta_A \cdot \left(\frac{\mu}{\rho}\right)_A (E) - \eta_B \cdot \left(\frac{\mu}{\rho}\right)_B (E)\right] dE. \qquad (12.23)$$

Here, subscripts "H" and "L" symbolize the high- and low-kVp scans, respectively. With the measurements, I_L and I_H, and Eqs. (12.22) and (12.23), we can solve for the equivalent density projections, η_A and η_B, of the unknown material. The process of solving η_A and η_B from the measured projection values is often called the basis material decomposition. If we have a monoenergetic x-ray source, integration over energy E disappears in Eqs. (12.22) and (12.23), and the basis material decomposition process is a simple solution to a set of linear equations. For a clinical CT system, polynomial expansion can be used to arrive at an approximate solution. A material other than A or B will contribute to both η_A and η_B in a specific fashion, and the contribution values can be interpreted as components in a 2D vector space with basis vectors being the basis materials.[93]

Note that the equivalent density projections, η_A and η_B, are the line integrals of the local densities of the scanned object (Eq. 12.21). For each high- and low-kVp sampling pair, I_L and I_H, we obtain a unique density projection pair, η_A and η_B, representing two complete sets of tomographic projection datasets of the equivalent densities. When the tomographic reconstruction process is applied on these density projections, we obtain the local equivalent densities in g/cm^3.

Similarly to the previous discussion, a material other than A or B will contribute to both reconstructed local equivalent density images of A and B. Typically, there are more than two materials inside a human body, especially if iodine contrast is injected to enhance the vascular structures. If we select water and iodine as the material bases, the portion of the object that contains bone will be represented in both "water-equivalent-density" and "iodine-equivalent-density" images. For illustration, the two images in the left column of Fig. 12.55 depict the conventional CT images collected and reconstructed with 80-kVp and 140-kVp settings. The phantom is a chest phantom filled with iodine contrast to simulate a cardiac scan. With the process described above, the two kVp projections were mapped to "water-equivalent density" and "iodine-equivalent density" projections, and a cone-beam reconstruction algorithm identical to that used in the conventional CT scans was applied on the material decomposition projections to produce "water-equivalent-density" and "iodine-equivalent-density" images shown in the right column of the figure. Note that the iodine contrast present in the conventional CT images is completely removed in the water-equivalent-density image. Similarly, the soft-tissue (water) portion of the phantom does not show up in the iodine-equivalent-density image. The bone portion of the image (ribs and spinal cord), on the other hand, is depicted in both images.

In a stricter sense, the water-equivalent-density image in the water-iodine pair should be called a "non-iodine-density image." Similarly, the iodine-equivalent-density image should be called a non-water density image." A word of caution is needed when interpreting the material-equivalent-density images shown in Fig. 12.55; these images are no longer depicted in the conventional CT HU, but merely represent the equivalent densities in g/cm^3 in terms of the basis material of the scanned object; thus, the interpretation of such images often causes confusions to the radiologist.

An alternative way of utilizing the equivalent density projections, η_A and η_B, is to synthesize a CT system employing a monoenergetic x-ray source instead of the wide-spectrum source of a clinical CT system. If we replace the energy variable E in Eq. (12.20) by a specific photon energy E_0, Eq. 12.20 becomes

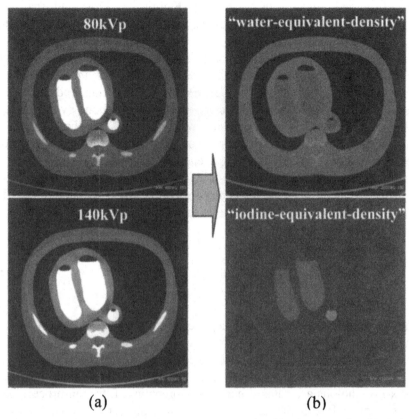

(a)   (b)

Figure 12.55 Dual-energy images and material decomposition images of a chest phantom study. (a) Conventional CT images in HU, and (b) equivalent-density images of a basis material in g/cm³.

$$\int \mu(\mathbf{r}, E_0)ds = \eta_A \cdot \left(\frac{\mu}{\rho}\right)_A (E_0) + \eta_B \cdot \left(\frac{\mu}{\rho}\right)_B (E_0). \qquad (12.24)$$

The left side of the equation represents the line integral of the linear attenuation coefficients of the scanned object at a specific energy E_0, and therefore represents a monoenergetic projection. For given basis materials A and B, their mass attenuation coefficients, $(\mu/\rho)_A$ and $(\mu/\rho)_B$, are known and are readily available from the National Institute of Standard and Technology (NIST).[96] Equation 12.24 indicates that we can readily derive a set of monoenergetic projections from the equivalent density projections and mass attenuation coefficients of materials A and B. When a regular tomographic reconstruction algorithm is applied to the monoenergetic projections, we arrive at a reconstructed image that synthesizes a monoenergetic CT acquisition. The reconstructed image is now in the HU, instead of the density unit of g/cm³. Figure 12.56 shows a human skull phantom scanned on a Discovery™ HD 750 scanner (GE Healthcare, Waukesha, WI) capable of collecting fast kV-switching projections. (The data collection details will be discussed later in this section.). Figures 12.56(a) and (b) are the conventional 140-kVp and 80-kVp images, while Fig. 12.56(c) was generated at 80 keV with the process outlined above.

The physics principle behind the seemingly strange representation of a third material with two other basis materials is actually fairly straightforward. Figure 12.57 plots the mass attenuation coefficients of three materials: iodine, bone, and soft tissue. Iodine is represented by the curve with circles, bone by the curve with triangles, and soft tissue by the curve with crosses. The three curves look significantly different in terms of their shapes, especially with the K edge of the iodine. However, if we ignore the low-energy range ($E > 40$ keV), the three curves are much more similar, as shown by the nonshaded region in the figure. If the least-square fit of these curves is performed, the mass attenuation coefficient of bone at the higher energy range can be approximated by the mass attenuations of iodine and soft tissue:

$$\left(\frac{\mu}{\rho}\right)_{bone} = 0.88\left(\frac{\mu}{\rho}\right)_{soft\ tissue} + 0.018\left(\frac{\mu}{\rho}\right)_{iodine}. \quad (12.25)$$

The fitted bone mass attenuation is depicted by the curve with diamonds. Over the energy range of 40 keV $< E <$ 150 keV, the measured and fitted bone mass attenuation curves nearly overlap each other. A significant difference does exist, however, for the energy range below the K edge of iodine. Because most low-energy x-ray photons are filtered out by the prepatient filtration process and the patient's body before reaching the CT detector, the error caused by the low-energy photons has little impact on the final result. This clearly justifies the use of two basis materials to represent other Z materials.

The discussion so far has only presented the ways in which projections and images can be generated from dual kVp scans, but other data acquisition technologies can be explored. Generally speaking, the approaches to acquire a

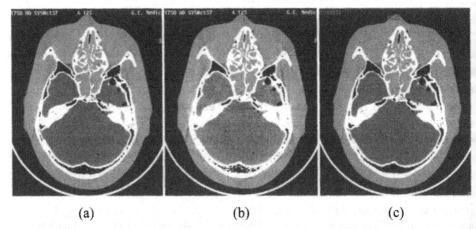

      (a)   (b)   (c)

Figure 12.56 A human skull phantom collected on a Discovery™ HD 750 scanner. (a) 140-kVp image, (b) 80-kVp image, and (c) synthesized monoenergetic 80-keV image.

DECT dataset can be classified into two major approaches: the source-driven approach and the detector-driven approach. In the source-driven approach, the CT detector does not have any energy discrimination capability. Instead, we rely solely on the x-ray source to provide different ranges of x-ray energy photons at either different times or different orientations, as illustrated in Fig. 12.58(a) and (b). In Fig. 12.58(a), the x-ray tube voltage is first set to the high-kVp (or low-kVp) level to complete the initial data acquisition. The tube voltage is then quickly switched to the low-kVp (or high-kVp) setting for the second scan. Because two projection datasets are perfectly aligned in terms of their projection angles, material decomposition can be carried out in either the projection space or the image space. The most significant potential issue for this approach is patient motion, since two projections are collected sequentially. Any patient motion occurring between two scans can lead to misregistration. The approach depicted in Fig. 12.58(b), on the other hand, is performed simultaneously in order to avoid patient motion, since one tube–detector pair images the patient with a high-kVp setting and the other with a low-kVp setting. The potential drawbacks of this approach are the increased cross-scatter (since both x-ray tubes are turned on at the same time), and CT number instability if the second tube–detector pair has a significantly smaller FOV in the x-y plane.

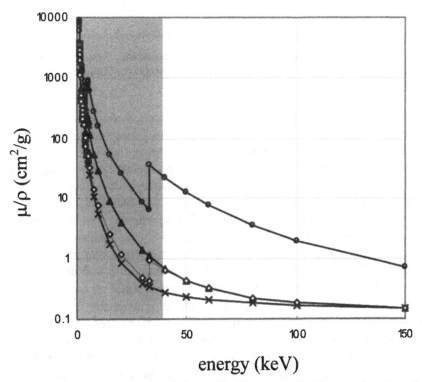

Figure 12.57 Mass attenuation coefficients for different materials as a function of x-ray energy. Line with circles: iodine; line with crosses: soft-tissue; line with triangles: bone; line with diamonds: fitted bone with iodine and soft tissue.

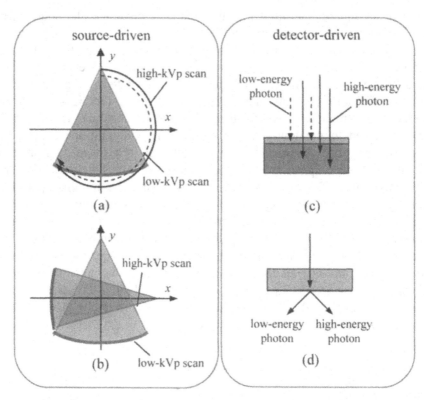

Figure 12.58 Different approaches for dual energy acquisition. Source driven: (a) same orientation, different time; (b) same time, different orientation. Detector driven: (c) dual layer and (d) photon counting.

As its name implies, the detector-driven approach relies on the detector's energy sensitivity to differentiate high-energy versus low-energy x-ray photons, as illustrated by Figs. 12.58(c) and (d). In Fig. 12.58(c), the detector is formed by two scintillating layers, with the top layer having a much smaller stopping power than the bottom layer. X-ray photons with lower energies will likely be absorbed by the top layer, and photons with higher energies will likely pass through it. The bottom layer will interact mostly with the higher-energy photons. This design allows the x-ray tube to produce x-ray photons in a wide spectrum and relies on the detector to produce two signals at different energy ranges. Clearly, both datasets are collected simultaneously, so patient motion is not an issue. On the other hand, the two datasets may contain a higher level of overlapping x-ray energies, which in turn may reduce the effectiveness of the material decomposition. The approach illustrated in Fig. 12.58(d) uses a photon-counting detector similar to the SPECT concept. As each x-ray photon interacts with the detector, the photon energy is determined based on the generated signal, and the reading is placed into different bins based on the photon energy. Like the approach described for Fig. 12.58(c), the data collection takes place simultaneously and patient motion is not an issue. In addition, since the energy level of each event can be determined, there is a potential for a better separation between the high- and low-energy datasets, and a potential for providing more than two energy levels beyond the dual-energy approach. However, the technical difficulty of this approach is high, because typical CT applications demand a high x-ray flux, which in turn requires the high count-rate capability of the detector to avoid pileup and other nonideal phenomena.

An alternative method is the so-called fast-kVp switching approach, in which the x-ray tube voltage is switched quickly between the high-kVp and low-kVp settings so the kVp setting changes from view to view, as illustrated in Fig. 12.59. Because two adjacent views with different energies are collected within a fraction of a millisecond, patient motion is negligible. Similarly, the orientations between adjacent views differ by a small fraction of a degree, and the material decomposition can be carried out in either the projection or the image space. Of course, the design of such systems requires more advanced technologies since the high-voltage generator must switch quickly between two voltage settings to avoid cross-contamination between the high- and low-voltage samples. A similar requirement exists for the detector's primary speed in order to keep the energy separation as wide as possible.

Despite the long history of DECT, its clinical and technical research is still in its infancy. In the past few years, there have been increased clinical research interests to take advantage of the recent technological development in dual-energy and explore ways to leverage the additional information provided by dual-energy for the identification and treatment of different pathologies.[100-108] These applications include the removal of bony objects for better visualization of the vascular structures, identification and removal of calcified plaques to investigate the integrity of the vessels, separation of benign and malignant tumors based on the iodine contrast uptake, differentiation of various lung nodule types, separation of fatty content in the liver, and improved characterization of renal stones. The number of clinical applications is likely to grow as dual-energy CT becomes more widely available.

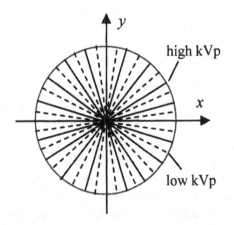

Figure 12.59 Fast-kVp switching approach.

New hardware technologies are continuously being developed to enable the application of dual energy to more clinical studies while minimizing the dose. More advanced reconstruction technologies are also underway to take full advantage of the extra energy information.[109–117] At the same time, many clinical experiments are being conducted to determine the clinical utility of the technology. Although DECT promises to significantly reduce beam hardening for artifact reduction and improved quantitative results, separate different material types to allow tissue characterization, and remove undesirable objects and materials to allow better visualization, the "killer application" has yet to be determined that can revolutionize the 30+-year-old x-ray CT.

12.7 Problems

12-1   Calculate the z coverage of a cardiac scan in the helical mode data acquisition over one cardiac cycle. Assume that the source-to-iso-center distance is 550 mm, the source-to-detector distance is 950 mm, and the cardiac FOV is 25 cm in diameter. The helical pitch is 0.2 used with a detector z coverage of 40 mm at the iso-center.

12-2   For the CT system described in problem 12-1, the patient heart is not centered in the scan FOV. Therefore, in order to completely cover the heart, the reconstruction FOV of 25 cm in diameter is centered (at 40 mm, 50 mm). Calculate the system z coverage in one cardiac cycle.

12-3   One potential issue of multicycle data acquisition and reconstruction is that the gantry may not be at the desired angle during successive cardiac cycles. The worst-case scenario is when the gantry rotation is in sync with the patient's heart rate, and projections collected during the successive heart cycles have projection angles identical to those collected in the first cycle. Describe two different approaches to overcome this problem.

12-4   Derive a pitch selection range [see Eq. (12.1)] when extrapolation is allowed during the reconstruction. Assume a zeroth-order extrapolation is used if the ray intersection with the detector (during the backprojection process) is within one detector row width beyond the physical detector z coverage of a 64-slice configuration.

12-5   Given the cardiac-induced motion in the lung, as shown in Fig. 12.21, do you recommend the use of EKG gating when performing lung scans? What are the pros and cons of scanning with and without EKG gating?

12-6   Figure 12.22 shows that the centrifugal force increases with the distance to the iso-center. One student argues that an optimal CT system should be designed so that both the x-ray tube and the detector are placed just outside the scan FOV to minimize the G-force impact. What are other considerations for determining the tube and detector locations?

12-7   Figure 12.25(b) shows a system configuration in which full-size detectors are used in both tube–detector pairs. Assuming that the source-to-iso-center distance is 550 mm and the reconstruction FOV is 50 cm in diameter, what is the offset angle between the two tube–detector pairs and the minimum distance between the detector and the iso-center that will make this arrangement possible? Ignore the size of the x-ray tube and supporting structures of the detector.

12-8   For the system configuration discussed in problem 12-7, calculate the percentage improvement in terms of temporal resolution compared to the temporal resolution of a single tube–detector half-scan.

12-9   Assume that the entire heart and coronary artery is contained inside a 20-cm-diameter sphere centered at the iso-center. Calculate the minimum detector z coverage (at iso-center) if the entire heart must be imaged in one gantry rotation, assuming that the source-to-iso-center distance is 550 mm.

12-10   Design a recursive reconstruction approach that uses a half-scan weighting function in CT fluoroscopy. Fan-to-parallel rebinning is not allowed.

12-11   After row-wise fan-to-parallel rebinning, the original cone-beam projections can be converted into a set of tilted parallel projections. Given the fan-angle-independent nature of the half-scan weighting in the parallel geometry, a recursive reconstruction implementation is straightforward. List the pros and cons of using this approach in CT fluoroscopy applications.

12-12   One student argues that based on the Nyquist sampling theory, for perfusion applications we need to reconstruct images at 0.5-sec time intervals for a continuous perfusion acquisition with a 1-sec gantry rotation speed. Do you agree with this argument? What other factors need to be considered to determine the reconstruction time interval?

12-13   You are asked to determine the minimum sampling intervals (in time) for CT perfusion in the liver. Is it sufficient to examine only the arterial contrast uptake curves?

12-14   List at least three approaches to combat patient motion in CT perfusion. Consider different methodologies for the brain, liver, and heart.

12-15   Section 12.5.1 discussed the impact of the reconstruction algorithm and detector aperture on lung nodule volume calculation. Do you think the helical pitch impacts the volume calculation? Justify your answer.

12-16   The fly-through technique shown in Fig. 12.52(c) can sometimes fail to detect polyps. Describe a scenario in which this phenomenon might occur.

12-17   For quantitative CT, we discussed the importance of having a reference phantom placed close to the object of interest. Do you think it is important for the reference material to be similar or identical to the object of interest? If it is important, how do you design a reference phantom given the wide range of BMDs of bones? How important is the shape of the reference phantom (sphere, cube, cylinder, etc.) in regards to the accuracy of the results?

12-18   An engineer argues that the separation between high- and low-kVp settings for DECT applications must be as wide as possible. Do you agree with this argument from a pure physics point of view? Do you agree with this argument from a CT system design point of view?

12-19   An engineer proposes to add two x-ray tube voltage settings for DECT applications: 60 kVp and 160 kVp. Discuss the pros and cons of this proposal.

12-20   For a liver scan with iodine-contrast injection, two images are generated with DECT: a soft-tissue-equivalent-density image and an iodine-equivalent-density image. In the soft-tissue image, all iodine components are removed. Do you expect the soft-tissue image to be identical to the image reconstructed from a precontrast scan of the liver, since the iodine contrast is effectively removed?

References

1.   C. H. Harris, "CT: It's not just for diagnosis anymore," *Med. Imag.* **16**(8), 54–61 (2001).

2.   M. J. Budoff and J. S. Shimbane, *Cardiac CT Imaging: Diagnosis of Cardiovascular Disease*, Springer-Verlag, London (2006).

3.   J. M. Siegel, *Pediatric Body CT*, Lippincott Williams & Wilkins (2007).

4.   D. J. Anschel, A. Mazumdar, P. Romanelli, *Clinical Neuroimaging: Cases and Key Points*, McGraw Hill, East Syracuse, NY (2007).

5.   W. R. Webb, W. Brant, N. Major, *Fundamentals of Body CT*, 3rd ed. Elsevier, Philadelphia (2005).

6.   S. H. Fox, L. N. Tanenbaum, S. Ackelsberg, H. D. He, J. Hsieh, and H. Hu, "Future directions in CT technology," in *Neuroimaging Clinics of North America: CT in Neuroimaging Revisited*, Vol. 8, No. 3, L. N. Tanenbaum, Ed., W. B. Saunders Co., Philadelphia (1998).

7.   M. F. Oliver, E. Samuel, P. Morley, G. P. Young, and P. L. Kapur, "Detection of coronary artery calcification during life," *Lancet* **1**, 891–895 (1964).

8.   J. R. Margolis, J. Chen, and Y. Kong, "The diagnostic and prognostic significance of coronary artery calcification: A report of 800 cases," *Radiology* **137**, 609–616 (1980).

9. R. M. Benitez and R. A. Vogel, "Assessment of subclinical atherosclerosis and cardiovascular risk," *Clin. Cardiol.* **24**, 642–650 (2001).

10. C. R. Conti, "Detecting coronary artery calcification," *Clin. Cardiol.* **23**, 717–718 (2000).

11. J. A. Rumberger, D. B. Simons, L. A. Fitzpatrick, P. F. Sheedy, and R. S. Schwartz, "Coronary artery calcium area by electron beam computed tomography and coronary atherosclerotic plaque area: A histopathologic correlative study," *Circulation* **15**, 2157–2162 (1995).

12. D. A. Laudon, L. F. Yukov, J. F. Breen, J. A. Rumberger, P. C. Wollan, and P. F. Sheedy, "Use of electron-beam computed tomography in the evaluation of chest pain patients in the emergency department," *Amer. Emerg. Med.* **33**, 15–21 (1999).

13. H. C. Yoon, L. E. Greaser III, R. Mather, S. Sinha, M. F. NcNitt-Gray, and J. G. Goldin, "Coronary artery calcium: alternate methods for accurate and reproducible quantitation," *Acad. Radiol.* **4**, 666–673 (1997).

14. Y. Arad, L. A. Spadaro, K. Goodman, A. Lledo-Perez, S. Sherman, G. Lerner, and A. D. Guerci, "Predictive value of electron beam computed tomography of the coronary arteries: 19-month follow-up of 1173 asymptomatic subjects," *Circulation* **93**, 1951–1953 (1996).

15. A. S. Agatston, W. R. Janowitz, F. J. Hildner, N. R. Zusmer, M. Viamonte, Jr., and R. Detrano, "Quantification of coronary artery calcium using ultrafast computed tomography," *J. Amer. Coll. Cardiol.* **15**, 827–832 (1990).

16. T. Q. Callister, B. Cooil, S. P. Raya, N. J. Lippolis, D. J. Russo, and P. Raggi, "Coronary artery disease: Improved reproducibility of calcium scoring with an electron-beam CT volumetric method," *Radiology* **208**(3), 807–814 (1998).

17. C. H. McCollough, R. B. Kaufmann, B. M. Cameron, D. J. Katz, P. F. Sheedy, and P. A. Peyser, "Electron-beam CT: Use of a calibration phantom to reduce variability in calcium quantification," *Radiology* **196**, 159–165 (1995).

18. D. P. Boyd, Presentation entitled "Calcium as a biomarker for atherosclerosis progression and regression in coronary artery disease," *Conference on Biomarkers and Surrogate Endpoints: Advancing Clinical Research and Applications,* National Institutes of Health and U.S. Food and Drug Administration (1999). http://www4.od.nih.gov/biomarkers/agenda.htm.

19. M. Kachelriess and W. A. Kalender, "Electrocardiogram-correlated image reconstruction from sub-second spiral computed tomography scans of the heart," *Med. Phys.* **25**, 2417–2431 (1998).

20. M. Kachelriess, S. Ulzheimer, and W. A. Kalender, "ECG-correlated imaging of the heart with subsecond multislice spiral CT," *IEEE Trans. Med. Imag.* **19**, 888–901 (2000).

21. J. Hsieh, T. Pan, K. C. Acharya, Y. Shen, and M. Woodford, "Non-uniform phase coded image reconstruction for cardiac CT," *Radiology* **213**, 401 (1999).

22. J. Hsieh, J. Mayo, K. C. Archarya, and T. Pan, "Adaptive phase-coded reconstruction for cardiac CT," *Proc. SPIE* **3978**, 501–508 (2000).

23. C. E. Woodhouse, W. R. Janowitz, and M. Viamonte, "Coronary arteries: Retrospective cardiac gating technique to reduce cardiac motion artifact at spiral CT," *Radiology* **204**, 566–569 (1997).

24. T. Pan and S. Yun, "Cardiac CT with variable gantry speeds and multi-sector reconstruction," *Radiology* **217**, 438 (2000).

25. T. Flohr, B. Ohnesorge, C. Becker, A. Kopp, S. Hailliburton, and A. Knez, "A reconstruction concept for ECG-gated multi-slice spiral CT of the heart with pulse-rate adaptive optimization of spiral and temporal resolution," *Radiology* **217**, 2438 (2000).

26. M. Hiraoka, A. Adachi, K. Taguchi, and H. Anno, "Evaluation of cardiac volumetric imaging methods for multi-slice helical CT," *Radiology* **217**, 438 (2000).

27. J. Hsieh and R. F. Senzig, "Dual cardiac CT scanner," U.S. Patent No. 6,421,412 (2002).

28. H. Seifarth, S. Wienbeck, M. Pusken, K. Juergens, D. Maintz, C. Vahlhaus, W. Heindel, and R. Fischbach, "Optimal systolic and diastolic reconstruction windows for coronoray CT angiography using dual-source CT," *Am. J. Roentgen.* **189**, 1317–1323 (2007).

29. C. H. McCollough, A. N. Primak, O. Saba, H. Bruder, K. Stierstorfer, R. Raupach, C. Suess, B. Schmidt, B. M. Ohnesorge, and T. G. Flohr, "Dose performance of a 64-channel dual-source CT scanner," *Radiology* **243**, 775–784 (2007).

30. Y. Kyriakou and W. A. Kalender, "Intensity distribution and impact of scatter for dual-energy CT," *Phys. Med. Biol.* **52**, 6969–6989 (2007).

31. J. Hsieh, J. Londt, M. Vass, J. Li, X. Tang, and D. Okerlund, "Step-and-shoot data acquisition and reconstruction for cardiac x-ray computed tomography," *Med. Phys.* **33**(11), 4236–4248 (2006).

32. J. P. Earls, E. L. Berman, B. A. Urban, C. A. Curry, J. L. Lane, R. S. Jennings, C. C. McCulloch, J. Hsieh, and J. H. Londt, "Prospectively gated transverse coronary CT angiography versus retrospectively gated helical technique: improved image quality and reduced radiation dose," *Radiology* **246**, 742–753 (2008).

33. J. Hsieh, "An optimization reconstruction algorithm for temporal resolution improvement in CT fluoroscopy," *Radiology* **209**, 435 (1998).

34. J. Hsieh, "Reconstruction optimization for temporal response improvement in CT fluoroscopy," *Proc. IEEE Intern. Conf. Imag.*, **2**, 672–676 (1999).

35. J. Hsieh, S. W. Metz, G. Saligram, G. M. Besson, H. Hu, S. Dutta, R. F. Senzig, and X. Min, "Image reconstruction in a CT fluoroscopy system," U.S. Patent No. 5,907,593 (1999).

36. J. Hsieh, "Fluoroscopy image reconstruction," U.S. Patent No. 6,061,423 (2000).

37. J. Hsieh, "Efficient reconstruction algorithm for CT fluoroscopy," *Proc. SPIE* **4322**, 79–86 (2001).

38. F. Ali, J. Hsieh, G. Saligram, S. P. Faessler, C. J. Mussack, D. M. Deaven, S. H. Fox, and A. Zavaljevski, "Apparatus and method for displaying computed tomography fluoroscopy images," U.S. Patent No. 6,101,234 (2000).

39. K. Taguchi and M. S. Otawara, "Improvement of CT fluoroscopy image quality by feathering the data to cover the lag between acquisition times," *Radiology* **209**, 435 (1998).

40. J. Astrup, B. Siesjo, and L. Symon, "Thresholds in cerebral ischemia—The ischemic penumbra," *Stroke* **12**, 723–725 (1981).

41. K. A. Hossmann, "Viability thresholds and the penumbra of focal ischemia," *Ann. Neurol.* **36**, 557–565 (1994).

42. N. A. Lassen and J. Astrup, "Ischemic penumbra," in *Cerebral Blood Flow: Physiologic and Clinical Aspects*, J. H. Wood, Ed., McGraw-Hill, New York (1987).

43. W. D. Heiss and G. Rosner, "Functional recovery of cortical neurons as related to degree and duration of ischemia," *Ann. Neurol.* **14**, 294–301 (1983).

44. P. Meier and K. L. Zierler, "On the theory of the indicator-dilution method for measurement of blood flow and volume," *J. Appl. Physiol.* **6**, 731–744 (1954).

45. L. Axel, "Cerebral blood flow determination by rapid sequence computed tomography," *Radiology* **137**, 679–686 (1980).

46. A. M. Peters, R. D. Gunasekera, B. L. Henderson, J. Brown, J. P. Lavender, M. De Souza, J. M. Ash, and D. L. Gilday, "Non-invasive measurements of blood flow and extraction fraction," *Nucl. Med. Commun.* **8**, 827–837 (1987).

47. K. A. Miles, "Measurement of tissue perfusion by dynamic computed tomography," *Brit. J. Radiol.* **64**, 409–412 (1991).

48. K. A. Miles, M. P. Hayball, and A. K. Dixon, "Measurement of human pancreatic perfusion using dynamic computed tomography with perfusion imaging," *Brit. J. Radiol.* **68**, 471–475 (1995).

49. E. Klotz and M. Konig, "Perfusion measurements of the brain: Using dynamic CT for the quantitative assessment of cerebral ischemia in acute stroke," *Eur. J. Radiol.* **30**, 170–184 (1999).

50. M. J. K. Blomley, R. Coulden, C. Bufkin, M. J. Lipton, and P. Dawson, "Contrast bolus dynamic computed tomography for the measurement of solid organ perfusion," *Invest. Radiol.* **28**, 72–77 (1993).

51. W. R. Jaschke, R. G. Gould, M. G. Cogan, R. Sievers, and M. J. Lipton, "Cine-CT measurement of cortical renal blood flow," *J. Comput. Assist. Tomogr.* **11**, 779–784 (1987).

52. A. Cenic, D. G. Nabavi, R. A. Craen, A. W. Gelb, and T.-Y. Lee, "Dynamic CT measurement of cerebral blood flow: A validation study," *Amer. J. Neuroradiol.* **20**, 63–73 (1999).

53. D. G. Nabavi, A, Cenic, J. Dool, R. M. L. Smith, F. Espinosa, R. A. Craen, A. W. Gelb, and T.-Y. Lee, "Quantitative assessment of cerebral hemodynamics using CT: stability, accuracy, and precision studies in dogs," *J. Comput. Assist. Tomogra.* **23**, 506–515 (1999).

54. D. G. Nabavi, A. Cenic, R. A. Craen, A. W. Gelb, J. D. Bennett, R. Kozak, and T.-Y. Lee, "CT assessment of cerebral perfusion: experimental validation and initial clinical experience," *Radiology* **213**, 141–149 (1999).

55. T.-Y. Lee, D. G. Nabavi, R. A. Craen, et al., "Validation of cerebral blood flow maps from contrast-enhanced cine CT scanning: Comparison of infarct sizes from TTC staining and blood flow threshold in a rabbit focal cerebral ischemia model," *Radiology* **213**, 1305 (1999).

56. D. G. Nabavi, L. LeBlanc, B. Baxter, A. J. Fox, D. H. Lee, S. P. Lownie, G. G. Ferguson, R. A. Craen, A. W. Gelb, and T.-Y. Lee, "Monitoring of cerebral blood flow after subarachnoid hemorrhage using computed tomography," *Neuroradiology* **43**, 7–16 (2001).

57. D. G. Nabavi, A. Cenic, S. Henderson, A. W. Gelb, and T.-Y. Lee, "Perfusion mapping using computed tomography allows accurate prediction of cerebral infarction in experimental brain ischemia," *Stroke* **32**, 175–183 (2001).

58. T. G. Purdie, E. Henderson, and T.-Y. Lee, "Functional CT imaging of angiogenesis in rabbit VX2 soft-tissue tumor," *Phys. Med. Biol.* **46**, 3161–3176 (2001).

59. J. B. Bassingthwaighte, F. P. Chinard, C. Crone, N. A. Lassen, and W. Perl, "Definitions and terminology for indicator dilution methods," in *Capillary Permeability*, C. Crone and N. A. Lassen, Eds., Munksgaard, Copenhagen, 664–669 (1970).

60. P. Licato, I. Stevic, B. D. Murphy, and T. Lee, "Investigation of the effects of sampling interval on parameter estimates from CT perfusion," *RSNA Scientific Assembly and Annual Meeting* **462** (2006).

61. P. Montes and G. Lauritsch, "Noise reduction by temporal estimation in perfusion computed tomography," in *Proc. IEEE Med. Imag. Conf. 2005*, 2747–2751 (2005).

62. Z. M. Lin, S. Pohlman, A. J. Cook II, S. Chandra, "CT perfusion, comparison of gamma-variate fit and deconvolution," *Proc. SPIE* **4683**, 102–109 (2002).

63. K. Nieman, M. D. Shapiro, M. Ferencik, C. H. Nomura, S. Abbara, U. Hoffmann, H. K. Gold, I. Jang, T. J. Brady, and R. C. Cury, "Reperfused myocardial infarction: contrast-enhanced 64-section CT in comparison to MR imaging," *Radiology* **247**(1), 49–56 (2008).

64. S. H. Landis, T. Murray, S. Bolden, and P. A. Wingo, "Cancer statistics 1998," *CA Cancer J. Clin.* **48**, 6–29 (1998).

65. S. H. Landis, T. Murray, S. Bolden, and P. A. Wingo, "Cancer statistics 1999," *CA Cancer J. Clin.* **49**, 8–31 (1999).

66. R. T. Greenlee, T. Murray, S. Bolden, and P. A. Wingo, "Cancer statistics 2000," *CA Cancer J. Clin.* **50**, 7–33 (2000).

67. "Lung Cancer Trends," Centers for Disease Control and Prevention, http://www.cdc.gov/cancer/lung/statistics/trends.htm.

68. A. Jernal, R. Siegel, E. Ward, et al., "Cancer statistics, 2008," *CA Cancer J. Clin.* **58**(2), 71–96 (2008).

69. B. J. Flehinger, M. Kimmel, and M. R. Melamed, "Survival from early lung cancer: implications for screening," *Chest* **101**, 1013–1018 (1992).

70. C. I. Henschke, D. I. McDauley, D. F. Yankelevitz, D. P. Naidich, G. McGuinness, O. S. Miettinen, D. M. Libby, M. W. Pasmantier, J. Koizumi, N. K. Altorki, and J. P. Smith, "Early lung cancer action project: Overall design and findings from baseline screening," *Lancet* **354**, 99–105 (1999).

71. B. Zhao, A. Reeves, D. Yankelevitz, and C. Henschke, "Two-dimensional multi-criterion segmentation of pulmonary nodules on helical CT images," *Med. Phys.* **26**, 889–895 (1999).

72. S. G. Armato, R. M. Engelmann, M. L. Giger, K. Doi, and H. MacMahon, "A computer-aided diagnostic methods for detection of lung nodules in CT scans," *Radiology* **217**(p), 243 (2000).

73. P. F. Judy and F. L. Jacobson, "CT lung nodule volumes: Comparison 2D and 3D image analysis," *Radiology* **217**(p), 129 (2000).

74. L. Fan, C. L. Novak, J. Qian, G. Kohl, and D. P. Naidich, "Automatic detection of lung nodules from multi-slice low-dose CT images," *Proc. SPIE* **4322**, 1828–1835 (2001).

75. C. E. Cann, H. K. Genant, "Precise measurement of vertebral mineral content using computed tomography," *J. Comput. Assist. Tomogr.* **4**, 493–500 (1980).

76. M. Bligh, L. Bidaut, R. A. White, W. A. Murphy, Jr., D. M. Stevens, and D. D. Cody, "Helical multidetector row quantitative computed tomography (QCT) precision," *Academic Radiology* **16**(2), 150–159 (2009).

77. R. F. Thoeni and I. Laufer, "Polyps and cancer," in *Textbook of Gastrointestinal Radiology*, Vol. 1, R. M. Gore, M. S. Levine, and I. Laufer, Eds., W. B. Saunders Co., Philadelphia, 1160–1199 (1994).

78. R. K. Hung, J. Yee, J. Resayo, and C. Henry, "CT colonography at the San Francisco VAMC: Pictorial tour," RSNA E-Journal:*Radioactive Archives*, **3** 27 (1999).

79. A. K. Hara, C. D. Johnson, J. E. Reed, R. L. Ehman, and D. M. Ilstrup, "Colorectal poly detection with CT colography: Two- versus three-dimensional techniques," *Radiology* **200**, 49–54 (1996).

80. A. K. Hara, C. D. Johnson, J. E. Reed, D. A. Ahlquist, H. Nelson, R. L. MacCarty, W. S. Harmsen, and D. M. Ilstrup, "Detection of colorectal polyps with CT colography: initial assessment of sensitivity and specificity," *Radiology* **205**, 59–65 (1997).

81. H. Yoshida, M. Yoshitaka, P. MacEnaney, and A. Dachman, "Computer-aided detection of polyps in CT colonography based on geometric features," *Proc. SPIE* **4321**, 53–57 (2001).

82. K. D. Hopper, M. Khandelwal, and C. Thompson, "CT colonoscopy: Experience of 100 cases using volumetric rendering," *Proc. SPIE* **4321**, 489–494 (2001).

83. G. Wang, E. McFarland, B. Brown, and M. Vannier, "GI tract unraveling with curved cross sections," *IEEE Trans. Med. Imag.* **17**, 318–322 (1998).

84. S. Haker, S. Angenent, A. Tannenbaum, and R. Kikinis, "Conformal 3D visualization for virtual colonoscopy," *Proc. SPIE* **3978**, 154–164 (2000).

85. R. E. Alvarez and A. Macovski, "Energy-selective reconstructions in x-ray computerized tomography," *Phys. Med. Biol.* **21**(5), 733–744 (1976).

86. F. A. Rutherford, B. R. Pullan, and I. Isherwood, "Measurement of effective atomic number and electron density using an EMI scanner," *Neuroradiology* **11**(1), 15–21 (1976).

87. G. Christ, "Exact treatment of the dual-energy method in CT using polyenergetic x-ray spectra,"*Phys. Med. Biol.* **29**(12), 1501–1510 (1984).

88. R. J. Johnson, X. Zhu, and I. Isherwood, A. I. Morris, B. A. McVerry, D. R. Triger, F. E. Preston, and S. B. Lucas, "Computed tomography: qualitative and quantitative recognition of liver disease in haemophilia," *J. Comput. Assist. Tomogr.* **7**(6), 1000–1006 (1983).

89. W. A. Kalender, W. H. Perman, J. R. Vetter, and E. Klotz, "Evaluation of a prototype dual-energy computed tomographic apparatus, I. Phantom studies," *Med. Phys.* **13**(3), 340–343 (1986).

90. K. Frederick, P. M. Joseph, and S. K. Hilal, "Noise considerations in dual energy CT scanning," *Med. Phys.* **6**(5), 418–425 (1979).

91. P. B. Dunscombe, D. E. Katz, A. J. Stacey, and M. Phil, "Some practical aspects of dual-energy CT scanning," *British Journal of Radiology* **57**, 82–87 (1984).

92. W. H. Marshall, R. E. Alvarez, and A. Macovski, "Initial results with prereconstruction dual-energy computed tomography (PREDECT)," *Body CT* **140**, 421–430 (1981).

93. L. A. Lehmann, R. E. Alvarez, A. Macovski, and W. R. Brody, "Generalized image combination in dual kVp digital radiography," Med. *Phys.* **8**(5), 659–667 (1981).

94. F. L. Roder, "Principles, history, and status of dual-energy computerized tomographic explosives detection," *J. Test. Eval.* **13**(3), 211–216 (1985).

95. J. Hsieh, "Dual energy imaging enhancement with fuzzy logic," *Proc. SPIE* **2434**, 290–299 (1995).

96. J. H. Hubbell and S. M. Seltzer, "Tables of x-ray mass attenuation coefficients and mass energy-absorption coefficients," http://physics.nist.gov/PhysRefData/XrayMassCoef/cover.html.

97. Z. Chen, R. Ning, D. Conover, Y. Yu, and X. Lu, "Dual-basis-material decomposition for dual-kVp cone-beam CT breast imaging," *Proc. SPIE* **5745**, 1322–1333 (2005).

98. D. Walter, X. Wu, Y. Du, E. Tkaczyk, and W. Ross, "Dual kVp material decomposition using flat-panel detectors," *Proc. SPIE* **5368**, 29–39 (2004).

99. K. Chuang and H. K. Huang, "Comparison of four dual energy image decomposition methods," *Phys. Med. Biol.* **33**(4), 455–456 (1988).

100. E. T. D. Hoey, D. Gopalan, and N. J. Screaton, "Dual-energy CT pulmonary angiography: a new horizon in the imaging of acute pulmonary thromboembolism," *Am. J. Roentgen.* **192**, 341–342 (2009).

101. K. Deng, C. Liu, R. Ma, C. Sun, X. Wang, Z. Ma, and X. Sun, "Clinical evaluation of dual-energy bone removal in CT angiography of the head and neck: comparison with conventional bone-subtraction CT angiography," *Clin. Radiol.* **64**(5), 534–541 (2009).

102. Y. Chen, H. D. Xue, Z. Y. Jin, W. Liu, H. Sun, X. Wang, W. M. Zhao, Y. Wang, and W. B. Mu, "Dual-energy CT angiography for evaluation of internal carotid artery stenosis and occlusion," *Zhongguo Yi Xue Ke Xue Yuan Xue Bao*, **31**(2), 215–220 (2009) (in Chinese).

103. D. T. Boll, N. A. Patil, E. K. Paulson, E. M. Merkle, W. N. Simmons, S. A. Pierre, and G. M. Preminger, "Renal stone assessment with dual-energy multidetector CT and advanced postprocessing techniques: improved characterization of renal stone composition-pilot study," *Radiol.* **250**(3), 813–820 (2009).

104. C. L. Brown, R. P. Hartman, O. P. Dzyubak, N. Takahashi, A. Kawashima, C. H. McCollough, M. R. Bruesewitz, A. M. Primak, and J. G. Fletcher,

"Dual-energy CT iodine overlay technique for characterization of renal masses as cyst or solid: a phantom feasibility study," *Euro. Radiol.* **19**(5), 1289–1295 (2009).

105. F. Schwarz, B. Ruzsics, U. J. Schoepf, G. Bastarrika, S. A. Chiaramida, J. A. Abro, R. L. Brothers, S. Vogt, B. Schmidt, P. Costello, and P. L. Zwerner, "Dual-energy CT of the heart-principles and protocols," *Euro. J. Radiol.* **68**(3), 423–433 (2008).

106. V. Raptopoulos, A. Karellas, J. Bernstein, F. R. Reale, C. Constantinou, and J. K. Zawacki, "Value of dual-energy CT in differentiating focal fatty infiltration of the liver from low-density masses," *Amer. J. Roentgen.* **157**, 721–725 (1991).

107. S. Sengupta, S. Jha, D. Walter, Y. Du, and E. J. Tkaczyk, "Dual energy for material differentiation in coronary arteries using electron-beam CT," *Proc. SPIE* **5745**, 1306–1316 (2005).

108. S. J. Swensen, K. Yamashita, C. H. McCollough, R. W. Viggiano, D. E. Midthun, E. F. Patz, J. R. Muhm, and A. L. Weaver, "Lung nodules: dual-kilovolt peak analysis with CT-multicenter study," *Radiology* **214**, 81–85 (2000).

109. N. Pelc, "Dual energy CT: physics principles," *Med. Phys.* **35**(6), 2861 (2008).

110. M. Bazalova, J. Carrier, L. Beaulieu, and F. Verhaegen, "Dual-energy CT-based material extraction for tissue segmentation in Monte Carlo dose calculations," *Phys. Med. Biol.* **53**, 2439–2456 (2008).

111. P. E. Kinahan, A. M. Alessio, and J. A. Fessler, "Dual energy CT attenuation correction methods for quantitative assessment of response to cancer therapy with PET/CT imaging," *Technology in Cancer Research and Treatment* **5**(4), 319–327 (2006).

112. L. Yu, A. N. Primak, X. Liu, and C. H. McCollough, "Image quality optimization and evaluation of linearly-mixed images in dual-source, dual-energy CT," *Med. Phys.* **36**(3), 1019–1024 (2009).

113. R. Carmi, G. Naveh, and A. Altman, "Material separation with dual-layer CT," *2005 IEEE Nuclear Science Symposium Conference Record*, 1876–1878 (2005).

114. D. J. Walter, E. J. Tkaczyk, and X. Wu, "Accuracy and precision of dual energy CT imaging for the quantification of tissue fat content," *Proc. SPIE* **6142**, 61421G (2006).

115. J. P. Schlomka, E. Roessl, R. Dorscheid, S. Dill, G. Martens, T. Istel, G. Baumer, C. Herrmann, R. Steadman, G. Zeitler, A. Livne, and R. Proksa, "Experimental feasibility of multi-energy photon-counting k-edge imaging

in pre-clinical computed tomography," *Phys. Med. Biol.* **53**, 4031–4047 (2008).

116. Z. Chen, R. Ning, D. Conover, Y. Yu, and X. Lu, "Dual-basis-material decomposition for dual-kVp cone-beam CT breast imaging," *Proc. SPIE* **5745**, 1322–1333 (2005).

117. E. Y. Sidky, Y. Zou, and X. Pan, "Effect of noise in dual-energy helical cone-beam computed tomography," *Proc. SPIE* **5368**, 396–402 (2004).

Glossary

air calibration. A calibration procedure that uses CT scans with no object placed in the gantry to normalize detector gain variations.

ALARA (as low as reasonably achievable) principle. A guiding principle to reduce radiation dose to patient.

aliasing. An occurrence in which signals of higher frequencies affect the signals of lower frequencies as a result of a violation of the Nyquist sampling criteria.

anode. The positively biased target of the x-ray tube where x-ray photons are generated by the bombardment of high-speed electrons.

artifact. A discrepancy between the reconstructed values in an image and the true attenuation coefficients of the object.

attenuation. The reduction of x-ray intensities when passing through matter.

axial plane. A plane that is parallel to both the left-right and anterior-posterior axes in the "patient-oriented" coordinate system.

backprojection. A step in the reconstruction to calculate the contribution of each projection sample to the reconstructed image along the sample path.

beam hardening. The disproportional reduction of low-energy (soft) photon intensities when polychromatic radiation passes through matter; the radiation becomes richer in high-energy (harder) photons.

Beer–Lambert law. For monochromatic x rays, the x-ray intensity passing through a uniform object falls off exponentially to the product of the path length and the linear attenuation coefficient of the object.

bowtie filter. A bowtie-shaped prepatient filter with a path length that increases with the angle from the iso-ray.

Bremsstrahlung radiation (general radiation). X-ray photon generated by the deceleration of a high-speed electron by the electric field of the target nuclei.

calcium scores. A scoring system in which the amount of calcium present in the coronary artery is mapped to the risk of a coronary event.

cathode. The negatively biased electrode of the x-ray tube from which high-speed electrons are emitted.

central slice theorem (Fourier slice theorem). A fundamental theorem of tomographic reconstruction that states that the Fourier transform of a parallel projection of an object is equal to a slice taken at the same angle in the 2D Fourier transform of the object.

central volume principle. A biological relationship between cerebral blood flow, cerebral blood volume, and mean transit time.

cerebral blood flow (CBF). The rate of blood flow through an organ.

cerebral blood volume (CBV). The total volume of blood in the large conductance vessels, arteries, arterioles, capillaries, venules, and sinuses.

characteristic radiation. X-ray photon generated by a high-speed electron's collision with and liberation of an inner shell electron of the target atom.

collimator. A mechanical device to define or shape the x-ray beam.

Compton scatter. The interaction of an x-ray photon with matter in which the photon is deflected to a new direction and retains part of its original energy.

cone angle. The angle formed by a projection sample and the detector center plane.

cone beam. The shape of a projection formed by a single focal spot and a 2D detector.

conjugate sampling pair. Two projection samples that are 180 deg apart with the same path in the x-y plane (but not necessarily in z).

coronal plane. A plane that is parallel to both the left-right and superior-inferior axes in the "patient-oriented" coordinate system.

coronary artery imaging (CAI). Imaging of the heart typically performed with the injection of iodine contrast.

CT colonography. A virtual colonographic procedure performed with reconstructed 3D-volume CT images of the colon.

CT dose index (CTDI). The ratio of the integrated dose profile over the total slice collimation. Ideally the integration is carried out from $-\infty$ to ∞. In practice, either a 7-slice thickness or 50 mm on each side of the plane is used. The total slice collimation is the product of the number of tomograms and the nominal tomographic section thickness.

CT fluoroscopy. Continuous imaging with CT to provide real-time or near-real-time feedback for interventional procedures.

CT number. An intensity scale for CT images, defined as $\dfrac{\mu - \mu_{water}}{\mu_{water}} \times 1000$. The unit is often called the Hounsfield unit (HU) in honor of Dr. G. N. Hounsfield.

CT perfusion. A CT procedure with intravenous contrast injection to characterize the physiological functions of tissues. The procedure typically covers the contrast uptake phase through the washout phase.

detector center plane. A plane formed by the detector "center row" and the focal spot. The detector center row is the real or imaginary row that represents the center of the detector in z.

detector quarter offset. A technique used to combat aliasing artifacts by aligning the detector center channel one-quarter cell width from the iso-ray.

display window width and level. A remapping of the original CT numbers between $(L_w - W_w/2)$ and $(L_w + W_w/2)$ to grayscale intensities of the display device, where W_w is the window width and L_w is the window level.

dose. A quantitative measure of the energy released in matter as a result of radiation exposure.

dose length product (DLP). A measurement of the total amount of dose for a series of scans, defined as the product of the $CTDI_v$ and the scan range in z.

dual-energy CT. A method of scanning the same object with two different x-ray energy spectrums.

effective dose. A measurement of the x-ray dose based on the radiation dose to individual organs and the relative radiation risk assigned to each organ.

electrocardiogram (EKG or ECG). A device to record the bioelectric signal of the heart.

electronic noise. Noise in the measured projection due to the random fluctuation of electronic devices (not due to the x-ray photon statistics).

equivalent density image. An image generated in dual-energy CT using the material decomposition technique to map the scanned object to equivalent densities of a basis material. The image is in g/cm^3, not in Hounsfield units.

fan angle. The angle γ formed by a projection sample and the iso-ray.

fan beam. The shape of a projection formed by a single focal spot and a 1D detector.

Fick principle. A theory in physiology stating that the rate of accumulation of contrast in an organ equals the difference between the influx rate and the efflux rate of contrast to the organ.

filtered backprojection. An image reconstruction process that includes the convolution of the measured projection with a reconstruction kernel, and the backprojection of the filtered projection.

focal spot. A small region on the surface of the x-ray tube anode that emits x-ray photons.

Fourier slice theorem. See central slice theorem.

full width at half maximum (FWHM). The distance on the abscissa of an intensity profile between two points whose intensities reach one-half of the maximum value.

full width at tenth maximum (FWTM). The distance on the abscissa of an intensity profile between two points whose intensities reach one-tenth of the maximum value.

gantry tilt. The tilt of the CT gantry with respect to the plane that is perpendicular to the patient long axis (z); 0-deg tilt occurs when the gantry is perpendicular to the z axis.

geometric detection efficiency (GDE). The ratio of the x-ray photons interacting with the active part of the detector over the total x-ray photons passing through the patient.

grays (Gy). A measurement unit for dose to patient described as the absorption of 1 Joule of energy per kilogram of matter.

helical pitch. The ratio of the distance that the patient table travels in one gantry rotation over the total slice collimation.

helical (spiral) scan. A CT scan mode in which the patient table travels continuously during the data acquisition.

high-contrast resolution (spatial resolution). The ability of a CT system to resolve a closely placed small object whose density is significantly different from the background.

Hounsfield unit (HU). The unit for the CT number scale (see also CT number).

hysteresis (radiation damage). The change in the detector gain due to its radiation exposure history.

iso-ray. The ray emitting from the focal spot of the x-ray tube and passing through the iso-center of the CT system.

iterative reconstruction. A reconstruction technique in which the final image is obtained by iteratively refining intermediate results using the synthesized projection and the measured projection.

line focus principle. A technique used in the x-ray tube design to increase the tube's thermal loading by setting the target surface at a small angle relative to the x-ray beam plane.

linear attenuation coefficient. The probability per centimeter thickness of matter that an x-ray photon will be attenuated.

low-contrast detectability (LCD). The ability of a CT system to detect a certain sized object whose density is slightly different from its background under certain dose conditions.

mA modulation. See tube-current modulation.

mass attenuation coefficient. The linear attenuation coefficient divided by the material density.

material decomposition. A theory stating that an unknown material can be characterized by the density projections of two basis materials.

maximum intensity projection (MIP). A technique used to represent a 3D volume with a 2D image by assigning the maximum intensity values along mathematical rays through the 3D volume to be the resulting intensities.

mean transit time. The average time it takes for blood to travel from the arterial inlet to the venous outlet.

modulation transfer function (MTF). The measurement of a system's high-contrast spatial resolution. The MTF can be obtained by taking the magnitude of the Fourier transform of the point-spread function (PSF).

monoenergetic (monochromatic) image. An image generated using dual-energy CT to synthesize a CT image when a monochromatic x-ray source is used.

monoenergetic (monochromatic) x ray. A condition in which all of the emitted x-ray photons have an identical energy level.

Monte Carlo simulation. A computer simulation technique in which the interactions of each x-ray photon with matter are traced and the final result is the ensemble of a large number of photons.

multislice CT. A CT system that employs a detector with two or more detector rows.

noise power spectrum. A method to characterize noise by taking the Fourier transform of the noise.

Nyquist sampling criteria. Mathematical criteria that state that a continuous band-limited function can be faithfully recovered from discretely sampled values if the sampling frequency is at least twice the highest-frequency present in the original function.

off-focal radiation. The x-ray photons emitted outside the focal spot region.

overbeam. A helical scan phenomenon in which the x-ray exposure length in z is longer than the range of reconstruction.

partial volume effect. An artifact caused by an object's inhomogeneity across the width of the x-ray beam of a detector element.

penumbra. The part of an x-ray beam that is contributed by only a portion of the x-ray focal spot.

photoelectric effect. The interaction between an x-ray photon and matter in which the photon energy is completely absorbed by an atomic electron that is ejected from the atom.

physiological gating. A modification of the CT scanning technique based on physiological information such as the respiratory or cardiac phase.

point-spread function (PSF). A system's response to a Dirac delta input. For a CT system, this is often obtained by scanning an object whose physical dimension is significantly smaller than the limiting resolution of the system.

polyenergetic (polychromatic) x-ray. X-ray photons that have a wide range (spectrum) of energy.

primary speed and afterglow. The residual or decayed signals of a solid state scintillator generated after x-ray radiation exposure.

prospective gating. A motion reduction approach in which the physiological gating is performed during the data acquisition.

quantum detection efficiency. The fraction of the incident x-ray photons that are attenuated by a detector.

rad. A measurement unit for patient dose defined as the absorption of 1×10^{-5} Joule of energy per gram of matter.

radiation damage. See hysteresis.

Rayleigh scattering (coherent scatter). The interaction between an x-ray photon and matter in which the radiation undergoes a change in direction without a change in energy.

rebinning (fan beam to parallel beam). A process by which the original fan-beam or cone-beam projection dataset is converted to a parallel-beam or tilted parallel-beam dataset.

reconstruction field of view (FOV). An area defined by the operator in which to reconstruct the scanned object.

reconstruction kernel. The convolution kernel used on a measured projection to compensate for the blurring caused by the backprojection process in the filtered backprojection (FBP) reconstruction. The reconstruction kernel can be adjusted to trade off spatial resolution versus noise.

region of interest (ROI). An area on the CT image defined by the operator.

retrospective gating. A motion reduction approach in which the physiological gating is performed during the image reconstruction.

roentgen (R). A measurement unit for radiation exposure of air that is defined as the amount of x-ray or gamma-ray energy required to produce an electrostatic charge of 2.58×10^{-4} coulomb in 1 kg of air.

sagittal plane. A plane that is parallel to both the anterior-posterior and superior-inferior axes in the "patient-oriented" coordinate system.

scan field of view. An area inside the CT gantry in which a complete projection dataset is acquired.

scanogram (scout view, projection radiograph). A projection image similar to an x-ray radiograph acquired with a stationary gantry and a continuously moving patient table.

scatter-to-primary ratio. The ratio of the signal that is due to scattered radiation over the signal generated from the primary beam.

sieverts (Sv). A dose-equivalent measurement unit named in honor of a Swedish scientist, defined as 1 Joule of energy per kilogram of matter.

sinogram. A 2D intensity plot of projection measurements as a function of the detector channel and projection view.

slice sensitivity profile (SSP). The CT system response to a Dirac delta function in z.

step-and-shoot (axial) scan. A CT scan mode in which the patient table remains stationary during the data acquisition.

temporal resolution. The ability of the CT system to resolve changes in CT values over time.

tube-current modulation (mA modulation). A dose reduction technique in which the x-ray tube current is modified based on the patient anatomical information.

umbra. The portion of the x-ray beam that is contributed by the entire x-ray focal spot.

volume rendering (VR). A technique used to represent a 3D volume with a 2D image by accumulating contributions from all voxels along mathematical rays through the 3D volume via opacity and color functions.

Index

A

adaptive array detector (see multislice CT)
afterglow (see x-ray detector)
Agatston method (see cardiac imaging)
air scans, 212, 246–247, 254, 517
algebraic reconstruction technique (ART), 59–60, 105–106
aliasing (see artifact)
area score (see cardiac imaging)
artifacts (see also helical CT and multislice CT)
    aliasing (projection and view), 17, 72, 188, 214–225, 264, 312–313
    alignment, 315, 364–367, 420–421
    beam-hardening, 44, 167, 270–281
    cone-beam, 378, 388–389, 392, 394
    gain, 195, 246–247, 254–256, 517
    hysteresis (radiation damage), 45, 195, 247
    incomplete projection, 283–288
    linearity, 45, 196, 247–248
    longitudinal truncation, 378, 389–391, 413, 496
    metal, 110, 280–283
    noise-induced, 214, 235–236, 289, 360–362, 416
    off-focal radiation, 45, 146, 239–242
    offset, 244–247
    partial volume, 45, 152, 169, 226–231, 278–280, 290, 291, 475
    patient motion, 14, 169, 258–270, 283, 289, 304, 375, 511, 529–531
    primary speed and afterglow, 192, 248–251
    ring and band, 210–214
    rotor wobble, 45, 244
    scatter, 167, 231–235, 319–321, 491
    shading, 44, 45, 66, 67, 75, 209, 210–211, 231, 233, 235, 239, 250–252, 261, 274, 276, 277, 278, 281, 284, 290
    SSP distortion, 419–420
    streak, 209–210, 211, 238
    tilt, 394–395, 416–418
    tube arcing (tube-spit), 45, 242–243

    uniformity, 253–257
    z aliasing, 410–411
as low as reasonably achievable (ALARA), 433, 434, 458, 470, 514
atherosclerosis, 472, 476
attenuation, 17, 39, 40–44, 58
attenuation coefficient, 9, 10, 40–43, 55, 57, 102, 104, 270, 271, 276–277, 316–318, 522–528
axial scan (see scan mode)
axial transverse tomography, 5, 6
azimuthal direction, 1, 147, 148, 149, 218, 219

B

backprojection, 7, 9, 59–61, 70, 81–82, 85, 92–93
    fan-beam implementation, 94–95
    parallel-beam implementation, 84–85
    pixel-driven, 68, 82–83, 86, 92
    ray-driven, 82–83
backprojection-filtering approach, 61
basis material (see dual-energy CT)
basis material decomposition (see dual-energy CT)
beam hardening, 44, 146, 270–280, 316–319
Beer–Lambert law (see x ray)
bone-mineral density (BMD), 517
bowtie filter (see x ray)
bremsstrahlung radiation, 35, 36

C

calcium scores (see cardiac imaging)
calibration (see preprocessing)
cardiac imaging, 80, 452–453, 471–496
    Agatston method, 474
    area scoring, 474–475
    artifacts, 476, 485–490
    calcium scores, 474–476
    coronary artery calcification (CAC), 472–476
    coronary artery imaging (CAI), 153, 476–478, 480, 493

 **Jiang Hsieh**, Ph.D., is the Chief Scientist at GE Healthcare and an adjunct professor in the Department of Medical Physics at University of Wisconsin-Madison. He has over 26 years of experience in medical imaging, holds over 160 U.S. patents, and has authored or coauthored more than 200 research articles, conference papers, and book chapters. He teaches short courses on x-ray computed tomography at SPIE Medical Imaging conferences, Radiology Society of North American annual meetings (RSNA), the IEEE Nuclear Science Symposium and Medical Imaging Conference, AAPM annual meetings, and AAPM summer school. His research interests include tomographic reconstruction, medical image processing, image artifact identification and correction, and advanced CT applications.

Printed in the United States
By Bookmasters